ELECTRICITY AND MAGNETISM

BERKELEY PHYSICS COURSE – VOLUME 2

SECOND EDITION

ELECTRICITY AND MAGNETISM

BERKELEY PHYSICS COURSE – VOLUME 2

EDWARD M. PURCELL
Gerhard Gade University Professor Emeritus
Harvard University

McGraw-Hill, Inc.
New York St. Louis San Francisco Auckland Bogotá
Caracas Lisbon London Madrid Mexico City Milan
Montreal New Delhi San Juan Singapore
Sydney Tokyo Toronto

This book was set in Times Roman by University Graphics, Inc.
The editors were Stephen Zlotnick and James W. Bradley;
the designer was Nicholas Krenitsky;
the production supervisor was Charles Hess.
New drawings were done by Felix Cooper.

ELECTRICITY AND MAGNETISM
Berkeley Physics Course
Volume 2

Printed and bound by Book-mart Press, Inc.

10 11 12 13 14 15 BKM BKM 9 9 8 7 6 5 4

ISBN 0-07-004908-4

The first edition of the Berkeley Physics Course, *Electricity and Magnetism,* Volume 2, copyright © 1965, 1964, 1963 by Education Development Center, Inc. (successor by merger to Educational Services Incorporated) was supported by a grant from the National Science Foundation to EDC. In accordance with the National Science Foundation's policies concerning curriculum revision material developed under its auspices, material in the first edition is available to publishers and authors on a royalty-free basis by applying to the Education Development Center. Inquiries should be addressed to the Managing Director, Education Development Center, 55 Chapel Street, Newton, Massachusetts, 02160.

Library of Congress Cataloging in Publication Data

Purcell, Edward M.
Electricity and magnetism

Includes bibliographical references and index.
1. Electricity. 2. Magnetism. I. Title.
QC522.P85 1985 537 83-19590
ISBN 0-07-004908-4

CONTENTS

PREFACE TO THE SECOND EDITION OF VOLUME 2

This revision of "Electricity and Magnetism," Volume 2 of the Berkeley Physics Course, has been made with three broad aims in mind. First, I have tried to make the text clearer at many points. In years of use teachers and students have found innumerable places where a simplification or reorganization of an explanation could make it easier to follow. Doubtless some opportunities for such improvements have still been missed; not too many, I hope.

A second aim was to make the book practically independent of its companion volumes in the Berkeley Physics Course. As originally conceived it was bracketed between Volume 1, which provided the needed special relativity, and Volume 3, "Waves and Oscillations," to which was allocated the topic of electromagnetic waves. As it has turned out, Volume 2 has been rather widely used alone. In recognition of that I have made certain changes and additions. A concise review of the relations of special relativity is included as Appendix A. Some previous introduction to relativity is still assumed. The review provides a handy reference and summary for the ideas and formulas we need to understand the fields of moving charges and their transformation from one frame to another. The development of Maxwell's equations for the vacuum has been transferred from the heavily loaded Chapter 7 (on induction) to a new Chapter 9, where it leads naturally into an elementary treatment of plane electromagnetic waves, both running and standing. The propagation of a wave in a dielectric medium can then be treated in Chapter 10 on Electric Fields in Matter.

A third need, to modernize the treatment of certain topics, was most urgent in the chapter on electrical conduction. A substantially rewritten Chapter 4 now includes a section on the physics of homogeneous semiconductors, including doped semiconductors. Devices are not included, not even a rectifying junction, but what is said about bands, and donors and acceptors, could serve as a starting point for development of such topics by the instructor. Thanks to solid-state electronics the physics of the voltaic cell has become even more relevant to daily life as the number of batteries in use approaches in order of magnitude the world's population. In the first edition of this book I unwisely chose as the example of an electrolytic cell the one cell—the Weston standard cell—which advances in physics were soon to render utterly obsolete. That section has been replaced by an analysis, with new diagrams, of the lead-acid storage battery—ancient, ubiquitous, and far from obsolete.

One would hardly have expected that, in the revision of an elementary text in classical electromagnetism, attention would have to be paid to new developments in particle physics. But that is the case for two questions that were discussed in the first edition, the significance of charge quantization, and the apparent absence of magnetic monopoles. Observation of proton decay would profoundly affect our view of the first question. Assiduous searches for that, and also for magnetic monopoles, have at this writing yielded no confirmed events, but the possibility of such fundamental discoveries remains open.

Three special topics, optional extensions of the text, are introduced in short appendixes: Appendix B: Radiation by an Accelerated Charge; Appendix C: Superconductivity; and Appendix D: Magnetic Resonance.

Our primary system of units remains the Gaussian CGS system. The SI units, ampere, coulomb, volt, ohm, and tesla are also introduced in the text and used in many of the problems. Major formulas are repeated in their SI formulation with explicit directions about units and conversion factors. The charts inside the back cover summarize the basic relations in both systems of units. A special chart in Chapter 11 reviews, in both systems, the relations involving magnetic polarization. The student is not expected, or encouraged, to memorize conversion factors, though some may become more or less familiar through use, but to look them up whenever needed. There is no objection to a "mixed" unit like the ohm-cm, still often used for resistivity, providing its meaning is perfectly clear.

The definition of the meter in terms of an assigned value for the speed of light, which has just become official, simplifies the exact relations among the units, as briefly explained in Appendix E.

There are some 300 problems, more than half of them new.

It is not possible to thank individually all the teachers and students who have made good suggestions for changes and corrections. I fear that some will be disappointed to find that their suggestions have not been followed quite as they intended. That the net result is a substantial improvement I hope most readers familiar with the first edition will agree. Mistakes both old and new will surely be found. Communications pointing them out will be gratefully received.

It is a pleasure to thank Olive S. Rand for her patient and skill-full assistance in the production of the manuscript.

Edward M. Purcell

PREFACE TO THE FIRST EDITION OF VOLUME 2

The subject of this volume of the Berkeley Physics Course is electricity and magnetism. The sequence of topics, in rough outline, is not unusual: electrostatics; steady currents; magnetic field; electromagnetic induction; electric and magnetic polarization in matter. However, our approach is different from the traditional one. The difference is most conspicuous in Chaps. 5 and 6 where, building on the work of Vol. I, we treat the electric and magnetic fields of moving charges as manifestations of relativity and the invariance of electric charge. This approach focuses attention on some fundamental questions, such as: charge conservation, charge invariance, the meaning of field. The only formal apparatus of special relativity that is really necessary is the Lorentz transformation of coordinates and the velocity-addition formula. It is essential, though, that the student bring to this part of the course some of the ideas and attitudes Vol. I sought to develop—among them a readiness to look at things from different frames of reference, an appreciation of invariance, and a respect for symmetry arguments. We make much use also, in Vol. II, of arguments based on superposition.

Our approach to electric and magnetic phenomena in matter is primarily "microscopic," with emphasis on the nature of atomic and molecular dipoles, both electric and magnetic. Electric conduction, also, is described microscopically in the terms of a Drude-Lorentz model. Naturally some questions have to be left open until the student takes up quantum physics in Vol. IV. But we freely talk in a matter-of-fact way about molecules and atoms as electrical structures with size, shape, and stiffness, about electron orbits, and spin. We try to treat carefully a question that is sometimes avoided and sometimes beclouded in introductory texts, the meaning of the macroscopic fields **E** and **B** inside a material.

In Vol. II, the student's mathematical equipment is extended by adding some tools of the vector calculus—gradient, divergence, curl, and the Laplacian. These concepts are developed as needed in the early chapters.

In its preliminary versions, Vol. II has been used in several classes at the University of California. It has benefited from criticism by many people connected with the Berkeley Course, especially from contributions by E. D. Commins and F. S. Crawford, Jr., who taught the first classes to use the text. They and their students discovered numerous places where clarification, or something more drastic, was needed; many of the revisions were based on their suggestions. Students' criticisms of the last preliminary version were collected by Robert Goren, who also helped to organize the problems. Valuable criticism has come also from J. D. Gavenda, who used the preliminary version at the University of Texas, and from E. F. Taylor, of Wesleyan University. Ideas were contributed by Allan Kaufman at an early stage of the writing. A. Felzer worked through most of the first draft as our first "test student."

The development of this approach to electricity and magnetism was encouraged, not only by our original Course Committee, but by colleagues active in a rather parallel development of new course material at the Massachusetts Institute of Technology. Among the latter, J. R. Tessman, of the MIT Science Teaching Center and Tufts University, was especially helpful and influential in the early formulation of the strategy. He has used the preliminary version in class, at MIT, and his critical reading of the entire text has resulted in many further changes and corrections.

Publication of the preliminary version, with its successive revisions, was supervised by Mrs. Mary R. Maloney. Mrs. Lila Lowell typed most of the manuscript. The illustrations were put into final form by Felix Cooper.

The author of this volume remains deeply grateful to his friends in Berkeley, and most of all to Charles Kittel, for the stimulation and constant encouragement that have made the long task enjoyable.

Edward M. Purcell

PREFACE TO THE BERKELEY PHYSICS COURSE

This is a two-year elementary college physics course for students majoring in science and engineering. The intention of the writers has been to present elementary physics as far as possible in the way in which it is used by physicists working on the forefront of their field. We have sought to make a course which would vigorously emphasize the foundations of physics. Our specific objectives were to introduce coherently into an elementary curriculum the ideas of special relativity, of quantum physics, and of statistical physics.

This course is intended for any student who has had a physics course in high school. A mathematics course including the calculus should be taken at the same time as this course.

There are several new college physics courses under development in the United States at this time. The idea of making a new course has come to many physicists, affected by the needs both of the advancement of science and engineering and of the increasing emphasis on science in elementary schools and in high schools. Our own course was conceived in a conversation between Philip Morrison of Cornell University and C. Kittel late in 1961. We were encouraged by John Mays and his colleagues of the National Science Foundation and by Walter C. Michels, then the Chairman of the Commission on College Physics. An informal committee was formed to guide the course through the initial stages. The committee consisted originally of Luis Alvarez, William B. Fretter, Charles Kittel, Walter D. Knight, Philip Morrison, Edward M. Purcell, Malvin A. Ruderman, and Jerrold R. Zacharias. The committee met first in May 1962, in Berkeley; at that time it drew up a provisional outline of an entirely new physics course. Because of heavy obligations of several of the original members, the committee was partially reconstituted in January 1964, and now consists of the undersigned. Contributions of others are acknowledged in the prefaces of the individual volumes.

The provisional outline and its associated spirit were a powerful influence on the course material finally produced. The outline covered in detail the topics and attitudes which we believed should and could be taught to beginning college students of science and engineering. It was never our intention to develop a course limited to honors students or to students with advanced standing. We have sought to present the principles of physics from fresh and unified viewpoints, and parts of the course may therefore seem almost as new to the instructor as to the students.

The five volumes of the course as planned will include:

1. Mechanics (Kittel, Knight, Ruderman)
2. Electricity and Magnetism (Purcell)
3. Waves and Oscillations (Crawford)
4. Quantum Physics (Wichmann)
5. Statistical Physics (Reif)

The authors of each volume have been free to choose that style and method of presentation which seemed to them appropriate to their subject.

The initial course activity led Alan M. Portis to devise a new elementary physics laboratory, now known as the Berkeley Physics Laboratory. Because the course emphasizes the principles of physics, some teachers may feel that it does not deal sufficiently with experimental physics. The laboratory is rich in important experiments, and is designed to balance the course.

The financial support of the course development was provided by the National Science Foundation, with considerable indirect support by the University of California. The funds were administered by Educational Services Incorporated, a nonprofit organization established to administer curriculum improvement programs. We are particularly indebted to Gilbert Oakley, James Aldrich, and William Jones, all of ESI, for their sympathetic and vigorous support. ESI established in Berkeley an office under the very competent direction of Mrs. Mary R. Maloney to assist the development of the course and the laboratory. The University of California has no official connection with our program, but it has aided us in important ways. For this help we thank in particular two successive Chairmen of the Department of Physics, August C. Helmholz and Burton J. Moyer; the faculty and nonacademic staff of the Department; Donald Coney, and many others in the University. Abraham Olshen gave much help with the early organizational problems.

Your corrections and suggestions will always be welcome.

Eugene D. Commins
Frank S. Crawford, Jr.
Walter D. Knight
Philip Morrison
Alan M. Portis
Edward M. Purcell
Frederick Reif
Malvin A. Ruderman
Eyvind H. Wichmann

Berkeley, California
Charles Kittel, *Chairman*

ELECTRICITY AND MAGNETISM

BERKELEY PHYSICS COURSE – VOLUME 2

1

ELECTROSTATICS: CHARGES AND FIELDS

ELECTRIC CHARGE

1.1 Electricity appeared to its early investigators as an extraordinary phenomenon. To draw from bodies the "subtle fire," as it was sometimes called, to bring an object into a highly electrified state, to produce a steady flow of current, called for skillful contrivance. Except for the spectacle of lightning, the ordinary manifestations of nature, from the freezing of water to the growth of a tree, seemed to have no relation to the curious behavior of electrified objects. We know now that electrical forces largely determine the physical and chemical properties of matter over the whole range from atom to living cell. For this understanding we have to thank the scientists of the nineteenth century, Ampère, Faraday, Maxwell, and many others, who discovered the nature of electromagnetism, as well as the physicists and chemists of the twentieth century who unraveled the atomic structure of matter.

Classical electromagnetism deals with electric charges and currents and their interactions as if all the quantities involved could be measured independently, with unlimited precision. Here *classical* means simply "nonquantum." The quantum law with its constant h is ignored in the classical theory of electromagnetism, just as it is in ordinary mechanics. Indeed, the classical theory was brought very nearly to its present state of completion before Planck's discovery. It has survived remarkably well. Neither the revolution of quantum physics nor the development of special relativity dimmed the luster of the electromagnetic field equations Maxwell wrote down 100 years ago.

Of course the theory was solidly based on experiment, and because of that was fairly secure within its original range of application—to coils, capacitors, oscillating currents, and eventually radio waves and light waves. But even so great a success does not guarantee validity in another domain, for instance, the inside of a molecule.

Two facts help to explain the continuing importance in modern physics of the classical description of electromagnetism. First, special relativity required no revision of classical electromagnetism. Historically speaking, special relativity *grew out of* classical electromagnetic theory and experiments inspired by it. Maxwell's field equations, developed long before the work of Lorentz and Einstein, proved to be entirely compatible with relativity. Second, quantum modifications of the electromagnetic forces have turned out to be unimportant down to distances less than 10^{-10} centimeters (cm), 100 times smaller than the atom. We can describe the repulsion and attraction of particles in the atom using the same laws that apply to the leaves of an electroscope, although we need quantum mechanics to predict how the particles will behave under those forces. For still smaller distances, a fusion of electromagnetic theory and quantum theory, called *quantum electrodynamics,* has been remarkably successful. Its predictions are confirmed by experiment down to the smallest distances yet explored.

It is assumed that the reader has some acquaintance with the elementary facts of electricity. We are not going to review all the experiments by which the existence of electric charge was demonstrated, nor shall we review all the evidence for the electrical constitution of matter. On the other hand, we do want to look carefully at the experimental foundations of the basic laws on which all else depends. In this chapter we shall study the physics of stationary electric charges—*electrostatics*.

Certainly one fundamental property of electric charge is its existence in the two varieties that were long ago named *positive* and *negative*. The observed fact is that all charged particles can be divided into two classes such that all members of one class repel each other, while attracting members of the other class. If two small electrically charged bodies A and B, some distance apart, attract one another, and if A attracts some third electrified body C, then we always find that B repels C. Contrast this with gravitation: There is only one kind of gravitational mass, and every mass attracts every other mass.

One may regard the two kinds of charge, positive and negative, as opposite manifestations of one quality, much as *right* and *left* are the two kinds of handedness. Indeed, in the physics of elementary particles, questions involving the sign of the charge are sometimes linked to a question of handedness, and to another basic symmetry, the relation of a sequence of events, *a,* then *b,* then *c,* to the temporally reversed sequence *c,* then *b,* then *a.* It is only the duality of electric charge that concerns us here. For every kind of particle in nature, as far as we know, there can exist an *antiparticle,* a sort of electrical "mirror image." The antiparticle carries charge of the opposite sign. If any other intrinsic quality of the particle has an opposite, the antiparticle has that too, whereas in a property which admits no opposite, such as mass, the antiparticle and particle are exactly alike. The electron's charge is negative; its antiparticle, called a *positron,* has a positive charge, but its mass is precisely the same as that of the electron. The proton's antiparticle is called simply an *antiproton;* its electric charge is negative. An electron and a proton combine to make an ordinary hydrogen atom. A positron and an antiproton could combine in the same way to make an atom of antihydrogen. Given the building blocks, positrons, antiprotons, and antineutrons,† there could be built up the whole range of antimatter, from antihydrogen to antigalaxies. There is a practical difficulty, of course. Should a positron meet an electron or an antiproton meet a proton, that pair of particles will quickly vanish in a burst of radiation. It is therefore not surprising that even positrons and antiprotons, not to speak of antiatoms, are exceedingly rare and short-lived in our world. Perhaps the universe contains,

†Although the electric charge of each is zero, the neutron and its antiparticle are not interchangeable. In certain properties that do not concern us here, they are opposite.

somewhere, a vast concentration of antimatter. If so, its whereabouts is a cosmological mystery.

The universe around us consists overwhelmingly of matter, not antimatter. That is to say, the abundant carriers of negative charge are electrons, and the abundant carriers of positive charge are protons. The proton is nearly 2000 times heavier than the electron and very different, too, in some other respects. Thus matter at the atomic level incorporates negative and positive electricity in quite different ways. The positive charge is all in the atomic nucleus, bound within a massive structure no more than 10^{-12} cm in size, while the negative charge is spread, in effect, through a region about 10^4 times larger in dimensions. It is hard to imagine what atoms and molecules—and all of chemistry—would be like, if not for this fundamental electrical asymmetry of matter.

What we call negative charge, by the way, could just as well have been called positive. The name was a historical accident. There is nothing essentially negative about the charge of an electron. It is not like a negative integer. A negative integer, once multiplication has been defined, differs essentially from a positive integer in that its square is an integer of opposite sign. But the product of two charges is not a charge; there is no comparison.

Two other properties of electric charge are essential in the electrical structure of matter: Charge is *conserved,* and charge is *quantized.* These properties involve *quantity* of charge and thus imply a measurement of charge. Presently we shall state precisely how charge can be measured in terms of the force between charges a certain distance apart, and so on. But let us take this for granted for the time being, so that we may talk freely about these fundamental facts.

FIGURE 1.1
Charged particles are created in pairs with equal and opposite charge.

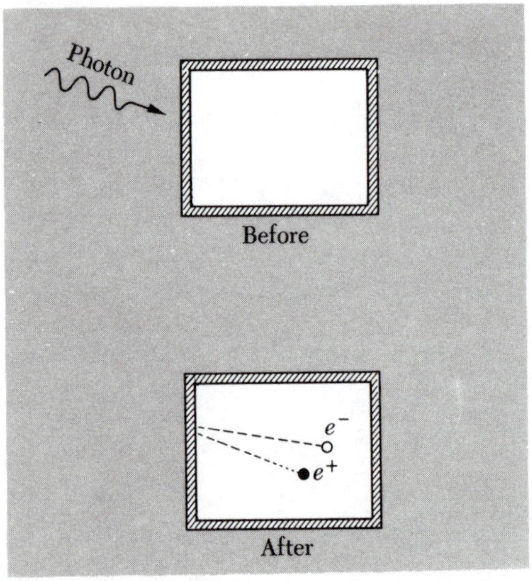

CONSERVATION OF CHARGE

1.2 The total charge in an isolated system never changes. By *isolated* we mean that no matter is allowed to cross the boundary of the system. We could let light pass into or out of the system, since the "particles" of light, called *photons,* carry no charge at all. Within the system charged particles may vanish or reappear, but they always do so in pairs of equal and opposite charge. For instance, a thin-walled box in a vacuum exposed to gamma rays might become the scene of a "pair-creation" event in which a high-energy photon ends its existence with the creation of an electron and a positron (Fig. 1.1). Two electrically charged particles have been newly created, but the net change in total charge, in and on the box, is zero. An event that *would* violate the law we have just stated would be the creation of a positively charged particle *without* the simultaneous creation of a negatively charged particle. Such an occurrence has never been observed.

Of course, if the electric charges of an electron and a positron were not precisely equal in magnitude, pair creation would still violate the strict law of charge conservation. That equality is a manifestation of the particle–antiparticle duality already mentioned, a universal symmetry of nature.

One thing will become clear in the course of our study of electromagnetism: Nonconservation of charge would be quite incompatible with the structure of our present electromagnetic theory. We may therefore state, either as a postulate of the theory or as an empirical law supported without exception by all observations so far, the charge conservation law:

> The total electric charge in an isolated system, that is, the algebraic sum of the positive and negative charge present at any time, never changes.

Sooner or later we must ask whether this law meets the test of relativistic invariance. We shall postpone until Chapter 5 a thorough discussion of this important question. But the answer is that it does, and not merely in the sense that the statement above holds in any given inertial frame but in the stronger sense that observers in different frames, measuring the charge, obtain the same number. In other words the total electric charge of an isolated system is a relativistically invariant number.

QUANTIZATION OF CHARGE

1.3 The electric charges we find in nature come in units of one magnitude only, equal to the amount of charge carried by a single electron. We denote the magnitude of that charge by e. (When we are paying attention to sign, we write $-e$ for the charge on the electron itself.) We have already noted that the positron carries precisely that amount of charge, as it must if charge is to be conserved when an electron and a positron annihilate, leaving nothing but light. What seems more remarkable is the apparently exact equality of the charges carried by all other charged particles—the equality, for instance, of the positive charge on the proton and the negative charge on the electron.

That particular equality is easy to test experimentally. We can see whether the net electric charge carried by a hydrogen molecule, which consists of two protons and two electrons, is zero. In an experiment carried out by J. G. King,† hydrogen gas was compressed into

†J. G. King, *Phys. Rev. Lett.* **5**:562 (1960). References to previous tests of charge equality will be found in this article and in the chapter by V. W. Hughes in "Gravitation and Relativity," H. Y. Chieu and W. F. Hoffman (eds.), W. A. Benjamin, New York, 1964, chap. 13.

a tank that was electrically insulated from its surroundings. The tank contained about 5×10^{24} molecules [approximately 17 grams (gm)] of hydrogen. The gas was then allowed to escape by means which prevented the escape of any ion—a molecule with an electron missing or an extra electron attached. If the charge on the proton differed from that on the electron by, say, one part in a billion, then each hydrogen molecule would carry a charge of $2 \times 10^{-9}e$, and the departure of the whole mass of hydrogen would alter the charge of the tank by $10^{16}e$, a gigantic effect. In fact, the experiment could have revealed a residual molecular charge as small as $2 \times 10^{-20}e$, and none was observed. This proved that the proton and the electron do not differ in magnitude of charge by more than 1 part in 10^{20}.

Perhaps the equality is really *exact* for some reason we don't yet understand. It may be connected with the possibility, suggested by recent theories, that a proton can, *very* rarely, decay into a positron and some uncharged particles. If that were to occur, even the slightest discrepancy between proton charge and positron charge would violate charge conservation. Several experiments designed to detect the decay of a proton have not yet, as this is written in 1983, registered with certainty a single decay. If and when such an event is observed, it will show that exact equality of the magnitude of the charge of the proton and the charge of the electron (the positron's antiparticle) can be regarded as a corollary of the more general law of charge conservation.

That notwithstanding, there is now overwhelming evidence that the *internal* structure of all the strongly interacting particles called *hadrons*—a class which includes the proton and the neutron—involves basic units called *quarks,* whose electric charges come in multiples of $e/3$. The proton, for example, is made with three quarks, two of charge $\frac{2}{3}e$ and one with charge $-\frac{1}{3}e$. The neutron contains one quark of charge $\frac{2}{3}e$ and two quarks with charge $-\frac{1}{3}e$.

Several experimenters have searched for single quarks, either free or attached to ordinary matter. The fractional charge of such a quark, since it cannot be neutralized by any number of electrons or protons, should betray the quark's presence. So far no fractionally charged particle has been conclusively identified. There are theoretical grounds for suspecting that the liberation of a quark from a hadron is impossible, but the question remains open at this time.

The fact of charge quantization lies outside the scope of classical electromagnetism, of course. We shall usually ignore it and act as if our point charges q could have any strength whatever. This will not get us into trouble. Still, it is worth remembering that classical theory cannot be expected to explain the structure of the elementary particles. (It is not certain that present quantum theory can either!) What holds the electron together is as mysterious as what fixes the precise value of its charge. Something more than electrical forces must be involved, for the electrostatic forces between different parts of the electron would be repulsive.

In our study of electricity and magnetism we shall treat the charged particles simply as carriers of charge, with dimensions so small that their extension and structure is for most purposes quite insignificant. In the case of the proton, for example, we know from high-energy scattering experiments that the electric charge does not extend appreciably beyond a radius of 10^{-13} cm. We recall that Rutherford's analysis of the scattering of alpha particles showed that even heavy nuclei have their electric charge distributed over a region smaller than 10^{-11} cm. For the physicist of the nineteenth century a "point charge" remained an abstract notion. Today we are on familiar terms with the atomic particles. The graininess of electricity is so conspicuous in our modern description of nature that we find a point charge less of an artificial idealization than a smoothly varying distribution of charge density. When we postulate such smooth charge distributions, we may think of them as averages over very large numbers of elementary charges, in the same way that we can define the macroscopic density of a liquid, its lumpiness on a molecular scale notwithstanding.

COULOMB'S LAW

1.4 As you probably already know, the interaction between electric charges at rest is described by Coulomb's law: Two stationary electric charges repel or attract one another with a force proportional to the product of the magnitude of the charges and inversely proportional to the square of the distance between them.

We can state this compactly in vector form:

$$\mathbf{F}_2 = k\,\frac{q_1 q_2 \hat{\mathbf{r}}_{21}}{r_{21}^2} \qquad (1)$$

Here q_1 and q_2 are numbers (scalars) giving the magnitude and sign of the respective charges, $\hat{\mathbf{r}}_{21}$ is the unit vector in the direction† from charge 1 to charge 2, and $\mathbf{F}_2$ is the force acting on charge 2. Thus Eq. 1 expresses, among other things, the fact that like charges repel and unlike attract. Also, the force obeys Newton's third law; that is, $\mathbf{F}_2 = -\mathbf{F}_1$.

The unit vector $\hat{\mathbf{r}}_{21}$ shows that the force is parallel to the line joining the charges. It could not be otherwise unless space itself has some built-in directional property, for with two point charges alone in empty and isotropic space, no other direction could be singled out.

†The convention we adopt here may not seem the natural choice, but it is more consistent with the usage in some other parts of physics and we shall try to follow it throughout this book.

If the point charge itself had some internal structure, with an axis defining a direction, then it would have to be described by more than the mere scalar quantity q. It is true that some elementary particles, including the electron, do have another property, called *spin.* This gives rise to a magnetic force between two electrons in addition to their electrostatic repulsion. This magnetic force does not, in general, act in the direction of the line joining the two particles. It decreases with the inverse fourth power of the distance, and at atomic distances of 10^{-8} cm the Coulomb force is already about 10^4 times stronger than the magnetic interaction of the spins. Another magnetic force appears if our charges are moving—hence the restriction to stationary charges in our statement of Coulomb's law. We shall return to these magnetic phenomena in later chapters.

Of course we must assume, in writing Eq. 1, that both charges are well localized, each occupying a region small compared with r_{21}. Otherwise we could not even define the distance r_{21} precisely.

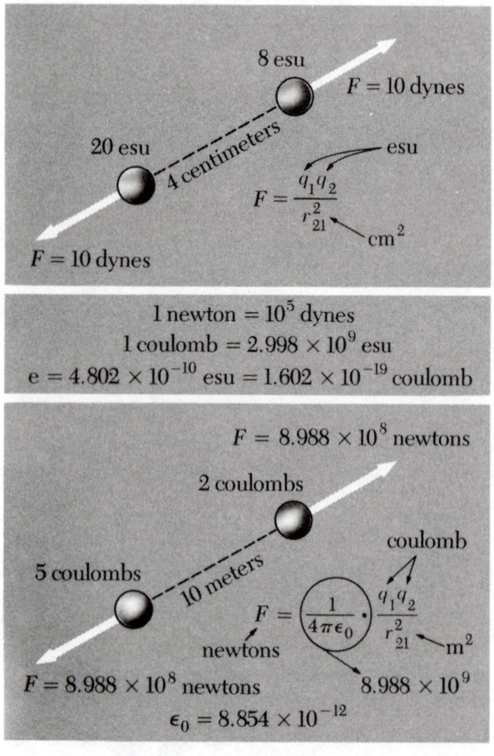

FIGURE 1.2
Coulomb's law expressed in CGS electrostatic units (top) and in SI units (bottom). The constant ϵ_0 and the factor relating coulombs to esu are connected, as we shall learn later, with the speed of light. We have rounded off the constants in the figure to four-digit accuracy. The precise values are given in Appendix E.

The value of the constant k in Eq. 1 depends on the units in which r, **F**, and q are to be expressed. Usually we shall choose to measure r_{21} in cm, **F** in dynes, and charge in electrostatic units (esu). Two like charges of 1 esu each repel one another with a force of 1 dyne when they are 1 cm apart. Equation 1, with $k = 1$, is the definition of the unit of charge in CGS electrostatic units, the dyne having already been defined as the force that will impart an acceleration of one centimeter per second per second to a one-gram mass. Figure 1.2*a* is just a graphic reminder of the relation. The magnitude of e, the fundamental quantum of electric charge, is 4.8023×10^{-10} esu.

We want to be familiar also with the unit of charge called the *coulomb.* This is the unit for electric charge in the *Système Internationale* (SI) family of units. That system is based on the meter, kilogram, and second as units of length, mass, and time, and among its electrical units are the familiar volt, ohm, ampere, and watt.

The SI unit of force is the newton, equivalent to exactly 10^5 dynes, the force that will cause a one-kilogram mass to accelerate at one meter per second per second. The coulomb is defined by Eq. 1 with **F** in newtons, r_{21} in meters, charges q_1 and q_2 in coulombs, and $k = 8.988 \times 10^9$. A charge of 1 coulomb equals 2.998×10^9 esu. Instead of k, it is customary to introduce a constant ϵ_0, which is just $(4\pi k)^{-1}$, with which the same equation is written

$$\mathbf{F} = \frac{1}{4\pi\epsilon_0} \frac{q_1 q_2 \hat{\mathbf{r}}_{21}}{r_{21}^2} \qquad (1')$$

Refer to Fig. 1.2*b* for an example. The constant ϵ_0 will appear in several SI formulas that we'll meet in the course of our study. The exact value of ϵ_0 and the exact relation of the coulomb to the esu can be found in Appendix E. For our purposes the following approximations are quite accurate enough: $k = 9 \times 10^9$; 1 coulomb $= 3 \times 10^9$ esu.

Fortunately the electronic charge e is very close to an easily remembered approximate value in either system: $e = 4.8 \times 10^{-10}$ esu $= 1.6 \times 10^{-19}$ coulomb.

The only way we have of detecting and measuring electric charges is by observing the interaction of charged bodies. One might wonder, then, how much of the apparent content of Coulomb's law is really only definition. As it stands, the significant physical content is the statement of inverse-square dependence and the implication that electric charge is *additive* in its effect. To bring out the latter point, we have to consider *more* than two charges. After all, if we had only two charges in the world to experiment with, q_1 and q_2, we could never measure them separately. We could verify only that F is proportional to $1/r_{21}^2$. Suppose we have *three* bodies carrying charges q_1, q_2, and q_3. We can measure the force on q_1 when q_2 is 10 cm away from q_1 and q_3 is very far away, as in Fig. 1.3*a*. Then we can take q_2 away, bring q_3 into q_2's former position, and again measure the force on q_1. Finally, we bring q_2 and q_3 very close together and locate the combination 10 cm from q_1. We find by measurement that the force on q_1 is equal to the sum of the forces previously measured. This is a significant result that could *not* have been predicted by logical arguments from symmetry like the one we used above to show that the force between two point charges *had* to be along the line joining them. *The force with which two charges interact is not changed by the presence of a third charge.*

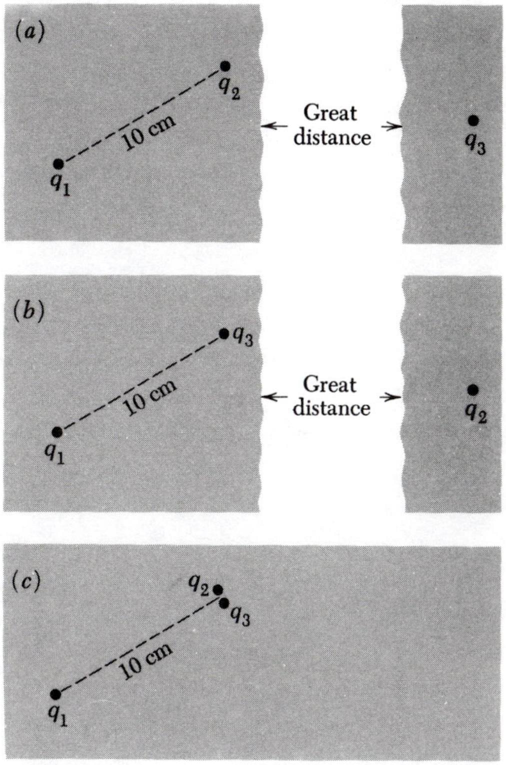

FIGURE 1.3

The force on q_1 in (*c*) is the sum of the forces on q_1 in (*a*) and (*b*).

No matter how many charges we have in our system. Coulomb's law (Eq. 1) can be used to calculate the interaction of every pair. This is the basis of the principle of *superposition,* which we shall invoke again and again in our study of electromagnetism. Superposition means combining two sets of sources into one system by adding the second system "on top of" the first without altering the configuration of either one. Our principle ensures that the force on a charge placed at any point in the combined system will be the vector sum of the forces that each set of sources, acting alone, causes to act on a charge at that point. This principle must not be taken lightly for granted. There may well be a domain of phenomena, involving very small distances or very intense forces, where superposition *no longer holds.* Indeed, we know of quantum phenomena in the electromagnetic field which do represent a failure of superposition, seen from the viewpoint of the classical theory.

Thus the physics of electrical interactions comes into full view only when we have *more* than two charges. We can go beyond the explicit statement of Eq. 1 and assert that, with the three charges in Fig. 1.3 occupying any positions whatever, the force on any one of them, such as q_3, is correctly given by this equation:

$$\mathbf{F}_3 = \frac{q_3 q_1 \hat{\mathbf{r}}_{31}}{r_{31}^2} + \frac{q_3 q_2 \hat{\mathbf{r}}_{32}}{r_{32}^2} \qquad (2)$$

The experimental verification of the inverse-square law of electrical attraction and repulsion has a curious history. Coulomb himself announced the law in 1786 after measuring with a torsion balance the force between small charged spheres. But 20 years earlier Joseph Priestly, carrying out an experiment suggested to him by Benjamin Franklin, had noticed the absence of electrical influence within a hollow charged container and made an inspired conjecture: "May we not infer from this experiment that the attraction of electricity is subject to the same laws with that of gravitation and is therefore according to the square of the distances; since it is easily demonstrated that were the earth in the form of a shell, a body in the inside of it would not be attracted to one side more than the other."† The same idea was the basis of an elegant experiment in 1772 by Henry Cavendish. Cavendish charged a spherical conducting shell which contained within it, and temporarily connected to it, a smaller sphere. The outer shell was then separated into two halves and carefully removed, the inner sphere having been first disconnected. This sphere was tested for charge, the absence of which would confirm the inverse-square law. Assuming that a deviation from the inverse-square law could be expressed as a difference in the exponent, $2 + \delta$, say, instead of 2, Cavendish concluded that δ must be less than 0.03. This experiment of Cavendish remained largely unknown until Maxwell discovered and published Cavendish's notes a century later (1876). At that time also Maxwell repeated the experiment with improved apparatus, pushing the limit down to $\delta < 10^{-6}$. The latest of several modern versions of the Cavendish experiment,‡ if interpreted the same way, yielded the fantastically small limit $\delta < 10^{-15}$.

During the second century after Cavendish, however, the question of interest changed somewhat. Never mind how perfectly Coulomb's law works for charged objects in the laboratory—is there a range of distances where it completely breaks down? There are two domains in either of which a breakdown is conceivable. The first is the domain of very small distances, distances less than 10^{-14} cm where electromagnetic theory as we know it may not work at all. As for very large distances, from the geographical, say, to the astronomical, a test of Coulomb's law by the method of Cavendish is obviously not feasible. Nevertheless we do observe certain large-scale electromagnetic phenomena which prove that the laws of classical electromagnetism work over very long distances. One of the most stringent tests is provided by planetary magnetic fields, in particular, the magnetic field of the giant planet Jupiter, which was surveyed in the mission of Pioneer

†Joseph Priestly, "The History and Present State of Electricity," vol. II, London, 1767.

‡E. R. Williams, J. G. Faller, and H. Hill. *Phys. Rev. Lett.* **26**:721 (1971).

10. The spatial variation of this field was carefully analyzed† and found to be entirely consistent with classical theory out to a distance of at least 10^5 kilometers (km) from the planet. This is tantamount to a test, albeit indirect, of Coulomb's law over that distance.

To summarize, we have every reason for confidence in Coulomb's law over the stupendous range of 24 decades in distance, from 10^{-14} to 10^{10} cm, if not farther, and we take it as the foundation of our description of electromagnetism.

ENERGY OF A SYSTEM OF CHARGES

1.5 In principle, Coulomb's law is all there is to electrostatics. Given the charges and their locations we can find all the electrical forces. Or given that the charges are free to move under the influence of other kinds of forces as well, we can find the equilibrium arrangement in which the charge distribution will remain stationary. In the same sense, Newton's laws of motion are all there is to mechanics. But in both mechanics and electromagnetism we gain power and insight by introducing other concepts, most notably that of energy.

Energy is a useful concept here because electrical forces are *conservative.* When you push charges around in electric fields, no energy is irrecoverably lost. Everything is perfectly reversible. Consider first the work which must be done *on* the system to bring some charged bodies into a particular arrangement. Let us start with two charged bodies or particles very far apart from one another, as indicated at the top of Fig. 1.4, carrying charges q_1 and q_2. Whatever energy may have been needed to create these two concentrations of charge originally we shall leave entirely out of account. Bring the particles slowly together until the distance between them is r_{12}. How much work does this take?

It makes no difference whether we bring q_1 toward q_2 or the other way around. In either case the work done is the integral of the product: force times displacement in direction of force. The force that has to be applied to move one charge toward the other is equal to and opposite the Coulomb force.

$$W = \int \text{force} \times \text{distance} = \int_{r=\infty}^{r_{12}} \frac{q_1 q_2(-dr)}{r^2} = \frac{q_1 q_2}{r_{12}} \qquad (3)$$

Because r is changing from ∞ to r_{12}, the increment of displacement is $-dr$. We know the work done on the system must be positive for charges of like sign; they have to be pushed together. With q_1 and q_2 in esu, and r_{12} in cm, Eq. 3 gives the work in ergs.

†L. Davis, Jr., A. S. Goldhaber, M. M. Nieto, *Phys. Rev. Lett.* **35**:1402 (1975). For a review of the history of the exploration of the outer limit of classical electromagnetism, see A. S. Goldhaber and M. M. Nieto, *Rev. Mod. Phys.* **43**:277 (1971).

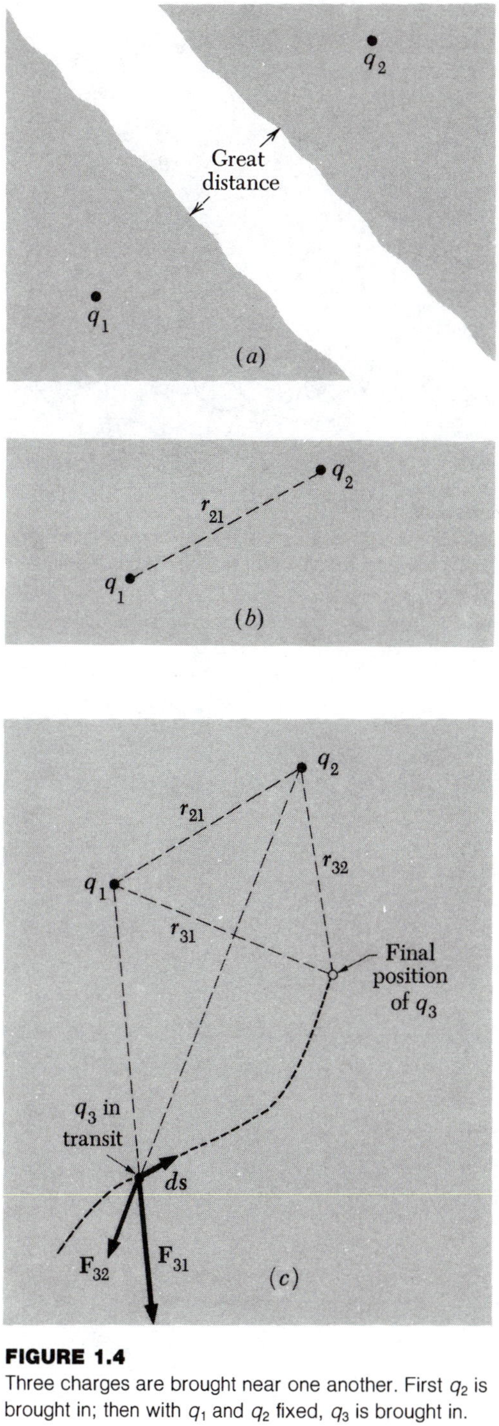

FIGURE 1.4
Three charges are brought near one another. First q_2 is brought in; then with q_1 and q_2 fixed, q_3 is brought in.

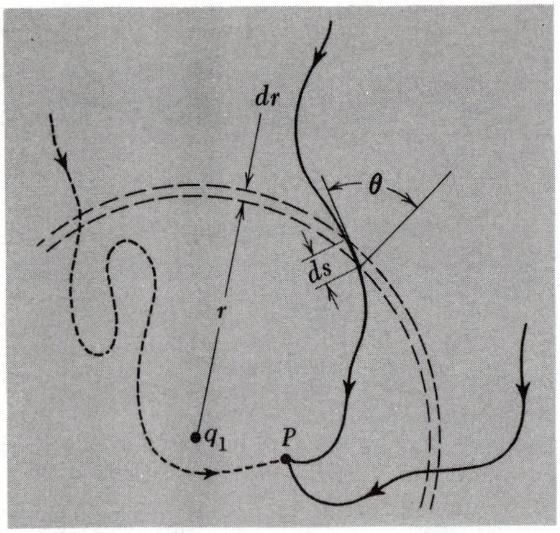

FIGURE 1.5
Because the force is central, the sections of different paths between $r + dr$ and r require the same amount of work.

This work is the same whatever the path of approach. Let's review the argument as it applies to the two charges q_1 and q_2 in Fig. 1.5. There we have kept q_1 fixed, and we show q_2 moved to the same final position along two different paths. Every spherical shell such as the one indicated between r and $r + dr$ must be crossed by both paths. The increment of work involved, $-\mathbf{F} \cdot d\mathbf{s}$ in this bit of path, is the same for the two paths.† The reason is that $\mathbf{F}$ has the same magnitude at both places and is directed radially from q_1, while $ds = dr/\cos\theta$; hence $\mathbf{F} \cdot d\mathbf{s} = F\,dr$. Each increment of work along one path is matched by a corresponding increment on the other, so the sums must be equal. Our conclusion holds even for paths that loop in and out, like the dotted path in Fig. 1.5. (Why?)

Returning now to the two charges as we left them in Fig. 1.4*b*, let us bring in from some remote place a third charge q_3 and move it to a point P_3 whose distance from charge 1 is r_{31} cm, and from charge 2, r_{32} cm. The work required to effect this will be

$$W_3 = -\int_{\infty}^{P_3} \mathbf{F}_3 \cdot d\mathbf{s} \tag{4}$$

Thanks to the additivity of electrical interactions, which we have already emphasized,

$$\begin{aligned} -\int \mathbf{F}_3 \cdot d\mathbf{s} &= -\int (\mathbf{F}_{31} + \mathbf{F}_{32}) \cdot d\mathbf{s} \\ &= -\int \mathbf{F}_{31} \cdot d\mathbf{r} - \int \mathbf{F}_{32} \cdot d\mathbf{r} \end{aligned} \tag{5}$$

That is, the work required to bring q_3 to P_3 is the sum of the work needed when q_1 is present alone and that needed when q_2 is present alone.

$$W_3 = \frac{q_1 q_3}{r_{31}} + \frac{q_2 q_3}{r_{32}} \tag{6}$$

The total work done in assembling this arrangement of three charges, which we shall call U, is therefore

$$U = \frac{q_1 q_2}{r_{12}} + \frac{q_1 q_3}{r_{13}} + \frac{q_2 q_3}{r_{23}} \tag{7}$$

We note that q_1, q_2, and q_3 appear symmetrically in the expression above, in spite of the fact that q_3 was brought up last. We would have reached the same result if q_3 had been brought in first. (Try it.) Thus U is independent of the *order* in which the charges were assem-

†Here we use for the first time the scalar product, or "dot product," of two vectors. A reminder: the scalar product of two vectors $\mathbf{A}$ and $\mathbf{B}$, written $\mathbf{A} \cdot \mathbf{B}$, is the number $A\,B\cos\theta$. A and B are the magnitudes of the vectors $\mathbf{A}$ and $\mathbf{B}$, and θ is the angle between them. Expressed in terms of cartesian components of the two vectors, $\mathbf{A} \cdot \mathbf{B} = A_x B_x + A_y B_y + A_z B_z$.

bled. Since it is independent also of the route by which each charge was brought in, U must be a unique property of the final arrangement of charges. We may call it the *electrical potential energy* of this particular system. There is a certain arbitrariness, as always, in the definition of a potential energy. In this case we have chosen the zero of potential energy to correspond to the situation with the three charges already in existence but infinitely far apart from one another. The potential energy *belongs to the configuration as a whole.* There is no meaningful way of assigning a certain fraction of it to one of the charges.

It is obvious how this very simple result can be generalized to apply to any number of charges. If we have N different charges, in any arrangement in space, the potential energy of the system is calculated by summing over all pairs, just as in Eq. 7. The zero of potential energy, as in that case, corresponds to all charges far apart.

As an example, let us calculate the potential energy of an arrangement of eight negative charges on the corners of a cube of side b, with a positive charge in the center of the cube, as in Fig. 1.6*a*. Suppose each negative charge is an electron with charge $-e$, while the central particle carries a double positive charge, $2e$. Summing over all pairs, we have

$$U = \frac{8(-2e^2)}{(\sqrt{3}/2)b} + \frac{12e^2}{b} + \frac{12e^2}{\sqrt{2}b} + \frac{4e^2}{\sqrt{3}b} = \frac{4.32e^2}{b} \qquad (8)$$

Figure 1.6*b* shows where each term in this sum comes from. The energy is positive, indicating that work had to be done on the system to assemble it. That work could, of course, be recovered if we let the charges move apart, exerting forces on some external body or bodies. Or if the electrons were simply to fly apart from this configuration, the *total kinetic energy* of all the particles would become equal to U. This would be true whether they came apart simultaneously and symmetrically, or were released one at a time in any order. Here we see the power of this simple notion of the total potential energy of the system. Think what the problem would be like if we had to compute the resultant vector force on every particle at every stage of assembly of the configuration! In this example, to be sure, the geometrical symmetry would simplify that task; even so, it would be more complicated than the simple calculation above.

One way of writing the instruction for the sum over pairs is this:

$$U = \frac{1}{2}\sum_{j=1}^{N}\sum_{k\neq j}\frac{q_j q_k}{r_{jk}} \qquad (9)$$

The double-sum notation, $\Sigma_{j=1}^{N}\Sigma_{k\neq j}$, says: Take $j = 1$ and sum over $k = 2, 3, 4, \ldots, N$; then take $j = 2$ and sum over $k = 1, 3, 4, \ldots, N$; and so on, through $j = N$. Clearly this includes every pair *twice,* and to correct for that we put in front the factor ½.

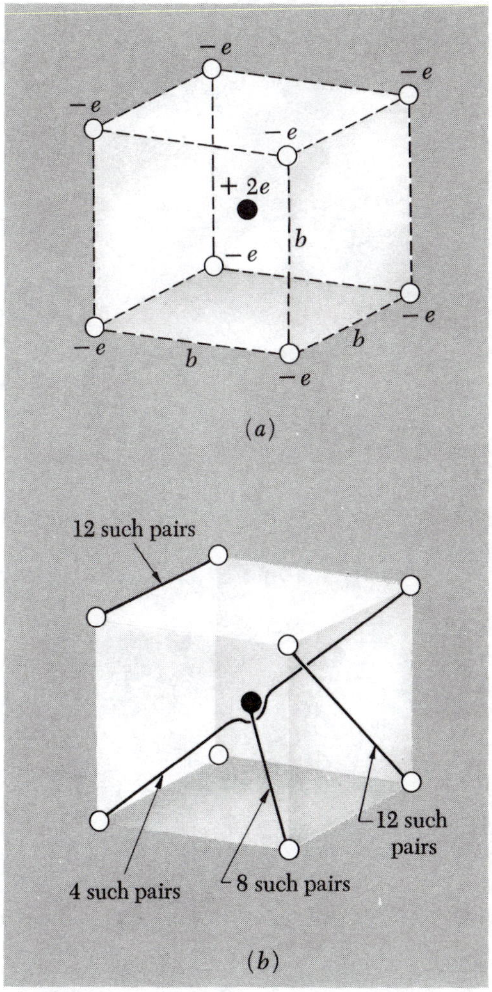

FIGURE 1.6
(a) The potential energy of this arrangement of nine point charges is given by Eq. 9. *(b)* Four types of pairs are involved in the sum.

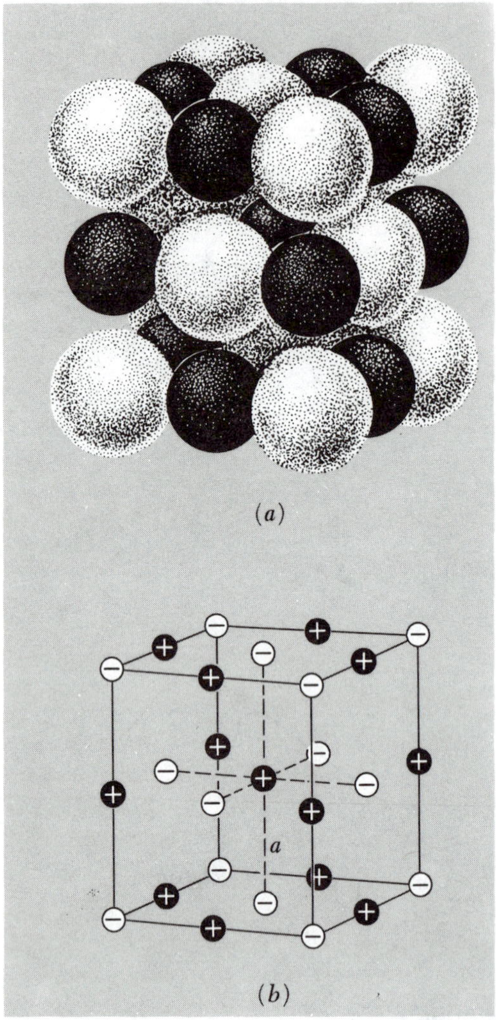

FIGURE 1.7
A portion of a sodium chloride crystal, with the ions Na^+ and $C1^-$ shown in about the right relative proportions (*a*), and replaced by equivalent point charges (*b*).

ELECTRICAL ENERGY IN A CRYSTAL LATTICE

1.6 These ideas have an important application in the physics of crystals. We know that an ionic crystal like sodium chloride can be described, to a very good approximation, as an arrangement of positive ions (Na^+) and negative ions (Cl^-) alternating in a regular three-dimensional array or lattice. In sodium chloride the arrangement is that shown in Fig. 1.7*a*. Of course the ions are not point charges, but they are nearly spherical distributions of charge and therefore (as we shall presently prove) the electrical forces they exert on one another are the same as if each ion were replaced by an equivalent point charge at its center. We show this electrically equivalent system in Fig. 1.7*b*. The electrostatic potential energy of the lattice of charges plays an important role in the explanation of the stability and cohesion of the ionic crystal. Let us see if we can estimate its magnitude.

We seem to be faced at once with a sum that is enormous, if not doubly infinite, for any macroscopic crystal contains 10^{20} atoms at least. Will the sum converge? Now what we hope to find is the potential energy per unit volume or mass of crystal. We confidently expect this to be independent of the size of the crystal, based on the general argument that one end of a macroscopic crystal can have little influence on the other. Two grams of sodium chloride ought to have twice the potential energy of 1 gm, and the shape should not be important so long as the surface atoms are a small fraction of the total number of atoms. We would be *wrong* in this expectation if the crystal were made out of ions of one sign only. Then, 1 gm of crystal would carry an enormous electric charge, and putting two such crystals together to make a 2-gm crystal would take a fantastic amount of energy. (You might estimate how much!) The situation is saved by the fact that the crystal structure is an alternation of equal and opposite charges, so that any macroscopic bit of crystal is very nearly neutral.

To evaluate the potential energy we first observe that every positive ion is in a position equivalent to that of every other positive ion. Furthermore, although it is perhaps not immediately obvious from Fig. 1.7, the arrangement of positive ions around a negative ion is exactly the same as the arrangement of negative ions around a positive ion, and so on. Hence we may take one ion as a center, it matters not which kind, sum over *its* interactions with all the others, and simply multiply by the total number of ions of both kinds. This reduces the double sum in Eq. 9, to a single sum and a factor N; we must still apply the factor ½ to compensate for including each pair twice. That is, the energy of a sodium chloride lattice composed of a total of N ions is

$$U = \frac{1}{2} N \sum_{k=2}^{N} \frac{q_1 q_k}{r_{1k}} \tag{10}$$

Taking the positive ion at the center as in Fig. 1.7*b,* our sum runs over all its neighbors near and far. The leading terms start out as follows:

$$U = \frac{1}{2} N\left(-\frac{6e^2}{a} + \frac{12e^2}{\sqrt{2}a} - \frac{8e^2}{\sqrt{3}a} + \cdots \right) \tag{11}$$

The first term comes from the 6 nearest chlorine ions, at distance a, the second from the 12 sodium ions on the cube edges, and so on. It is clear, incidentally, that this series does not converge *absolutely;* if we were so foolish as to try to sum all the positive terms first, that sum would diverge. To evaluate such a sum, we should arrange it so that as we proceed outward, including ever more distant ions, we include them in groups which represent nearly neutral shells of material. Then if the sum is broken off, the more remote ions which have been neglected will be such an even mixture of positive and negative charges that we can be confident their contribution would have been small. This is a crude way to describe what is actually a somewhat more delicate computational problem. The numerical evaluation of such a series is easily accomplished with a computer. The answer in this example happens to be

$$U = \frac{-0.8738Ne^2}{a} \tag{12}$$

Here N, the number of ions, is twice the number of NaCl molecules.

The negative sign shows that work would have to be *done* to take the crystal apart into ions. In other words, the electrical energy helps to explain the cohesion of the crystal. If this were the whole story, however, the crystal would collapse, for the potential energy of the charge distribution is obviously *lowered* by shrinking all the distances. We meet here again the familiar dilemma of classical—that is, non-quantum—physics. No system of stationary particles can be in stable equilibrium, according to classical laws, under the action of electrical forces alone. Does this make our analysis useless? Not at all. Remarkably, and happily, in the quantum physics of crystals the electrical potential energy can still be given meaning, and can be computed very much in the way we have learned here.

THE ELECTRIC FIELD

1.7 Suppose we have some arrangement of charges, $q_1, q_2, \ldots, q_N$, fixed in space, and we are interested not in the forces they exert on one another but only in their effect on some other charge q_0 which might be brought into their vicinity. We know how to calculate the

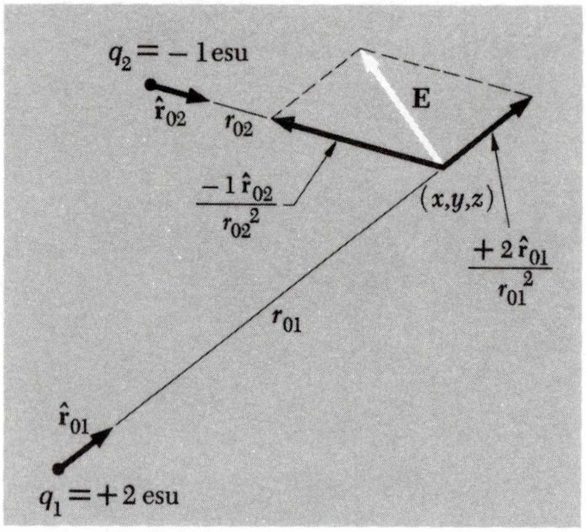

FIGURE 1.8
The field at a point is the vector sum of the fields of each of the charges in the system.

resultant force on this charge, given its position which we may specify by the coordinates x, y, z. The force on the charge q_0 is

$$\mathbf{F}_0 = \sum_{j=1}^{N} \frac{q_0 q_j \hat{\mathbf{r}}_{0j}}{r_{0j}^2} \tag{13}$$

where $\mathbf{r}_{0j}$ is the vector from the jth charge in the system to the point (x, y, z). The force is proportional to q_0, so if we divide out q_0 we obtain a vector quantity which depends only on the structure of our original system of charges, $q_1, \ldots, q_N$, and on the position of the point (x, y, z). We call this vector function of x, y, z the *electric field* arising from the $q_1, \ldots, q_N$ and use the symbol **E** for it. The charges $q_1, \ldots, q_N$ we call *sources* of the field. We may take as the *definition* of the electric field **E** of a charge distribution, at the point (x, y, z)

$$\mathbf{E}(x, y, z) = \sum_{j=1}^{N} \frac{q_j \hat{\mathbf{r}}_{0j}}{r_{0j}^2} \tag{14}$$

Figure 1.8 illustrates the vector addition of the field of a point charge of 2 esu to the field of a point charge of -1 esu, at a particular point in space. In the CGS system of units, electric field strength is expressed in dynes per unit charge, that is, dynes/esu.

In SI units with the coulomb as the unit of charge and the newton as the unit of force, the electric field strength **E** can be expressed in newtons/coulomb, and Eq. 14 would be written like this:

$$\mathbf{E} = \frac{1}{4\pi\epsilon_0} \sum_{j=1}^{N} \frac{q_j \hat{\mathbf{r}}_{0j}}{r_{0j}^2} \tag{14'}$$

each distance r_{0j} being measured in meters.

After the introduction of the electric potential in the next chapter, we shall have another, and completely equivalent, way of expressing the unit of electric field strength; namely, statvolts/cm in the CGS system of units and volts/meter in SI units.

So far we have nothing really new. The electric field is merely another way of describing the system of charges; it does so by giving the force per unit charge, in magnitude and direction, that an exploring charge q_0 would experience at any point. We have to be a little careful with that interpretation. Unless the source charges are really immovable, the introduction of some finite charge q_0 may cause the source charges to shift their positions, so that the field itself, as defined by Eq. 14, is different. That is why we assumed fixed charges to begin our discussion. People sometimes define the field by requiring q_0 to be an "infinitesimal" test charge, letting **E** be the limit of $\mathbf{F}/q_0$ as $q_0 \to 0$. Any flavor of rigor this may impart is illusory. Remember that in the real world we have never observed a charge smaller than e! Actually, if we take Eq. 14 as our *definition* of **E**, without reference to a test charge, no problem arises and the sources need not be fixed.

If the introduction of a new charge causes a shift in the source charges, then it has indeed brought about a change in the electric field, and if we want to predict the force on the new charge, we must use the new electric field in computing it.

Perhaps you still want to ask, what *is* an electric field? Is it something real, or is it merely a name for a factor in an equation which has to be multiplied by something else to give the numerical value of the force we measure in an experiment? Two observations may be useful here. First, since it works, it doesn't make any difference. That is not a frivolous answer, but a serious one. Second, the fact that the electric field vector at a point in space is all we need know to predict the force that will act on *any* charge at that point is by no means trivial. It might have been otherwise! If no experiments had ever been done, we could imagine that, in two different situations in which unit charges experience equal force, test charges of strength 2 units might experience different forces, depending on the nature of the other charges in the system. If that were true, the field description wouldn't work. The electric field attaches to every point in a system a *local property,* in this sense: If we know **E** in some small neighborhood, we know, *without further inquiry,* what will happen to any charges in that neighborhood. We don't need to ask what produced the field.

To visualize an electric field, you need to associate a vector, that is, a magnitude and direction, with every point in space. We shall use various schemes, none of them wholly satisfactory, to depict vector fields in this book.

It is hard to draw in two dimensions a picture of a vector func-

FIGURE 1.9
(*a*) Field of a charge $q_1 = 3$. (*b*) Field of a charge $q_2 = -1$. Both representations are necessarily crude and only roughly quantitative.

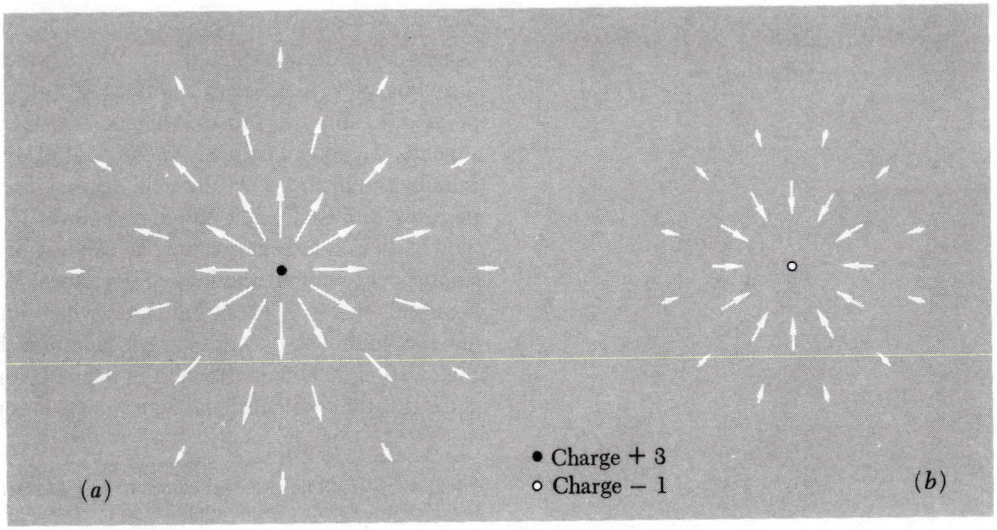
• Charge + 3
∘ Charge − 1
(a)
(b)

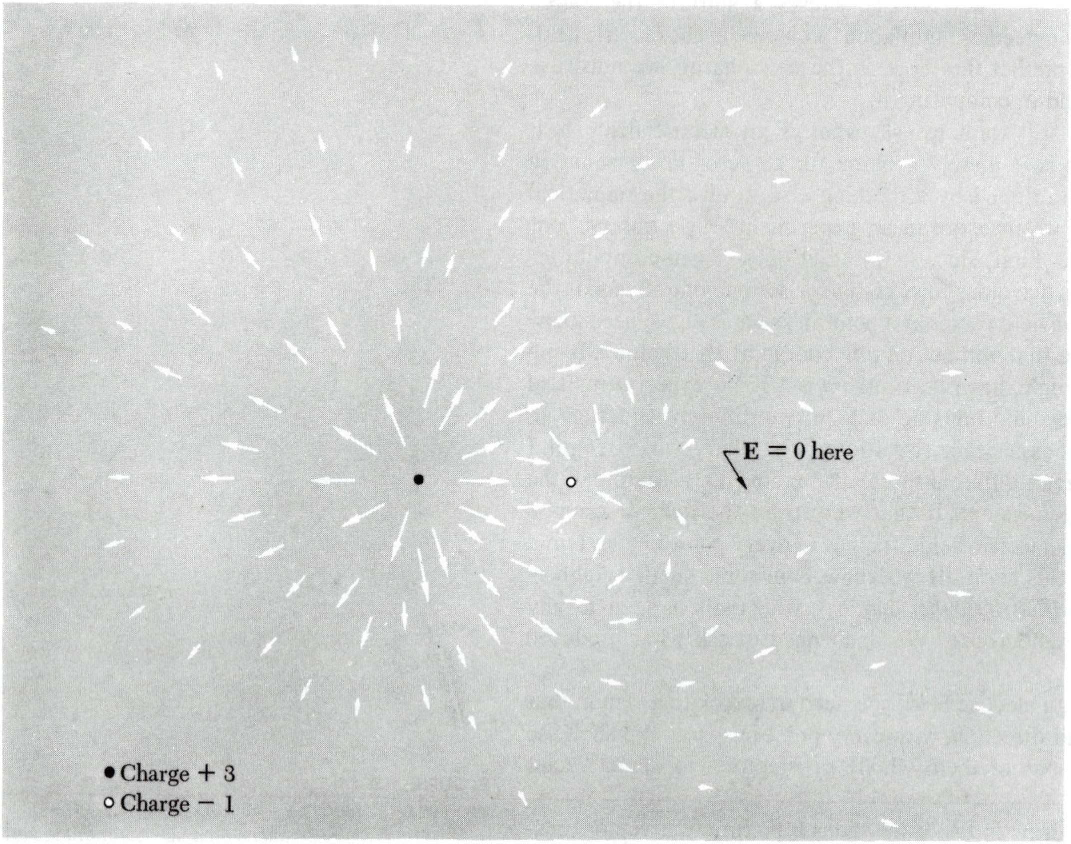
E = 0 here
● Charge + 3
○ Charge − 1

FIGURE 1.10
The field in the vicinity of two charges, $q_1 = +3$, $q_2 = -1$, is the superposition of the fields in Fig. 1.9*a* and *b*.

tion in three-dimensional space. We can indicate the magnitude and direction of **E** at various points by drawing little arrows near those points, making the arrows longer where E is larger.† Using this scheme, we show in Fig. 1.9*a* the field of an isolated point charge of 3 units and in Fig. 1.9*b* the field of a point charge of -1 unit. These pictures admittedly add nothing whatever to our understanding of the field of an isolated charge; anyone can imagine a simple radial inverse-square field without the help of a picture. We show them in order to combine the two fields in Fig. 1.10, which indicates in the same manner the field of two such charges separated by a distance a. All that Fig. 1.10 can show is the field in a plane containing the charges. To get a full three-dimensional representation one must imagine the fig-

†Such a representation is rather clumsy at best. It is hard to indicate the point in space to which a particular vector applies, and the range of magnitudes of E is usually so large that it is impracticable to make the lengths of the arrows proportional to E.

ure rotated around the symmetry axis. In Fig. 1.10 there is one point in space where **E** is zero. How far from the nearest charge must this point lie? Notice also that toward the edge of the picture the field points more or less radially outward all around. One can see that at a very large distance from the charges the field will look very much like the field from a positive point charge. This is to be expected because the separation of the charges cannot make very much difference for points far away, and a point charge of 2 units is just what we would have left if we superimposed our two sources at one spot.

Another way to depict a vector field is to draw *field lines*. These are simply curves whose tangent, at any point, lies in the direction of the field at that point. Such curves will be smooth and continuous

FIGURE 1.11
Some field lines in the electric field around two charges, $q_1 = +3$, $q_2 = -1$.

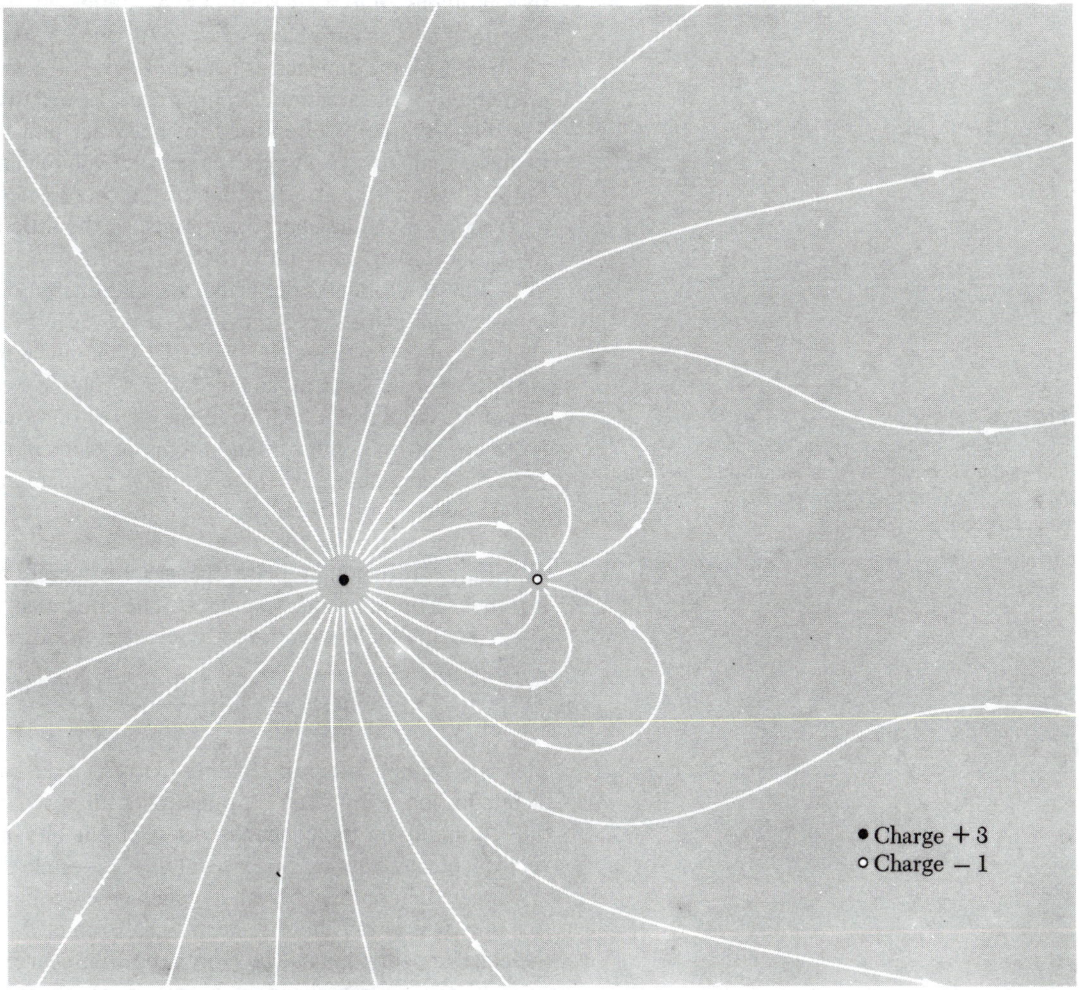
• Charge + 3
○ Charge − 1

except at singularities such as point charges, or points like the one in the example of Fig. 1.10 where the field is zero. A field line plot does not directly give the magnitude of the field, although we shall see that, in a general way, the field lines converge as we approach a region of strong field and spread apart as we approach a region of weak field. In Fig. 1.11 are drawn some field lines for the same arrangement of charges as in Fig. 1.10, a positive charge of 3 units and a negative charge of 1 unit. Again, we are restricted by the nature of paper and ink to a two-dimensional section through a three-dimensional bundle of curves.

CHARGE DISTRIBUTIONS

1.8 This is as good a place as any to generalize from *point charges* to *continuous charge distributions.* A volume distribution of charge is described by a scalar charge-density function ρ, which is a function of position, with the dimensions *charge/volume.* That is, ρ times a volume element gives the amount of charge contained in that volume element. The same symbol is often used for mass per unit volume, but in this book we shall always give charge per unit volume first call on the symbol ρ. If we write ρ as a function of the coordinates x, y, z, then $\rho(x, y, z)\,dx\,dy\,dz$ is the charge contained in the little box, of volume $dx\,dy\,dz$, located at the point (x, y, z).

On an atomic scale, of course, the charge density varies enormously from point to point; even so, it proves to be a useful concept in that domain. However, we shall use it mainly when we are dealing with large-scale systems, so large that a volume element $dv = dx\,dy\,dz$ can be quite small relative to the size of our system, although still large enough to contain many atoms or elementary charges. As we have remarked before, we face a similar problem in defining the ordinary mass density of a substance.

FIGURE 1.12
Each element of the charge distribution $\rho(x', y', z')$ makes a contribution to the electric field **E** at this point (x, y, z). The total field at this point is the sum of all such contributions (Eq. 15).

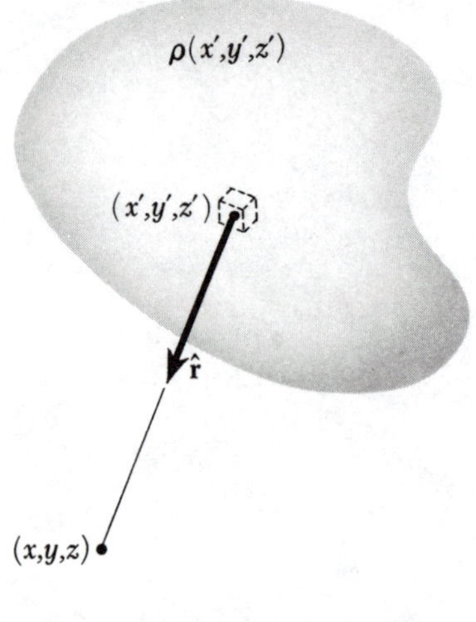

If the source of the electric field is to be a continuous charge distribution rather than point charges, we merely replace the sum in Eq. 14 with the appropriate integral. The integral gives the electric field at (x, y, z), which is produced by charges at other points (x', y', z').

$$\mathbf{E}(x, y, z) = \int \frac{\rho(x', y', z')\,\hat{\mathbf{r}}\,dx'\,dy'\,dz'}{r^2} \tag{15}$$

This is a volume integral. Holding (x, y, z) fixed we let the variables of integration, x', y', and z', range over all space containing charge, thus summing up the contributions of all the bits of charge. The unit vector $\hat{\mathbf{r}}$ points from (x', y', z') to (x, y, z)—unless we want to put a minus sign before the integral, in which case we may reverse the direction of $\hat{\mathbf{r}}$. It is always hard to keep signs straight. Let's remember that the electric field points *away* from a positive source (Fig. 1.12).

In the neighborhood of a true point charge the electric field grows infinite like $1/r^2$ as we approach the point. It makes no sense to talk about the field *at* the point charge. As our ultimate physical sources of field are not, we believe, infinite concentrations of charge in zero volume but instead finite structures, we simply ignore the mathematical singularities implied by our point-charge language and rule out of bounds the interior of our elementary sources. A continuous charge distribution $\rho(x', y', z')$ which is nowhere infinite gives no trouble at all. Equation 15 can be used to find the field at any point within the distribution. The integrand doesn't blow up at $r = 0$ because the volume element in the numerator is in that limit proportional to $r^2\,dr$. That is to say, so long as ρ remains finite, the field will remain finite everywhere, even in the interior or on the boundary of a charge distribution.

FLUX

1.9 The relation between the electric field and its sources can be expressed in a remarkably simple way, one that we shall find very useful. For this we need to define a quantity called *flux*.

Consider some electric field in space and in this space some arbitrary closed surface, like a balloon of any shape. Figure 1.13 shows such a surface, the field being suggested by a few field lines. Now divide the whole surface into little patches which are so small that over any one patch the surface is practically flat and the vector field does not change appreciably from one part of a patch to another. In other words, don't let the balloon be too crinkly, and don't let its surface pass right through a singularity† of the field such as a point charge. The area of a patch has a certain magnitude in cm^2, and a patch defines a unique direction—the outward-pointing normal to its surface. (Since the surface is closed, you can tell its inside from its outside; there is no ambiguity.) Let this magnitude and direction be represented by a vector. Then for every patch into which the surface has been divided, such as patch number j, we have a vector $\mathbf{a}_j$ giving its area and orientation. The steps we have just taken are pictured in Fig. 1.13*b* and *c*. Note that the vector $\mathbf{a}_j$ does not depend at all on the shape of the patch; it doesn't matter how we have divided up the surface, as long as the patches are small enough.

Let $\mathbf{E}_j$ be the electric field vector at the location of patch number j. The scalar product $\mathbf{E}_j \cdot \mathbf{a}_j$ is a number. We call this number the *flux* through that bit of surface. To understand the origin of the name,

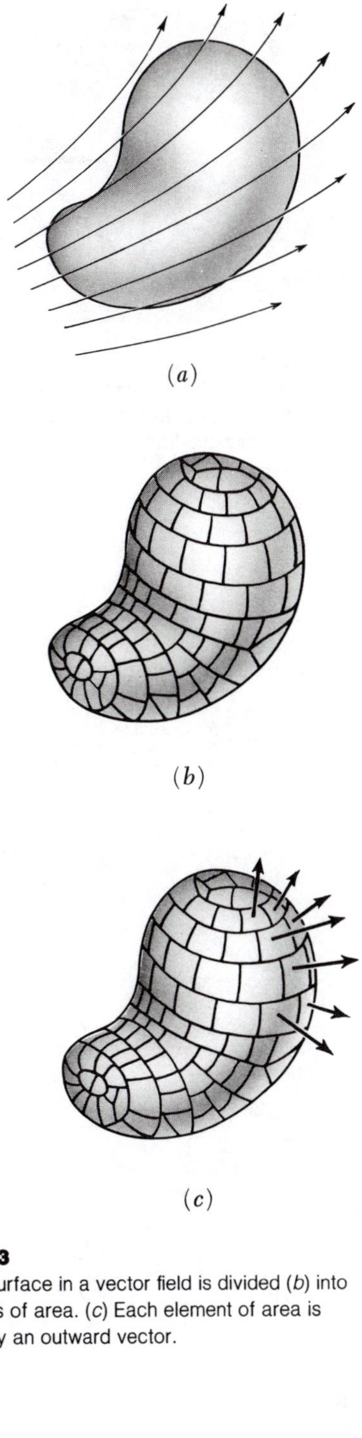

FIGURE 1.13
(*a*) A closed surface in a vector field is divided (*b*) into small elements of area. (*c*) Each element of area is represented by an outward vector.

†By a singularity of the field we would ordinarily mean not only a point source where the field approaches infinity, but any place where the field changes magnitude or direction discontinuously, such as an infinitesimally thin layer of concentrated charge. Actually this latter, milder, kind of singularity would cause no difficulty here unless our balloon's surface were to coincide with the surface of discontinuity over some finite area.

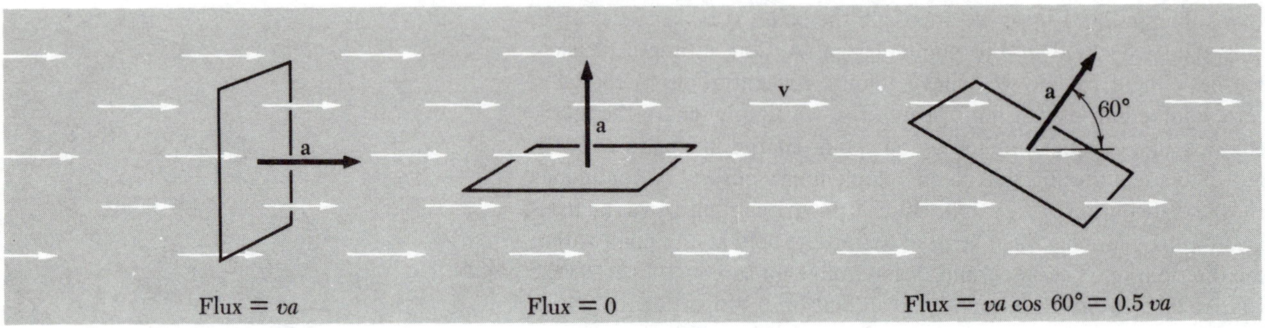
v
a
a
a
60°
Flux = va
Flux = 0
Flux = va cos 60° = 0.5 va

FIGURE 1.14
The flux through the frame of area **a** is **v** · **a**, where **v** is the velocity of the fluid. The flux is the volume of fluid passing through the frame, per unit time.

imagine a vector function which represents the velocity of motion in a fluid—say in a river, where the velocity varies from one place to another but is constant in time at any one position. Denote this vector field by **v**, measured, say, in meters/sec. Then, if **a** is the oriented area in square meters of a frame lowered into the water, **v** · **a** is the *rate of flow* of water through the frame in cubic meters per second (Fig. 1.14). We must emphasize that our definition of flux is applicable to any vector function, whatever physical variable it may represent.

Now let us add up the flux through all the patches to get the flux through the entire surface, a scalar quantity which we shall denote by Φ:

$$\Phi = \sum_{\text{All } j} \mathbf{E}_j \cdot \mathbf{a}_j \tag{16}$$

Letting the patches become smaller and more numerous without limit, we pass from the sum in Eq. 16 to a surface integral:

$$\Phi = \int_{\substack{\text{Entire}\\ \text{surface}}} \mathbf{E} \cdot d\mathbf{a} \tag{17}$$

A surface integral of any vector function **F**, over a surface S, means just this: Divide S into small patches, each represented by a vector outward, of magnitude equal to the patch area; at every patch, take the scalar product of the patch area vector and the local **F**; sum all these products, and the limit of this sum, as the patches shrink, is the surface integral. Do not be alarmed by the prospect of having to perform such a calculation for an awkwardly shaped surface like the one in Fig. 1.13. The surprising property we are about to demonstrate makes that unnecessary!

GAUSS'S LAW

1.10 Take the simplest case imaginable; suppose the field is that of a single isolated positive point charge q and the surface is a sphere of radius r centered on the point charge (Fig. 1.15). What is the flux Φ through this surface? The answer is easy because the magnitude of **E** at every point on the surface is q/r^2 and its direction is the same as that of the outward normal at that point. So we have

$$\Phi = E \times \text{total area} = \frac{q}{r^2} \times 4\pi r^2 = 4\pi q \qquad (18)$$

The flux is independent of the size of the sphere.

Now imagine a second surface, or balloon, enclosing the first, but *not* spherical, as in Fig. 1.16. We claim that the total flux through this surface is the same as that through the sphere. To see this, look at a cone, radiating from q, which cuts a small patch **a** out of the sphere and continues on to the outer surface where it cuts out a patch **A** at a distance R from the point charge. The area of the patch **A** is larger than that of the patch **a** by two factors: first, by the ratio of the distance squared $(R/r)^2$; and second, owing to its inclination, by the factor $1/\cos\theta$. The angle θ is the angle between the outward normal and the radial direction (see Fig. 1.16). The electric field in that neighborhood is reduced from its magnitude on the sphere by the factor $(r/R)^2$ and is still radially directed. Letting $\mathbf{E}_{(R)}$ be the field at the outer patch and $\mathbf{E}_{(r)}$ be the field at the sphere, we have

$$\text{Flux through outer patch} = \mathbf{E}_{(R)} \cdot \mathbf{A} = E_{(R)} A \cos\theta \qquad (19)$$
$$\text{Flux through inner patch} = \mathbf{E}_{(r)} \cdot \mathbf{a} = E_{(r)} a$$
$$E_{(R)} A \cos\theta = \left[E_{(r)}\left(\frac{r}{R}\right)^2\right]\left[a\left(\frac{R}{r}\right)^2 \frac{1}{\cos\theta}\right]\cos\theta = E_{(r)} a$$

This proves that the flux through the two patches is the same.

Now every patch on the outer surface can in this way be put into correspondence with part of the spherical surface, so the total flux must be the same through the two surfaces. That is, the flux through the new surface must be just $4\pi q$. But this was a surface of *arbitrary* shape and size.† We conclude: The flux of the electric field through *any* surface enclosing a point charge q is $4\pi q$. As a corollary we can say that the total flux through a closed surface is *zero* if the charge lies *outside* the surface. We leave the proof of this to the reader, along with Fig. 1.17 as a hint of one possible line of argument.

There is a way of looking at all this which makes the result seem obvious. Imagine at q a source which emits particles—such as bullets or photons—in all directions at a steady rate. Clearly the flux of particles through a window of unit area will fall off with the inverse square of the window's distance from q. Hence we can draw an analogy between the electric field strength E and the intensity of particle

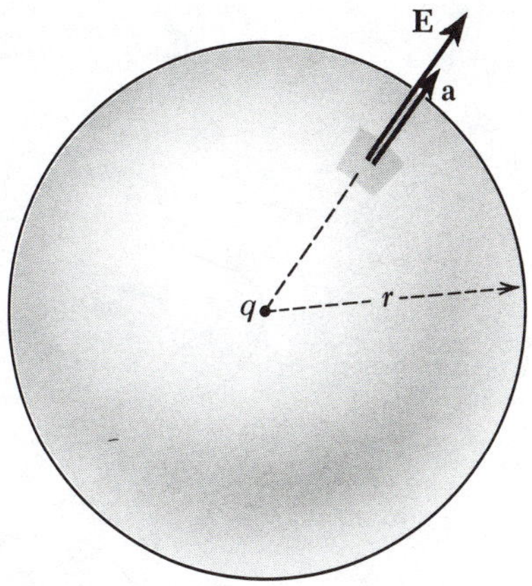

FIGURE 1.15
In the field **E** of a point charge q, what is the outward flux over a sphere surrounding q?

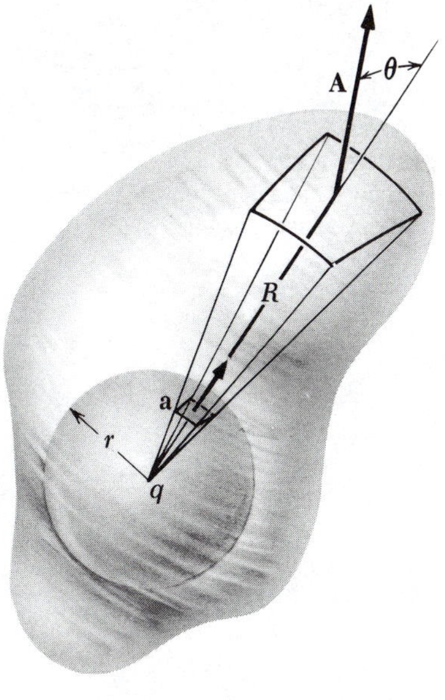

FIGURE 1.16
Showing that the flux through any closed surface around q is the same as the flux through the sphere.

†To be sure, we had the second surface enclosing the sphere, but it didn't have to, really. Besides, the sphere can be taken as small as we please.

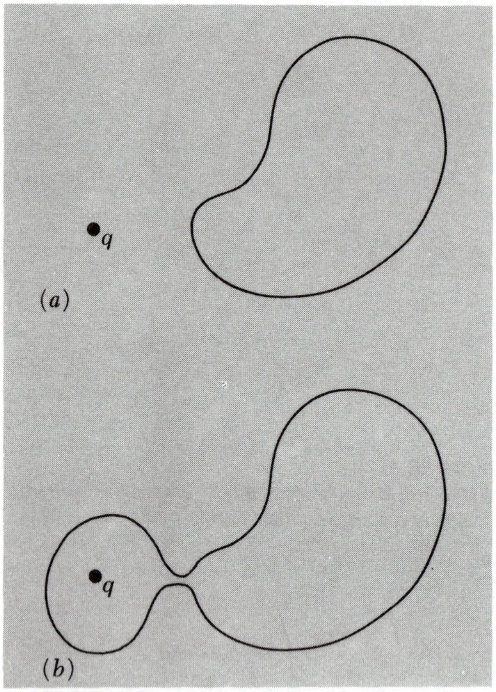

FIGURE 1.17
To show that the flux through the closed surface in (*a*) is zero, you can make use of (*b*).

flow in bullets per unit area per unit time. It is pretty obvious that the flux of bullets through any surface completely surrouding q is independent of the size and shape of that surface, for it is just the total number emitted per unit time. Correspondingly, the flux of E through the closed surface must be independent of size and shape. The common feature responsible for this is the inverse-square behavior of the intensity.

The situation is now ripe for superposition! Any electric field is the sum of the fields of its individual sources. This property was expressed in our statement, Eq. 13, of Coulomb's law. Clearly flux is an additive quantity in the same sense, for if we have a number of sources, $q_1, q_2, \ldots, q_N$, the fields of which, if each were present alone, would be $\mathbf{E}_1, \mathbf{E}_2 \ldots, \mathbf{E}_N$, the flux Φ through some surface S in the actual field can be written

$$\Phi = \int_S \mathbf{E} \cdot d\mathbf{a} = \int_S (\mathbf{E}_1 + \mathbf{E}_2 + \cdots + \mathbf{E}_N) \cdot d\mathbf{a} \qquad (20)$$

We have just learned that $\int_S \mathbf{E}_n \cdot d\mathbf{a}$ equals $4\pi q_n$ if the charge q_n is inside S and equals zero otherwise. So every charge q inside the surface contributes exactly $4\pi q$ to the surface integral of Eq. 20 and all charges outside contribute nothing. We have arrived at Gauss's law:

> The flux of the electric field $\mathbf{E}$ through any closed surface, that is, the integral $\int \mathbf{E} \cdot d\mathbf{a}$ over the surface, equals 4π times the total charge enclosed by the surface:
>
> $$\int \mathbf{E} \cdot d\mathbf{a} = 4\pi \sum_i q_i = 4\pi \int \rho \, dv \qquad (21)$$

We call the statement in the box a *law* because it is equivalent to Coulomb's law and it could serve equally well as the basic law of electrostatic interactions, after charge and field have been defined. Gauss's law and Coulomb's law are not two independent physical laws, but the same law expressed in different ways.†

†There is one difference, inconsequential here, but relevant to our later study of the fields of moving charges. Gauss' law is obeyed by a wider class of fields than those represented by the electrostatic field. In particular, a field that is inverse-square in r but not spherically symmetrical can satisfy Gauss' law. In other words, Gauss' law alone does not imply the symmetry of the field of a point source which is implicit in Coulomb's law.

Looking back over our proof, we see that it hinged on the inverse-square nature of the interaction and of course on the additivity of interactions, or superposition. Thus the theorem is applicable to any inverse-square field in physics, for instance, to the gravitational field.

It is easy to see that Gauss's law would *not* hold if the law of force were, say, inverse-cube. For in that case the flux of electric field from a point charge q through a sphere of radius R centered on the charge would be

$$\Phi = \int \mathbf{E} \cdot d\mathbf{a} = \frac{q}{R^3} \cdot 4\pi R^2 = \frac{4\pi q}{R} \tag{22}$$

By making the sphere large enough we could make the flux through it as small as we pleased, while the total charge inside remained constant.

This remarkable theorem enlarges our grasp in two ways. First, it reveals a connection between the field and its sources that is the converse of Coulomb's law. Coulomb's law tells us how to derive the electric field if the charges are given; with Gauss's law we can determine how much charge is in any region if the field is known. Second, the mathematical relation here demonstrated is a powerful analytic tool; it can make complicated problems easy, as we shall see.

FIGURE 1.18
A charge distribution with spherical symmetry.

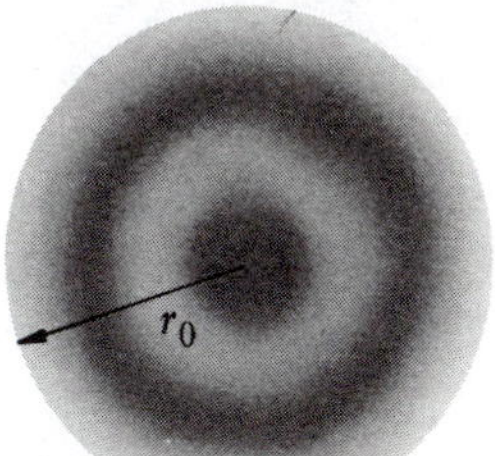

FIELD OF A SPHERICAL CHARGE DISTRIBUTION

1.11 We can use Gauss's law to find the electric field of a spherically symmetrical distribution of charge, that is, a distribution in which the charge density ρ depends only on the radius from a central point. Figure 1.18 depicts a cross section through some such distribution. Here the charge density is high at the center, and is zero beyond r_0. What is the electric field at some point such as P_1 outside the distribution, or P_2 inside it (Fig. 1.19)? If we could proceed only from Coulomb's law, we should have to carry out an integration which would sum the electric field vectors at P_1 arising from each elementary volume in the charge distribution. Let's try a different approach which exploits both the symmetry of the system and Gauss's law.

FIGURE 1.19
The electric field of a spherical charge distribution.

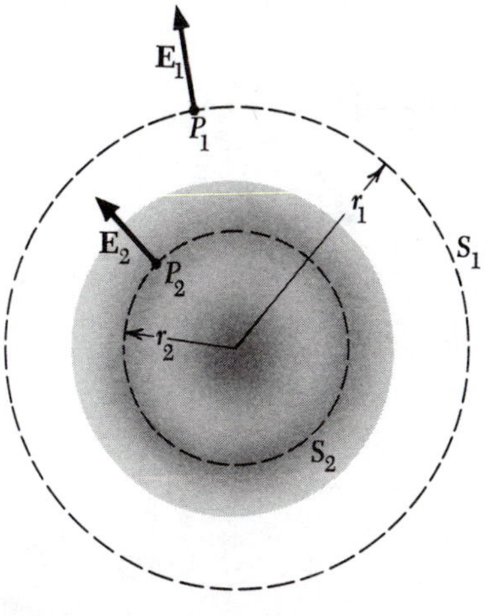

Because of the spherical symmetry, the electric field at any point must be radially directed—no other direction is unique. Likewise, the field magnitude E must be the same at all points on a spherical surface S_1 of radius r_1, for all such points are equivalent. Call this field magnitude E_1. The flux through this surface S_1 is therefore simply $4\pi r_1^2 E_1$, and by Gauss's law this must be equal to 4π times the charge enclosed by the surface. That is, $4\pi r_1^2 E_1 = 4\pi$ (charge inside S_1) or

$$E_1 = \frac{\text{charge inside } S_1}{r_1^2} \tag{23}$$

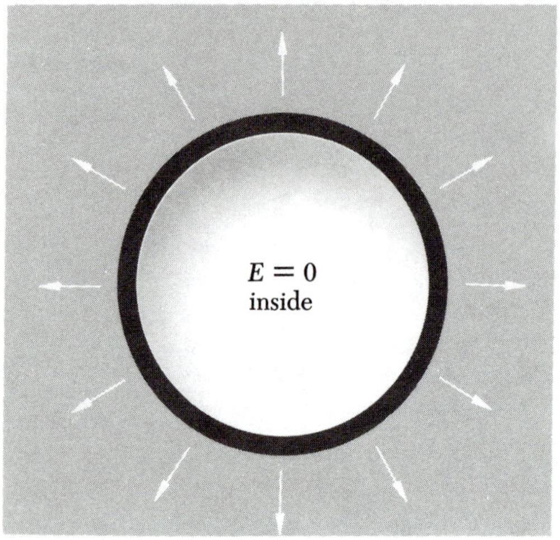

FIGURE 1.20
The field is zero inside a spherical shell of charge.

Comparing this with the field of a point charge, we see that *the field at all points on S_1 is the same as if all the charge within S_1 were concentrated at the center.* The same statement applies to a sphere drawn *inside* the charge distribution. The field at any point on S_2 is the same as if all charge within S_2 were at the center, and all charge *outside* S_2 absent. Evidently the field inside a "hollow" spherical charge distribution is zero (Fig. 1.20).

The same argument applied to the gravitational field would tell us that the earth, assuming it is spherically symmetrical in its mass distribution, attracts outside bodies as if its mass were concentrated at the center. That is a rather familiar statement. Anyone who is inclined to think the principle expresses an obvious property of the center of mass must be reminded that the theorem is not even true, in general, for other shapes. A perfect cube of uniform density does *not* attract external bodies as if its mass were concentrated at its geometrical center.

Newton didn't consider the theorem obvious. He needed it as the keystone of his demonstration that the moon in its orbit around the earth and a falling body on the earth are responding to similar forces. The delay of nearly 20 years in the publication of Newton's theory of gravitation was apparently due, in part at least, to the trouble he had in proving this theorem to his satisfaction. The proof he eventually devised and published in the *Principia* in 1686 (Book I, Section XII, Theorem XXXI) is a marvel of ingenuity in which, roughly speaking, a tricky volume integration is effected without the aid of the integral calculus as we know it. The proof is a good bit longer than our whole preceding discussion of Gauss's law, and more intricately reasoned. You see, with all his mathematical resourcefulness and originality, Newton lacked Gauss's theorem—a relation which, once it has been shown to us, seems so obvious as to be almost trivial.

FIELD OF A LINE CHARGE

1.12 A long, straight, charged wire, if we neglect its thickness, can be characterized by the amount of charge it carries per unit length. Let λ, measured in esu/cm, denote this *linear charge density*. What is the electric field of such a line charge, assumed infinitely long and with constant linear charge density λ? We'll do the problem in two ways, first by an integration starting from Coulomb's law.

To evaluate the field at the point P, shown in Fig. 1.21, we must add up the contributions from all segments of the line charge, one of which is indicated as a segment of length dx. The charge dq on this element is given by $dq = \lambda\, dx$. Having oriented our x axis along the line charge, we may as well let the y axis pass through P, which is r cm from the nearest point on the line. It is a good idea to take advantage of symmetry at the outset. Obviously the electric field at P must

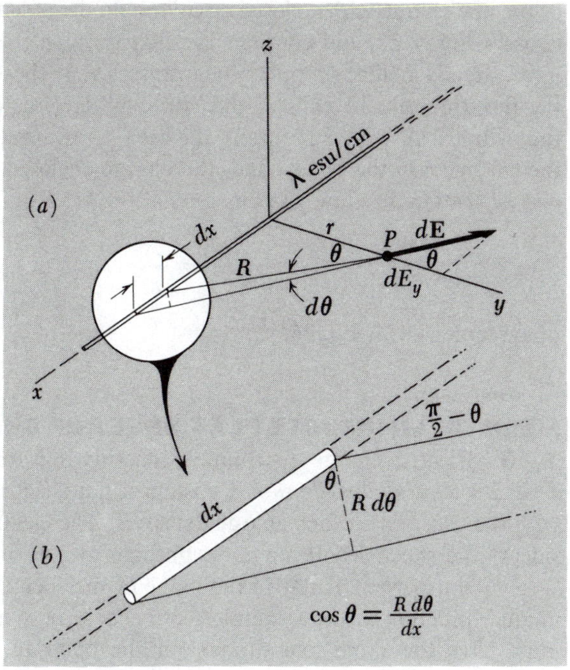

FIGURE 1.21
(*a*) The field at *P* is the vector sum of contributions from each element of the line charge. (*b*) Detail of (*a*).

point in the y direction, so that E_x and E_z are both zero. The contribution of the charge dq to the y component of the electric field at P is

$$dE_y = \frac{dq}{R^2}\cos\theta = \frac{\lambda\, dx}{R^2}\cos\theta \tag{24}$$

where θ is the angle the vector field of dq makes with the y direction. The total y component is then

$$E_y = \int dE_y = \int_{-\infty}^{\infty} \frac{\lambda \cos\theta}{R^2}\, dx \tag{25}$$

It is convenient to use θ as the variable of integration. Since $R = r/\cos\theta$ and $dx = R\, d\theta/\cos\theta$, the integral becomes

$$E_y = \int_{-\pi/2}^{\pi/2} \frac{\lambda \cos\theta\, d\theta}{r} = \frac{\lambda}{r}\int_{-\pi/2}^{\pi/2} \cos\theta\, d\theta = \frac{2\lambda}{r} \tag{26}$$

We see that the field of an infinitely long, uniformly dense line charge is proportional to the reciprocal of the distance from the line. Its direction is of course radially outward if the line carries a positive charge, inward if negative.

Gauss' law leads directly to the same result. Surround a segment

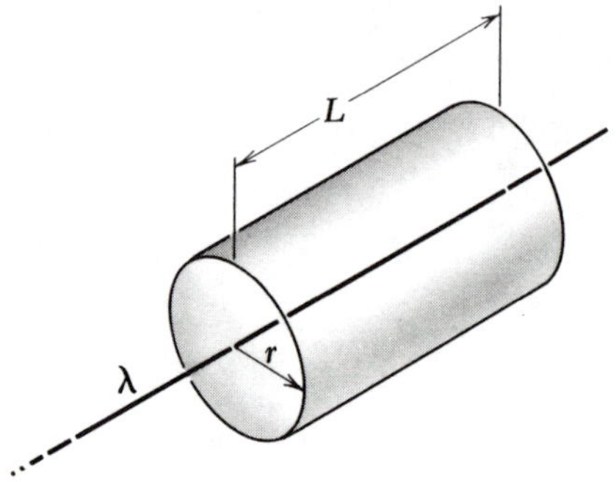

FIGURE 1.22
Using Gauss' law to find the field of a line charge.

of the line charge with a closed circular cylinder of length L and radius r, as in Fig. 1.22, and consider the flux through this surface. As we have already noted, symmetry guarantees that the field is radial, so the flux through the ends of the "tin can" is zero. The flux through the cylindrical surface is simply the area, $2\pi rL$, times E_r, the field at the surface. On the other hand, the charge enclosed by the surface is just λL, so Gauss's law gives us $2\pi rLE_r = 4\pi\lambda L$ or

$$E_r = \frac{2\lambda}{r} \tag{27}$$

in agreement with Eq. 26.

FIELD OF AN INFINITE FLAT SHEET OF CHARGE

1.13 Electric charge distributed smoothly in a thin sheet is called a *surface charge distribution.* Consider a flat sheet infinite in extent, with the constant surface charge density σ. The electric field on either side of the sheet, whatever its magnitude may turn out to be, must surely point perpendicular to the plane of the sheet; there is no other unique direction in the system. Also because of symmetry, the field must have the same magnitude and the opposite direction at two points P and P' equidistant from the sheet on opposite sides. With these facts established, Gauss's law gives us at once the field intensity, as follows: Draw a cylinder, as in Fig. 1.23, with P on one side and P' on the other, of cross-section area A. The outward flux is found only at the ends, so that if E_P denotes the magnitude of the field at P, and $E_{P'}$ the magnitude of P', the outward flux is $AE_P + AE_{P'} = 2AE_P$. The charge enclosed is σA. Hence $2AE_P = 4\pi\sigma A$, or

$$E_P = 2\pi\sigma \tag{28}$$

We see that the field strength is independent of r, the distance from the sheet. Equation 28 could have been derived more laboriously by calculating the vector sum of the contributions to the field at P from all the little elements of charge in the sheet.

The field of an infinitely long line charge, we found, varies inversely as the distance from the line, while the field of an infinite sheet has the same strength at all distances. These are simple consequences of the fact that the field of a point charge varies as the inverse square of the distance. If that doesn't yet seem compellingly obvious, look at it this way: Roughly speaking, the part of the line charge that is mainly responsible for the field at P, in Fig. 1.21, is the near part—the charge within a distance of order of magnitude r. If we lump all this together and forget the rest, we have a concentrated charge of magnitude $q \approx \lambda r$, which ought to produce a field proportional to q/r^2, or λ/r. In the case of the sheet, the amount of charge that is "effective," in this sense, increases proportionally to r^2 as we go out

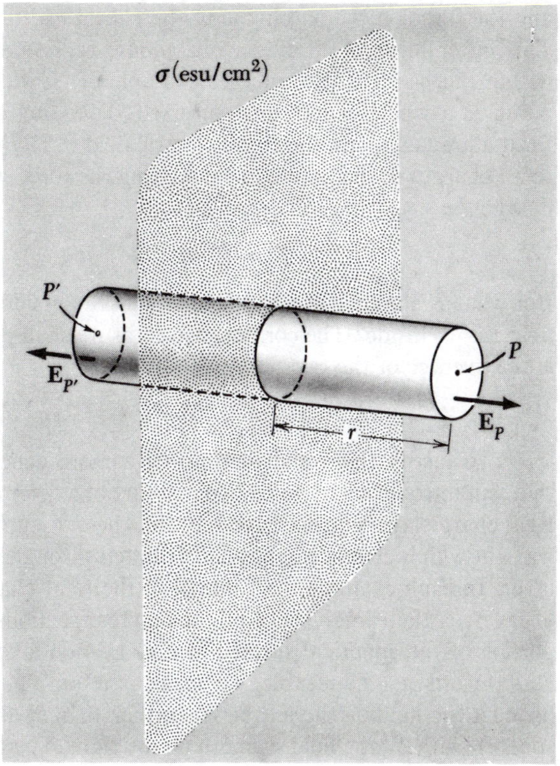

FIGURE 1.23
Using Gauss' law to find the field of an infinite flat sheet of charge.

from the sheet, which just offsets the $1/r^2$ decrease in the field from any given element of charge.

THE FORCE ON A LAYER OF CHARGE

1.14 The sphere in Fig. 1.24 has a charge distributed over its surface with the uniform density σ, in esu/cm^2. Inside the sphere, as we have already learned, the electric field of such a charge distribution is zero. Outside the sphere the field is Q/r^2, where Q is the total charge on the sphere, equal to $4\pi r_0^2\sigma$. Just outside the surface of the sphere the field strength is $4\pi\sigma$. Compare this with Eq. 28 and Fig. 1.23. In both cases Gauss' law is obeyed: The *change* in E, from one side of the layer to the other, is equal to $4\pi\sigma$.

What is the electrical force experienced by the charges that make up this distribution? The question may seem puzzling at first because the field **E** arises from these very charges. What we must think about is the force on some small element of charge dq, such as a small patch of area dA with charge $dq = \sigma\, dA$. Consider, separately, the force on dq due to all the other charges in the distribution,

FIGURE 1.24
A spherical surface with uniform charge density σ.

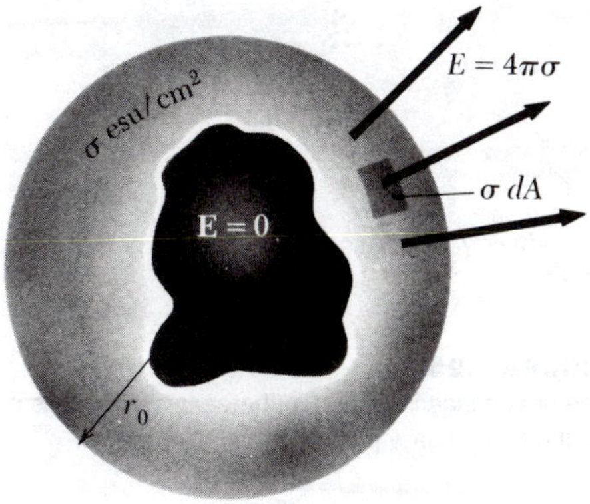

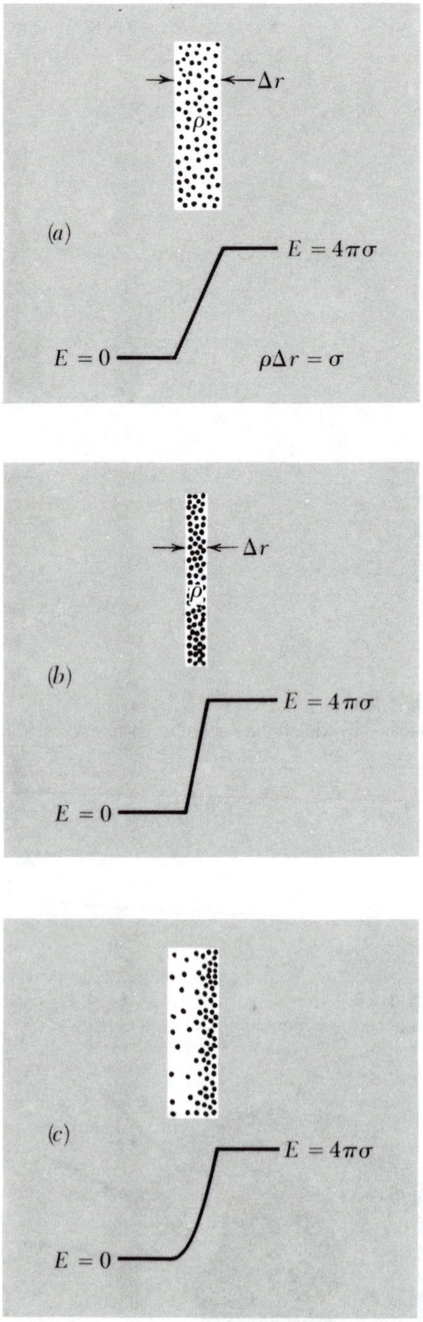

FIGURE 1.25
The net change in field at a charge layer depends only on the total charge per unit area.

and the force on the patch due to the charges within the patch itself. This latter force is surely zero. Coulomb repulsion between charges within the patch is just another example of Newton's third law; the patch as a whole cannot push on itself. That simplifies our problem, for it allows us to use the entire electric field **E**, *including* the field due to all charges in the patch, in calculating the force $d\mathbf{F}$ on the patch of charge dq:

$$d\mathbf{F} = \mathbf{E}\,dq = \mathbf{E}\,\sigma\,dA \tag{29}$$

But what E shall we use, the field $E = 4\pi\sigma$ outside the sphere or the field $E = 0$ inside? The correct answer, as we shall prove in a moment, is the *average* of the two fields.

$$dF = \tfrac{1}{2}(4\pi\sigma + 0)\,\sigma\,dA = 2\pi\sigma^2\,dA \tag{30}$$

To justify this we shall consider a more general case, and one that will introduce a more realistic picture of a layer of surface charge. Real charge layers do not have zero thickness. Figure 1.25 shows some ways in which charge might be distributed through the thickness of a layer. In each example the value of σ, the total charge per unit area of layer, is the same. These might be cross sections through a small portion of the spherical surface in Fig. 1.24 on a scale such that the curvature is not noticeable. To make it more general, however, we have let the field on the left be E_1 (rather than 0, as it was inside the sphere), with E_2 the field strength on the right. The condition imposed by Gauss's law, for given σ, is in each case

$$E_2 - E_1 = 4\pi\sigma \tag{31}$$

Now let us look carefully within the layer where the field is changing continuously from E_1 to E_2 and there is a volume charge density $\rho(x)$ extending from $x = 0$ to $x = x_0$, the thickness of the layer (Fig. 1.26). Consider a much thinner slab, of thickness $dx \ll x_0$, which contains per unit area an amount of charge $\rho\,dx$. The force on it is

$$dF = E\rho\,dx \tag{32}$$

Thus the total force per unit area of our charge layer is

$$F = \int_0^{x_0} E\rho\,dx \tag{33}$$

But Gauss's law tells us that dE, the change in E through the thin slab, is just $4\pi\rho\,dx$. Hence $\rho\,dx$ in Eq. 33 can be replaced by $dE/4\pi$, and the integral becomes

$$F = \frac{1}{4\pi}\int_{E_1}^{E_2} E\,dE = \frac{1}{8\pi}(E_2^2 - E_1^2) \tag{34}$$

Since $E_2 - E_1 = 4\pi\sigma$, the result in Eq. 34, after being factored, can be expressed as

$$F = \tfrac{1}{2}(E_1 + E_2)\sigma \tag{35}$$

We have shown, as promised, that for given σ the force per unit area on a charge layer is determined by the mean of the external field on one side and that on the other.† This is independent of the thickness of the layer, as long as it is small compared to the total area, and of the variation $\rho(x)$ in charge density within the layer.

The direction of the electrical force on an element of the charge on the sphere is, of course, outward whether the surface charge is positive or negative. If the charges do not fly off the sphere, that outward force must be balanced by some inward force, not included in our equations, which can hold the charge carriers in place. To call such a force "nonelectrical" would be misleading, for electrical attractions and repulsions are the dominant forces in the structure of atoms and in the cohesion of matter generally. The difference is that these forces are effective only at short distances, from atom to atom, or from electron to electron. Physics on that scale is a story of individual particles. Think of a charged rubber balloon, say, 10 cm in radius, with 20 esu of negative charge spread as uniformly as possible on its outer surface. It forms a surface charge of density $\sigma = 20/400\pi = 0.016$ esu/cm^2. The resulting outward force, per cm^2 of surface charge, is $2\pi\sigma^2$, or 0.0016 dynes/cm^2. In fact our charge consists of about 4×10^{10} electrons attached to the rubber film. As there are about 30 million extra electrons per cm^2, "graininess" in the charge distribution is hardly apparent. However, if we could look at one of these extra electrons, we would find it roughly 10^{-4} cm—an enormous distance on an atomic scale—from its nearest neighbor. This electron would be stuck, electrically stuck, to a local molecule of rubber. The rubber molecule would be attached to adjacent rubber molecules, and so on. If you pull on the electron, the force is transmitted in this way to the whole piece of rubber. Unless, of course, you pull hard enough to tear the electron loose from the molecule to which it is attached. That would take an electric field many thousands of times stronger than the field in our example.

ENERGY ASSOCIATED WITH THE ELECTRIC FIELD

1.15 Suppose our spherical shell of charge is compressed slightly, from an initial radius of r_0 to a smaller radius, as in Fig. 1.27. This requires that work be done against the repulsive force, $2\pi\sigma^2$ dynes for

†Note that this is *not* necessarily the same as the average field within the layer, a quantity of no special interest or significance.

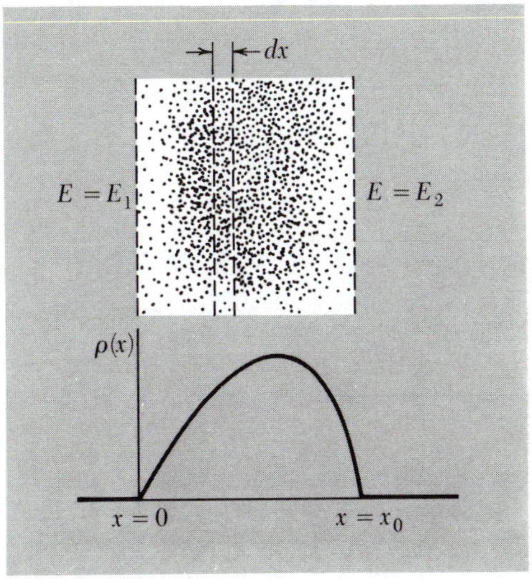

FIGURE 1.26
Within the charge layer of density $\rho(x)$, $E(x + dx) - E(x) = 4\pi\rho\, dx$.

FIGURE 1.27
Shrinking a spherical shell or charged balloon.

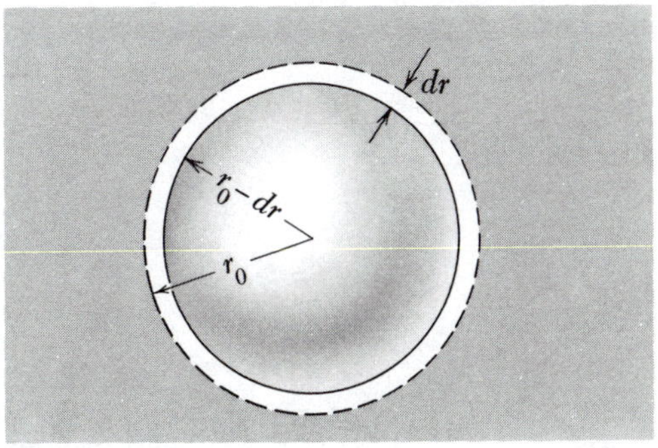

each square centimeter of surface. The displacement being dr, the total work done is $(4\pi r_0^2)(2\pi\sigma^2)\,dr$, or $8\pi^2 r_0^2 \sigma^2\,dr$. This represents an *increase* in the energy required to assemble the system of charges, the energy U we talked about in Section 1.5:

$$dU = 8\pi^2 r_0^2 \sigma^2\,dr \tag{36}$$

Notice how the electric field E has been changed. Within the shell of thickness dr the field was zero and is now $4\pi\sigma$. Beyond r_0 the field is unchanged. In effect we have created a field of strength $E = 4\pi\sigma$ filling a region of volume $4\pi r_0^2\,dr$. We have done so by investing an amount of energy given by Eq. 36 which, if we substitute $E/4\pi$ for σ, can be written like this:

$$dU = \frac{E^2}{8\pi} 4\pi r_0^2\,dr \tag{37}$$

This is an instance of a general theorem which we shall not prove now: *The potential energy U of a system of charges, which is the total work required to assemble the system, can be calculated from the electric field itself simply by assigning an amount of energy $(E^2/8\pi)\,dv$ to every volume element dv and integrating over all space where there is electric field.*

$$\boxed{U = \frac{1}{8\pi}\int_{\substack{\text{Entire}\\ \text{field}}} E^2\,dv} \tag{38}$$

E^2 is a scalar quantity, of course: $E^2 \equiv \mathbf{E}\cdot\mathbf{E}$.

One may think of this energy as "stored" in the field. The system being conservative, that amount of energy can of course be recovered by allowing the charges to go apart; so it is nice to think of the energy as "being somewhere" meanwhile. Our accounting comes out right if we think of it as stored in space with a density of $E^2/8\pi$, in ergs/cm^3. There is no harm in this, but in fact we have no way of identifying, quite independently of anything else, the energy stored in a particular cubic centimeter of space. Only the total energy is physically measurable, that is, the work required to bring the charge into some configuration, starting from some other configuration. Just as the concept of electric field serves in place of Coulomb's law to explain the behavior of electric charges, so when we use Eq. 38 rather than Eq. 9 to express the total potential energy of an electrostatic system, we are merely using a different kind of bookkeeping. Sometimes a change in viewpoint, even if it is at first only a change in bookkeeping, can stimulate new ideas and deeper understanding. The notion of the electric field as an independent entity will take form when we study the dynamical behavior of charged matter and electromagnetic radiation.

We run into trouble if we try to apply Eq. 38 to a system that contains a point charge, that is, a finite charge q of zero size. Locate q at the origin of the coordinates. Close to the origin E^2 will approach q^2/r^4. With $dv = 4\pi r^2\, dr$, the integrand $E^2\, dv$ will behave like dr/r^2, and our integral will blow up at the limit $r = 0$. That simply tells us that it would take infinite energy to pack finite charge into zero volume—which is true but not helpful. In the real world we deal with particles like electrons and protons. They are so small that for most purposes we can ignore their dimensions and think of them as point charges when we consider their electrical interaction with one another. How much energy it took to make such a particle is a question that goes beyond the range of classical electromagnetism. We have to regard the particles as supplied to us ready-made. The energy we are concerned with is the work done in moving them around.

The distinction is usually clear. Consider two charged particles, a proton and a negative pion, for instance. Let $\mathbf{E}_p$ be the electric field of the proton, $\mathbf{E}_\pi$ that of the pion. The total field is $\mathbf{E} = \mathbf{E}_p + \mathbf{E}_\pi$, and $\mathbf{E} \cdot \mathbf{E}$ is $E_p^2 + E_\pi^2 + 2\mathbf{E}_p \cdot \mathbf{E}_\pi$. According to Eq. 38 the total energy in the electric field of this two-particle system is

$$\begin{aligned} U &= \frac{1}{8\pi}\int E^2\, dv \\ &= \frac{1}{8\pi}\int E_p^2\, dv + \frac{1}{8\pi}\int E_\pi^2\, dv + \frac{1}{4\pi}\int \mathbf{E}_p \cdot \mathbf{E}_\pi\, dv \end{aligned} \tag{39}$$

The value of the first integral is a property of any isolated proton. It is a constant of nature which is not changed by moving the proton around. The same goes for the second integral, involving the pion's electric field alone. It is the third integral that directly concerns us, for it expresses the energy required to assemble the system *given* a proton and a pion as constituents.

The distinction could break down if the two particles interact so strongly that the electrical structure of one is distorted by the presence of the other. Knowing that both particles are in a sense composite (the proton consisting of three quarks, the pion of two), we might expect that to happen during a close approach. In fact, nothing much happens down to a distance of 10^{-13} cm. At shorter distances, for strongly interacting particles like the proton and the pion, nonelectrical forces dominate the scene anyway.

That explains why we do not need to include "self-energy" terms like the first two integrals in Eq. 39 in our energy accounts for a system of elementary charged particles. Indeed, we want to omit them. We are doing just that, in effect, when we replace the actual distribution of discrete elementary charges (the electrons on the rubber balloon) by a perfectly continuous charge distribution.

PROBLEMS

1.1 In the domain of elementary particles, a natural unit of mass is the mass of a *nucleon,* that is, a proton or a neutron, the basic massive building blocks of ordinary matter. Given the nucleon mass as 1.6×10^{-24} gm and the gravitational constant G as 6.7×10^{-8} cm^3/gm-sec^2, compare the gravitational attraction of two protons with their electrostatic repulsion. This shows why we call gravitation a very *weak* force. The distance between the two protons in the helium nucleus could be at one instant as much as 10^{-13} cm. How large is the force of electrical repulsion between two protons at that distance? Express it in newtons, and in pounds. Even stronger is the *nuclear* force that acts between any pair of hadrons (including neutrons and protons) when they are that close together.

1.2 On the utterly unrealistic assumption that there are no other charged particles in the vicinity, at what distance below a proton would the upward force on an electron (electron mass $\simeq 10^{-27}$ gm) equal the electron's weight?

1.3 Two volley balls, mass 0.3 kilogram (kg) each, tethered by nylon strings and charged with an electrostatic generator, hang as shown in the diagram. What is the charge on each in coulombs, assuming the charges are equal? (*Reminder:* the weight of a 1-kg mass on earth is 9.8 newtons, just as the weight of a 1-gm mass is 980 dynes.)

PROBLEM 1.3

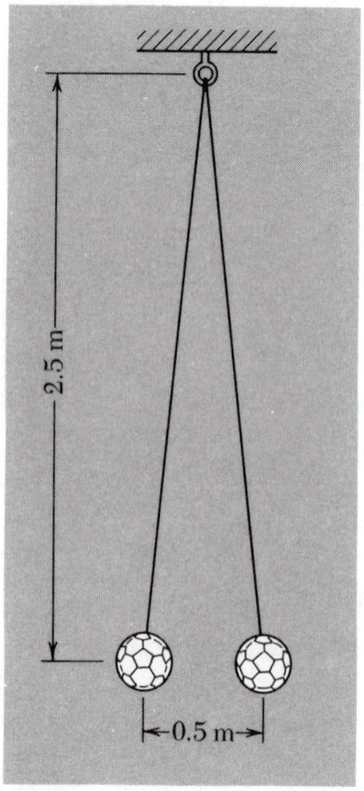

1.4 At each corner of a square is a particle with charge q. Fixed at the center of the square is a point charge of opposite sign, of magnitude Q. What value must Q have to make the total force on each of the four particles zero? With Q set at that value, the system, in the absence of other forces, is in equilibrium. Do you think the equilibrium is stable?

Ans. $Q = 0.957q$.

1.5 A thin plastic rod bent into a semicircle of radius R has a charge of Q, in esu, distributed uniformly over its length. Find the strength of the electric field at the center of the semicircle.

1.6 Three positive charges, A, B, and C, of 3×10^{-6}, 2×10^{-6}, and 2×10^{-6} coulombs, respectively, are located at the corners of an equilateral triangle of side 0.2 meter.

(*a*) Find the magnitude in newtons of the force on each charge.

(*b*) Find the magnitude in newtons/coulomb of the electric field at the center of the triangle.

Ans. (*a*) 2.34 newtons on A, 1.96 newtons on B and C; (*b*) 6.74×10^5 newtons/coulomb.

1.7 Find a geometrical arrangement of one proton and two electrons such that the potential energy of the system is exactly zero. How many such arrangements are there with the three particles on the same straight line?

1.8 Calculate the potential energy, per ion, for an infinite one-dimensional ionic crystal, that is, a row of equally spaced charges of magnitude e and alternating sign. [*Hint:* The power-series expansion of $\ln(1 + x)$ may be of use.]

1.9 A spherical volume of radius a is filled with charge of uniform density ρ. We want to know the potential energy U of this sphere of charge, that is, the work done in assembling it. Calculate it by building the sphere up layer by layer, making use of the fact that the field outside a spherical distribution of charge is the same as if all the charge were at the center. Express the result in terms of the total charge Q in the sphere.

Ans. $U = \frac{3}{5}(Q^2/a)$.

1.10 At the beginning of the century the idea that the rest mass of the electron might have a purely electrical origin was very attractive, especially when the equivalence of energy and mass was revealed by special relativity. Imagine the electron as a ball of charge, of constant volume density out to some maximum radius r_0. Using the result of Problem 1.9, set the potential energy of this system equal to mc^2 and see what you get for r_0. One defect of the model is rather obvious: Nothing is provided to hold the charge together!

1.11 A charge of 1 esu is at the origin. A charge of -2 esu is at $x = 1$ on the x axis.

(*a*) Find a point on the x axis where the electric field is zero.

(*b*) Locate, at least approximately, a point on the y axis where the electric field is parallel to the x axis. [A calculator should help with (*b*).]

1.12 Another problem for your calculator: Two positive ions and one negative ion are fixed at the vertices of an equilateral triangle. Where can a fourth ion be placed so that the force on it will be zero? Is there more than one such place?

1.13 The passage of a thundercloud overhead caused the vertical electric field strength in the atmosphere, measured at the ground, to rise to 0.1 statvolt/cm.

(*a*) How much charge did the thundercloud contain, in esu per cm^2 of horizontal area?

(*b*) Suppose there was enough water in the thundercloud in the form of 1-millimeter (mm)-diameter drops to make 0.25 cm of rainfall, and that it was those drops which carried the charge. How large was the electric field strength at the surface of one of the drops?

1.14 A charge Q is distributed uniformly around a thin ring of radius b which lies in the xy plane with its center at the origin. Locate the point on the positive z axis where the electric field is strongest.

1.15 Consider a spherical charge distribution which has a constant density ρ from $r = 0$ out to $r = a$ and is zero beyond. Find the electric field for all values of r, both less than and greater than a. Is there a discontinuous change in the field as we pass the surface of the charge distribution at $r = a$? Is there a discontinuous change at $r = 0$?

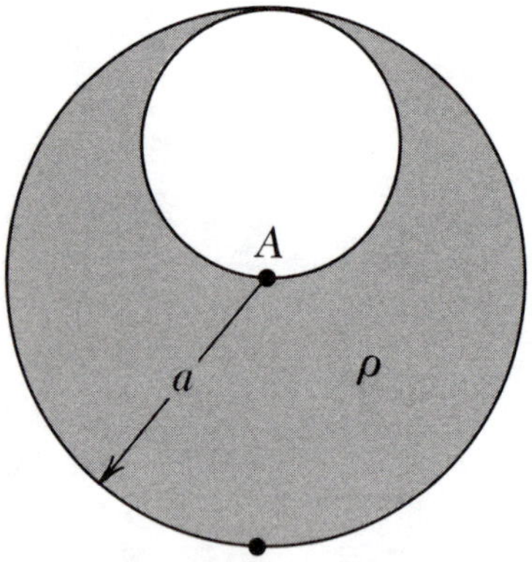

PROBLEM 1.16

1.16 The sphere of radius a was filled with positive charge at uniform density ρ. Then a smaller sphere of radius $a/2$ was carved out, as shown in the figure, and left empty. What are the direction and magnitude of the electric field at A? At B?

1.17 (a) A point charge q is located at the center of a cube of edge length d. What is the value of $\int \mathbf{E} \cdot d\mathbf{a}$ over one face of the cube?

(b) The charge q is moved to one corner of the cube. What is now the value of the flux of $\mathbf{E}$ through each of the faces of the cube?

1.18 Two infinite plane sheets of surface charge, of density $\sigma = 6$ esu/cm^2 and $\sigma = -4$ esu/cm^2, are located 2 cm apart, parallel to one another. Discuss the electric field of this system. Now suppose the two planes, instead of being parallel, intersect at right angles. Show what the field is like in each of the four regions into which space is thereby divided.

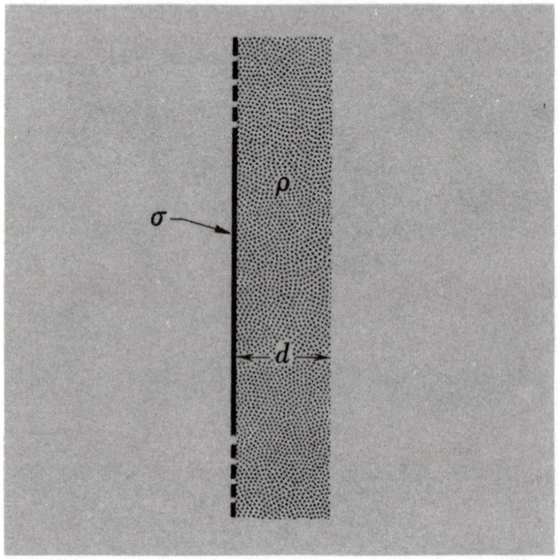

PROBLEM 1.19

1.19 An infinite plane has a uniform surface charge distribution σ on its surface. Adjacent to it is an infinite parallel layer of charge of thickness d and uniform volume charge density ρ. All charges are fixed. Find $\mathbf{E}$ everywhere.

1.20 Consider a distribution of charge in the form of a circular cylinder, like a long charged pipe. Prove that the field inside the pipe is zero. Prove that the field outside is the same as if the charge were all on the axis. Is either statement true for a pipe of square cross section on which the charge is distributed with uniform surface density?

1.21 The neutral hydrogen atom in its normal state behaves in some respects like an electric charge distribution which consists of a point charge of magnitude e surrounded by a distribution of negative charge whose density is given by $-\rho(r) = Ce^{-2r/a_0}$. Here a_0 is the *Bohr radius*, 0.53×10^{-8} cm, and C is a constant with the value required to make the total amount of negative charge exactly e. What is the net electric charge inside a sphere of radius a_0? What is the electric field strength at this distance from the nucleus?

1.22 Consider three plane charged sheets, A, B, and C. The sheets are parallel with B below A and C below B. On each sheet there is surface charge of uniform density: -4 esu/cm^2 on A, 7 esu/cm^2 on B, and -3 esu/cm^3 on C. (The density given includes charge on both sides of the sheet.) What is the magnitude of the electrical force on each sheet, in dynes/cm^2? Check to see that the total force on the three sheets is zero.

Ans. 32π dynes/cm^2 on A; 14π dynes/cm^2 on B; 18π dynes/cm^2 on C.

1.23 A sphere of radius R has a charge Q distributed uniformly over its surface. How large a sphere contains 90 percent of the energy stored in the electrostatic field of this charge distribution?

Ans. Radius: $10R$.

1.24 A thin rod 10 cm long carries a total charge of 8 esu uniformly distributed along its length. Find the strength of the electric field at each of the two points A and B located as shown in the diagram.

PROBLEM 1.24

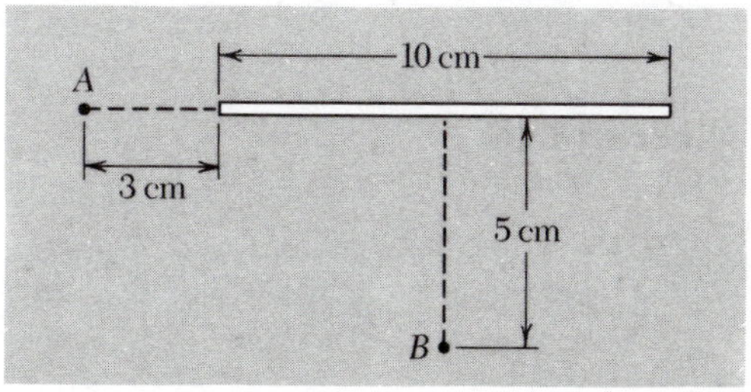

1.25 The relation in Eq. 27 expressed in SI units becomes

$$E = \frac{1}{4\pi\epsilon_0}\frac{2\lambda}{r}$$

with r in meters, λ in coulombs/meter, and E in newtons/coulomb. Consider a high-voltage direct current power line which consists of two parallel conductors suspended 3 meters apart. The lines are oppositely charged. If the electric field strength halfway between them is 15,000 newtons/coulomb, how much excess positive charge resides on a 1-km length of the positive conductor?

Ans. 6.26×10^{-4} coulomb.

1.26 Two long, thin parallel rods, a distance $2b$ apart, are joined by a semicircular piece of radius b, as shown. Charge of uniform linear density λ is deposited along the whole filament. Show that the field $\mathbf{E}$ of this charge distribution vanishes at the point C. Do this by comparing the contribution of the element at A to that of the element at B which is defined by the same values of θ and $d\theta$.

PROBLEM 1.26

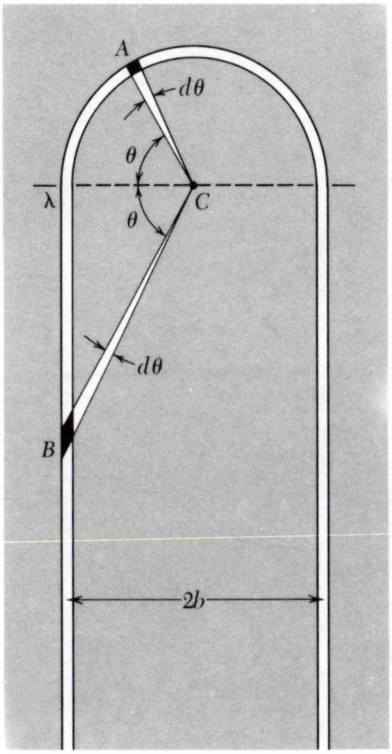

1.27 An infinite chessboard with squares of side s has a charge e at the center of every white square and a charge $-e$ at the center of every black square. We are interested in the work W required to transport one charge from its position on the board to an infinite distance from the board, along a path perpendicular to the plane of the board. Given that W is finite (which is plausible but not so easy to prove), do you think it is positive or negative? To calculate an approximate value for W, consider removing the charge from the central square of a 7×7 board. (Only 9 different terms are involved in that sum.) Or write a program and compute the work to remove the central charge from a much larger array, for instance a 101×101 board. Comparing the result for the 101×101 board with that for a 99×99 board, and for a 103×103 board, should give some idea of the rate of convergence toward the value for the infinite array.

1.28 Three protons and three electrons are to be placed at the vertices of a regular octahedron of edge length a. We want to find the energy of the system, that is, the work required to assemble it starting with the particles very far apart. There are two essentially different arrangements. What is the energy of each?

Ans. $-3.879e^2/a$; $-2.121e^2/a$.

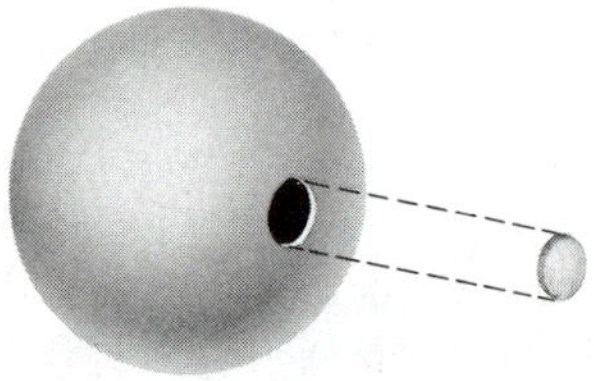

PROBLEM 1.29

1.29 The figure shows a spherical shell of charge, of radius a and surface density σ, from which a small circular piece of radius $b \ll a$ has been removed. What is the direction and magnitude of the field at the midpoint of the aperture? There are two ways to get the answer. You can integrate over the remaining charge distribution to sum the contributions of all elements to the field at the point in question. Or, remembering the superposition principle, you can think about the effect of replacing the piece removed, which itself is practically a little disk. Note the connection of this result with our discussion of the force on a surface charge—perhaps that is a third way in which you might arrive at the answer.

1.30 Concentric spherical shells of radius a and b, with $b > a$, carry charge Q and $-Q$, respectively, each charge uniformly distributed. Find the energy stored in the electric field of this system.

1.31 Like the charged rubber balloon described on page 31, a charged soap bubble experiences an outward electrical force on every bit of its surface. Given the total charge Q on a bubble of radius R, what is the magnitude of the resultant force tending to pull any hemispherical half of the bubble away from the other half? (Should this force divided by $2\pi R$ exceed the surface tension of the soap film interesting behavior might be expected!)

Ans. $Q^2/8R^2$.

1.32 Suppose three positively charged particles are constrained to move on a fixed circular track. If the charges were all equal, an equilibrium arrangement would obviously be a symmetrical one with the particles spaced 120° apart around the circle. Suppose that two of the charges are equal and the equilibrium arrangement is such that these two charges are 90° apart rather than 120°. What is the relative magnitude of the third charge?

Ans. 3.154.

1.33 Imagine a sphere of radius a filled with negative charge of uniform density, the total charge being equivalent to that of two electrons. Imbed in this jelly of negative charge two protons and assume that in spite of their presence the negative charge distribution remains uniform. Where must the protons be located so that the force on each of them is zero? (This is a surprisingly realistic caricature of a hydrogen molecule; the magic that keeps the electron cloud in the molecule from collapsing around the protons is explained by quantum mechanics!)

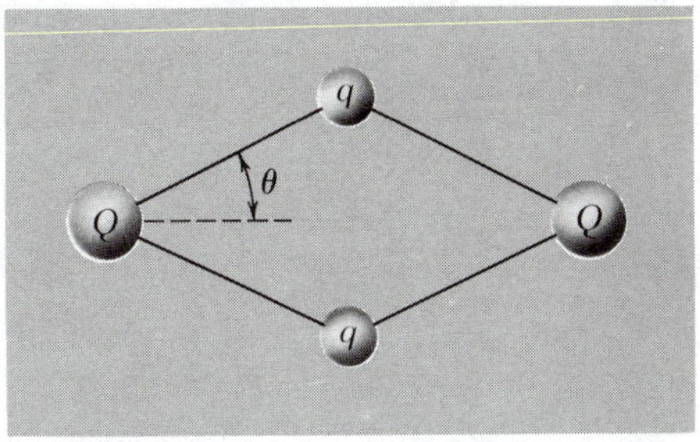

PROBLEM 1.34

1.34 Four positively charged bodies, two with charge Q and two with charge q, are connected by four unstretchable strings of equal length. In the absence of external forces they assume the equilibrium configuration shown in the diagram. Show that $\tan^3 \theta = q^2/Q^2$. This can be done in two ways. You could show that this relation must hold if the total force on each body, the vector sum of string tension and electrical repulsion, is zero. Or you could write out the expression for the energy U of the assembly (like Eq. 7 but for four charges instead of three) and minimize it.

1.35 Consider the electric field of two protons b cm apart. According to Eq. 1.38 (which we stated but did not prove) the potential energy of the system ought to be given by

$$U = \frac{1}{8\pi}\int \mathbf{E}^2\, dv = \frac{1}{8\pi}\int (\mathbf{E}_1 + \mathbf{E}_2)^2\, dv$$

$$= \frac{1}{8\pi}\int \mathbf{E}_1^2\, dv + \frac{1}{8\pi}\int \mathbf{E}_2^2\, dv + \frac{1}{4\pi}\int \mathbf{E}_1 \cdot \mathbf{E}_2\, dv$$

where $\mathbf{E}_1$ is the field of one particle alone and $\mathbf{E}_2$ that of the other. The first of the three integrals on the right might be called the "electrical self-energy" of one proton; an intrinsic property of the particle, it depends on the proton's size and structure. We have always disregarded it in reckoning the potential energy of a system of charges, on the assumption that it remains constant; the same goes for the second integral. The third integral involves the distance between the charges. The third integral is not hard to evaluate if you set it up in spherical polar coordinates with one of the protons at the origin and the other on the polar axis, and perform the integration over r first. Thus, by direct calculation, you can show that the third integral has the value e^2/b, which we already know to be the work required to bring the two protons in from an infinite distance to positions a distance b apart. So you will have proved the correctness of Eq. 38 for this case, and by invoking superposition you can argue that Eq. 38 must then give the energy required to assemble any system of charges.

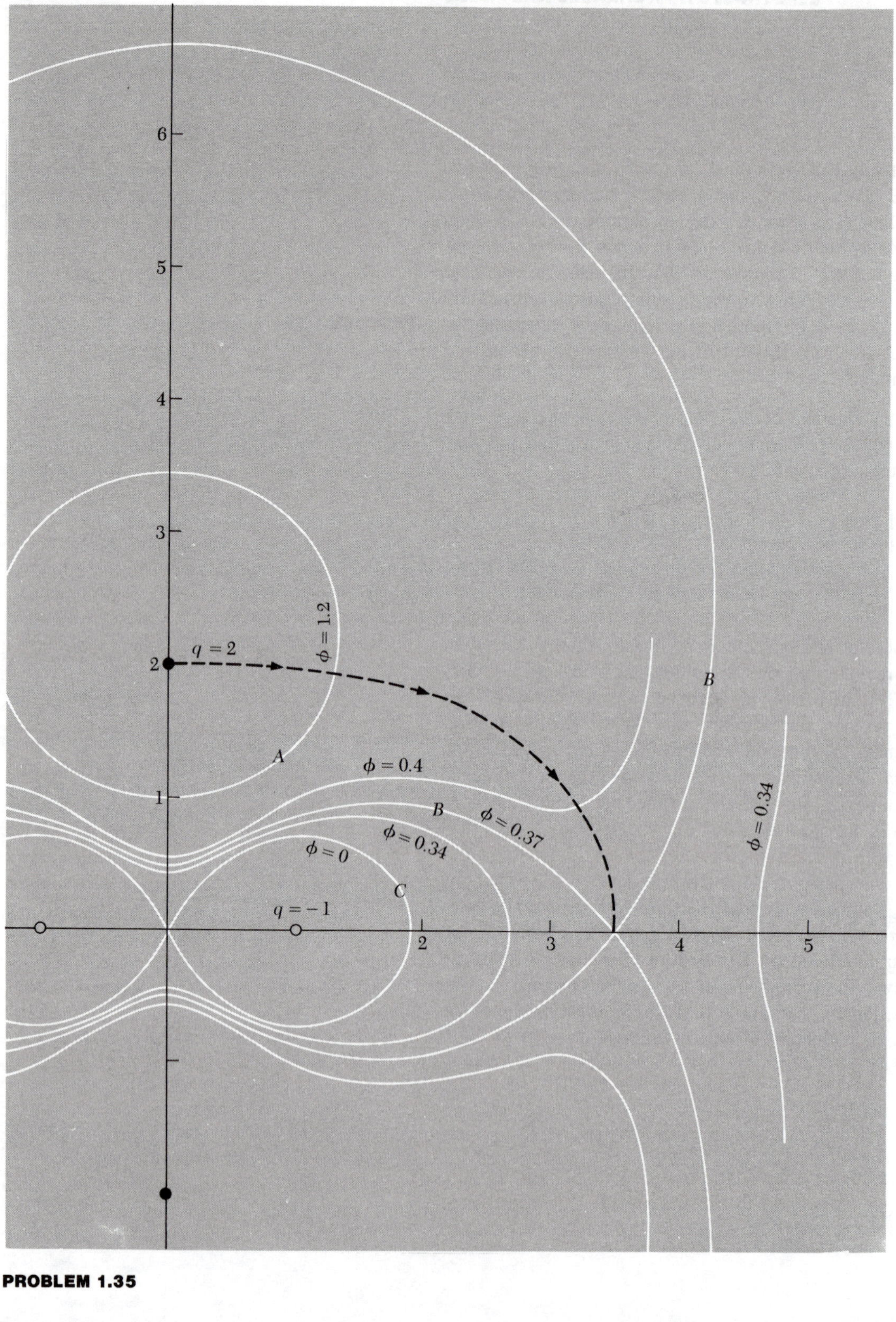

6
5
4
3
2
1
$\phi = 1.2$
$q = 2$
B
A
$\phi = 0.4$
B
$\phi = 0.37$
$\phi = 0.34$
$\phi = 0.34$
$\phi = 0$
C
$q = -1$
2
3
4
5

PROBLEM 1.35

2

THE ELECTRIC POTENTIAL

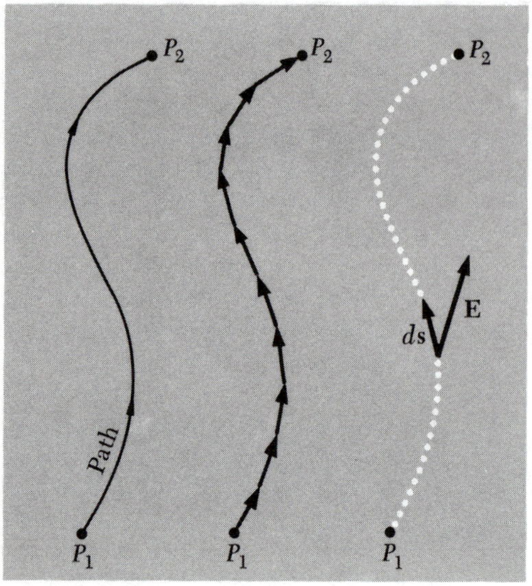

FIGURE 2.1
Showing the division of the path into path elements $d\mathbf{s}$.

LINE INTEGRAL OF THE ELECTRIC FIELD

2.1 Suppose that **E** is the field of some stationary distribution of electric charges. Let P_1 and P_2 denote two points anywhere in the field. The line integral of E between the two points is $\int_{P_1}^{P_2} \mathbf{E} \cdot d\mathbf{s}$, taken along some path that runs from P_1 to P_2, as in Fig. 2.1. This means: Divide the chosen path into short segments, each segment being represented by a vector connecting its ends; take the scalar product of the path-segment vector with the field **E** at that place; add these products up for the whole path. The integral as usual is to be regarded as the limit of this sum as the segments are made shorter and more numerous without limit.

Let's consider the field of a point charge q and some paths running from point P_1 to point P_2 in that field. Two different paths are shown in Fig. 2.2. It is easy to compute the line integral of **E** along path A, which is made up of a radial segment running outward from P_1 and an arc of radius r_2. Along the radial segment of path A, **E** and $d\mathbf{s}$ are parallel, the magnitude of **E** is q/r^2, and $\mathbf{E} \cdot d\mathbf{s}$ is simply $(q/r^2)\,ds$. Thus the line integral on that segment is

$$\int_{r_1}^{r_2} \frac{q\,dr}{r^2} = q\left(\frac{1}{r_1} - \frac{1}{r_2}\right) \tag{1}$$

The second leg of path A, the circular segment, gives zero because **E** is perpendicular to $d\mathbf{s}$ everywhere on that arc. The entire line integral is therefore

$$\int_{P_1}^{P_2} \mathbf{E} \cdot d\mathbf{s} = q\left(\frac{1}{r_1} - \frac{1}{r_2}\right) \tag{2}$$

Now look at path B. Because **E** is radial with magnitude q/r^2, $\mathbf{E} \cdot d\mathbf{s} = (q/r^2)\,dr$ even when $d\mathbf{s}$ is not radially oriented. The corresponding pieces of path A and path B indicated in the diagram make identical contributions to the integral. The part of path B that loops beyond r_2 makes a net contribution of zero; contributions from corresponding outgoing and incoming parts cancel. For the entire line integral, path B will give the same result as path A. As there is nothing special about path B, Eq. 1 must hold for *any* path running from P_1 to P_2.

Here we have essentially repeated, in different language, the argument in Section 1.5, illustrated in Fig. 1.5, concerning the work done in moving one point charge near another. But now we are interested in the total electric field produced by any distribution of charges. One more step will bring us to an important conclusion. The line integral of the sum of fields equals the sum of the line integrals of the

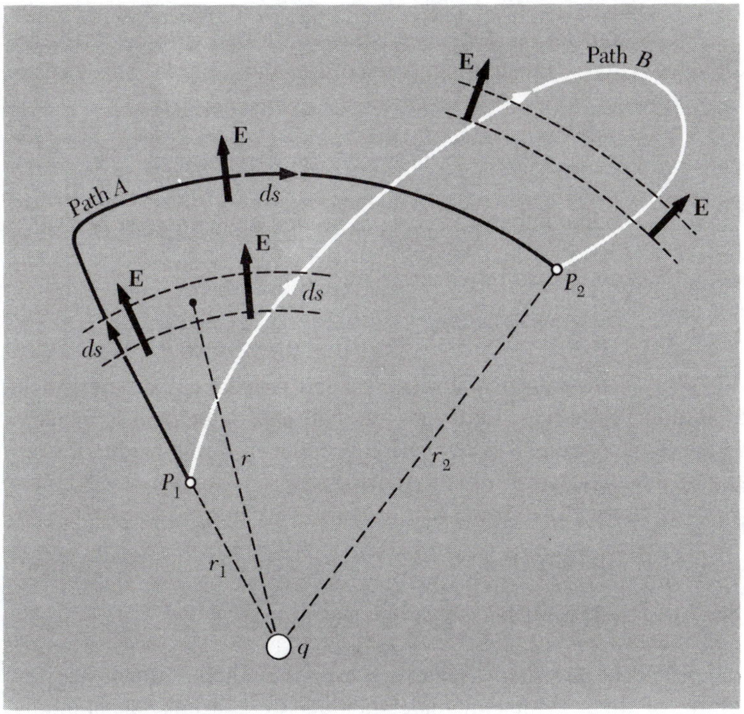

FIGURE 2.2
The electric field **E** is that of a positive point charge q. The line integral of **E** from P_1 to P_2 along path A has the value $q(1/r_1 - 1/r_2)$. It will have exactly the same value if calculated for path B, or for any other path from P_1 to P_2.

fields calculated separately. Or, stated more carefully, if $\mathbf{E} = \mathbf{E}_1 + \mathbf{E}_2 + \ldots$, then

$$\int_{P_1}^{P_2} \mathbf{E} \cdot d\mathbf{s} = \int_{P_1}^{P_2} \mathbf{E}_1 \cdot d\mathbf{s} + \int_{P_1}^{P_2} \mathbf{E}_2 \cdot d\mathbf{s} + \cdots \quad (3)$$

where the same path is used for all the integrations. Now any electrostatic field can be regarded as the sum of a number (possibly enormous) of point-charge fields, as expressed in Eq. 1.14 or 1.15. Therefore if the line integral from P_1 to P_2 is independent of path for each of the point-charge fields $\mathbf{E}_1, \mathbf{E}_2, \ldots$, the total field **E** must have this property:

> The line integral $\int_{P_1}^{P_2} \mathbf{E} \cdot d\mathbf{s}$ for any electrostatic field **E** has the same value for all paths from P_1 to P_2 (4)

The points P_2 and P_1 may coincide. In that case the paths are all closed curves, among them paths of vanishing length. This leads to the corollary:

$$\text{The line integral } \int \mathbf{E} \cdot d\mathbf{s} \text{ around any closed path in an electrostatic field is zero} \tag{5}$$

By *electrostatic field* is meant, strictly speaking, the electric field of stationary charges. Later on, we shall encounter electric fields in which the line integral is *not* path-independent. Those fields will usually be associated with rapidly moving charges. For our present purposes we can say that, if the source charges are moving slowly enough, the field $\mathbf{E}$ will be such that $\int \mathbf{E} \cdot d\mathbf{s}$ is practically path-independent. Of course, if $\mathbf{E}$ itself is varying in time, the $\mathbf{E}$ in $\int \mathbf{E} \cdot d\mathbf{s}$ must be understood as the field that exists over the whole path at a given instant of time. With that understanding we can talk meaningfully about the line integral in a changing electrostatic field.

POTENTIAL DIFFERENCE AND THE POTENTIAL FUNCTION

2.2 Because the line integral in the electrostatic field is path-independent, we can use it to define a scalar quantity ϕ_{21}, without specifying any particular path:

$$\phi_{21} = -\int_{P_1}^{P_2} \mathbf{E} \cdot d\mathbf{s} \tag{6}$$

Thus ϕ_{21} is the work per unit charge done in moving a positive charge from P_1 to P_2 in the field $\mathbf{E}$. Thus ϕ_{21} is a single-valued scalar function of the two positions P_1 and P_2. We call it the *electric potential difference* between the two points.

In our CGS system of units, potential difference is measured in erg/esu. This unit has a name of its own, the statvolt ("stat" comes from "electrostatic"). The volt is the unit of potential difference in SI units, the system in which the coulomb is the unit of charge and the joule the unit of energy. One joule (10^7 ergs) of work is required to move a charge of one coulomb through a potential difference of one volt. The exact relations between CGS and SI electrical units are given in Appendix E, taking into account the very recent official redefinition of the meter in terms of the speed of light. Those exact relations need not concern us now. Two approximate relations are all we shall usually need: One coulomb is equivalent to 3×10^9 esu. One volt is equivalent to $\frac{1}{300}$ statvolt. These are accurate to better than 0.1 percent, thanks to the accident that c is that close to 3×10^8 meters/sec.

Suppose we hold P_1 fixed at some reference position. Then ϕ_{21} becomes a function of P_2 only, that is, a function of the spatial coordinates x, y, z. We can write it simply $\phi(x, y, z)$, without the subscript, if we remember that its definition still involves agreement on a reference point P_1. We can say that ϕ is the potential associated with the vector field **E**. It is a scalar function of position, or a scalar field (they mean the same thing). Its value at a point is simply a number (in units of work per unit charge) and has no direction associated with it. Once the vector field **E** is given, the potential function ϕ is determined, except for an arbitrary additive constant allowed by the arbitrariness in our choice of P_1.

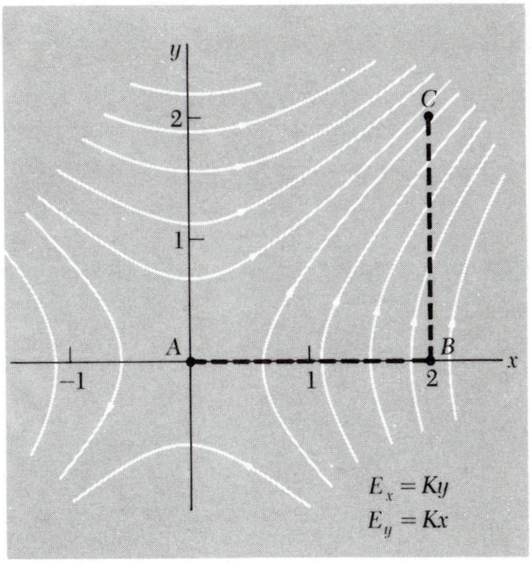

FIGURE 2.3
(*a*) A particular path, *ABC*, in the electric field $E_x = Ky$, $E_y = Kx$. Some field lines are shown.

As an example, let us find the potential associated with the electric field described in Fig. 2.3, the components of which are: $E_x = Ky$, $E_y = Kx$, $E_z = 0$, with K a constant. This is a possible electrostatic field. Some field lines are shown. Since $E_z = 0$, the potential will be independent of z and we need consider only the xy plane. Let x_1, y_1 be the coordinates of P_1, and x_2, y_2 the coordinates of P_2. It is convenient to locate P_1 at the origin: $x_1 = 0$, $y_1 = 0$. To evaluate $-\int \mathbf{E} \cdot d\mathbf{s}$ from this reference point to a general point (x_2, y_2) it is easiest to use a path like the dotted path *ABC* in Fig. 2.3.

$$\phi(x_2, y_2) = -\int_{(0,0)}^{(x_2,y_2)} \mathbf{E} \cdot d\mathbf{s}$$

$$= -\int_{(0,0)}^{(x_2,0)} E_x\, dx - \int_{(x_2,0)}^{(x_2,y_2)} E_y\, dy \qquad (7)$$

The first of the two integrals on the right is zero because E_x is zero along the x axis. The second integration is carried out at constant x, with $E_y = Kx_2$:

$$-\int_{(x_2,0)}^{(x_2,y_2)} E_y\, dy = -\int_0^{y_2} Kx_2\, dy = -Kx_2y_2 \qquad (8)$$

There was nothing special about the point (x_2, y_2) so we can drop the subscripts:

$$\phi = -Kxy \qquad (9)$$

for the potential at any point (x, y) in this field, with zero potential at the origin. Any constant could be added to this. That would only mean that the reference point to which zero potential is assigned had been located somewhere else.

We must be careful not to confuse the potential ϕ associated with a given field **E** with the potential energy of a system of charges. The potential energy of a system of charges is the total work required to assemble it, starting with all the charges far apart. In Eq. 1.8, for example, we expressed U, the potential energy of the charge system in Fig. 1.6. The electric potential $\phi(x, y, z)$ associated with the field in Fig. 1.6 would be the work per unit charge required to move a unit positive test charge from some chosen reference point to the point (x, y, z) in the field of that structure of eight charges.

GRADIENT OF A SCALAR FUNCTION

2.3 Given the electric field, we can find the electric potential function. But we can also proceed in the other direction; from the potential we can derive the field. It appears from Eq. 6 that the field is in some sense the *derivative* of the potential function. To make this idea precise we introduce the *gradient* of a scalar function of position. Let $f(x, y, z)$ be some continuous, differentiable function of the coordinates. With its partial derivatives $\partial f/\partial x$, $\partial f/\partial y$, and $\partial f/\partial z$ we can construct at every point in space a vector, the vector whose x, y, z components are equal to the respective partial derivatives.† This vector we call the *gradient* of f, written "grad f," or ∇f.

$$\nabla f = \hat{\mathbf{x}} \frac{\partial f}{\partial x} + \hat{\mathbf{y}} \frac{\partial f}{\partial y} + \hat{\mathbf{z}} \frac{\partial f}{\partial z} \tag{10}$$

∇f is a vector that tells how the function f varies in the neighborhood of a point. Its x component is the partial derivative of f with respect to x, a measure of the rate of change of f as we move in the x direction. The direction of the vector ∇f at any point is the direction in which one must move from that point to find the most rapid increase in the function f. Suppose we were dealing with a function of two variables only, x and y, so that the function could be represented by a surface in three dimensions. Standing on that surface at some point, we see the surface rising in some direction, sloping downward in the opposite direction. There is a direction in which a short step will take us higher than a step of the same length in any other direction. The gradient of

†We remind the reader that a partial derivative with respect to x, of a function of x, y, z, written simply $\partial f/\partial x$, means the rate of change of the function with respect to x with the other variables y and z held constant. More precisely,

$$\frac{\partial f}{\partial x} = \lim_{\Delta x \to 0} \frac{f(x + \Delta x, y, z) - f(x, y, z)}{\Delta x}$$

As an example, if $f = x^2yz^3$,

$$\frac{\partial f}{\partial x} = 2xyz^3 \qquad \frac{\partial f}{\partial y} = x^2z^3 \qquad \frac{\partial f}{\partial z} = 3x^2yz^2$$

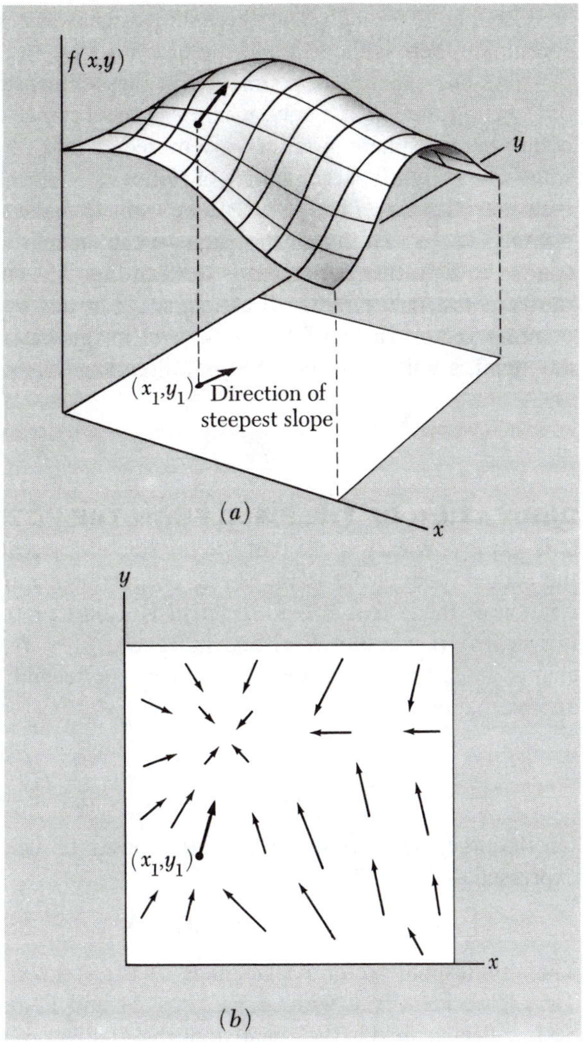

FIGURE 2.4
The scalar function $f(x, y)$ is represented by the surface in (*a*). The arrows in (*b*) represent the vector function, grad f.

the function is a vector in that direction of steepest ascent, and its magnitude is the slope measured in that direction.

Figure 2.4 may help you to visualize this. Suppose some particular function of two coordinates x and y is represented by the surface $f(x, y)$ sketched in Fig. 2.4*a*. At the location (x_1, y_1) the surface rises most steeply in a direction that makes an angle of about 80° with the positive x direction. The gradient of $f(x, y)$, ∇f, is a vector function of x and y. Its character is suggested in Fig. 2.4*b* by a number of vectors at various points in the two-dimensional space, including the point (x_1, y_1). The vector function ∇f defined in Eq. 10 is simply an extension of this idea to three-dimensional space. [Be careful not to

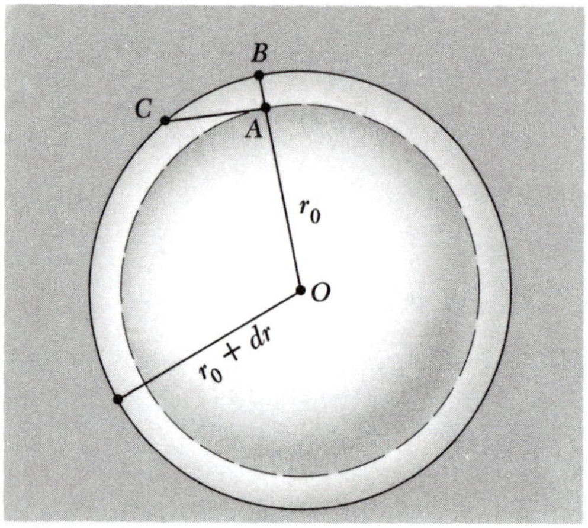

FIGURE 2.5
The shortest step for a given change in f is the radial step AB, if f is a function of r only.

confuse Fig. 2.4*a* with real three-dimensional xyz space; the third coordinate there is the value of the function $f(x, y)$.]

As one example of a function in three-dimensional space, suppose f is a function of r only, where r is the distance from some fixed point O. On a sphere of radius r_0 centered about O, $f = f(r_0)$ is constant. On a slightly larger sphere of radius $r_0 + dr$ it is also constant, with the value $f = f(r_0 + dr)$. If we want to make the change from $f(r_0)$ to $f(r_0 + dr)$, the shortest step we can make is to go radially (as from A to B) rather than from A to C, in Fig. 2.5. The "slope" of f is thus greatest in the radial direction, so ∇f at any point is a radially pointing vector. In fact $\nabla f = \hat{\mathbf{r}}\,(df/dr)$ in this case, $\hat{\mathbf{r}}$ denoting, for any point, a unit vector in the radial direction.

DERIVATION OF THE FIELD FROM THE POTENTIAL

2.4 It is now easy to see that the relation of the scalar function f to the vector function ∇f is the same, except for a minus sign, as the relation of the potential φ to the field $\mathbf{E}$. Consider the value of φ at two nearby points, (x, y, z) and $(x + dx, y + dy, z + dz)$. The change in φ, going from the first point to the second, is in first-order approximation

$$d\varphi = \frac{\partial \varphi}{\partial x}\,dx + \frac{\partial \varphi}{\partial y}\,dy + \frac{\partial \varphi}{\partial z}\,dz \tag{11}$$

On the other hand, from the definition of φ, the change can also be expressed as

$$d\varphi = -\mathbf{E} \cdot d\mathbf{s} \tag{12}$$

The infinitesimal vector displacement $d\mathbf{s}$ is just $\hat{\mathbf{x}}\,dx + \hat{\mathbf{y}}\,dy + \hat{\mathbf{z}}\,dz$. Thus if we identify $\mathbf{E}$ with $-\nabla\varphi$, Eqs. 11 and 12 become identical. So the electric field is the negative of the gradient of the potential:

$$\mathbf{E} = -\nabla\varphi \tag{13}$$

The minus sign came in because the electric field points from a region of positive potential toward a region of negative potential, whereas the vector $\nabla\varphi$ is defined so that it points in the direction of increasing φ.

To show how this works, we go back to the example of the field in Fig. 2.3. From the potential given by Eq. 9, $\varphi = -Kxy$, we can recover the electric field we started with:

$$\begin{aligned}\mathbf{E} &= -\nabla(-Kxy) \\ &= -\left(\hat{\mathbf{x}}\frac{\partial}{\partial x} + \hat{\mathbf{y}}\frac{\partial}{\partial y}\right)(-Kxy) = K(\hat{\mathbf{x}}y + \hat{\mathbf{y}}x)\end{aligned} \tag{14}$$

POTENTIAL OF A CHARGE DISTRIBUTION

2.5 We already know the potential that goes with a single point charge, because we calculated the work required to bring one charge into the neighborhood of another in Eq. 3 of Chapter 1. The potential at any point, in the field of an isolated point charge q, is just q/r, where r is the distance from the point in question to the source q, and where we have assigned zero potential to points infinitely far from the source.

Superposition must work for potentials as well as fields. If we have several sources, the potential function is simply the sum of the potential functions that we would have for each of the sources present alone—*providing* we make a consistent assignment of the zero of potential in each case. If all the sources are contained in some finite region, it is always possible, and usually the simplest choice, to put zero potential at infinite distance. If we adopt this rule, the potential of any charge distribution can be specified by the integral:

$$\varphi(x, y, z) = \int_{\substack{\text{All} \\ \text{sources}}} \frac{\rho(x', y', z')\, dx'\, dy'\, dz'}{r} \tag{15}$$

where r is the distance from the volume element $dx'\, dy'\, dz'$ to the point (x, y, z) at which the potential is being evaluated (Fig. 2.6). That is, $r = [(x - x')^2 + (y - y')^2 + (z - z')^2]^{1/2}$. Notice the difference between this and the integral giving the electric field of a charge distribution (Eq. 15 of Chapter 1). Here we have r in the denominator, not r^2, and the integral is a scalar not a vector. From the scalar potential function $\varphi(x, y, z)$ we can always find the electric field by taking the negative gradient of φ, according to Eq. 13.

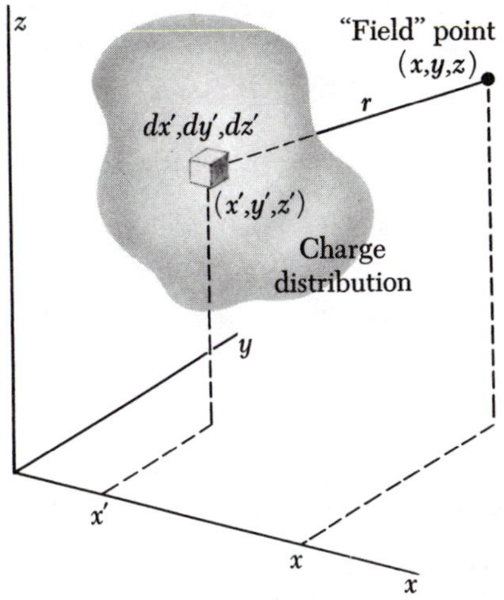

FIGURE 2.6
Each element of the charge distribution $\rho\,(x', y', z')$ contributes to the potential ϕ at the point (x, y, z). The potential at this point is the sum of all such contributions (Eq. 15).

Potential of two point charges. Consider a very simple example, the potential of the two point charges shown in Fig. 2.7. A positive charge of 12 esu is located 3 cm away from a negative charge, -6 esu. The potential at any point in space is the sum of the potentials due to each charge alone. The potentials for some selected points in space are given in the diagram. No vector addition is involved here, only the algebraic addition of scalar quantities. For instance, at the point on the far right which is 6 cm from the positive charge and 5 cm from the negative charge, the potential has the value $\frac{12}{6} + (-\frac{6}{5}) = 0.8$. The unit here comes out esu/cm, which is the same as erg/esu, or *statvolts*. The potential approaches zero at infinite distance. It would take 0.8 erg of work to bring a unit positive charge in from infinity to a point where $\varphi = 0.8$ statvolt. Note that two of the points shown on the diagram have $\varphi = 0$. The net work done in bringing in any charge to one of these points would be zero. You can see that there must be an infinite number of such points, forming a surface in space surrounding the negative charge. In fact the locus of points

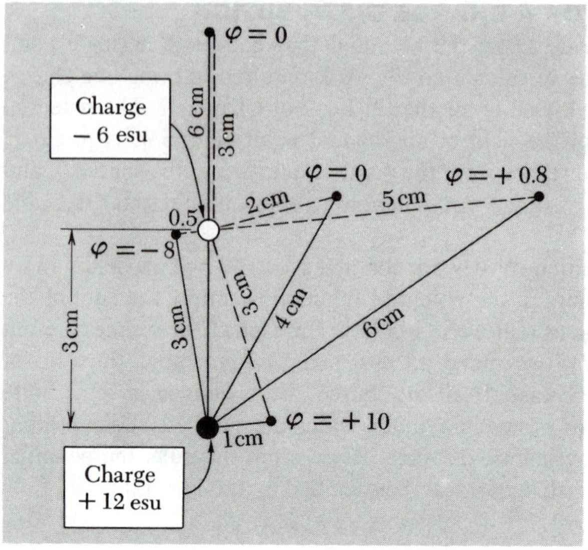

FIGURE 2.7
The electric potential ϕ at various points in a system of two point charges. ϕ goes to zero at infinite distance. ϕ is given in units of statvolts, or ergs per unit charge.

with any particular value of φ is a surface—an *equipotential surface*—which would show on our two-dimensional diagram as a curve.

Potential of a long charged wire. There is one restriction on the use of Eq. 15: It may not work unless all sources are confined to some finite region of space. A simple example of the difficulty that arises with charges distributed out to infinite distance is found in the long charged wire whose field **E** we studied in Section 1.12. If we attempt to carry out the integration over the charge distribution indicated in Eq. 15, we find that the integral diverges—we get an infinite result. No such difficulty arose in finding the electric *field* of the infinitely long wire, because the contributions of elements of the line charge to the field decrease so rapidly with distance. Evidently we had better locate the zero of potential somewhere close to home, in a system which has charges distributed out to infinity. Then it is simply a matter of calculating the difference in potential φ_{21}, between the general point (x, y, z) and the selected reference point, using the fundamental relation, Eq. 6.

To see how this goes in the case of the infinitely long charged wire, let us arbitrarily locate the reference point P_1 at a distance r_1 from the wire. Then to carry a charge from P_1 to any other point P_2 at distance r_2 requires the work per unit charge

$$\varphi_{21} = -\int_{P_1}^{P_2} \mathbf{E} \cdot d\mathbf{s} = -\int_{r_1}^{r_2} \left(\frac{2\lambda}{r}\right) dr$$
$$= -2\lambda \ln r_2 + 2\lambda \ln r_1 \qquad (16)$$

This shows that the electrical potential for the charged wire can be taken as

$$\varphi = -2\lambda \ln r + \text{const} \tag{17}$$

The constant, $2\lambda \ln r_1$ in this case, has no effect when we take $-\text{grad}\,\varphi$ to get back to the field **E**. In this case

$$-\nabla\varphi = -\hat{\mathbf{r}}\frac{d\varphi}{dr} = \frac{2\lambda\hat{\mathbf{r}}}{r} \tag{18}$$

UNIFORMLY CHARGED DISK

2.6 Let us study, as a concrete example, the electric potential and field around a uniformly charged disk. This is a charge distribution like that discussed in Section 1.13, except that it has a limited extent. The flat disk of radius a in Fig. 2.8 carries a positive charge spread over its surface with the constant density σ, in esu/cm^2. (This is a single sheet of charge of infinitesimal thickness, not two layers of charge, one on each side. That is, the total charge in the system is $\pi a^2\sigma$.) We shall often meet surface charge distributions in the future, especially on metallic conductors. However, the object just described is *not* a conductor; if it were, as we shall soon see, the charge could not remain uniformly distributed but would redistribute itself, crowding more toward the rim of the disk. What we have is an insulating disk, like a sheet of plastic, upon which charge has been "sprayed" so that every square centimeter of the disk has received, and holds fixed, the same amount of charge.

As a start, let's find the potential at some point P_1 on the axis of symmetry, which we have made the y axis. All charge elements in a thin, ring-shaped segment of the disk lie at the same distance from P_1. If s denotes the radius of such an annular segment and ds is its width, its area is $2\pi s\, ds$. The amount of charge it contains, dq, is therefore $dq = \sigma\, 2\pi s\, ds$. All parts of this ring are the same distance away from P_1, namely, $r = \sqrt{y^2 + s^2}$, so the contribution of the ring to the potential at P_1 is dq/r, or $2\pi\sigma s\, ds/\sqrt{y^2 + s^2}$. To get the potential due to the whole disk, we have to integrate over all such rings:

$$\varphi(0, y, 0) = \int \frac{dq}{r} = \int_0^a \frac{2\pi\sigma s\, ds}{\sqrt{y^2 + s^2}}$$

$$= 2\pi\sigma\left[\sqrt{y^2 + s^2}\right]_{s=0}^{s=a} \tag{19}$$

The integral happened to be an elementary one; on substituting $u = y^2 + s^2$ it takes the form $\int u^{-1/2}\, du$. Putting in the limits, we obtain

$$\varphi(0, y, 0) = 2\pi\sigma\left(\sqrt{y^2 + a^2} - y\right) \qquad \text{for } y > 0 \tag{20}$$

FIGURE 2.8
Finding the potential at a point P_1 on the axis of a uniformly charged disk.

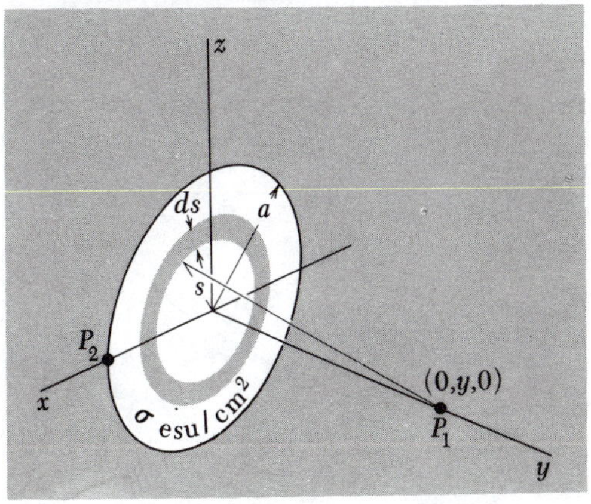

A minor point deserves a comment: The result we have written down in Eq. 20 holds for all points on the *positive* y axis. It is obvious from the physical symmetry of the system (there is no difference between one face of the disk and the other) that the potential must have the same value for negative and positive y, and this is reflected in Eq. 19, where only y^2 appears. But in writing Eq. 20 we made a choice of sign in taking the square root of y^2, with the consequence that it holds *only* for positive y. The correct expression for $y < 0$ is obtained by the other choice of root and is

$$\varphi(0, y, 0) = 2\pi\sigma(\sqrt{y^2 + a^2} + y) \qquad \text{for } y < 0 \qquad (21)$$

In view of this, we should not be surprised to find a singularity in $\varphi(0, y, 0)$ at $y = 0$. Indeed, the function has an abrupt change of slope there, as we see in Fig. 2.9, where we have plotted as a function of y the potential on the axis. The potential at the center of the disk is $\varphi(0, 0, 0) = 2\pi\sigma a$. That much work would be required to bring a unit positive charge in from infinity, by any route, and leave it sitting at the center of the disk.

The behavior of $\varphi(0, y, 0)$ for very large y is interesting. For $y \gg a$ we can approximate Eq. 20 as follows:

$$\begin{aligned} \sqrt{y^2 + a^2} - y &= y\left[\left(1 + \frac{a^2}{y^2}\right)^{1/2} - 1\right] \\ &= y\left[1 + \frac{1}{2}\left(\frac{a^2}{y^2}\right) \cdots - 1\right] \approx \frac{a^2}{2y} \end{aligned} \qquad (22)$$

FIGURE 2.9
A graph of the potential on the axis. The dashed curve is the potential of a point charge $q = \pi a^2\sigma$.

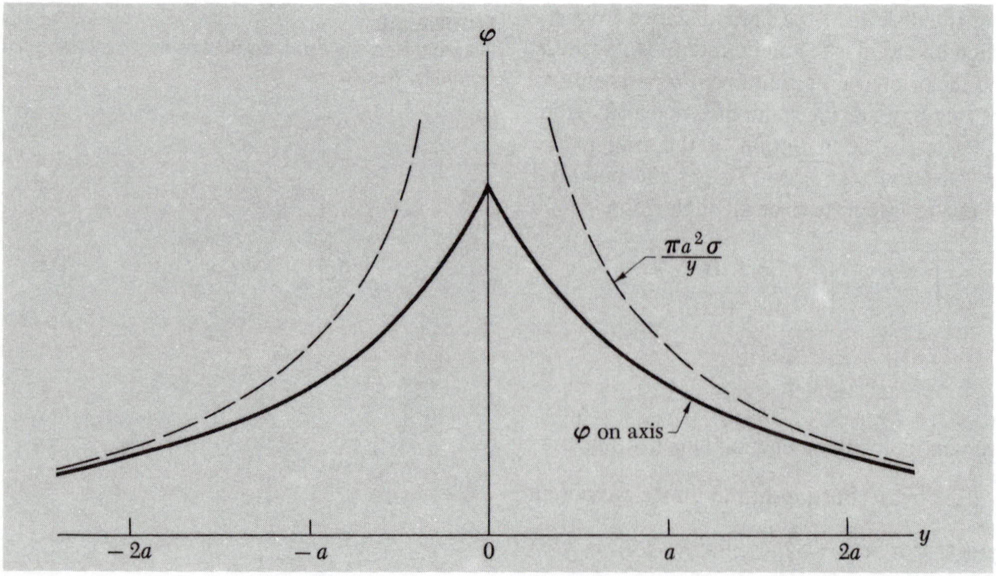
φ
πa²σ / y
φ on axis
y
– 2a
– a
0
a
2a

Hence

$$\varphi(0, y, 0) \approx \frac{\pi a^2 \sigma}{y} \qquad \text{for } y \gg a \tag{23}$$

Now $\pi a^2 \sigma$ is the total charge q on the disk, and Eq. 23 is just the expression for the potential due to a point charge of this magnitude. As we should expect, at a considerable distance from the disk (relative to its diameter), it doesn't matter much how the charge is shaped; only the total charge matters, in first approximation. In Fig. 2.9 we have drawn, as a dotted curve, the function $\pi a^2 \sigma / y$. You can see that the axial potential function approaches its asymptotic form pretty quickly.

It is not quite so easy to derive the potential for general points away from the axis of symmetry, because the definite integral isn't so simple. It proves to be something called an *elliptic integral*. These functions are well-known and tabulated, but there is no point in pursuing here mathematical details peculiar to a special problem. One further calculation, which is easy enough, may be instructive. We can find the potential at a point on the very edge of the disk, such as P_2 in Fig. 2.10.

To calculate the potential at P_2 we can consider first the thin wedge of length R and angular width $d\theta$ in Fig. 2.10. An element of the wedge, the black patch at distance r from P_2, contains an amount of charge $\sigma r \, d\theta \, dr$. Its contribution to the potential at P_2 is therefore just $\sigma \, d\theta \, dr$. The contribution of the entire wedge is then $\sigma \, d\theta \int_0^R dr = \sigma R \, d\theta$. Now R is $2a \cos\theta$ from the geometry of the right triangle, and the whole disk is swept out as θ ranges from $-\pi/2$ to $\pi/2$. Thus we find the potential at P_2:

$$\phi = \int_{\pi/2}^{\pi/2} 2\sigma a \cos\theta \, d\theta = 4\sigma a \tag{24}$$

FIGURE 2.10
Finding the potential at a point P_2 on the rim of a uniformly charged disk.

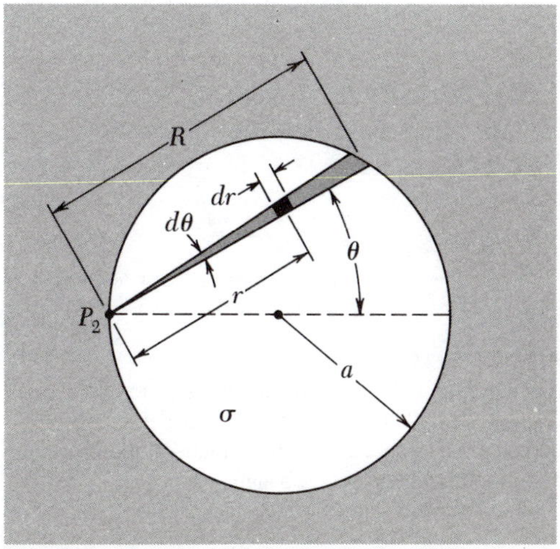

Comparing this with $2\pi\sigma a$, the potential at the center of the disk, we see that, as we should expect, the potential falls off from the center to the edge of the disk. The electric field, therefore, must have an *outward* component in the plane of the disk. That is why we remarked earlier that the charge, if free to move, would redistribute itself toward the rim. To put it another way, our uniformly charged disk is *not* a surface of constant potential, which any conducting surface must be unless charge is moving.†

The electric field on the symmetry axis can be computed directly from the potential function:

$$E_y = -\frac{\partial \varphi}{\partial y} = -\frac{d}{dy} 2\pi\sigma(\sqrt{y^2 + a^2} - y) \tag{25}$$

†The fact that conducting surfaces have to be equipotentials will be discussed thoroughly in Chapter 3.

giving

$$E_y = 2\pi\sigma\left[1 - \frac{y}{\sqrt{y^2 + a^2}}\right] \qquad y > 0 \qquad (26)$$

(To be sure, it is not hard to compute E_y directly from the charge distribution, for points on the axis.)

As y approaches zero from the positive side, E_y approaches $2\pi\sigma$. On the negative y side of the disk, which we shall call the back, **E** points in the other direction and its y component E_y is $-2\pi\sigma$. This is the same as the field of an infinite sheet of charge of density σ, derived in Section 1.13. It ought to be, for at points close to the center of the disk, the presence or absence of charge out beyond the rim can't make much difference. In other words, any sheet looks infinite if viewed from close up. Indeed, E_y has the value $2\pi\sigma$ not only at the center but all over the disk.

In Fig. 2.11 we show some field lines for this system and also, plotted as dashed curves, the intersections on the yz plane of the surfaces of constant potential. Near the center of the disk these are lens-like surfaces, while at distances much greater than a they approach the spherical form of equipotential surfaces around a point charge.

Figure 2.11 illustrates a general property of field lines and equipotential surfaces. A field line through any point and the equipotential surface through that point *are perpendicular to one another,* just as, on a contour map of hilly terrain, the slope is steepest at right angles to a contour of constant elevation. This must be so, because if the field at any point had a component parallel to the equipotential surface through that point, it would require work to move a test charge along a constant-potential surface.

The energy associated with this electric field could be expressed as the integral over all space of $E^2\,dv/8\pi$. It is equal to the work done in assembling this distribution, starting with infinitesimal charges far apart. In this particular example, as Problem 2.27 will demonstrate, that work is not hard to calculate directly if we know the potential at the rim of a uniformly charged disk.

There is a general relation between the work U required to assemble a charge distribution $\rho(x, y, z)$ and the potential $\phi(x, y, z)$ of that distribution:

$$U = \frac{1}{2}\int \rho\phi\,dv \qquad (27)$$

Equation 9 of Chapter 1, for the energy of a system of discrete point charges, could have been written in this way:

$$U = \frac{1}{2}\sum_{j=1}^{N} q_j \sum_{k\neq j} \frac{q_k}{r_{jk}} \qquad (28)$$

FIGURE 2.11

(Facing page.) The electric field of the uniformly charged disk. Solid curves are field lines. Dashed curves are intersections, with the plane of the figure, of surfaces of constant potential.

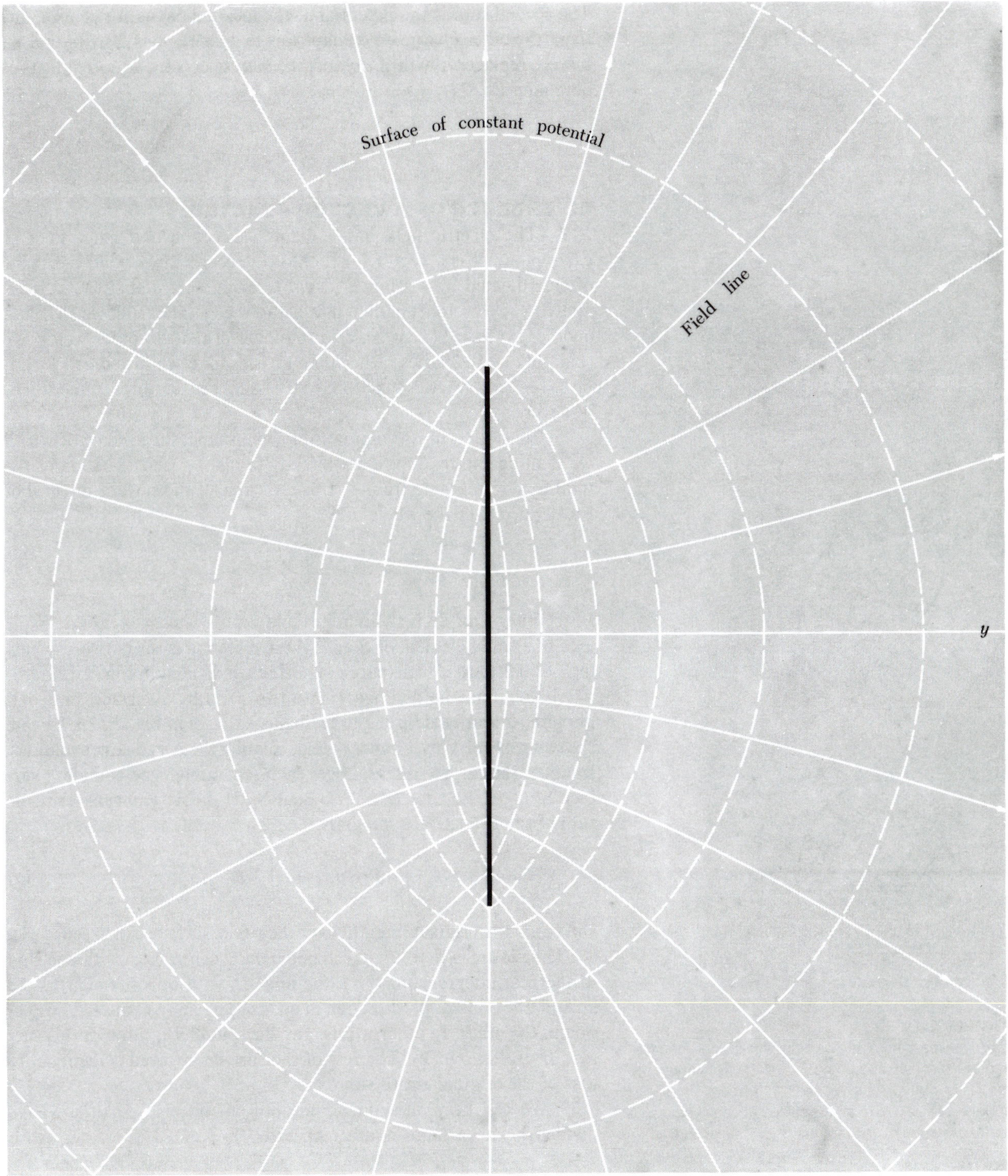
Surface of constant potential
Field line
y

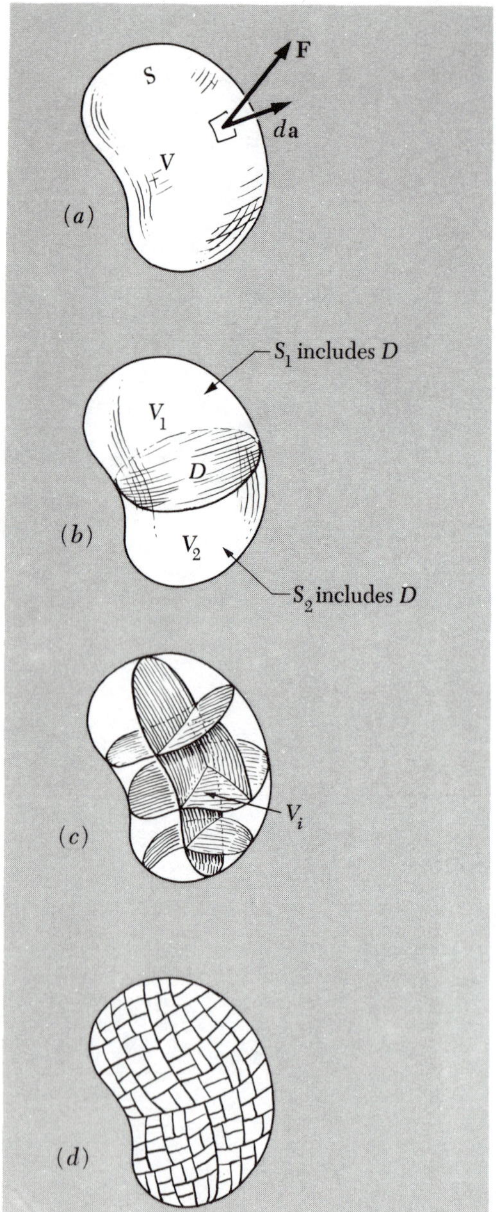

FIGURE 2.12
(*a*) A volume V enclosed by a surface S is divided (*b*) into two pieces enclosed by S_1 and S_2. No matter how far this is carried, as in (*c*) and (*d*), the sum of the surface integrals over all the pieces equals the original surface integral over S, for any vector function **F**.

The second sum is the potential at the location of the jth charge, due to all the other charges. To adapt this to a continuous distribution we merely replace q_j with $\rho\, dv$ and the sum over j by an integral, thus obtaining Eq. 27.

DIVERGENCE OF A VECTOR FUNCTION

2.7 The electric field has a definite direction and magnitude at every point. It is a vector function of the coordinates, which we have often indicated by writing $\mathbf{E}(x, y, z)$. What we are about to say can apply to any vector function, not just to the electric field; we shall use another symbol, $\mathbf{F}(x, y, z)$, as a reminder of that. In other words, we shall talk mathematics rather than physics for a while and call **F** simply a general vector function. We shall keep to three dimensions, however.

Consider a finite volume V of some shape, the surface of which we shall denote by S. We are already familiar with the notion of the total flux Φ emerging from S. It is the value of the surface integral of **F** extended over the whole of S:

$$\Phi = \int_S \mathbf{F} \cdot d\mathbf{a} \tag{29}$$

In the integrand $d\mathbf{a}$ is the infinitesimal vector whose magnitude is the area of a small element of S and whose direction is the outward-pointing normal to that little patch of surface, indicated in Fig. 2.12*a*.

Now imagine dividing V into two parts by a surface, or a diaphragm, D that cuts through the "balloon" S, as in Fig. 2.12*b*. Denote the two parts of V by V_1 and V_2 and, treating them as distinct volumes, compute the surface integral over each separately. The boundary surface S_1 of V_1 includes D, and so does S_2. It is pretty obvious that the sum of the two surface integrals

$$\int_{S_1} \mathbf{F} \cdot d\mathbf{a}_1 + \int_{S_2} \mathbf{F} \cdot d\mathbf{a}_2 \tag{30}$$

will equal the original integral over the whole surface expressed in Eq. 29. The reason is that any given patch on D contributes with one sign to the first integral and the same amount with opposite sign to the second, the "outward" direction in one case being the "inward" direction in the other. In other words, any flux *out* of V_1, through this surface D, is flux *into* V_2. The rest of the surface involved is identical to that of the original entire volume.

We can keep on subdividing until our internal partitions have divided V into a large number of parts, $V_1, \ldots, V_i, \ldots, V_N$, with surfaces $S_1, \ldots, S_i, \ldots, S_N$. No matter how far this is carried we can still be sure that

$$\sum_{i=1}^{N} \int_{S_i} \mathbf{F} \cdot d\mathbf{a}_i = \int_S \mathbf{F} \cdot d\mathbf{a} = \Phi \tag{31}$$

What we are after is this: In the limit as N becomes enormous we want to identify something which is characteristic of a particular small region—and ultimately, of the neighborhood of a point. Now the surface integral

$$\int_{S_i} \mathbf{F} \cdot d\mathbf{a}_i \tag{32}$$

over one of the small regions, is *not* such a quantity, for if we divide everything again, so that N becomes $2N$, this integral divides into two terms, each smaller than before since their sum is constant. In other words, as we consider smaller and smaller volumes in the same locality, the surface integral over one such volume gets steadily smaller. But we notice that, when we divide, the volume is also divided into two parts which sum to the original volume. This suggests that we look at the ratio of surface integral to volume for an element in the subdivided space:

$$\frac{\int_{S_i} \mathbf{F} \cdot d\mathbf{a}_i}{V_i} \tag{33}$$

It seems plausible that for N large enough, that is, for sufficiently fine-grained subdivision, we can halve the volume every time we halve the surface integral so that we shall find that with continuing subdivision of any particular region this ratio approaches a limit. If so, this limit is a property characteristic of the vector function $\mathbf{F}$ in that neighborhood. We call it the *divergence* of $\mathbf{F}$, written div $\mathbf{F}$. That is, the value of div $\mathbf{F}$ at any point is defined as

$$\text{div } \mathbf{F} \equiv \lim_{V_i \to 0} \frac{1}{V_i} \int_{S_i} \mathbf{F} \cdot d\mathbf{a}_i \tag{34}$$

where V_i is a volume including the point in question, and S_i, over which the surface integral is taken, is the surface of V_i. We must include the proviso that the limit exists and is independent of our method of subdivision. For the present we shall assume that this is true.

The meaning of div $\mathbf{F}$ can be expressed in this way: div $\mathbf{F}$ is the flux out of V_i, per unit of volume, in the limit of infinitesimal V_i. It is a scalar quantity, obviously. It may vary from place to place, its value at any particular location (x, y, z) being the limit of the ratio in Eq.

34 as V_i is chopped smaller and smaller while always enclosing the point (x, y, z). So div **F** is simply a scalar function of the coordinates.

GAUSS'S THEOREM AND THE DIFFERENTIAL FORM OF GAUSS'S LAW

2.8 If we know this scalar function of position div **F**, we can work our way right back to the surface integral over a large volume: We first write Eq. 31 in this way:

$$\int_S \mathbf{F} \cdot d\mathbf{a} = \sum_{i=1}^{N} \int_{S_i} \mathbf{F} \cdot d\mathbf{a}_i = \sum_{i=1}^{N} V_i \left[\frac{\int_{S_i} \mathbf{F} \cdot d\mathbf{a}_i}{V_i} \right] \tag{35}$$

In the limit $N \to \infty$, $V_i \to 0$, the term in brackets becomes the divergence of **F** and the sum goes into a volume integral:

$$\int_S \mathbf{F} \cdot d\mathbf{a} = \int_V \operatorname{div} \mathbf{F}\, dv \tag{36}$$

Equation 36 is called Gauss's theorem, or the divergence theorem. It holds for any vector field for which the limit involved in Eq. 34 exists.

Let us see what this implies for the electric field **E**. We have Gauss's law which assures us that

$$\int_S \mathbf{E} \cdot d\mathbf{a} = 4\pi \int_V \rho\, dv \tag{37}$$

If the divergence theorem holds for any vector field, it certainly holds for **E**:

$$\int_S \mathbf{E} \cdot d\mathbf{a} = \int_V \operatorname{div} \mathbf{E}\, dv \tag{38}$$

Both Eq. 37 and Eq. 38 hold for *any* volume we care to choose—of any shape, size, or location. Comparing them, we see that this can only be true if at every point,

$$\operatorname{div} \mathbf{E} = 4\pi\rho \tag{39}$$

If we adopt the divergence theorem as part of our regular mathematical equipment from now on, we can regard Eq. 39 simply as an alternative statement of Gauss's law. It is Gauss's law in differential form, that is, stated in terms of a local relation between charge density and electric field.

THE DIVERGENCE IN CARTESIAN COORDINATES

2.9 While Eq. 34 is the fundamental definition of *divergence,* independent of any system of coordinates, it is useful to know how to calculate the divergence of a vector function when we are given its explicit form. Suppose a vector function **F** is expressed as a function of cartesian coordinates x, y, and z. That means that we have three scalar functions, $F_x\,(x, y, z)$, $F_y\,(x, y, z)$, and $F_z\,(x, y, z)$. We'll take the region V_i in the shape of a little rectangular box, with one corner at the point (x, y, z) and sides Δx, Δy, and Δz, as in Fig. 2.13*a*. Whether some other shape will yield the same limit is a question we must face later.

Consider two opposite faces of the box, the top and bottom for instance, which would be represented by the vectors $\hat{\mathbf{z}}\,\Delta x\,\Delta y$ and $-\hat{\mathbf{z}}\,\Delta x\,\Delta y$. The flux through these faces involves only the z component of **F**, and the net contribution depends on the *difference* between F_z at the top and F_z at the bottom or, more precisely, on the difference between the average of F_z over the top face and the average of F_z over the bottom face of the box. To the first order in small quantities this difference is $(\partial F_z/\partial z)\,\Delta z$. Figure 2.13*b* will help to explain this. The average value of F_z on the bottom surface of the box, if we consider only first-order variations in F_z over this small rectangle, is its value at the center of the rectangle. That value is, to first order† in Δx and Δy,

$$F_z\,(x, y, z) + \frac{\Delta x}{2}\frac{\partial F_z}{\partial x} + \frac{\Delta y}{2}\frac{\partial F_z}{\partial y} \tag{40}$$

For the average of F_z over the top face we take the value at the center of the top face, which to first order in the small displacements is

$$F_z\,(x, y, z) + \frac{\Delta x}{2}\frac{\partial F_z}{\partial x} + \frac{\Delta y}{2}\frac{\partial F_z}{\partial y} + \Delta z\,\frac{\partial F_z}{\partial z} \tag{41}$$

†This is nothing but the beginning of a Taylor expansion of the scalar function F_z, in the neighborhood of (x, y, z). That is, $F_z(x + a, y + b, z + c) = F_z(x, y, z) + \left(a\frac{\partial}{\partial x} + b\frac{\partial}{\partial y} + c\frac{\partial}{\partial z}\right) F_z + \cdots + \left(\frac{1}{n!}\right)\left(a\frac{\partial}{\partial x} + b\frac{\partial}{\partial y} + c\frac{\partial}{\partial z}\right)^n F_z + \cdots$. The derivatives are all to be evaluated at (x, y, z). In our case $a = \Delta x/2$, $b = \Delta y/2$, $c = 0$, and we drop the higher-order terms in the expansion.

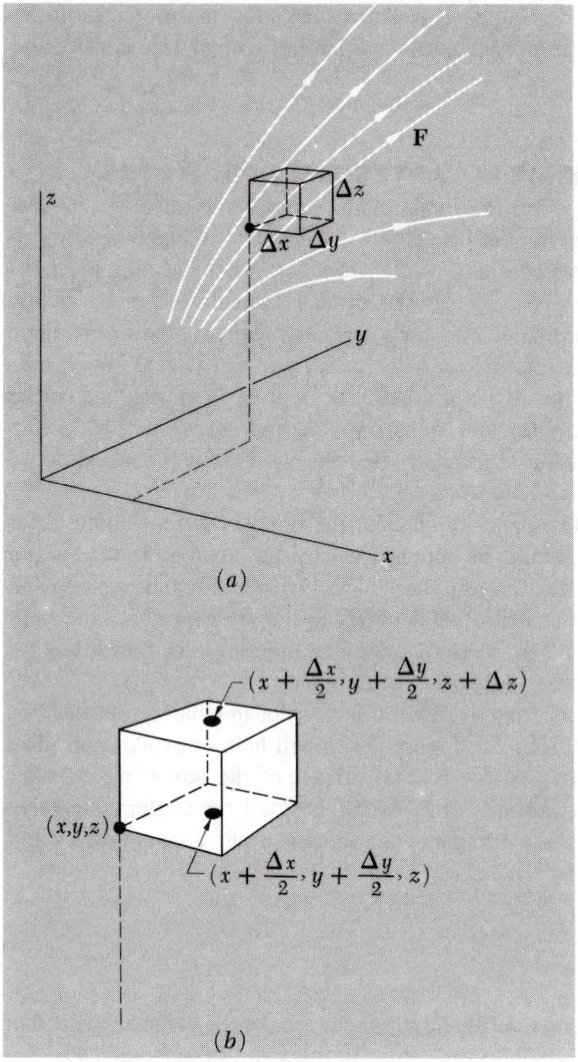

FIGURE 2.13
Calculation of flux from the box of volume $\Delta x\ \Delta y\ \Delta z$.

The net flux out of the box through these two faces, each of which has the area of $\Delta x\ \Delta y$, is therefore

$$\underbrace{\Delta x\, \Delta y \left[F_z\,(x, y, z) + \frac{\Delta x}{2}\frac{\partial F_z}{\partial x} + \frac{\Delta y}{2}\frac{\partial F_z}{\partial y} + \Delta z\,\frac{\partial F_z}{\partial z} \right]}_{\text{(flux out of box at top)}}$$

$$\underbrace{-\ \Delta x\, \Delta y \left[F_z\,(x, y, z) + \frac{\Delta x}{2}\frac{\partial F_z}{\partial x} + \frac{\Delta y}{2}\frac{\partial F_z}{\partial y} \right]}_{\text{(flux into box at bottom)}} \qquad (42)$$

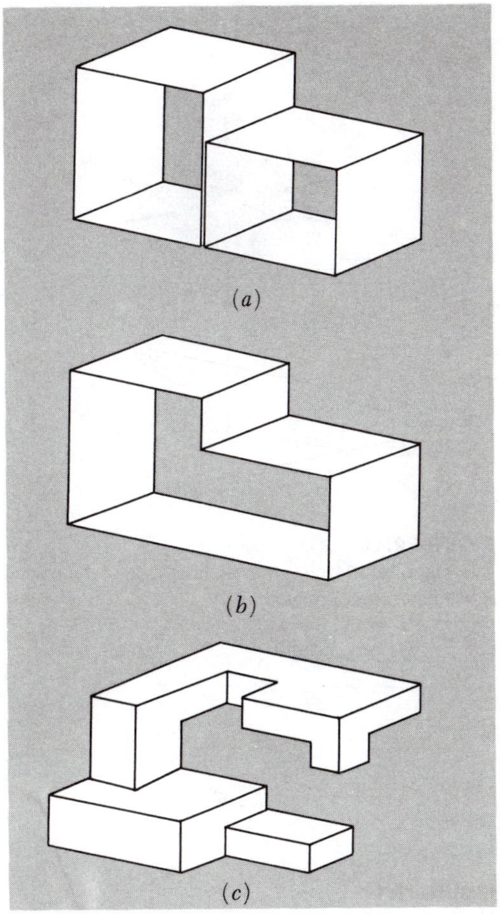

FIGURE 2.14
The limit of the flux/volume ratio is independent of the shape of the box.

which reduces to $\Delta x\, \Delta y\, \Delta z\, (\partial F_z/\partial z)$. Obviously, similar statements must apply to the other pairs of sides. That is, the net flux out of the box through the sides parallel to the yz plane is $\Delta y\, \Delta z\, \Delta x\, (\partial F_x/\partial x)$. Notice that the product $\Delta x\, \Delta y\, \Delta z$ occurs here too. Thus the total flux out of the little box is

$$\Phi = \Delta x\, \Delta y\, \Delta z \left(\frac{\partial F_x}{\partial x} + \frac{\partial F_y}{\partial y} + \frac{\partial F_z}{\partial z} \right) \tag{43}$$

The volume of the box is $\Delta x\, \Delta y\, \Delta z$, so the ratio of flux to volume is $\partial F_x/\partial x + \partial F_y/\partial y + \partial F_z/\partial z$, and as this expression does not contain the dimensions of the box at all, it remains as the limit when we let the box shrink. [Had we retained terms proportional to $(\Delta x)^2$, $(\Delta x\, \Delta y)$, etc., in the calculation of the flux, they would of course vanish on going to the limit.]

Now we can begin to see why this limit is going to be independent of the shape of the box. Obviously it is independent of the proportions of the rectangular box, but that isn't saying much. It is easy to see that it will be the same for any volume that we can make by sticking together little rectangular boxes of any size and shape. Consider the two boxes in Fig. 2.14. The sum of the flux Φ_1 out of box 1 and Φ_2 out of box 2 is not changed by removing the adjoining walls to make one box, for whatever flux went through that plane was negative flux for one and positive for the other. So we could have a bizarre shape like Fig. 2.14*c* without affecting the result. We leave it to the reader to generalize further. Tilted surfaces can be taken care of if you will first prove that the vector sum of the four surface areas of the tetrahedron in Fig. 2.15 is zero.

We conclude that, assuming only that the functions F_x, F_y, and F_z are differentiable, the limit does exist and is given by

$$\text{div}\, \mathbf{F} = \frac{\partial F_x}{\partial x} + \frac{\partial F_y}{\partial y} + \frac{\partial F_z}{\partial z} \tag{44}$$

If div **F** has a positive value at some point, we find—thinking of **F** as a velocity field—a net "outflow" in that neighborhood. For instance, if all three partial derivatives in Eq. 44 are positive at a point P, we might have a vector field in that neighborhood something like that suggested in Fig. 2.16. But the field could look quite different and still have positive divergence, for any vector function **G** such that div **G** $= 0$ could be superimposed. Thus one or two of the three partial derivatives could be negative, and we might still have div **F** > 0. The divergence is a quantity that expresses only one aspect of the spatial variation of a vector field.

FIGURE 2.15
You can prove that $\mathbf{a}_1 + \mathbf{a}_2 + \mathbf{a}_3 + \mathbf{a}_4 = 0$.

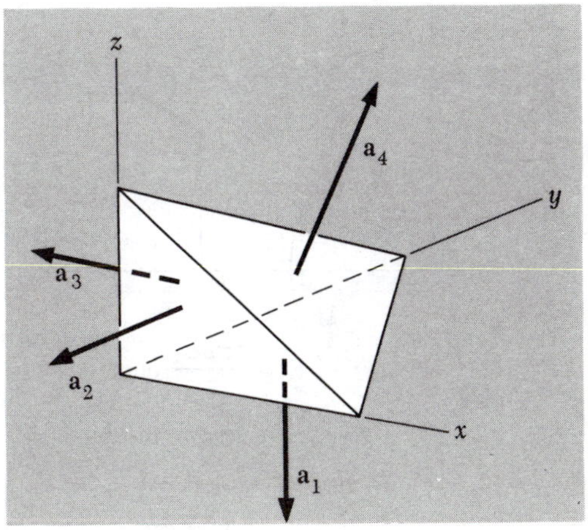

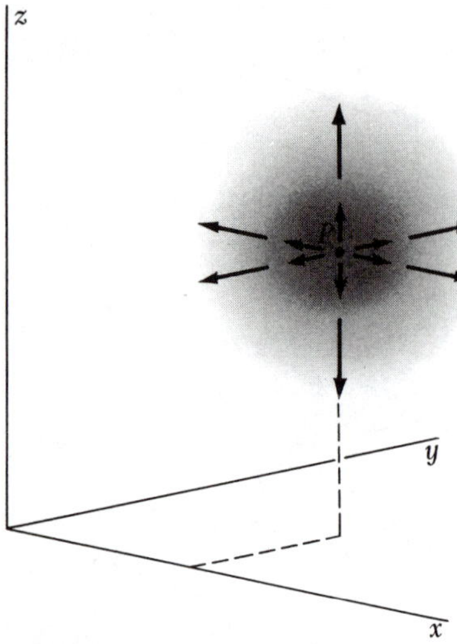
z
P
y
x

FIGURE 2.16
Showing a field which in the neighborhood of the point P has nonzero divergence.

Let's apply this to an electric field that is rather easy to visualize. An infinitely long circular cylinder of radius a is filled with a distribution of positive charge of density ρ. Outside the cylinder the electric field is the same as that of a line charge on the axis. It is a radial field with magnitude proportional to $1/r$. The field inside is found by applying Gauss' law to a cylinder of radius $r < a$. You can do this as an easy problem. You will find that the field inside is directly proportional to r, and of course it is radial also. The exact values are:

$$\begin{aligned} E &= \frac{2\pi\rho a^2}{r} \qquad \text{for } r > a \\ E &= 2\pi\rho r \qquad \text{for } r < a \end{aligned} \tag{45}$$

Figure 2.17 is a section perpendicular to the axis of the cylinder. Rectangular coordinates aren't the most natural choice here, but we'll use them anyway to get some practice with Eq. 44. With $r = \sqrt{x^2 + y^2}$, the field components are expressed as follows:

$$\begin{aligned} E_x &= \left(\frac{x}{r}\right) E = \frac{2\pi\rho a^2 x}{x^2 + y^2} \qquad \text{for } r > a \\ &= 2\pi\rho x \qquad \text{for } r < a \\ E_y &= \left(\frac{y}{r}\right) E = \frac{2\pi\rho a^2 y}{x^2 + y^2} \qquad \text{for } r > a \\ &= 2\pi\rho y \qquad \text{for } r < a \end{aligned} \tag{46}$$

E_z is zero, of course.

Outside the cylinder of charge, div **E** has the value given by

$$\frac{\partial E_x}{\partial x} + \frac{\partial E_y}{\partial y} = 2\pi\rho a^2 \left[\frac{1}{x^2 + y^2} - \frac{2x^2}{(x^2 + y^2)^2} + \frac{1}{x^2 + y^2} - \frac{2y^2}{(x^2 + y^2)^2}\right] = 0 \tag{47}$$

FIGURE 2.17
The field inside and outside a uniform cylindrical distribution of charge.

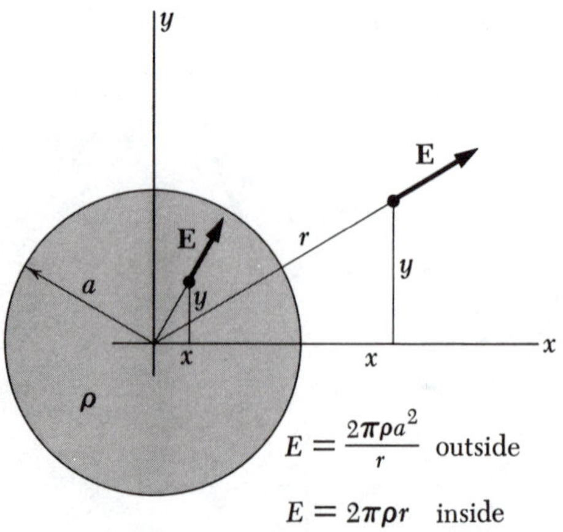
y
E
E
r
a
y
y
x
x
x
ρ
E = 2πρa²/r outside
E = 2πρr inside

Inside the cylinder, div **E** is

$$\frac{\partial E_x}{\partial x} + \frac{\partial E_y}{\partial y} = 2\pi\rho(1 + 1) = 4\pi\rho \tag{48}$$

We expected both results. Outside the cylinder where there is no charge, the net flux emerging from any volume—large or small—is zero, so the limit of the ratio *flux/volume* is certainly zero. Inside the cylinder we get the result required by the fundamental relation Eq. 39.

THE LAPLACIAN

2.10 We have now met two scalar functions related to the electric field, the potential function φ and the divergence, div **E**. In cartesian coordinates the relationships are expressed as

$$\mathbf{E} = -\text{grad}\,\varphi = -\left(\hat{\mathbf{x}}\frac{\partial\varphi}{\partial x} + \hat{\mathbf{y}}\frac{\partial\varphi}{\partial y} + \hat{\mathbf{z}}\frac{\partial\varphi}{\partial z}\right) \qquad (49)$$

$$\text{div}\,\mathbf{E} = \frac{\partial E_x}{\partial x} + \frac{\partial E_y}{\partial y} + \frac{\partial E_z}{\partial z} \qquad (50)$$

Equation 49 shows that the x component of **E** is $E_x = -\partial\varphi/\partial x$. Substituting this and the corresponding expressions for E_y and E_z into Eq. 50, we get a relation between div **E** and φ:

$$\text{div}\,\mathbf{E} = -\text{div grad}\,\varphi = -\left(\frac{\partial^2\varphi}{\partial x^2} + \frac{\partial^2\varphi}{\partial y^2} + \frac{\partial^2\varphi}{\partial z^2}\right) \qquad (51)$$

The operation on φ which is indicated by Eq. 51 except for the minus sign we could call "div grad," or "taking the divergence of the gradient of. . . ." The symbol used to represent this operation is ∇^2, called *the Laplacian operator,* or just *the Laplacian.* The expression

$$\frac{\partial^2}{\partial x^2} + \frac{\partial^2}{\partial y^2} + \frac{\partial^2}{\partial z^2}$$

is the prescription for the Laplacian in cartesian coordinates.

The notation ∇^2 is explained as follows. The gradient operator is often symbolized by ∇, called "del." Writing it out in cartesian coordinates,

$$\nabla = \hat{\mathbf{x}}\frac{\partial}{\partial x} + \hat{\mathbf{y}}\frac{\partial}{\partial y} + \hat{\mathbf{z}}\frac{\partial}{\partial z} \qquad (52)$$

If we handle this as a vector, then its square would be

$$\nabla \cdot \nabla = \frac{\partial^2}{\partial x^2} + \frac{\partial^2}{\partial y^2} + \frac{\partial^2}{\partial z^2} \qquad (53)$$

the same as the Laplacian in cartesian coordinates. So the Laplacian is often called "del squared," and we say "del squared φ," meaning "div grad φ." *Warning:* In other coordinate systems, spherical polar coordinates, for instance, the explicit forms of the gradient operator and the Laplacian operator are not so simply related. It is well to remember that the fundamental definition of the Laplacian operation is "divergence of the gradient of."

We can now express directly a *local* relation between the charge density at some point and the potential function in that immediate neighborhood. From Gauss' law in differential form, div $\mathbf{E} = 4\pi\rho$, we have

$$\nabla^2\varphi = -4\pi\rho \tag{54}$$

Equation 54, sometimes called *Poisson's equation,* relates the charge density to the second derivatives of the potential. Written out in cartesian coordinates it is

$$\frac{\partial^2\varphi}{\partial x^2} + \frac{\partial^2\varphi}{\partial y^2} + \frac{\partial^2\varphi}{\partial z^2} = -4\pi\rho \tag{55}$$

One may regard this as the differential expression of the relationship expressed by an integral in Eq. 15, which tells us how to find the potential at a point by summing the contributions of all sources near and far.†

LAPLACE'S EQUATION

2.11 Wherever $\rho = 0$, that is, in all parts of space containing no electric charge, the electric potential φ has to satisfy the equation

$$\nabla^2\varphi = 0 \tag{56}$$

This is called *Laplace's equation.* We run into it in many branches of physics. Indeed one might say that from a mathematical point of view the theory of classical fields is mostly a study of the solutions of this equation. The class of functions that satisfy Laplace's equation are called *harmonic functions.* They have some remarkable properties, one of which is this: *If $\varphi(x, y, z)$ satisfies Laplace's equation, then the average value of φ over the surface of any sphere (not necessarily a small sphere) is equal to the value of φ at the center of the sphere.* We can easily prove that this must be true of the electric potential ϕ in regions containing no charge. Consider a point charge q and a spherical surface S over which a charge q' is uniformly distributed. Let the charge q be brought in from infinity to a distance R from the center of the charged sphere, as in Fig. 2.18. The electric field of the sphere being the same as if its total charge q' were concentrated at its center,

†In fact, it can be shown that Eq. 55 is the *mathematical* equivalent of Eq. 15. This means, if you apply the Laplacian operator to the integral in Eq. 15, you will come out with $-4\pi\rho$. We shall not stop to show how this is done; you'll have to take our word for it or figure out how to do it.

the work required is qq'/R. Now suppose, instead, that the point charge q was there first and the charged sphere was later brought in from infinity. The work required for that is the product of q' and the average over the surface S of the potential due to the point charge q. Now the work is surely the same in the second case, namely, qq'/R, so the average over the sphere of the potential due to q must be q/R. That is indeed the potential at the center of the sphere due to the external point charge q. That proves the assertion for any single point charge outside the sphere. But the potential of many charges is just the sum of the potentials due to the individual charges, and the average of a sum is the sum of the averages. It follows that the assertion must be true for *any* system of sources lying wholly outside the sphere S.

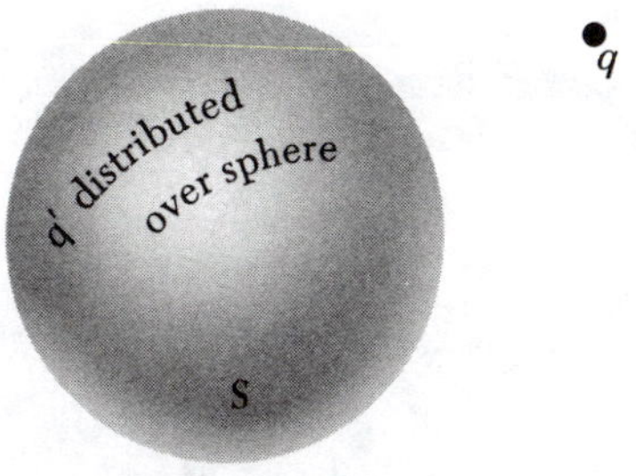

FIGURE 2.18
The work required to bring in q' and distribute it over the sphere is q' times the *average*, over the sphere, of the potential ϕ due to q.

This property of the potential, that its average over an empty sphere is equal to its value at the center, is closely related to a fact that you may find disappointing: You can't construct an electrostatic field that will hold a charged particle in *stable* equilibrium in empty space. This particular "impossibility theorem," like others in physics, is useful in saving fruitless speculation and effort. Let us see why it is true. Suppose we have an electric field in which, contrary to the theorem, there *is* a point P at which a positively charged particle would be in stable equilibrium. That means that *any* small displacement of the particle from P must bring it to a place where an electric field acts to push it back toward P. But that means that a little sphere around P must have $\mathbf{E}$ pointing inward *everywhere* on its surface. That contradicts Gauss's law, for there is no negative source charge within the region. (Our charged test particle doesn't count; besides, it's positive.) In other words, you can't have an empty region where the electric field points all inward or all outward, and that's what you would need for *stable* equilibrium. To express the same fact in terms of the electric potential, a stable position for a charged particle must be one where the potential φ is either lower than that at all neighboring points (if the particle is positively charged) or higher than that at all neighboring points (if the particle is negatively charged). Clearly neither is possible for a function whose average value over a sphere is always equal to its value at the center.

Of course one can have a charged particle in *equilibrium* in an electrostatic field, in the sense that the force on it is zero. The point where $\mathbf{E} = 0$ in Fig. 1.10 is such a location. The position midway between two equal positive charges is an equilibrium position for a third charge, either positive or negative. But the equilibrium is not stable. (Think what happens when the third charge is slightly displaced from its equilibrium position.) It *is* possible, by the way, to trap and hold stably an electrically charged particle by electric fields that vary in *time*.

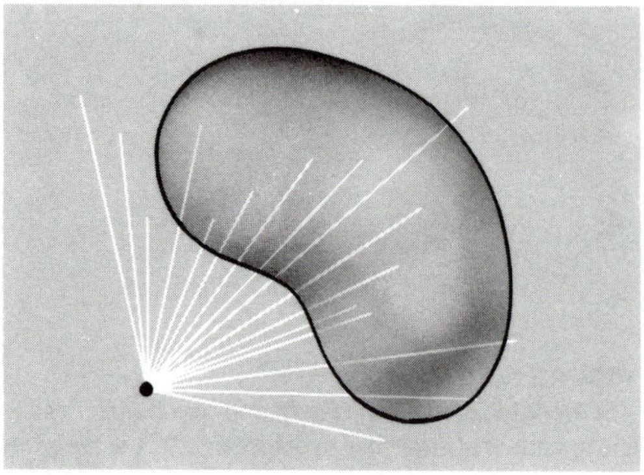

FIGURE 2.19
In a non-inverse-square field, the flux through a closed surface is not zero.

DISTINGUISHING THE PHYSICS FROM THE MATHEMATICS

2.12 In the last few sections we have been concerned with mathematical relations and new ways of expressing familiar facts. It may help to sort out physics from mathematics, and law from definition, if we try to imagine how things would be if the electric force were *not* a pure inverse-square force but instead a force with a finite range, for instance, a force varying like

$$\frac{e^{-\lambda r}}{r^2} \tag{57}$$

Then Gauss's law in the integral form expressed in Eq. 37 would surely fail, for by taking a very large surface enclosing some sources, we would find a vanishingly small field on this surface. The flux would go to zero as the surface expanded, rather than remain constant. However, we could still define a field at every point in space. We could calculate the divergence of that field, and Eq. 38, which describes a mathematical property of *any* vector field, would still be true. Is there a contradiction here? No, because Eq. 39 would also fail. The divergence of the field would no longer be the same as the source density. We can understand this by noting that a small volume empty of sources could still have a net flux through it owing to the effect of a source *outside* the volume, if the field has finite range. As suggested in Fig. 2.19, more flux would enter the side near the source than would leave the volume.

Thus we may say that Eqs. 37 and 39 express the same *physical law,* the inverse-square law that Coulomb established by direct measurement of the forces between charged bodies, while Eq. 38 is an expression of a *mathematical theorem* which enables us to translate our statement of this law from differential to integral form or the reverse. The relations that connect $\mathbf{E}$, ρ, and ϕ are gathered together in Fig. 2.20 and 2.20′.

How can we justify these differential relations between source and field in a world where electric charge is really not a smooth jelly but is concentrated on particles whose interior we know very little about? Actually, a statement like Eq. 54, Poisson's equation, is meaningful on a macroscopic scale only. The charge density ρ is to be interpreted as an average over some small but finite region containing many particles. Thus the function ρ cannot be continuous in the way a mathematician might prefer. When we let our region V_i shrink down in the course of demonstrating the differential form of Gauss's law, we know as physicists that we musn't let it shrink too far. That is awkward perhaps, but the fact is that we make out very well with the continuum model in large-scale electrical systems. In the atomic world we have the elementary particles, and vacuum. Inside the particles,

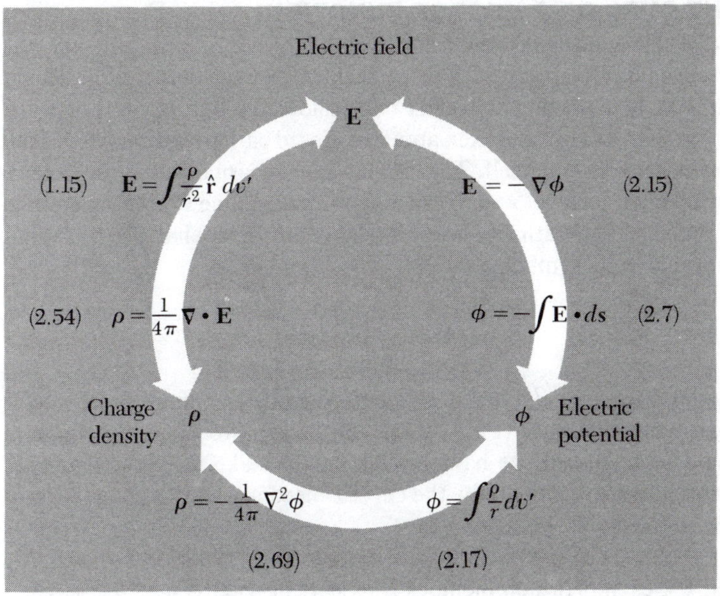

FIGURE 2.20
How electric charge density, electric potential, and electric field are related. The integral relations involve the line integral and the volume integral. The differential relations involve the gradient, the divergence, and div · grad or ∇^2, the Laplacian operator. Charge density ρ is in esu/cm^3, potential ϕ is in statvolts, field **E** is in statvolt/cm, and all lengths in cm.

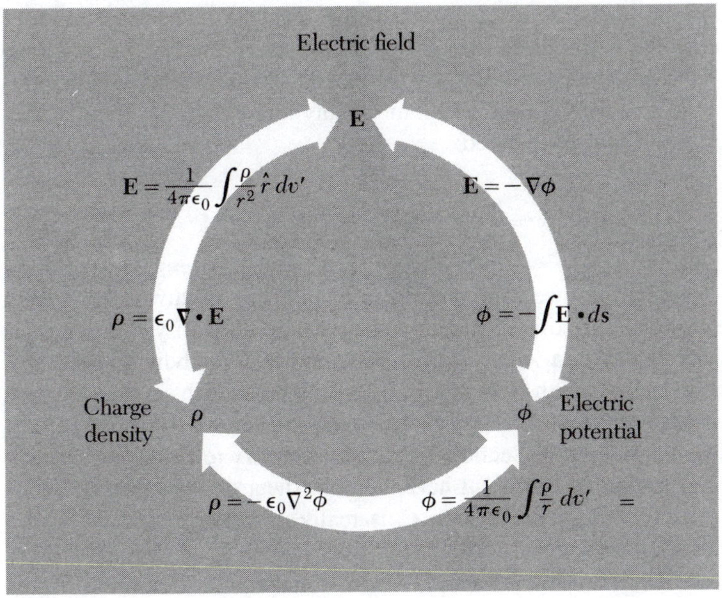

FIGURE 2.20′
The same relations in SI units. Charge density ρ is in coulomb/m^3, potential ϕ is in volts, field **E** is in volt/meter, and all lengths are in meters. ($\epsilon_0 = 8.854 \times 10^{-12}$ coulomb/volt-meter.)

even if Coulomb's law turns out to have some kind of meaning, much else is going on. The vacuum, so far as electrostatics is concerned, is ruled by Laplace's equation. Still, we cannot be sure that, even in the vacuum, passage to a limit of zero size has *physical* meaning.

THE CURL OF A VECTOR FUNCTION†

2.13 We developed the concept of divergence, a local property of a vector field, by starting from the surface integral over a large closed surface. In the same spirit, let us consider the line integral of some vector field $\mathbf{F}(x, y, z)$, taken around a closed path, some curve C which comes back to join itself. The curve C can be visualized as the boundary of some surface S which spans it. A good name for the magnitude of such a closed-path line integral is *circulation;* we shall use Γ (capital gamma) as its symbol:

$$\Gamma = \int_C \mathbf{F} \cdot d\mathbf{s} \tag{58}$$

FIGURE 2.21
For the subdivided loop, the sum of all the circulations Γ_i around the sections is equal to the circulation Γ around the original curve C.

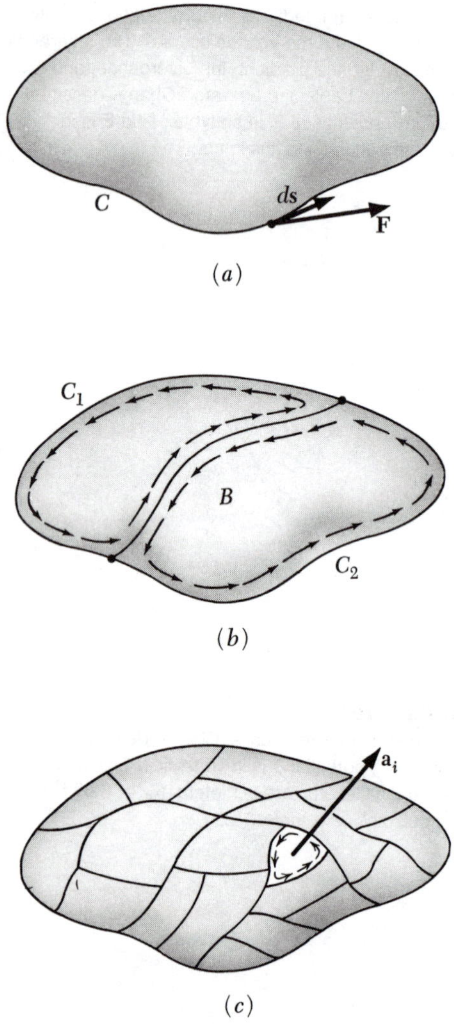

In the integrand $d\mathbf{s}$ is the element of path, an infinitesimal vector locally tangent to C (Fig. 2.21*a*). There are two senses in which C could be traversed; we have to pick one to make the direction of $d\mathbf{s}$ unambiguous. Incidentally, the curve C need not lie in a plane—it can be as crooked as you like.

Now bridge C with a new path B, thus making two loops, C_1 and C_2, each of which includes B as part of itself (Fig. 2.21*b*). Take the line integral around each of these, in the same directional sense. It is easy to see that the sum of the two circulations, Γ_1 and Γ_2, will be the same as the original circulation around C: The reason is that the bridge is traversed in opposite directions in the two integrations, leaving just the contributions which made up the original line integral around C. Further subdivision into many loops, $C_1, \ldots, C_i, \ldots, C_N$, leaves the sum unchanged:

$$\int_C \mathbf{F} \cdot d\mathbf{s} = \sum_{i=1}^{N} \int_{C_i} \mathbf{F} \cdot d\mathbf{s}_i \quad \text{or} \quad \Gamma = \sum_{i=1}^{N} \Gamma_i \tag{59}$$

Here, too, we can continue indefinitely to subdivide, by adding new bridges, seeking in the limit to arrive at a quantity characteristic of the field $\mathbf{F}$ in a local neighborhood. When we subdivide the loops, we make loops with smaller circulation, but also with smaller area. So it is natural to consider the ratio of *loop circulation* to *loop area,* just as we considered in Section 2.7 the ratio of *flux* to *volume.* However, things are a little different here, because the area $\mathbf{a}_i$ of the bit of surface that spans a small loop C_i is really a vector; a surface has an orientation in space. In fact, as we make smaller and smaller loops in some neighborhood, we can arrange to have a loop oriented in any direction we choose. (Remember, we are not committed to any partic-

†Study of this section and the remainder of Chapter 2 can be postponed until Chapter 6 is reached. Until then our only application of this vector derivative will be the demonstration that an electrostatic field is characterized by curl $\mathbf{E} = 0$, as explained in Section 2.16.

ular surface over the whole curve C.) Thus we can pass to the limit in essentially different ways, and we must expect the result to reflect this.

Let us choose some particular orientation for the patch as it goes through the last stages of subdivision. The unit vector $\hat{\mathbf{n}}$ will denote the normal to the patch, which is to remain fixed in direction as the patch surrounding a particular point P shrinks down toward zero size. The limit of the *ratio of circulation* to *patch area* will be written this way:

$$\lim_{a_i \to 0} \frac{\Gamma_i}{a_i} \quad \text{or} \quad \lim_{a_i \to 0} \frac{\int_{C_i} \mathbf{F} \cdot d\mathbf{s}}{a_i} \tag{60}$$

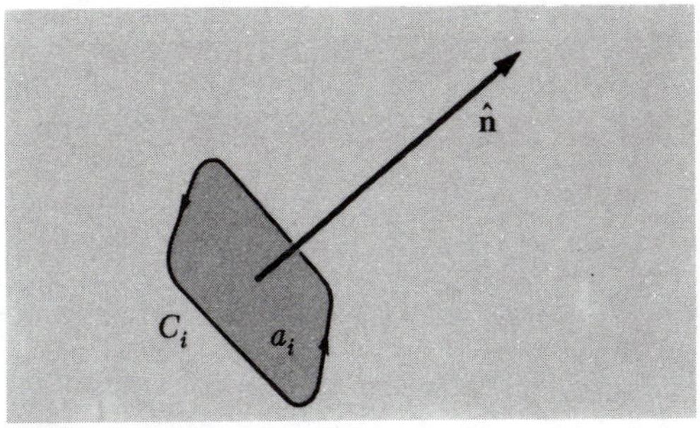

FIGURE 2.22
Right-hand-screw relation between the surface normal and the direction in which the circulation line integral is taken.

The rule for sign is that the direction of $\hat{\mathbf{n}}$ and the sense in which C_i is traversed in the line integral shall be related by a right-hand-screw rule, as in Fig. 2.22. The limit we obtain by this procedure is a scalar quantity which is associated with the point P in the vector field $\mathbf{F}$, and with the direction $\hat{\mathbf{n}}$. We could pick three directions, such as $\hat{\mathbf{x}}$, $\hat{\mathbf{y}}$, and $\hat{\mathbf{z}}$, and get three different numbers. It turns out that these numbers can be considered components of a vector. We call the vector curl $\mathbf{F}$. That is to say, the number we get for the limit with $\hat{\mathbf{n}}$ in a particular direction is the component, in that direction, of the vector curl $\mathbf{F}$. To state this in an equation,

$$(\text{curl } \mathbf{F}) \cdot \hat{\mathbf{n}} = \lim_{a_i \to 0} \frac{\Gamma_i}{a_i} = \lim_{a_i \to 0} \frac{\int_{C_i} \mathbf{F} \cdot d\mathbf{s}}{a_i} \tag{61}$$

For instance, the x component of curl $\mathbf{F}$ is obtained by choosing $\hat{\mathbf{n}} = \hat{\mathbf{x}}$, as in Fig. 2.23. As the loop shrinks down around the point P, we keep it in a plane perpendicular to the x axis. In general, the vector curl $\mathbf{F}$ will vary from place to place. If we let the patch shrink down around some other point, the ratio of circulation to area may have a different value, depending on the nature of the vector function $\mathbf{F}$. That is, curl $\mathbf{F}$ is itself a vector function of the coordinates. Its direction at each point in space is normal to the plane through this point in which the circulation is a maximum. Its magnitude is the limiting value of circulation per unit area, in this plane, around the point in question.

FIGURE 2.23
The patch shrinks around P, keeping its normal pointing in the x direction.

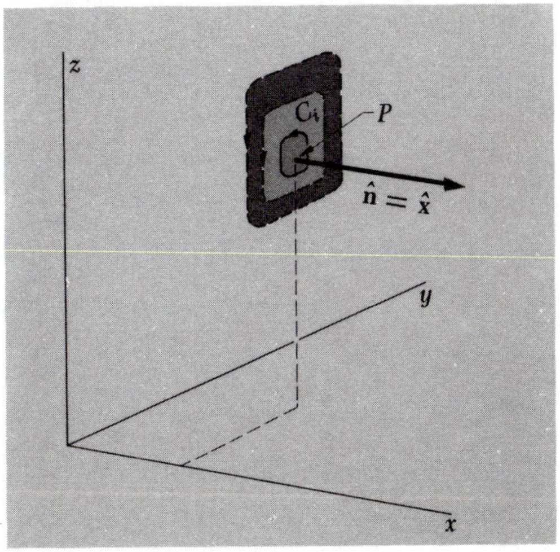

The last two sentences might be taken as a definition of curl $\mathbf{F}$. Like Eq. 61 they make no reference to a coordinate frame. We have not proved that the object so named and defined is a vector; we have only asserted it. Possession of direction and magnitude is not enough to make something a vector. The components as defined must behave like vector components. Suppose we have determined certain values for the x, y, and z components of curl $\mathbf{F}$ by applying Eq. 61 with $\hat{\mathbf{n}}$

chosen, successively, as $\hat{\mathbf{x}}$, $\hat{\mathbf{y}}$, and $\hat{\mathbf{z}}$. If curl **F** is a vector, it is uniquely determined by these three components. If some fourth direction is now chosen for $\hat{\mathbf{n}}$, the left side of Eq. 61 is fixed and the quantity on the right, the circulation in the plane perpendicular to the new $\hat{\mathbf{n}}$, had better agree with it! Indeed, until one is sure that curl **F** is a vector, it is not even obvious that there can be at most one direction for which the circulation per unit area at P is maximum—as was tacitly assumed in the latter definition. In fact, Eq. 61 does define a vector, but we shall not give a proof of that.

STOKES' THEOREM

2.14 From the circulation around an infinitesimal patch of surface we can now work back to the circulation around the original large loop C:

$$\Gamma = \int_C \mathbf{F} \cdot d\mathbf{s} = \sum_{i=1}^{N} \Gamma_i = \sum_{i=1}^{N} a_i \left(\frac{\Gamma_i}{a_i} \right) \tag{62}$$

In the last step we merely multiplied and divided by a_i. Now observe what happens to the right-hand side as N is made enormous and all the a_i's shrink. The quantity in parentheses becomes (curl **F**) $\cdot\ \hat{\mathbf{n}}_i$, where $\hat{\mathbf{n}}_i$ is the unit vector normal to the ith patch. So we have on the right the sum, over all patches that make up the entire surface S spanning C, of the product "patch area times normal component of (curl **F**)." This is nothing but the *surface integral,* over S, of the vector curl **F**:

$$\sum_{i=1}^{N} a_i \left(\frac{\Gamma_i}{a_i} \right) = \sum_{i=1}^{N} a_i \,(\text{curl } \mathbf{F}) \cdot \hat{\mathbf{n}}_i \rightarrow \int_S d\mathbf{a} \cdot \text{curl } \mathbf{F} \tag{63}$$

We thus find that

$$\int_C \mathbf{F} \cdot d\mathbf{s} = \int_S \text{curl } \mathbf{F} \cdot d\mathbf{a} \tag{64}$$

The relation expressed by Eq. 64 is a mathematical theorem called Stokes' theorem. Note how it resembles Gauss's theorem, the divergence theorem, in structure. Stokes' theorem relates the line integral of a vector to the surface integral of the curl of the vector. Gauss's theorem (Eq. 36) relates the surface integral of a vector to the volume integral of the divergence of the vector. Stokes' theorem involves a surface and the curve that bounds it. Gauss' theorem involves a volume and the surface that encloses it.

THE CURL IN CARTESIAN COORDINATES

2.15 Equation 61 is the fundamental definition of curl **F**, stated without reference to any particular coordinate system. In this respect it is like our fundamental definition of divergence, Eq. 34. As in that case, we should like to know how to calculate curl **F** when the vector function **F**(x, y, z) is explicitly given. To find the rule, we carry out the integration called for in Eq. 61, but we do it over a path of very simple shape, one that encloses a rectangular patch of surface parallel to the xy plane (Fig. 2.24). That is, we are taking $\hat{\mathbf{n}} = \hat{\mathbf{z}}$. In agreement with our rule about sign, the direction of integration around the rim

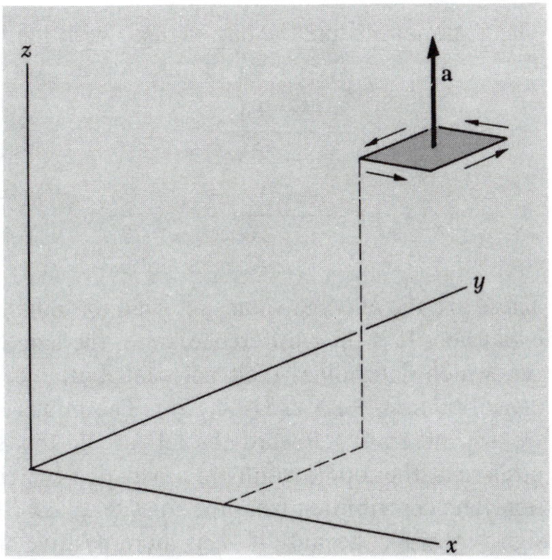

FIGURE 2.24
Circulation around a rectangular patch with $\hat{\mathbf{n}} = \hat{\mathbf{z}}$.

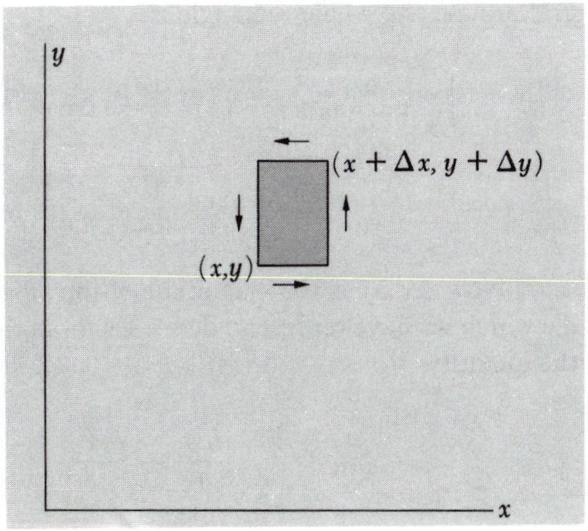

FIGURE 2.25
Looking down on the patch in Fig. 2.24.

must be clockwise as seen by someone looking up in the direction of $\hat{\mathbf{n}}$. In Fig. 2.25 we look down onto the rectangle from above.

The line integral of $\mathbf{A}$ around such a path depends on the variation of A_x with y and the variation of A_y with x. For if A_x had the same average value along the top of the frame, in Fig. 2.25, as along the bottom of the frame, the contribution of these two pieces of the whole line integral would obviously cancel. A similar remark applies to the side members. To the first order in the small quantities Δx and Δy, the difference between the average of A_x over the top segment of path at $y + \Delta y$ and its average over the bottom segment at y is

$$\left(\frac{\partial A_x}{\partial y}\right) \Delta y \tag{65}$$

The argument is like the one we used with Fig. 2.13*b*.

$$\begin{aligned} A_x &= A_x(x, y) + \frac{\Delta x}{2}\frac{\partial A_x}{\partial x} && \begin{pmatrix}\text{at midpoint of}\\ \text{bottom of frame}\end{pmatrix} \\ A_x &= A_x(x, y) + \frac{\Delta x}{2}\frac{\partial A_x}{\partial x} + \Delta y \frac{\partial A_x}{\partial y} && \begin{pmatrix}\text{at midpoint of}\\ \text{top of frame}\end{pmatrix} \end{aligned} \tag{66}$$

These are the average values referred to, to first order in the Taylor's expansion. It is their difference, times the length of the path segment Δx, which determines their net contribution to the circulation. This contribution is $-\Delta x\, \Delta y\, (\partial A_x/\partial y)$. The minus sign comes in because we are integrating toward the left at the top, so that if A_x is more positive at the top, it results in a negative contribution to the circulation. The contribution from the sides is $\Delta y\, \Delta x\, (\partial A_y/\partial x)$, and here the sign is positive, because if A_y is more positive on the right, the result is a positive contribution to the circulation.

Thus, neglecting any higher powers of Δx and Δy, the line integral around the whole rectangle is

$$\begin{aligned} \int_{\square} \mathbf{A} \cdot d\mathbf{s} &= (-\Delta x)\left(\frac{\partial A_x}{\partial y}\right) \Delta y + (\Delta y)\left(\frac{\partial A_y}{\partial x}\right) \Delta x \\ &= \Delta x\, \Delta y \left(\frac{\partial A_y}{\partial x} - \frac{\partial A_x}{\partial y}\right) \end{aligned} \tag{67}$$

Now $\Delta x\, \Delta y$ is the magnitude of the area of the enclosed rectangle which we have represented by a vector in the z direction. Evidently the quantity

$$\frac{\partial A_y}{\partial x} - \frac{\partial A_x}{\partial y} \tag{68}$$

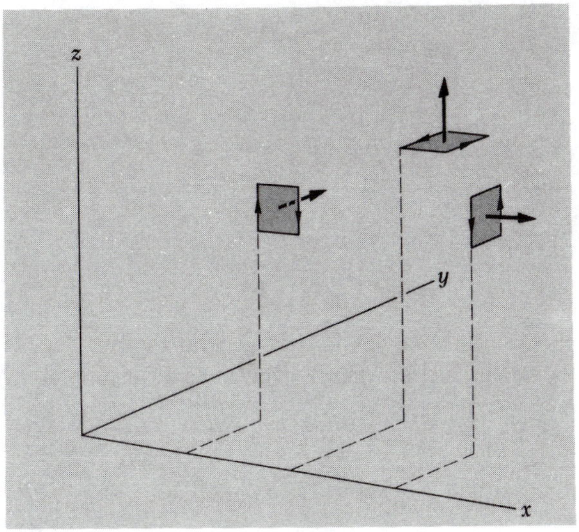

FIGURE 2.26
For each orientation, the limit of the ratio circulation/area determines a component of curl **A** at that point. To determine all components of the vector curl **A** at any point, the patches should all cluster around that point; here they are separated for clarity.

is the limit of the ratio

$$\frac{\text{Line integral around patch}}{\text{Area of patch}} \tag{69}$$

as the patch shrinks to zero size. If the rectangular frame had been oriented with its normal in the positive y direction, we would have found the expression

$$\frac{\partial A_x}{\partial z} - \frac{\partial A_z}{\partial x} \tag{70}$$

for the limit of the corresponding ratio, and if the frame had been oriented with its normal in the x direction, like the frame on the right in Fig. 2.26, we would have obtained

$$\frac{\partial A_z}{\partial y} - \frac{\partial A_y}{\partial z} \tag{71}$$

Although we have considered only rectangles, our result is actually independent of the shape of the little patch and its frame, for reasons much the same as in the case of the integrals involved in the divergence theorem. For instance, it is clear that we can freely join different rectangles to form other figures, because the line integrals along the merging sections of boundary cancel one another exactly (Fig. 2.27).

We conclude that, for any of these orientations, the limit of the ratio of circulation to area is independent of the shape of the patch we

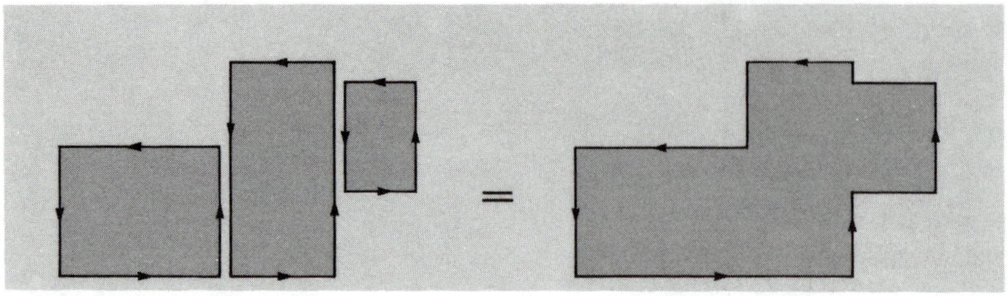

FIGURE 2.27
The circulation in the loop at the right is the sum of the circulations in the rectangles, and the area on the right is the sum of the rectangular areas. This diagram shows why the circulation/area ratio is independent of shape.

choose. Thus we obtain as a general formula for the components of the vector curl **F**, when **F** is given as a function of x, y, and z:

$$\text{curl}\,\mathbf{F} = \hat{\mathbf{x}}\left(\frac{\partial F_z}{\partial y} - \frac{\partial F_y}{\partial z}\right) + \hat{\mathbf{y}}\left(\frac{\partial F_x}{\partial z} - \frac{\partial F_z}{\partial x}\right) + \hat{\mathbf{z}}\left(\frac{\partial F_y}{\partial x} - \frac{\partial F_x}{\partial y}\right) \qquad (72)$$

You may find the following rule easier to remember than the formula itself: Make up a determinant like this:

$$\begin{vmatrix} \hat{\mathbf{x}} & \hat{\mathbf{y}} & \hat{\mathbf{z}} \\ \dfrac{\partial}{\partial x} & \dfrac{\partial}{\partial y} & \dfrac{\partial}{\partial z} \\ F_x & F_y & F_z \end{vmatrix} \qquad (73)$$

Expand it according to the rule for determinants, and you will get curl **F** as given by Eq. 72. Notice that the x component of curl **F** depends on the rate of change of F_z in the y direction and the negative of the rate of change of F_y in the z direction, and so on.

The symbol $\nabla\times$, read as "del cross," where ∇ is interpreted as the "vector"

$$\hat{\mathbf{x}}\frac{\partial}{\partial x} + \hat{\mathbf{y}}\frac{\partial}{\partial y} + \hat{\mathbf{z}}\frac{\partial}{\partial z} \qquad (74)$$

is often used in place of the name *curl*. If we write $\nabla\times\mathbf{F}$ and follow the rules for forming the components of a vector cross product, we get automatically the vector, curl **F**. So curl **F** and $\nabla\times\mathbf{F}$ mean the same thing.

THE PHYSICAL MEANING OF THE CURL

2.16 The name *curl* reminds us that a vector field with a nonzero curl has circulation, or vorticity. Maxwell used the name *rotation,* and in German a similar name is still used, abbreviated rot. Imagine a velocity vector field **G**, and suppose that curl **G** is not zero. Then the velocities in this field have something of this character: $\downarrow \overset{\leftarrow}{\underset{\rightarrow}{}} \uparrow$ or $\uparrow \overset{\rightarrow}{\underset{\leftarrow}{}} \downarrow$ superimposed, perhaps, on a general flow in one direction. For instance, the velocity field of water flowing out of a bathtub generally acquires a circulation. Its curl is not zero over most of the surface. Something floating on the surface rotates as it moves along. In the physics of fluid flow, hydrodynamics and aerodynamics, this concept is of central importance.

To make a "curlmeter" for an electric field—at least in our imagination—we could fasten positive charges to a hub by insulating spokes, as in Fig. 2.28. Exploring an electric field with this device, we would find, wherever curl **E** is not zero, a tendency for the wheel to turn around the shaft. With a spring to restrain rotation, the amount of twist could be used to indicate the torque, which would be proportional to the component of the vector curl **E** in the direction of the

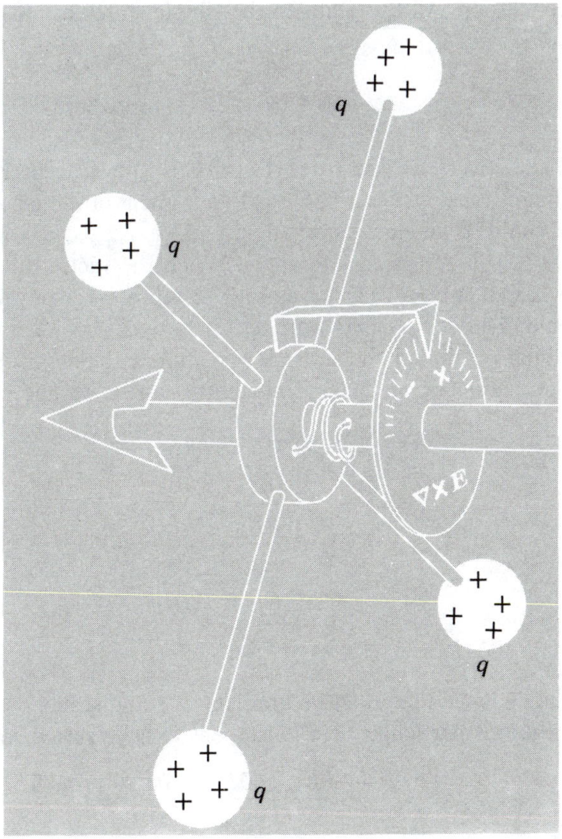

FIGURE 2.28
The curlmeter.

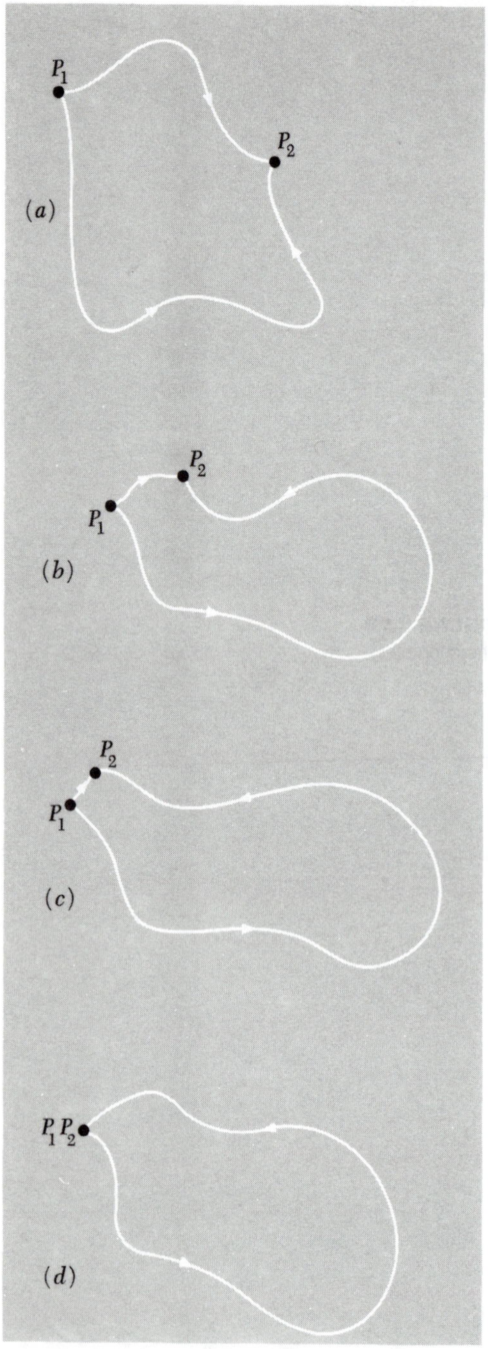

FIGURE 2.29
If the line integral between P_1 and P_2 is independent of path, the line integral around a closed loop must be zero.

shaft. If we can find the direction of the shaft for which the torque is maximum, and clockwise, that is the direction of the vector curl **E**. (Of course, we cannot trust the curlmeter in a field which varies greatly within the dimensions of the wheel itself.)

What can we say, in the light of all this, about the *electrostatic* field **E**? The conclusion we can draw is a simple one: The curlmeter will always read zero! That follows from a fact we have already learned; namely, in the electrostatic field the line integral of **E** around *any* closed path is zero. Just to recall why this is so, remember that the line integral of **E** between any two points such as P_1 and P_2 in Fig. 2.29 is independent of the path. As we bring the two points P_1 and P_2 close together, the line integral over the shorter path in the figure obviously vanishes—unless the final location is at a singularity such as a point charge, a case we can rule out. So the line integral must be zero over the closed loop in Fig. 2.29*d*. But now, if the circulation is zero around *any* closed path, it follows from Stokes' theorem that the surface integral of curl **E** is zero over a patch of any size, shape, or location. But then curl **E** must be zero *everywhere,* for if it were not zero somewhere we could devise a patch in that neighborhood to violate the conclusion. All this leads to the simple statement that in the electrostatic field **E**:

$$\text{curl } \mathbf{E} = 0 \qquad \text{(everywhere)} \tag{75}$$

The converse is also true. If curl **E** is known to be zero everywhere, then **E** must be describable as the gradient of some potential function; it could be an electrostatic field.

This test is easy to apply. When the vector function in Fig. 2.3 was first introduced, it was said to represent a possible electrostatic field. The components were specified by $E_x = Ky$ and $E_y = Kx$, to which we should add $E_z = 0$ to complete the description of a field in three-dimensional space. Calculating curl **E** we find

$$\begin{aligned}(\text{curl } \mathbf{E})_x &= \frac{\partial E_z}{\partial y} - \frac{\partial E_y}{\partial z} = 0 \\ (\text{curl } \mathbf{E})_y &= \frac{\partial E_x}{\partial z} - \frac{\partial E_z}{\partial x} = 0 \\ (\text{curl } \mathbf{E})_z &= \frac{\partial E_y}{\partial x} - \frac{\partial E_x}{\partial y} = K - K = 0\end{aligned} \tag{76}$$

This tells us that **E** is the gradient of some scalar potential. Incidentally, this particular field **E** happens to have zero divergence also:

$$\frac{\partial E_x}{\partial x} + \frac{\partial E_y}{\partial y} + \frac{\partial E_z}{\partial z} = 0 \tag{77}$$

It therefore represents an electrostatic field in a *charge-free* region.

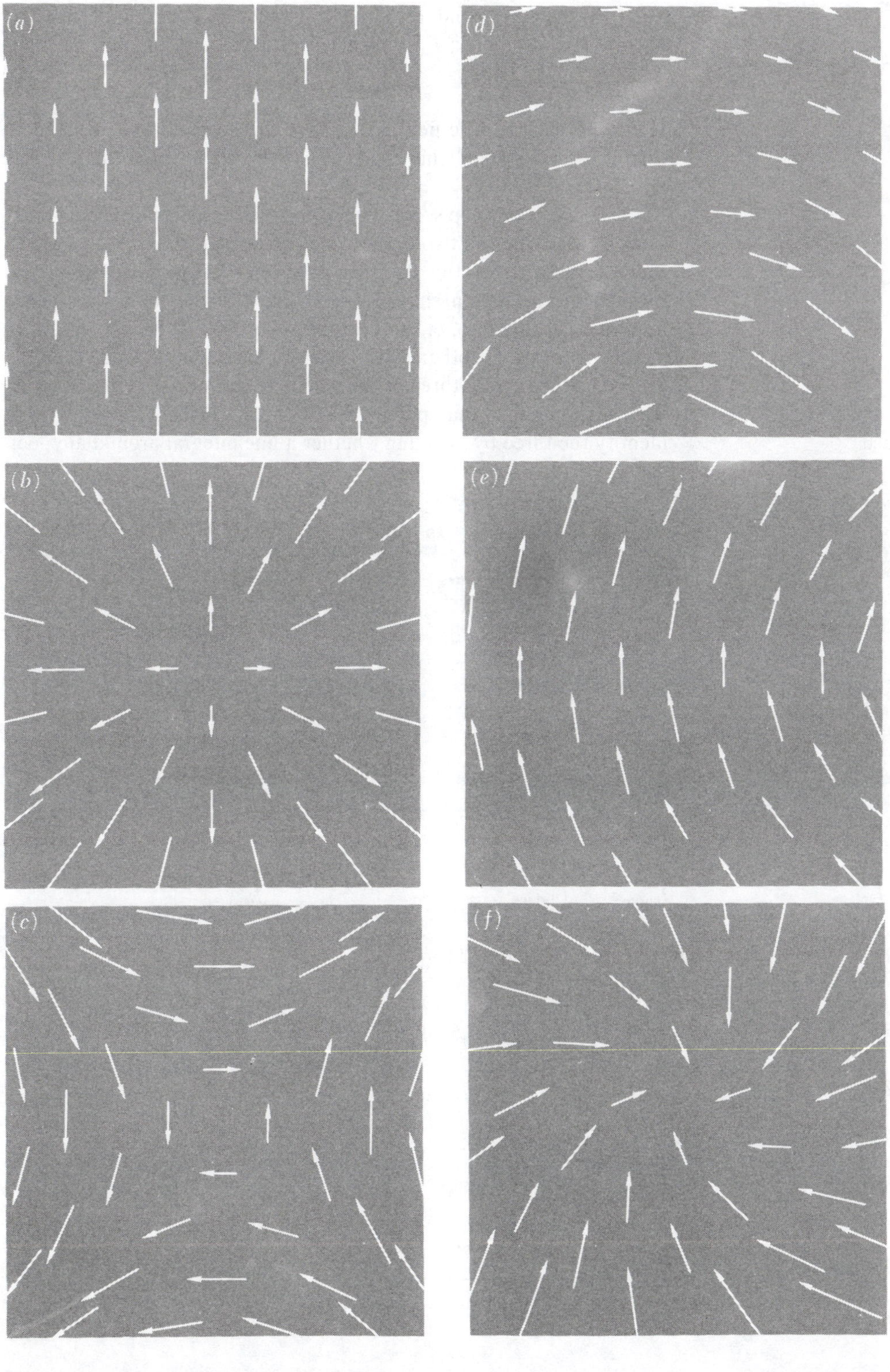

FIGURE 2.30
Four of these vector fields have zero divergence in the region shown. Three have zero curl. Can you spot them?

On the other hand, the equally simple vector function defined by $F_x = Ky$; $F_y = -Kx$; $F_z = 0$, does *not* have zero curl. Instead,

$$(\text{curl } \mathbf{F})_z = -2K \tag{78}$$

Hence no electrostatic field could have this form. If you will sketch roughly the form of this field, you will see at once that it has circulation.

You can develop some feeling for these aspects of vector functions by studying the two-dimensional fields pictured in Fig. 2.30. In four of these fields the divergence of the vector function is zero throughout the region shown. Try to identify the four. Divergence implies a net flux into, or out of, a neighborhood. It is easy to spot in certain patterns. In others you may be able to see at once that the divergence is zero. In three of the fields the curl of the vector function is zero throughout that portion of the field which is shown. Try to identify the three by deciding whether a line integral around any loop

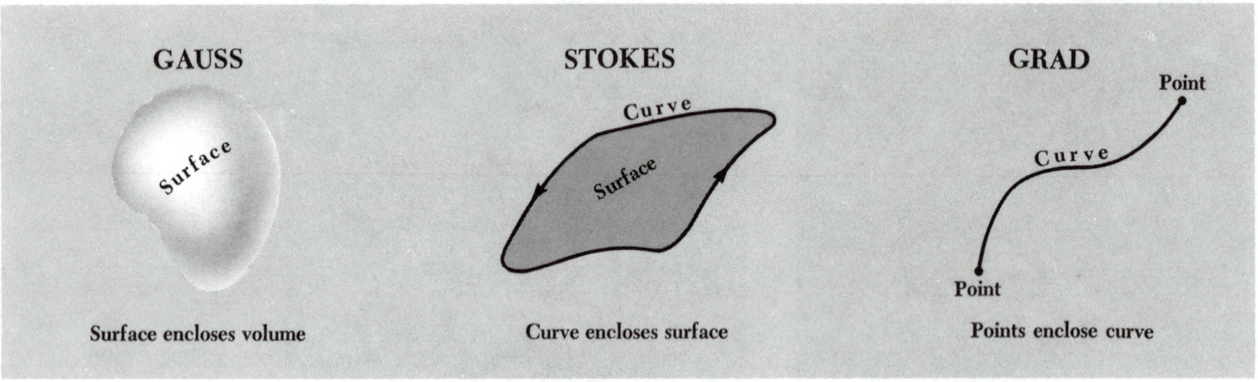
GAUSS
Surface
Surface encloses volume
STOKES
Curve
Surface
Curve encloses surface
GRAD
Point
Curve
Point
Points enclose curve

FIGURE 2.31
Some vector relations summarized.

$$\int_{\text{surface}} \mathbf{F} \cdot d\mathbf{a} = \int_{\text{volume}} \text{div}\,\mathbf{F}\, dv \qquad \int_{\text{curve}} \mathbf{A} \cdot d\mathbf{s} = \int_{\text{surface}} \text{curl}\,\mathbf{A} \cdot d\mathbf{a} \qquad \varphi_2 - \varphi_1 = \int_{\text{curve}} \text{grad}\,\varphi \cdot d\mathbf{s}$$

IN CARTESIAN COORDINATES

$$\text{div}\,\mathbf{F} = \frac{\partial F_x}{\partial x} + \frac{\partial F_y}{\partial y} + \frac{\partial F_z}{\partial z} = \nabla \cdot \mathbf{F}$$

$$\text{curl}\,\mathbf{A} = \hat{\mathbf{x}}\left(\frac{\partial A_z}{\partial y} - \frac{\partial A_y}{\partial z}\right) + \hat{\mathbf{y}}\left(\frac{\partial A_x}{\partial z} - \frac{\partial A_z}{\partial x}\right) + \hat{\mathbf{z}}\left(\frac{\partial A_y}{\partial x} - \frac{\partial A_x}{\partial y}\right) = \nabla \times \mathbf{A}$$

$$\text{grad}\,\varphi = \hat{\mathbf{x}}\frac{\partial \varphi}{\partial x} + \hat{\mathbf{y}}\frac{\partial \varphi}{\partial y} + \hat{\mathbf{z}}\frac{\partial \varphi}{\partial z} = \nabla \varphi$$

$$\nabla = \hat{\mathbf{x}}\frac{\partial}{\partial x} + \hat{\mathbf{y}}\frac{\partial}{\partial y} + \hat{\mathbf{z}}\frac{\partial}{\partial z}$$

would or would not be zero in each picture. That is the essence of *curl*. After you have studied the pictures, think about these questions before you compare your reasoning and your conclusions with the explanation given later in Fig. 2.32.

The curl of a vector field will prove to be a valuable tool later on when we deal with electric and magnetic fields whose curl is *not* zero. We have developed it at this point because the ideas involved are so close to those involved in the divergence. We may say that we have met two kinds of derivatives of a vector field. One kind, the divergence, involves the rate of change of a vector component in its own direction, $\partial F_x/\partial x$, and so on. The other kind, the curl, is a sort of "sideways derivative," involving the rate of change of F_x as we move in the y or z direction.

The relations called Gauss's theorem and Stokes' theorem are summarized in Fig. 2.31. The connection between the scalar potential function and the line integral of its gradient can also be looked on as a member of this family of theorems and is included in the third column.

PROBLEMS

2.1 The vector function which follows represents a possible electrostatic field:

$$E_x = 6xy \qquad E_y = 3x^2 - 3y^2 \qquad E_z = 0$$

Calculate the line integral of $\mathbf{E}$ from the point (0, 0, 0) to the point $(x_1, y_1, 0)$ along the path which runs straight from (0, 0, 0) to $(x_1, 0, 0)$ and thence to $(x_1, y_1, 0)$. Make a similar calculation for the path which runs along the other two sides of the rectangle, via the point $(0, y_1, 0)$. You ought to get the same answer if the assertion above is true. Now you have the potential function $\phi(x, y, z)$. Take the gradient of this function and see that you get back the components of the given field.

2.2 Consider the system of two charges shown in Fig. 2.7. Let z be the coordinate along the line on which the two charges lie, with $z = 0$ at the location of the positive charge. Make a plot of the potential ϕ along this line, plotting ϕ in statvolts against z in cm, from $z = -5$ to $z = 15$.

2.3 A charge of 2 esu is located at the origin. Two charges of -1 esu each are located at the point with x, y, z coordinates 1, 1, 0 and -1, 1, 0. It is easy to see that the potential ϕ is zero at the point (0, 1, 0) if it is zero at infinity. It follows that somewhere on the y axis beyond (0, 1, 0) the function $\phi(0, y, 0)$ must have a minimum or a maximum. At that point the electric field **E** must be zero. Why? Locate the point, at least approximately.

Ans. $y = 1.6207$.

2.4 Describe the electric field and the charge distribution that go with the following potential:

$$\phi = x^2 + y^2 + z^2 \qquad \text{for } x^2 + y^2 + z^2 < a^2$$

$$\phi = -a^2 + \frac{2a^3}{(x^2 + y^2 + z^2)^{1/2}} \qquad \text{for } a^2 < x^2 + y^2 + z^2$$

2.5 A sphere the size of a basketball is charged to a potential of -1000 volts. About how many extra electrons are on it, per cm^2 of surface?

Ans. 3×10^7.

2.6 A sphere the size of the earth has 1 coulomb of charge distributed evenly over its surface. What is the electric field strength just outside the surface, in volts/meter? What is the potential of the sphere, in volts, with zero potential at infinity?

Ans. 2.5×10^{-4} volt/meter; 1500 volts.

2.7 Designate the corners of a square, 5 cm on a side, in clockwise order, *A, B, C, D*. Put a charge 2 esu at *A*, -3 esu at *B*. Determine the value of the line integral of **E**, from point *C* to point *D*. (No actual integration needed!)

2.8 For the cylinder of uniform charge density in Fig. 2.17:

(*a*) Show that the expression there given for the field inside the cylinder follows from Gauss's law.

(*b*) Find the potential ϕ as a function of r, both inside and outside the cylinder, taking $\phi = 0$ at $r = 0$.

2.9 For the system in Fig. 2.10 sketch the equipotential surface that touches the rim of the disk. Find the point where it intersects the symmetry axis.

2.10 A thin rod extends along the z axis from $z = -d$ to $z = d$. The rod carries a charge uniformly distributed along its length with linear charge density λ. By integrating over this charge distribution calculate the potential at a point P_1 on the z axis with coordinates 0, 0, $2d$. By another integration find the potential at a point P_2 on the x

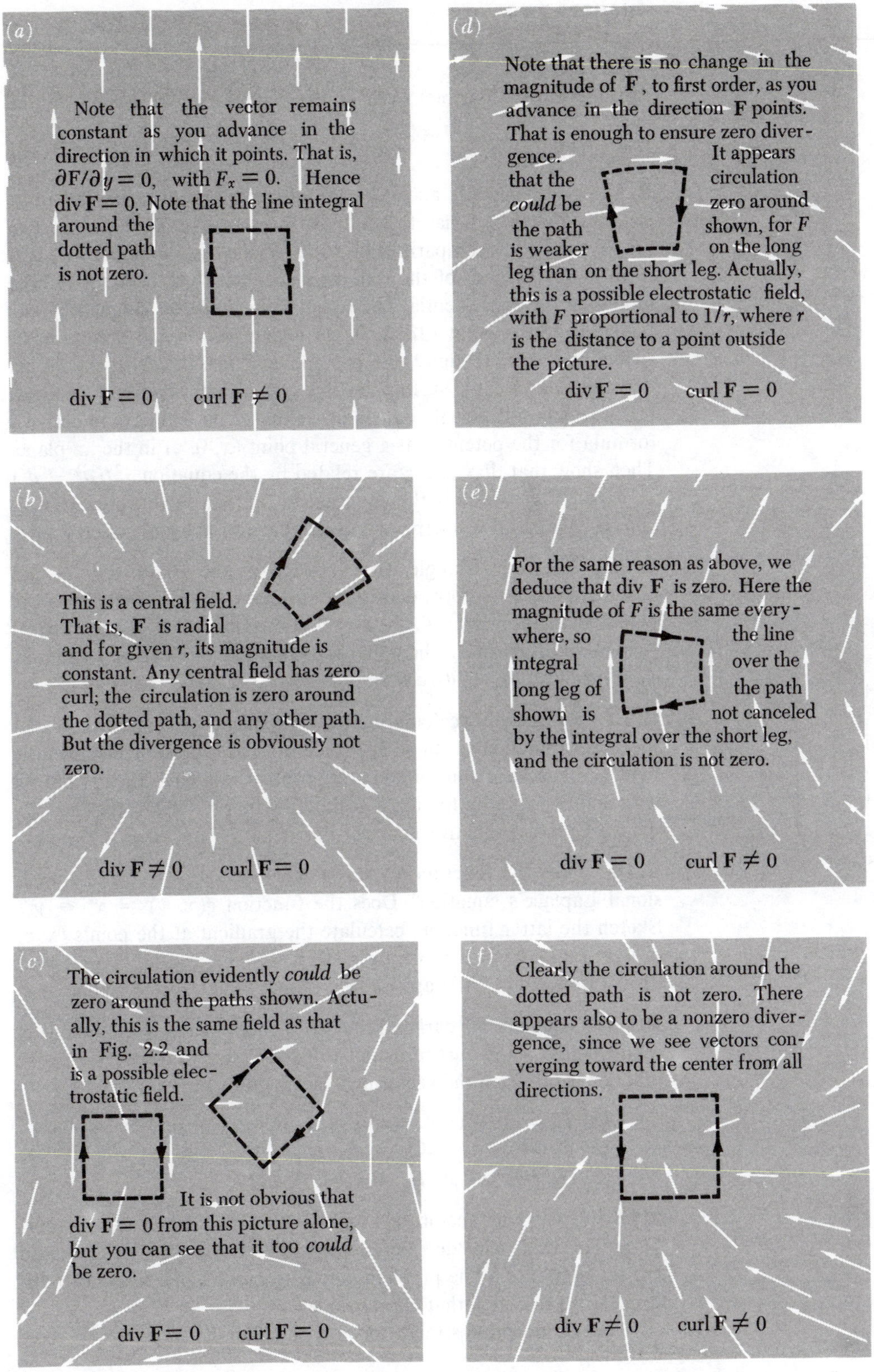
(a)
Note that the vector remains constant as you advance in the direction in which it points. That is, ∂F/∂y = 0, with $F_x = 0$. Hence div F = 0. Note that the line integral around the dotted path is not zero.
div F = 0 curl F ≠ 0
(d)
Note that there is no change in the magnitude of F, to first order, as you advance in the direction F points. That is enough to ensure zero divergence.
It appears that the circulation could be zero around the path shown, for F is weaker on the long leg than on the short leg. Actually, this is a possible electrostatic field, with F proportional to 1/r, where r is the distance to a point outside the picture.
div F = 0 curl F = 0
(b)
This is a central field. That is, F is radial and for given r, its magnitude is constant. Any central field has zero curl; the circulation is zero around the dotted path, and any other path. But the divergence is obviously not zero.
div F ≠ 0 curl F = 0
(e)
For the same reason as above, we deduce that div F is zero. Here the magnitude of F is the same everywhere, so the line integral over the long leg of the path shown is not canceled by the integral over the short leg, and the circulation is not zero.
div F = 0 curl F ≠ 0
(c)
The circulation evidently could be zero around the paths shown. Actually, this is the same field as that in Fig. 2.2 and is a possible electrostatic field.
It is not obvious that div F = 0 from this picture alone, but you can see that it too could be zero.
div F = 0 curl F = 0
(f)
Clearly the circulation around the dotted path is not zero. There appears also to be a nonzero divergence, since we see vectors converging toward the center from all directions.
div F ≠ 0 curl F ≠ 0

FIGURE 2.32
Discussion of Fig. 2.30.

axis and locate this point to make the potential equal to the potential at P_1.

Ans. $\lambda \ln 3$; $x = \sqrt{3}d$.

2.11 The points P_1 and P_2 in the preceding problem happen to lie on an ellipse which has the ends of the rod as its foci, as you can readily verify by comparing the sums of the distances from P_1 and from P_2 to the ends of the rod. This suggests that the whole ellipse might be an equipotential. Test that conjecture by calculating the potential at the point $(3d/2, 0, d)$ which lies on the same ellipse. Indeed it is true, though there is no obvious reason why it should be, that the equipotential surfaces of this system are a family of confocal prolate spheroids. See if you can prove that. You will have to derive a formula for the potential at a general point $(x, 0, z)$ in the xz plane. Then show that, if x and z are related by the equation $x^2/(a^2 - d^2) + z^2/a^2 = 1$, which is the equation for an ellipse with foci at $z = \pm d$, the potential will depend only on the parameter a, not on x or z.

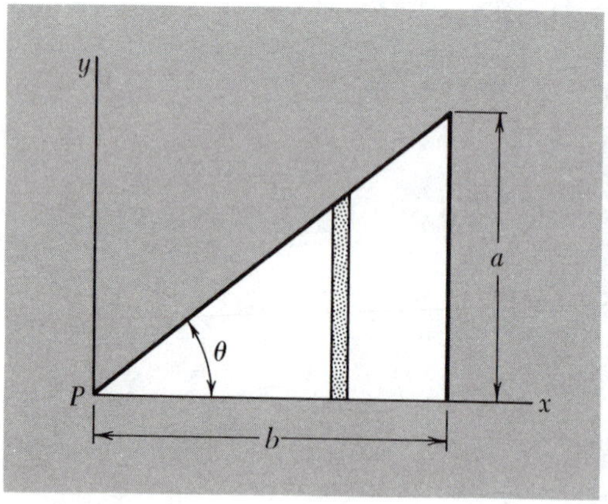

PROBLEM 2.12

2.12 The right triangle with vertex P at the origin, base b, and altitude a has a uniform density of surface charge σ. Determine the potential at the vertex P. First find the contribution of the vertical strip of width dx at x. Show that the potential at P can be written as $\phi_P = \sigma b \ln[(1 + \sin\theta)/\cos\theta]$.

2.13 By explicitly calculating the components of $\nabla \times \mathbf{E}$, show that the vector function specified in Problem 2.1 is a possible electrostatic field. (Of course, if you worked that problem, you have already proved it in another way by finding a scalar function of which it is the gradient.) Evaluate the divergence of this field.

2.14 Does the function $f(x, y) = x^2 + y^2$ satisfy the two-dimensional Laplace's equation? Does the function $g(x, y) = x^2 - y^2$? Sketch the latter function, calculate the gradient at the points $(x = 0, y = 1)$; $(x = 1, y = 0)$; $(x = 0, y = -1)$; and $(x = -1, y = 0)$ and indicate by little arrows how these gradient vectors point.

2.15 Calculate the curl and the divergence of each of the following vector fields. If the curl turns out to be zero, try to discover a scalar function ϕ of which the vector field is the gradient:

(*a*) $F_x = x + y$; $F_y = -x + y$; $F_z = -2z$.
(*b*) $G_x = 2y$; $G_y = 2x + 3z$; $G_z = 3y$.
(*c*) $H_x = x^2 - z^2$; $H_y = 2$; $H_z = 2xz$.

2.16 If $\mathbf{A}$ is any vector field with continuous derivatives, div (curl $\mathbf{A}$) = 0 or, using the "del" notation, $\nabla \cdot (\nabla \times \mathbf{A}) = 0$. We shall need this theorem later. The problem now is to prove it. Here are two different ways in which that can be done:

(*a*) (Uninspired straightforward calculation in a particular coordinate system): Using the formula for ∇ in cartesian coordinates, work out the string of second partial derivatives that $\nabla \cdot (\nabla \times \mathbf{A})$ implies.

(*b*) (With the divergence theorem and Stokes' theorem, no coordinates are needed): Consider the surface S in the figure, a balloon almost cut in two which is bounded by the closed curve C. Think about the line integral, over a curve like C, of any vector field. Then invoke Stokes and Gauss with suitable arguments.

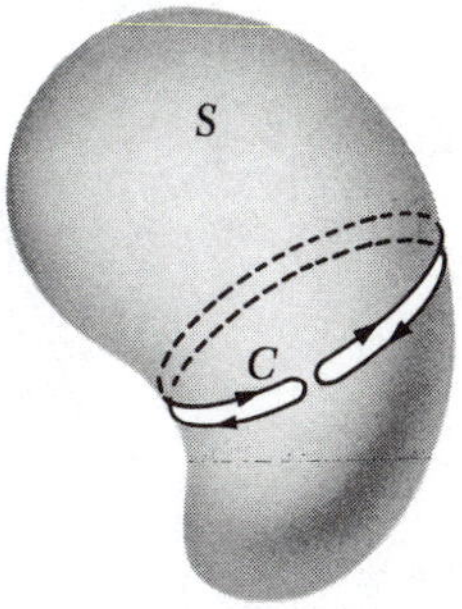

PROBLEM 2.16

2.17 Use the identity $\nabla (\phi \nabla \phi) = (\nabla \phi)^2 + \phi \nabla^2 \phi$ and the divergence theorem to prove that Eq. 38 of Chapter 1 and Eq. 27 of Chapter 2 are equivalent for any charge distribution of finite extent.

2.18 A hollow circular cylinder, of radius a and length b, with open ends, has a total charge Q uniformly distributed over its surface. What is the difference in potential between a point on the axis at one end and the midpoint of the axis? Show by sketching some field lines how you think the field of this thing ought to look.

2.19 We have two metal spheres, of radii R_1 and R_2, quite far apart from one another compared with these radii. Given a total amount of charge Q which we have to divide between the spheres, how should it be divided so as to make the potential energy of the resulting charge distribution as small as possible? To answer this, first calculate the potential energy of the system for an arbitrary division of the charge, q on one and $Q - q$ on the other. Then minimize the energy as a function of q. You may assume that any charge put on one of these spheres distributes itself uniformly over the sphere, the other sphere being far enough away so that its influence can be neglected. When you have found the optimum division of the charge, show that with that division the potential difference between the two spheres is zero. (Hence they could be connected by a wire, and there would still be no redistribution. This is a special example of a very general principle we shall meet in Chapter 3: on a conductor, charge distributes itself so as to minimize the total potential energy of the system.)

2.20 As a distribution of electric charge, the gold nucleus can be described as a sphere of radius 6×10^{-13} cm with a charge $Q = 79e$ distributed fairly uniformly through its interior. What is the potential ϕ_0 at the center of the nucleus, expressed in megavolts? (First derive a general formula for ϕ_0 for a sphere of charge Q and radius a. Do this by using Gauss's law to find the internal and external electric field and then integrating to find the potential.)

Ans. $\phi = 3Q/2a = 95{,}000$ statvolts $= 28.5$ megavolts.

2.21 Suppose eight protons are permanently fixed at the corners of a cube. A ninth proton floats freely near the center of the cube. There are no other charges around, and no gravity. Is the ninth proton trapped? Can it find an escape route that is all down hill in potential energy? Test it with your calculator. Many-digit accuracy will be needed!

2.22 An interstellar dust grain, roughly spherical with a radius of 3×10^{-7} meters, has acquired a negative charge such that its potential is -0.15 volt. How many extra electrons has it picked up? What is the strength of the electric field at its surface, expressed in volts/meter?

2.23 By means of a van de Graaff generator, protons are accelerated through a potential difference of 5×10^6 volts. The proton beam then passes through a thin silver foil. The atomic number of silver is 47, and you may assume that a silver nucleus is so massive compared with the proton that its motion may be neglected. What is the closest possible distance of approach, of any proton, to a silver nucleus? What will be the strength of the electric field acting on the proton at that position?

2.24 Which of the two boxed statements in Section 2.1 we regard as the corollary of the other is arbitrary. Show that, if the line integral $\int \mathbf{E} \cdot d\mathbf{s}$ is zero around any closed path, it follows that the line integral between two different points is path-independent.

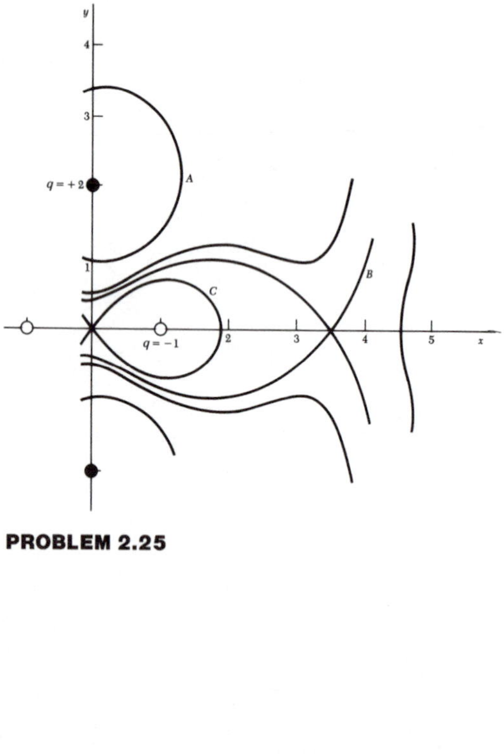

PROBLEM 2.25

2.25 Two point charges of strength 2 esu each, and two point charges of strength -1 esu each are symmetrically located in the xy plane as follows: The two positive charges are at (0, 2) and (0, -2), the two negative charges at (1, 0) and (-1, 0). Some of the equipotentials in the xy plane have been plotted in the figure. (Of course these curves are really the intersection of some three-dimensional equipotential surfaces with the xy plane.) Study this figure until you understand its general appearance. Now find the value of the potential ϕ on each of the curves A, B, and C, as usual taking $\phi = 0$ at infinite distance. Do this by calculating the potential at some point on the curve, a point chosen to make the calculation as easy as possible. Roughly sketch in some intermediate equipotentials.

2.26 Use the result for Problem 2.12 to answer this question: If a square with surface charge density σ and side s has the same potential at its center as a disk with the same surface charge density and diameter d, what must be the ratio s/d? Is your answer reasonable?

2.27 Use the result stated in Eq. 24 to calculate the energy stored in the electric field of the charged disk described in Section 2.6. (*Hint:* Consider the work done in building the disk of charge out from zero radius to radius a by adding successive rings of width dr. Express the total energy in terms of radius a and total charge $Q = \pi a^2 \sigma$.)

Ans. $8Q^2/3\pi a$.

2.28 A thin disk, radius 3 cm, has a circular hole of radius 1 cm in the middle. There is a uniform surface charge of -4 esu/cm^2 on the disk.

(*a*) What is the potential in statvolts at the center of the hole? (Assume zero potential at infinite distance.)

(*b*) An electron, starting from rest at the center of the hole, moves out along the axis, experiencing no forces except repulsion by the charges on the disk. What velocity does it ultimately attain? (Electron mass $= 9 \times 10^{-28}$ gm.)

2.29 One of two nonconducting spherical shells of radius a carries a charge Q uniformly distributed over its surface, the other a charge $-Q$, also uniformly distributed. The spheres are brought together until they touch. What does the electric field look like, both outside and inside the shells? How much work is needed to move them far apart?

2.30 Consider a charge distribution which has the constant density ρ everywhere inside a cube of edge b and is zero everywhere outside that cube. Letting the electric potential ϕ be zero at infinite distance from the cube of charge, denote by ϕ_0 the potential at the center of the cube and ϕ_1 the potential at a corner of the cube. Determine the ratio ϕ_0/ϕ_1. The answer can be found with very little calculation by combining a dimensional argument with superposition. (Think about the potential at the center of a cube with the same charge density and with twice the edge length.)

2.31 A flat nonconducting sheet lies in the xy plane. The only charges in the system are on this sheet. In the half-space above the sheet, $z > 0$, the potential is $\phi = \phi_0 e^{-kz} \cos kx$, where ϕ_0 and k are constants.

(*a*) Verify that ϕ satisfies Laplace's equation in the space above the sheet.

(*b*) What do the electric field lines look like?

(*c*) Describe the charge distribution on the sheet.

2.32 To show that it takes more than direction and magnitude to make a vector, let's try to define a vector which we'll name squrl **F** by a relation like Eq. 61 with the right-hand side squared:

$$(\text{squrl } \mathbf{F}) \cdot \hat{\mathbf{n}} = \left[\lim_{a_i \to 0} \frac{\int_{C_i} \mathbf{F} \cdot d\mathbf{s}}{a_i} \right]^2$$

Prove that this does *not* define a vector. (*Hint:* Consider reversing the direction of $\hat{\mathbf{n}}$.)

3

ELECTRIC FIELDS AROUND CONDUCTORS

CONDUCTORS AND INSULATORS

3.1 The earliest experimenters with electricity observed that substances differed in their power to hold the "Electrick Vertue." Some materials could be easily electrified by friction and maintained in an electrified state; others, it seemed, could not be electrified that way, or did not hold the Vertue if they acquired it. Experimenters of the early eighteenth century compiled lists in which substances were classified as "electricks" or "nonelectricks." Around 1730, the important experiments of Stephen Gray in England showed that the Electrick Vertue could be conducted from one body to another by horizontal string, over distances of several hundred feet, provided that the string was itself supported from above by silk threads.† Once this distinction between conduction and nonconduction had been grasped, the electricians of the day found that even a nonelectrick could be highly electrified if it were supported on glass or suspended by silk threads. A spectacular conclusion of one of the popular electric exhibitions of the time was likely to be the electrification of a boy suspended by many silk threads from the rafters; his hair stood on end and sparks could be drawn from the tip of his nose.

After the work of Gray and his contemporaries the elaborate lists of electricks and non-electricks were seen to be, on the whole, a division of materials into electrical *insulators* and electrical *conductors*. This distinction is still one of the most striking and extreme contrasts that nature exhibits. Common good conductors like ordinary metals differ in their electrical conductivity from common insulators like glass and plastics, by factors on the order of 10^{20}. To express it in a way the eighteenth-century experimenters like Gray or Benjamin Franklin would have understood, a metal globe on a metal post can lose its electrification in a millionth of a second; a metal globe on a glass post can hold its Vertue for many years. (To make good on the last assertion we would need to take some precautions beyond the capability of an eighteenth-century laboratory. Can you suggest some of them?)

The electrical difference between a good conductor and a good insulator is as vast as the mechanical difference between a liquid and a solid. That is not entirely accidental. Both properties depend on the *mobility* of atomic particles: in the electrical case, the mobility of the carriers of charge, electrons or ions; in the case of the mechanical properties, the mobility of the atoms or molecules that make up the structure of the material. To carry the analogy a bit further, we know of substances whose fluidity is intermediate between that of a solid and

†The "pack-thread" he used for his string was doubtless a rather poor conductor compared to metal wire, but good enough for transferring charge in electrostatic experiments. Gray found, too, that fine copper wire was a conductor, but mostly he used the pack-thread for the longer distances.

that of a liquid—substances such as tar or ice cream. Indeed some substances—glass is a good example—change gradually and continuously from a mobile liquid to a very permanent and rigid solid with a few hundred degrees' lowering of the temperature. In electrical conductivity, too, we find examples over the whole wide range from good conductor to good insulator, and some substances that can change conductivity over nearly as wide a range, depending on conditions such as their temperature. A fascinating and useful class of materials called semiconductors, which we shall meet in Chapter 4, have this property.

Whether we call a material solid or liquid sometimes depends on the time scale, and perhaps also on the scale of distances involved. Natural asphalt seems solid enough if you hold a chunk in your hand. Viewed geologically, it is a liquid, welling up from underground deposits and even forming lakes. We may expect that, for somewhat similar reasons, whether a material is to be regarded as an electrical insulator or a conductor will depend on the time scale of the phenomenon we are interested in.

CONDUCTORS IN THE ELECTROSTATIC FIELD

3.2 We shall look first at electrostatic systems involving conductors. That is, we shall be interested in the *stationary* state of charge and electric field that prevails after all redistributions of charge have taken place in the conductors. Any insulators present are assumed to be perfect insulators. As we have already mentioned, quite ordinary insulators come remarkably close to this idealization, so the systems we shall discuss are not too artificial. In fact, the air around us is an extremely good insulator. The systems we have in mind might be typified by some such example as this: Bring in two charged metal spheres, insulated from one another and from everything else. Fix them in positions relatively near one another. What is the resulting electric field in the whole space surrounding and between the spheres, and how is the charge that was on each sphere distributed? We begin with a more general question: After the charges have become stationary, what can we say about the electric field inside conducting matter?

In the static situation there is no further motion of charge. You might be tempted to say that the electric field must then be zero within conducting material. You might argue that, if the field were *not* zero, the mobile charge carriers would experience a force and would be thereby set in motion, and thus we would not have a static situation after all. Such an argument overlooks the possibility of *other* forces which may be acting on the charge carriers, and which would have to be counterbalanced by an electric force to bring about a stationary state. To remind ourselves that it is physically possible to have other than electrical forces acting on the charge carriers we need only think of gravity. A positive ion has weight; it experiences a steady force in a gravitational field and so does an electron; also, the forces they experience are not equal. This is a rather absurd example. We know that gravitational forces are utterly negligible on an atomic scale. There are other forces at work, however, which we may very loosely call "chemical." In a battery and in many, many other theaters of chemical reaction, including the living cell, charge carriers sometimes move *against* the general electric field; they do so because a reaction may thereby take place which yields more energy than it costs to buck the field. One hesitates to call these forces nonelectrical, knowing as we do that the structure of atoms and molecules and the forces between them can be explained in terms of Coulomb's law and quantum mechanics. Still, from the viewpoint of our *classical* theory of electricity, they must be treated as quite extraneous. Certainly they behave very differently from the inverse-square force upon which our theory is based. The general necessity for forces that are in this sense nonelectrical was already foreshadowed by our discovery in Chapter 2 that inverse-square forces alone cannot make a stable, static structure.

The point is simply this: We must be prepared to find, in some cases, unbalanced, non-Coulomb forces acting on charge carriers inside a conducting medium. When that happens, the electrostatic situation is attained when there *is* a finite electric field in the conductor that just offsets the influence of the other forces, whatever they may be.

Having issued this warning, however, we turn at once to the very familiar and important case in which there is no such force to worry about, the case of a homogeneous, isotropic conducting material. In the interior of such a conductor, in the static case, we can state confidently that the electric field must be zero.† If it weren't, charges would have to move. It follows that all regions inside the conductor, including all points just below its surface, must be at the same potential. Outside the conductor, the electric field is not zero. The surface of the conductor must be an equipotential surface of this field.

Imagine that we could change a material from insulator to conductor at will. (It's not impossible—glass becomes conducting when heated; any gas can be ionized by x-rays.) In Fig. 3.1*a* is shown an uncharged nonconductor in the electric field produced by two fixed layers of charge. The electric field is the same inside the body as outside. (A dense body such as glass would actually distort the field, an

†In speaking of the electric field inside matter, we mean an average field, averaged over a region large compared with the details of the atomic structure. We know, of course, that very strong fields exist in all matter, including the good conductors, if we search on a small scale near an atomic nucleus. The nuclear electric field does not contribute to the average field in matter, ordinarily, because it points in one direction on one side of a nucleus and in the opposite direction on the other side. Just how this average field ought to be defined, and how it could be measured, are questions we'll consider in Chapter 10.

effect we'll study in Chapter 10, but that is not important here.) Now, in one way or another, let mobile charges (or *ions*) be created, making the body a conductor. Positive ions are drawn in one direction by the field, negative ions in the opposite direction, as indicated in Fig. 3.1*b*. They can go no farther than the surface of the conductor. Piling up there, they begin themselves to create an electric field inside the body which tends to *cancel* the original field. And in fact the movement goes on until that original field is *precisely* canceled. The final distribution of charge at the surface, shown in Fig. 3.1*c*, is such that its field and the field of the fixed external sources combine to give *zero* electric field in the interior of the conductor. Because this "automatically" happens in every conductor, it is really only the surface of a conductor that we need to consider when we are concerned with the external fields.

With this in mind, let us see what can be said about a system of conductors, variously charged, in otherwise empty space. In Fig. 3.2 we see some objects. Think of them, if you like, as solid pieces of metal. They are prevented from moving by invisible insulators—perhaps by Stephen Gray's silk threads. The total charge of each object, by which we mean the net excess of positive over negative charge, is fixed because there is no way for charge to leak on or off. We denote it by Q_k, for the kth conductor. Each object can also be characterized by a particular value φ_k of the electric potential function φ. We say that conductor 2 is "at the potential φ_2." With a system like the one shown, where no physical objects stretch out to infinity, it is usually convenient to assign the potential zero to points infinitely far away. In that case φ_2 is the work per unit charge required to bring an infinitesimal test charge in from infinity and put it anywhere on conductor 2. (Notice, by the way, that this is just the kind of system in which the test charge needs to be kept small, a point raised in Section 1.7.)

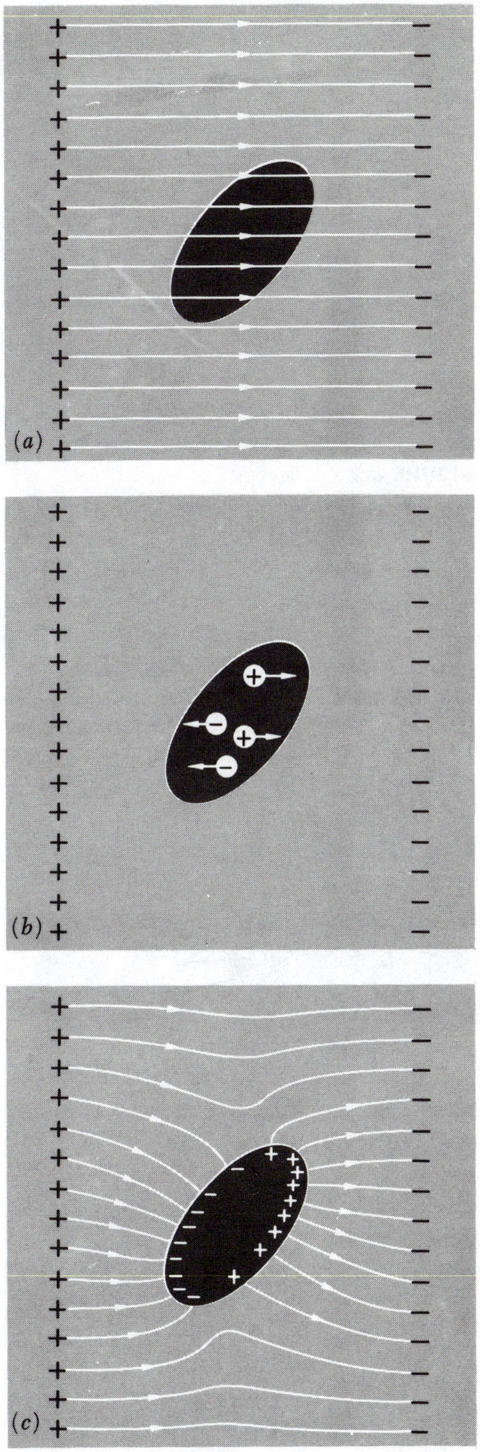

FIGURE 3.1
The object in (*a*) is a neutral nonconductor. The charges in it, both positive and negative, are immobile. In (*b*) the charges have been released and begin to move. They will move until the final condition, shown in (*c*), is attained.

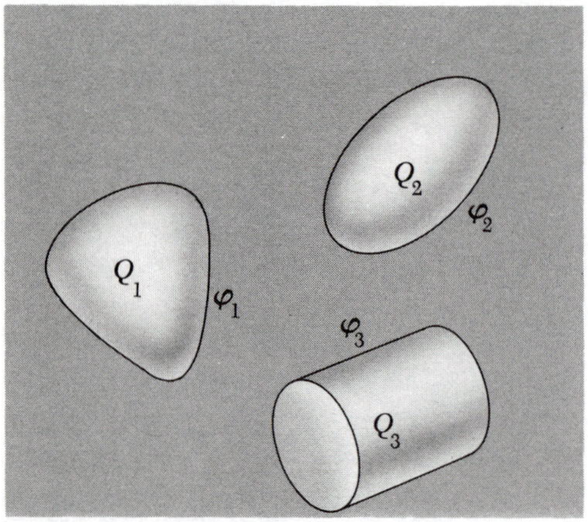

FIGURE 3.2
A system of three conductors. Q_1 is the charge on conductor 1, ϕ_1 is its potential, etc.

Because the surface of a conductor in Fig. 3.2 is necessarily a surface of constant potential, the electric field, which is $-\text{grad } \varphi$, must be *perpendicular* to the surface at every point on the surface. Proceeding from the interior of the conductor outward, we find at the surface an abrupt change in the electric field; **E** is not zero outside the surface, and it is zero inside. The discontinuity in **E** is accounted for by the presence of a surface charge, of density σ, which we can relate directly to **E** by Gauss's law. We can use a flat box enclosing a patch of surface (Fig. 3.3) like the one we used in analyzing the charged disk in Section 2.6. Here, there is *no* flux through the "bottom" of the box, which lies inside the conductor, and we conclude that $E_n = 4\pi\sigma$, where E_n is the component of electric field normal to the surface. As we have already seen, there *is* no other component in this case, the field being always perpendicular to the surface. The surface charge must account for the whole charge Q_k. That is, the surface integral of σ over the whole conductor must equal Q_k. In summary, we can make the following statements about *any* such system of conductors, whatever their shape and arrangement:

$$\varphi = \varphi_k \text{ at all points on the surface of the } k\text{th conductor} \tag{1}$$

At any point just outside the conductor, **E** is perpendicular to the surface, and $E = 4\pi\sigma$, where σ is the local density of surface charge (2)

$$Q_k = \int_{S_k} \sigma \, da = \frac{1}{4\pi} \int_{S_k} \mathbf{E} \cdot d\mathbf{a} \tag{3}$$

FIGURE 3.3
(*a*) Gauss' law relates the electric field strength at the surface of a conductor to the density of surface charge (Eq. 2). (*b*) Cross section through surface of conductor and box.

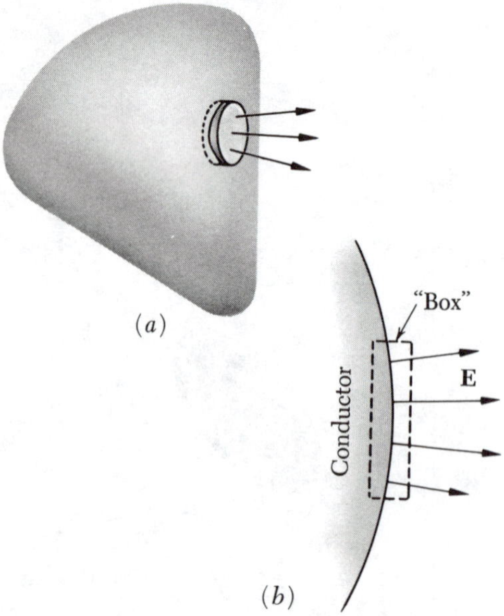

E is the total field arising from *all* the charges in the system, near and far, of which the surface charge is only a part. The surface charge on a conductor is obliged to "readjust itself" until relation (2) is fulfilled. That the conductor presents a special case, in contrast to other surface charge distributions, is brought out by the comparison in Fig. 3.4.

Figure 3.5 shows the field and charge distribution for a simple system like the one mentioned earlier. There are two conducting spheres, a sphere of unit radius carrying a total charge of + 1 unit, the other a somewhat larger sphere with total charge zero. Observe that the surface charge density is not uniform over either of the conductors. The sphere on the right, with total charge zero, has a negative surface charge density in the region which faces the other sphere, and a positive surface charge on the rearward portion of its surface. The dashed curves in Fig. 3.5 indicate the equipotential surfaces or, rather, their intersection with the plane of the figure. If we were to go a long way out, we would find the equipotential surfaces becoming nearly

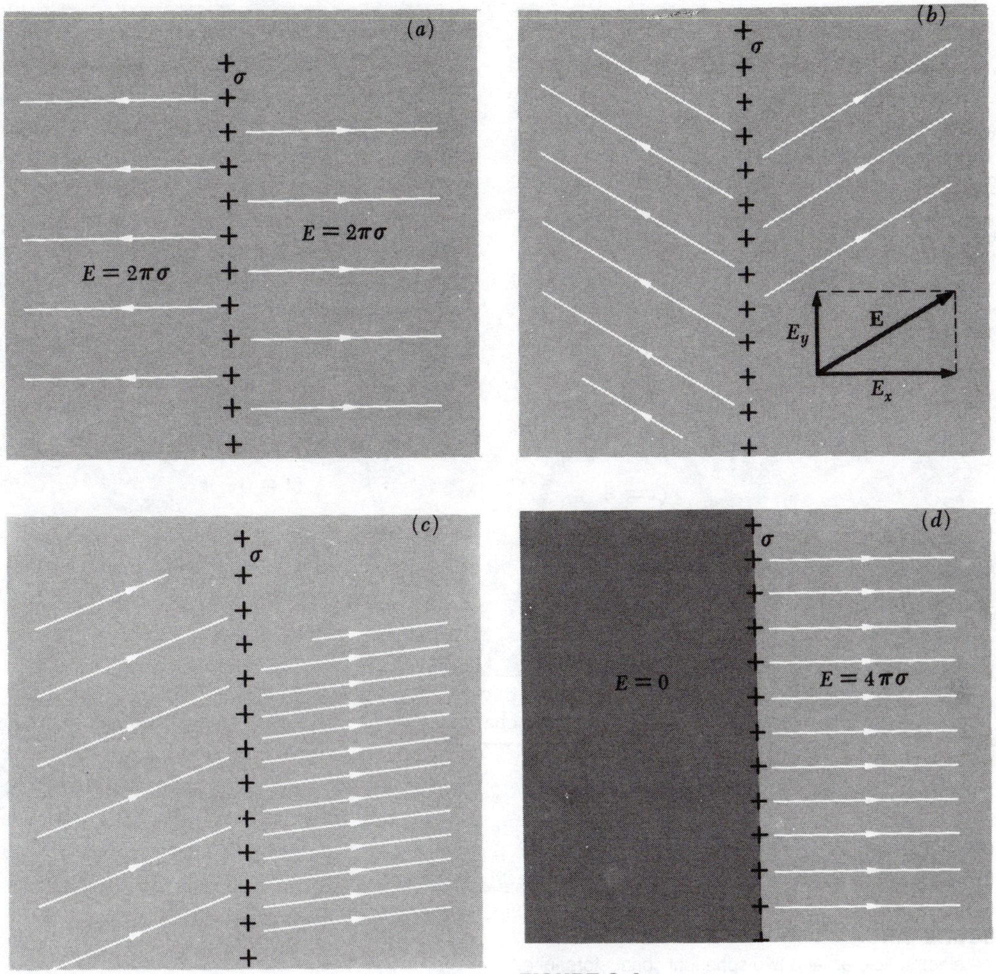

FIGURE 3.4
(*a*) An isolated sheet of surface charge with nothing else in the system. This was treated in Fig. 1.23. The field was determined as $2\pi\sigma$ on each side of the sheet by the assumption of symmetry. (*b*) If there are other charges in the system, we can say only that the change in E_x at the surface must be $4\pi\sigma$, with zero change in E_y. Many fields other than the field of (*a*) above could have this property. Two such are shown in (*b*) and (*c*). (*d*) If we know that the medium on one side of the surface is a conductor, we know that on the other side **E** must be perpendicular to the surface, with magnitude $E = 4\pi\sigma$. **E** could not have a component parallel to the surface without causing charge to move.

spherical and the field lines nearly radial, and the field would begin to look very much like that of a point charge of magnitude 1 and positive, which is the net charge on the entire system.

Figure 3.5 illustrates, at least qualitatively, all the features we anticipated, but we have an additional reason for showing it. Simple as the system is, the exact mathematical solution for this case cannot be obtained in a straightforward way. Our Fig. 3.5 was constructed from an approximate solution. In fact, the number of three-dimensional geometrical arrangements of conductors which permit a mathematical solution in closed form is lamentably small. One does not learn much physics by concentrating on the solution of the few neatly soluble examples. Let us instead try to understand the general nature of the mathematical problem such a system presents.

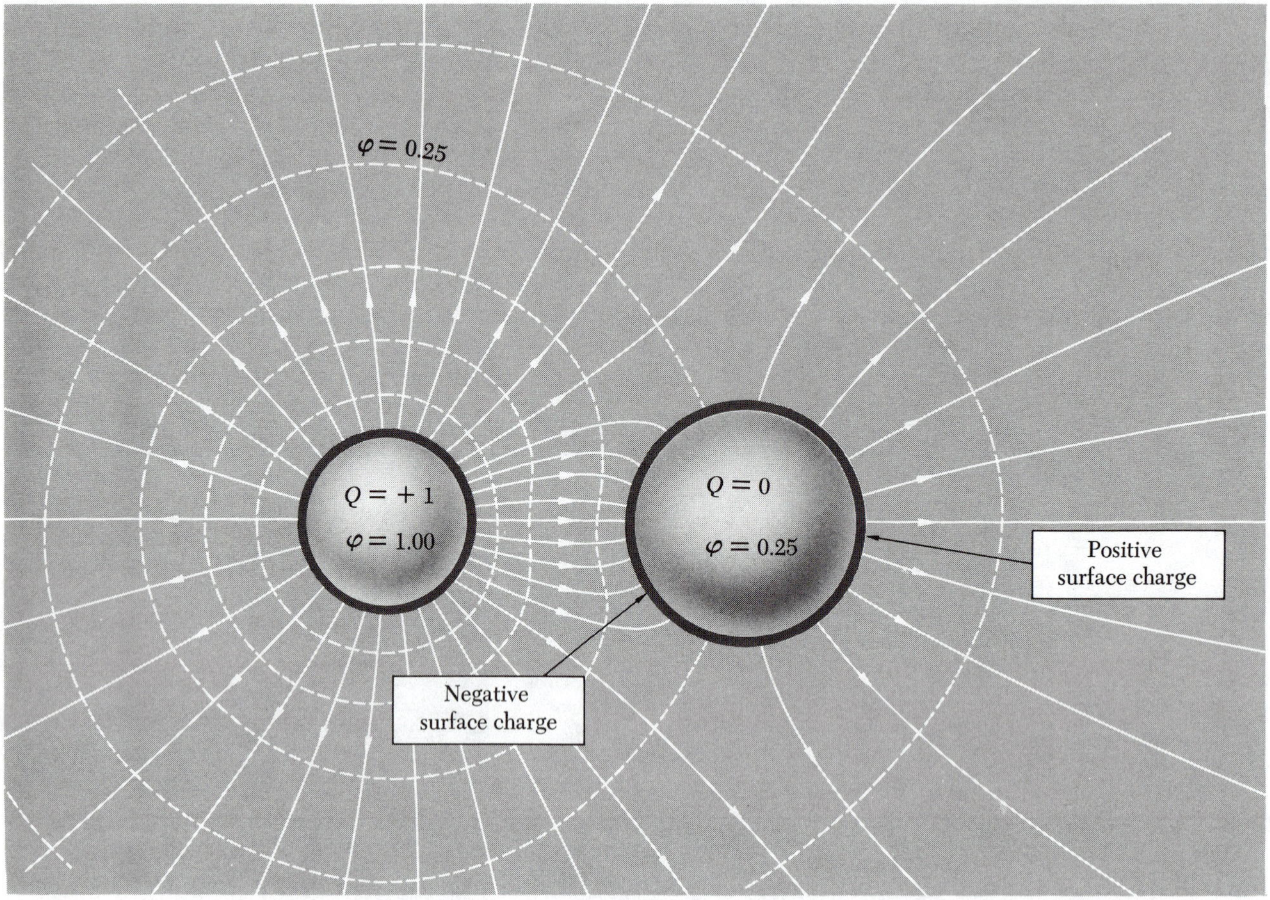

φ = 0.25
Q = +1
φ = 1.00
Q = 0
φ = 0.25
Positive surface charge
Negative surface charge

FIGURE 3.5
The electric field around two spherical conductors, one with total charge 1, and one with total charge zero. Dashed curves are intersections of equipotential surfaces with the plane of the figure. Zero potential is at infinity.

THE GENERAL ELECTROSTATIC PROBLEM; UNIQUENESS THEOREM

3.3 We can state the problem in terms of the potential function φ, for if φ can be found, we can at once get **E** from it. Everywhere outside the conductors φ has to satisfy the partial differential equation we met in Chapter 2, Laplace's equation: $\nabla^2\varphi = 0$. Written out in cartesian coordinates, Laplace's equation reads,

$$\frac{\partial^2\varphi}{\partial x^2} + \frac{\partial^2\varphi}{\partial y^2} + \frac{\partial^2\varphi}{\partial z^2} = 0 \tag{4}$$

The problem is to find a function that satisfies Eq. 4 and also meets the specified conditions on the conducting surfaces. These conditions might have been set in various ways. It might be that the potential of each conductor φ_k is fixed or known. (In a real system the potentials may be fixed by permanent connections to batteries or other constant-potential "power supplies.") Then our solution $\varphi(x, y, z)$ has to assume the correct value at all points on each of the surfaces. These surfaces in their totality *bound* the region in which φ is defined, if we include a large surface "at infinity," where we require φ to approach zero. Sometimes the region of interest is totally enclosed by a conducting surface; then, we can assign this conductor a potential and ignore anything outside it. In either case, we have a typical *boundary-value problem,* in which the value the function has to assume on the boundary is specified for the entire boundary.

One might, instead, have specified the total charge on each conductor, Q_k. (We could not specify arbitrarily all charges and potentials; that would overdetermine the problem.) With the charges specified, we have in effect fixed the value of the surface integral of grad φ over the surface of each conductor. This gives the mathematical problem a slightly different aspect. Or one can "mix" the two kinds of boundary conditions.

A general question of some interest is this: With the boundary conditions given in some way, does the problem have no solution, one solution, or more than one solution? We shall not try to answer this question in all the forms it can take, but one important case will show how such questions can be dealt with and will give us a useful result. Suppose the potential of each conductor, φ_k, has been specified, together with the requirement that φ approach zero at infinite distance, or on a conductor which encloses the system. We shall prove that this boundary-value problem has no more than one solution. It seems obvious, as a matter of physics, that it has *a* solution, for if we should actually arrange the conductors in the prescribed manner, connecting them by infinitesimal wires to the proper potentials, the system would have to settle down in *some* state. However, it is quite a different matter to prove mathematically that a solution always exists, and we shall not attempt it. Instead, we assume that there *is* a solution $\varphi(x, y, z)$ and show that it must be unique. The argument, which is typical of such proofs, runs as follows.

Assume there is another function $\psi(x, y, z)$ which is also a solution meeting the same boundary conditions. Now Laplace's equation is *linear*. That is, if φ and ψ satisfy Eq. 4, then so does $\varphi + \psi$ or any linear combination such as $c_1\varphi + c_2\psi$, where c_1 and c_2 are constants. In particular, the difference between our two solutions, $\varphi - \psi$, must satisfy Eq. 4. Call this function W:

$$W(x, y, z) = \varphi(x, y, z) - \psi(x, y, z) \tag{5}$$

Of course, W does *not* satisfy the boundary conditions. In fact, at the surface of every conductor W is zero, because ψ and φ take on the same value, φ_k, at the surface of a conductor k. Thus W is a solution of *another* electrostatic problem, one with the same conductors but with all conductors held at zero potential. We can now assert that, if this is so, W must be zero at all points in space. For if it is not, it must have either a maximum or a minimum somewhere—remember that W is zero at infinity as well as on all the conducting boundaries. If W has an extremum at some point P, consider a sphere centered on that point. As we saw in Chapter 2, the average over a sphere of a function that satisfies Laplace's equation is equal to its value at the center. This could not be true if the center is a maximum or minimum. Thus W cannot have a maximum or minimum; it must therefore be zero everywhere. It follows that $\psi = \varphi$ everywhere, that is, there can be only *one* solution of Eq. 4 that satisfies the prescribed boundary conditions.

We can now demonstrate easily another remarkable fact. *In the space inside a hollow conductor of any shape whatever, if that space itself is empty of charge, the electric field is zero.* This is true whatever the field may be outside the conductor. We are already familiar with the fact that the field is zero inside an isolated uniform spherical shell of charge, just as the gravitational field inside the shell of a hollow spherical mass is zero. The theorem we just stated is, in a way, more surprising. Consider the closed metal box shown partly cut away in Fig. 3.6. There are charges in the neighborhood of the box, and the external field is approximately as depicted. There is a highly nonuniform distribution of charge over the surface of the box. Now the field everywhere in space, *including the interior of the box,* is the sum of the field of this charge distribution and the fields of the external sources. It seems hardly credible that the surface charge has so cleverly arranged itself on the box that its field precisely *cancels* the field of the external sources at every point inside the box. Yet this must have happened, as we can prove in a few sentences.

The potential function inside the box, $\varphi(x, y, z)$, must satisfy Laplace's equation. The entire boundary of this region, namely, the box, is an equipotential, so we have $\varphi = \varphi_0$, a constant everywhere on the boundary. One solution is obviously $\varphi = \varphi_0$ throughout the volume. But there can be only one solution, according to our uniqueness theorem, so this is it. "φ = constant" implies $\mathbf{E} = 0$, because $\mathbf{E} = -\text{grad } \varphi$.

The absence of electric field inside a conducting enclosure is useful, as well as theoretically interesting. It is the basis for electrical shielding. For most practical purposes the enclosure does not need to be completely tight. If the walls are perforated with small holes, or made of metallic screen, the field inside will be extremely weak except in the immediate vicinity of a hole. A metal pipe with open ends, if it is a few diameters long, very effectively shields the space inside that is not close to either end. We are considering only static fields of course, but for slowly varying electric fields these remarks still hold. (A rapidly varying field can become a wave that travels through the

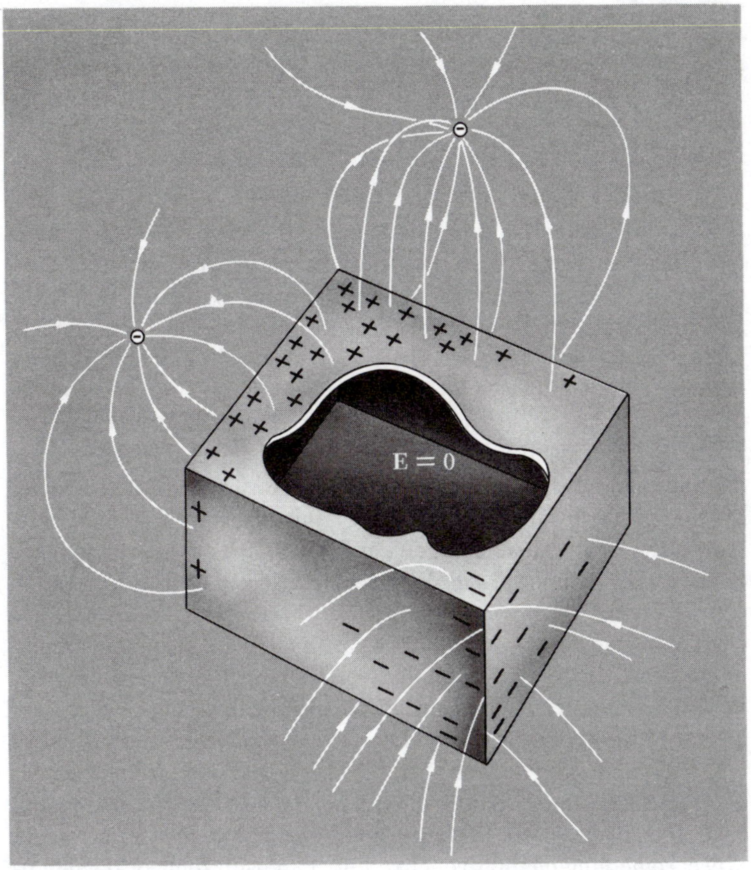
E = 0

FIGURE 3.6
The field is zero everywhere inside a closed conducting box.

pipe. *Rapidly* means here "in less time than light takes to travel a pipe diameter.")

FIGURE 3.7
With given charges Q_1 and Q_2 on the spherical shells, the potential of the inner shell is given by Eq. 6.

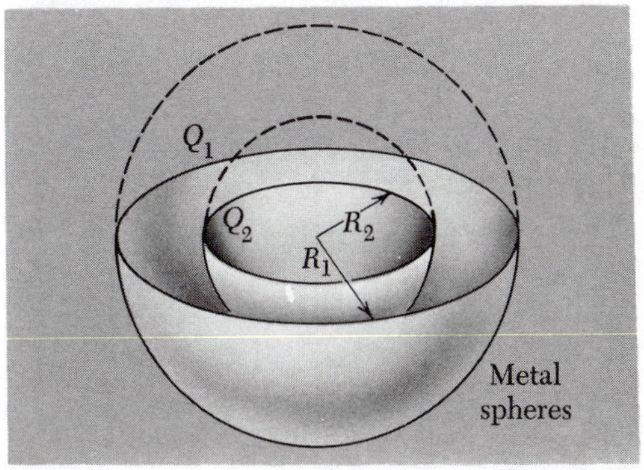
Q_1
Q_2
R_2
R_1
Metal spheres

SOME SIMPLE SYSTEMS OF CONDUCTORS

3.4 In this section we shall investigate a few particularly simple arrangements of conductors. We begin with two concentric metal spheres, of radii R_1 and R_2, carrying total charges Q_1 and Q_2, respectively (Fig. 3.7). This situation presents no new challenge. It is obvious from symmetry that the charge on each sphere must be distributed uniformly, so our example really belongs back in Chapter 1! Outside the larger sphere the field is that of a point charge of magnitude $Q_1 + Q_2$, so φ_1, the potential of the outer sphere, is

$$\frac{Q_1 + Q_2}{R_1}$$

The potential of the inner sphere is given by

$$\varphi_2 = \frac{Q_1 + Q_2}{R_1} + \int_{R_1}^{R_2} -\frac{Q_2}{r^2}\,dr = \frac{Q_1}{R_1} + \frac{Q_2}{R_1} + \frac{Q_2}{R_2} - \frac{Q_2}{R_1} \tag{6}$$
$$= \frac{Q_1}{R_1} + \frac{Q_2}{R_2}$$

φ_2 is also the potential at all points inside the inner sphere. We could have found $\varphi_2 = Q_1/R_1 + Q_2/R_2$ by simple superposition: Q_1/R_1 is the potential inside the larger sphere if it alone is present, Q_2/R_2 the potential inside the inner sphere if it alone is present. If the spheres carried equal and opposite charges, $Q_1 = -Q_2$, only the space between them will have a nonvanishing electric field.

About the simplest system in which the mobility of the charges in the conductor makes itself evident is the point charge near a conducting plane. Suppose the xy plane is the surface of a conductor extending out to infinity. Let's assign this plane the potential zero. Now bring in a positive charge Q and locate it h cm above the plane on the z axis, as in Fig. 3.8*a*. What sort of field and charge distribution can we expect? We expect the positive charge Q to attract negative charge, but we hardly expect the negative charge to pile up in an infinitely dense concentration at the foot of the perpendicular from Q. Why not? Also, we remember that the electric field is always perpendicular to the surface of a conductor, at the conductor's surface. Very near the point charge Q, on the other hand, the presence of the conducting plane can make little difference; the field lines must *start out* from Q as if they were leaving a point charge radially. So we might expect something qualitatively like Fig. 3.8*b*, with some of the details still a bit uncertain. Of course the whole thing is bound to be quite symmetrical about the z axis.

But how do we really solve the problem? The answer is, by a trick, but a trick that is both instructive and frequently useful. We find an easily soluble problem whose solution, or a piece of it, can be made to fit the problem at hand. Here the easy problem is that of two equal and opposite point charges, Q and $-Q$. On the plane which bisects the line joining the two charges, the plane indicated in cross section by the line AA in Fig. 3.8*c*, the electric field is everywhere perpendicular to the plane. If we make the distance of Q from the plane agree with the distance h in our original problem, the upper half of the field in Fig. 3.8*c* meets all our requirements: The field is per-

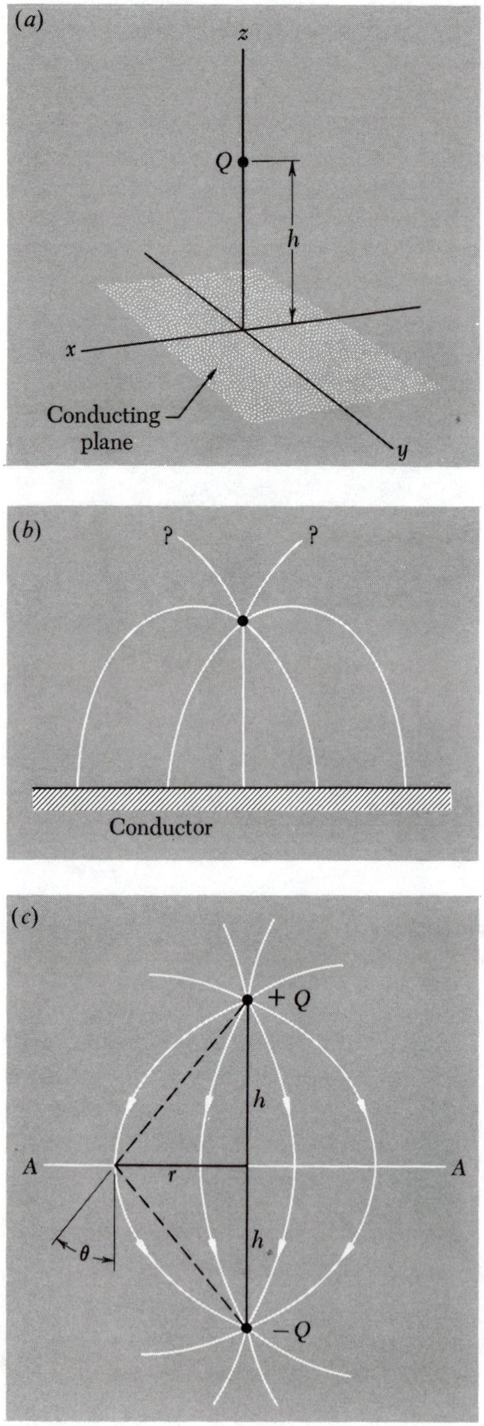
(a)
z
Q
h
x
Conducting plane
y
(b)
?
?
Conductor
(c)
+Q
h
A
r
A
θ
h
−Q

FIGURE 3.8
(*a*) A point charge *Q* above an infinite plane conductor. (*b*) The field must look something like this. (*c*) The field of a pair of opposite charges.

pendicular to the plane of the conductor, and in the neighborhood of Q it approaches the field of a point charge.

The boundary conditions here are not quite those that figured in our uniqueness theorem in the last section. The potential of the conductor is fixed, but we have in the system a point charge at which the potential approaches infinity. We can regard the point charge as the limiting case of a small, spherical conductor on which the total charge Q is fixed. For this mixed boundary condition—potentials given on some surfaces, total charge on others—a uniqueness theorem also holds. If our "borrowed" solution fits as well as this, it must be *the* solution.

Figure 3.9 shows the final solution for the field above the plane, with the density of the surface charge suggested. We can calculate the field strength and direction at any point by going back to the two-charge problem, Fig. 3.8*c*, and using Coulomb's law. Consider a point on the surface, a distance R from the origin. The square of its distance from Q is $r^2 + h^2$, and the z component of the field of Q, at this point, is $-Q \cos \theta/(r^2 + h^2)$. The "image charge," $-Q$, below the plane contributes an equal z component. Thus the electric field here is given by:

$$E_z = \frac{-2Q}{r^2 + h^2} \cos \theta = \frac{-2Q}{r^2 + h^2} \cdot \frac{h}{(r^2 + h^2)^{1/2}} = \frac{-2Qh}{(r^2 + h^2)^{3/2}} \quad (7)$$

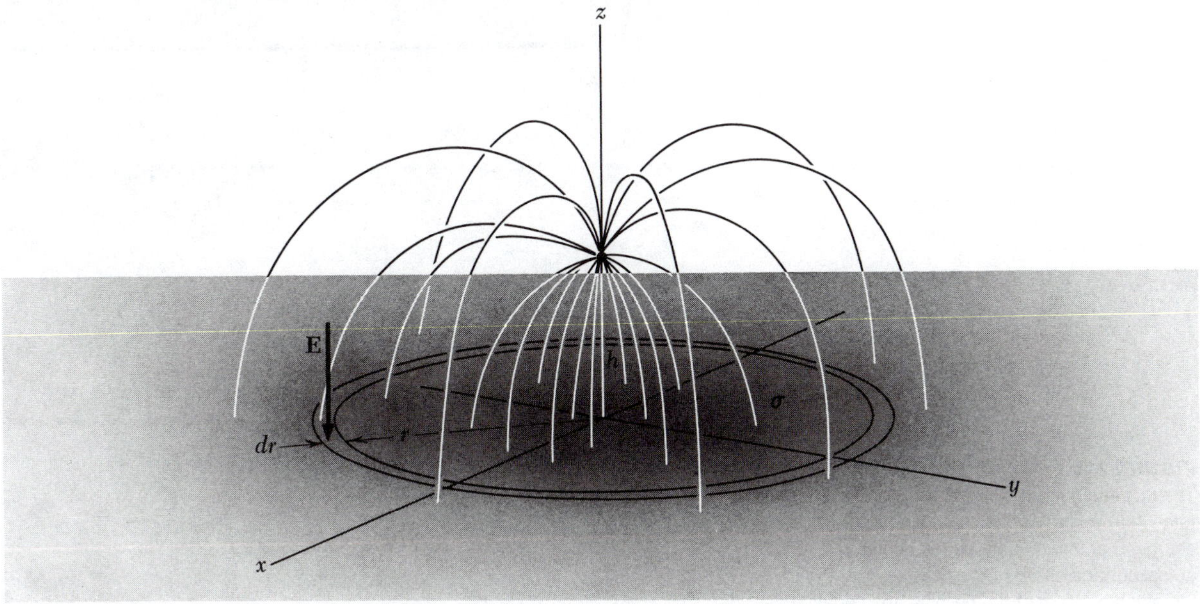

FIGURE 3.9
Some field lines for the charge above the plane. The field strength at the surface, given by Eq. 7, determines the surface charge density σ.

This tells us the surface charge density σ:

$$\sigma = \frac{E_z}{4\pi} = \frac{-Qh}{2\pi(r^2 + h^2)^{3/2}} \tag{8}$$

Let us calculate the total amount of charge on the surface by integrating over the distribution:

$$\int_0^\infty \sigma \cdot 2\pi r\, dr = -Q \int_0^\infty \frac{hr\, dr}{(h^2 + r^2)^{3/2}} = -Q \tag{9}$$

That result was to be expected. It means that all the flux leaving the charge Q ends on the conducting plane.

There is one puzzling point. What if the plane conductor had been completely uncharged before the charge Q was put in place above it? How can the conductor now exhibit a net charge $-Q$? The answer is that a compensating positive charge, $+Q$ in amount, must be distributed over the whole plane. To see what is going on here, imagine that the conducting plane is actually a metal disk, not infinite but finite and with a radius $R \gg h$. If a charge $+Q$ were to be spread uniformly over this disk, on *both* sides, the resulting surface density

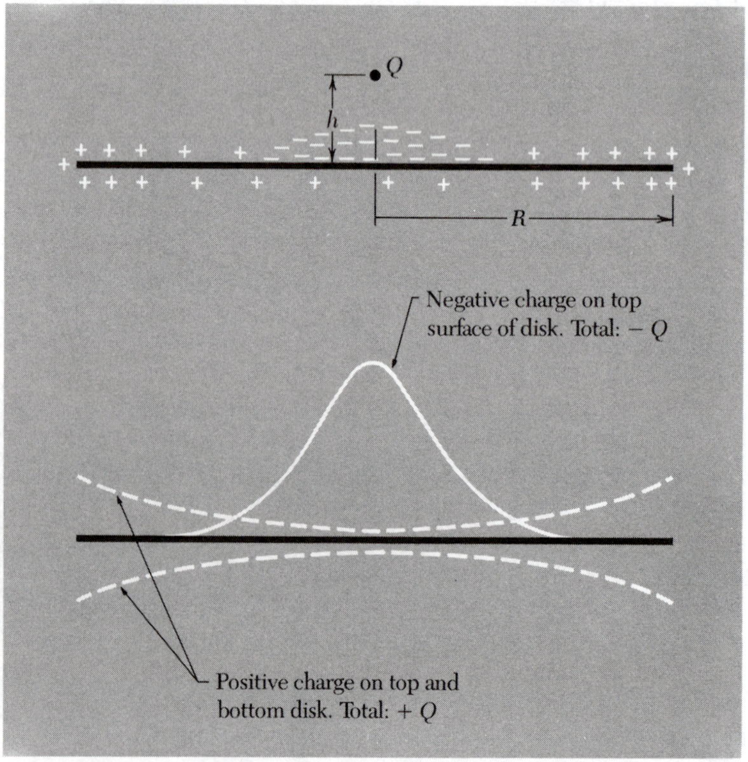

FIGURE 3.10
The distribution of charge on a conducting disk with total charge zero, in the presence of a positive point charge Q at height h above the center of the disk. The actual surface charge density at any point is of course the algebraic sum of the positive and negative densities shown.

would be $Q/2\pi R^2$, which would cause an electric field of strength $2Q/R^2$ normal to the plane of the disk. Since our disk is a conductor, on which charge can move, the charge density and the resulting field strength will be even *less* than $2Q/R^2$ near the center of the disk because of the tendency of the charge to spread out toward the rim. In any case the field of this distribution is smaller in order of magnitude by a factor h^2/R^2 than the field described by Eq. 7. As long as $R \gg h$ we were justified in ignoring it, and of course it vanishes completely for an unbounded conducting plane. Figure 3.10 shows in separate plots the surface charge density σ, given by Eq. 8, and the distribution of the compensating charge Q on the upper and lower surfaces of the disk. Here we have made R not very much larger than h, in order to show both distributions clearly in the same diagram. Notice that the compensating positive charge has arranged itself in exactly the same way on the top and bottom surfaces of the disk, as if it were utterly ignoring the pile of negative charge in the middle of the upper surface! Indeed, it is free to do so, for the field of that negative charge distribution *plus* that of the point charge Q that induced it has horizontal component zero at the surface of the disk, hence has no influence whatever on the distribution of the compensating positive charge.

The isolated conducting disk mentioned above belongs to another class of soluble problems, a class which includes any isolated conductor in the shape of a spheroid, an ellipsoid of revolution. Without going into the mathematics† we show in Fig. 3.11 some electric field lines and equipotential surfaces around the conducting disk. The field lines are hyperbolas. The equipotentials are oblate ellipsoids of revolution enclosing the disk. The potential ϕ of the disk itself, relative to infinity, turns out to be

$$\phi_0 = \frac{(\pi/2)\, Q}{a} \tag{10}$$

where Q is the total charge of the disk and a is its radius. Compare this picture with Fig. 2.11, the field of a *uniformly* charged *non*conducting disk. In that case the electric field at the surface was not normal to the surface; it had a radial component outward. If you could make that disk in Fig. 2.11 a conductor, the charge would flow outward until the field in Fig. 3.11 was established. According to the mathematical solution on which Fig. 3.11 is based, the charge density at the center of the disk would then be just half as great as it was at the center of the uniformly charged disk.

Figure 3.11 shows us the field not only of the conducting disk,

†Mathematically speaking, this class of problems is soluble because a spheroidal coordinate system happens to be one of those systems in which Laplace's equation takes on a particularly simple form.

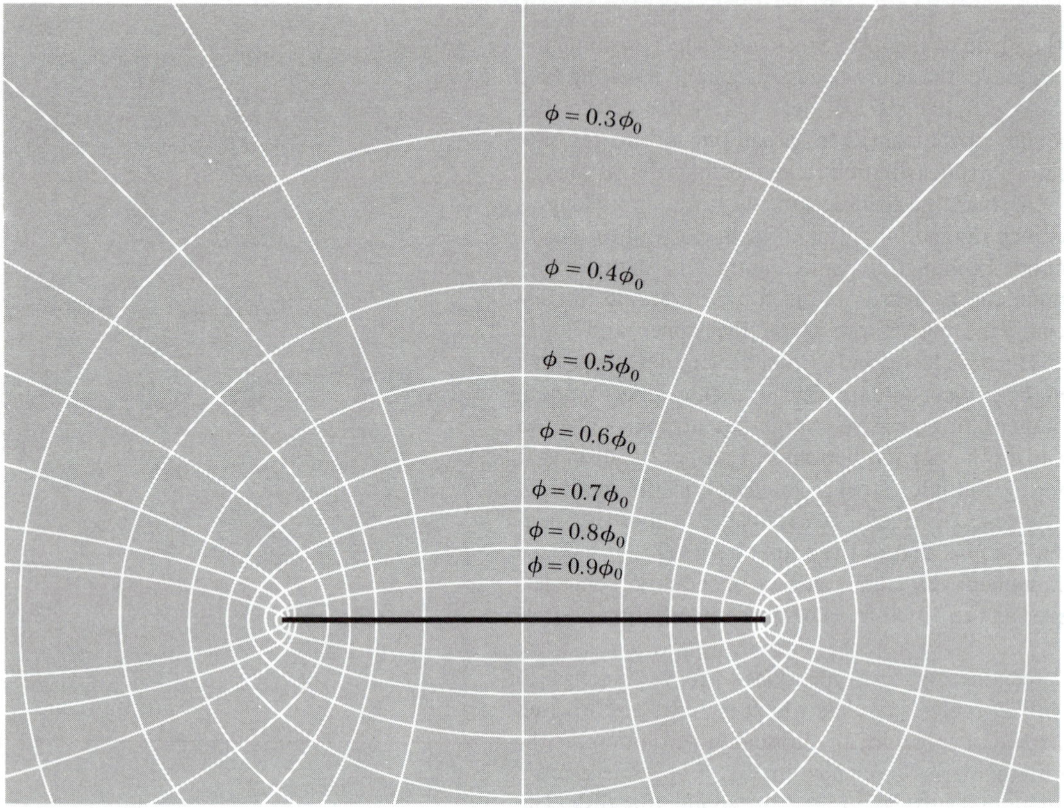

FIGURE 3.11
Equipotentials and field lines for a charged conducting disk.

but of any isolated oblate spheroidal conductor. To see that, choose one of the equipotential surfaces of revolution—say the one whose trace in the diagram is the ellipse marked $\phi = 0.6\phi_0$. Imagine that we could plate this spheroid with copper and deposit charge Q on it. Then the field shown outside it already satisfies the boundary conditions: electric field normal to surface; total flux $4\pi Q$. It is *a* solution, and in view of the uniqueness theorem it must be *the* solution for an isolated charged conductor of that particular shape. All we need to do is erase the field lines *inside* the conductor. Or imagine copperplating two of the spheroidal surfaces, putting charge Q on the inner surface, $-Q$ on the outer. The section of Fig. 3.11 between these two equipotentials shows us the field between two such concentric spheroidal conductors.

This suggests a general strategy. Given the solution for any electrostatic problem with the equipotentials located, we can extract from it the solution for any other system made from the first by copperplating one or more equipotential surfaces. Perhaps we should call the method "a solution in search of a problem." The situation was well described by Maxwell: "It appears, therefore, that what we should naturally call the inverse problem of determining the forms of the conductors when the expression for the potential is given is more manageable than the direct problem of determining the potential when the form of the conductors is given".†

If you worked Problem 2.11, you already possess the raw material for an important example. You found that a uniform line charge of finite length has equipotential surfaces in the shape of prolate ellipsoids of revolution. This solves the problem of the potential and field of any isolated charged conductor of prolate spheroidal shape, reducing it to the relatively easy calculation of the potential due to a line charge. You can try it in Problem 3.22.

CAPACITANCE AND CAPACITORS

3.5 An isolated conductor carrying a charge Q has a certain potential ϕ_0, with zero potential at infinity. Q is proportional to ϕ_0. The constant of proportionality depends only on the size and shape of the conductor. We call this factor the *capacitance* of that conductor and denote it by C.

$$Q = C\phi_0 \qquad (11)$$

Obviously the units for C depend on the units in which Q and ϕ_0 are expressed. Let us continue to measure Q in esu and ϕ_0 in statvolts. For an isolated spherical conductor of radius a we know that $\phi_0 = Q/a$. Hence the capacitance of the sphere, defined by Eq. 11, must be

$$C = \frac{Q}{\phi_0} = a \qquad (12)$$

For an isolated conducting disk of radius a, according to Eq. 10, $Q = (2/\pi)a\phi_0$, so the capacitance of such a conductor is $C = (2/\pi)a$. It is somewhat smaller than the capacitance of a sphere of the same radius, which seems reasonable. The CGS electrostatic unit of capacitance is the centimeter; it needs no other name. Since capacitance has the dimensions of length, for conductors of a given shape capacitance scales as a linear dimension of the object.

That applies to single, isolated conductors. The concept of capacitance is useful whenever we are concerned with charges on and potentials of conductors. By far the most common case of interest is that of two conductors oppositely charged, with Q and $-Q$, respec-

†James Clerk Maxwell, "Treatise on Electricity and Magnetism," vol. I, 3d ed., Oxford University Press, 1891, chap. VII; reprinted by Dover, New York, 1954. Every student of physics ought sometime to look into Maxwell's book. Chapter VII is a good place to dip in while we are on the present subject. At the end of Volume I you will find some beautiful diagrams of electric fields, and shortly beyond the quotation we have just given, Maxwell's reason for presenting these figures. One may suspect that he also took delight in their construction and their elegance.

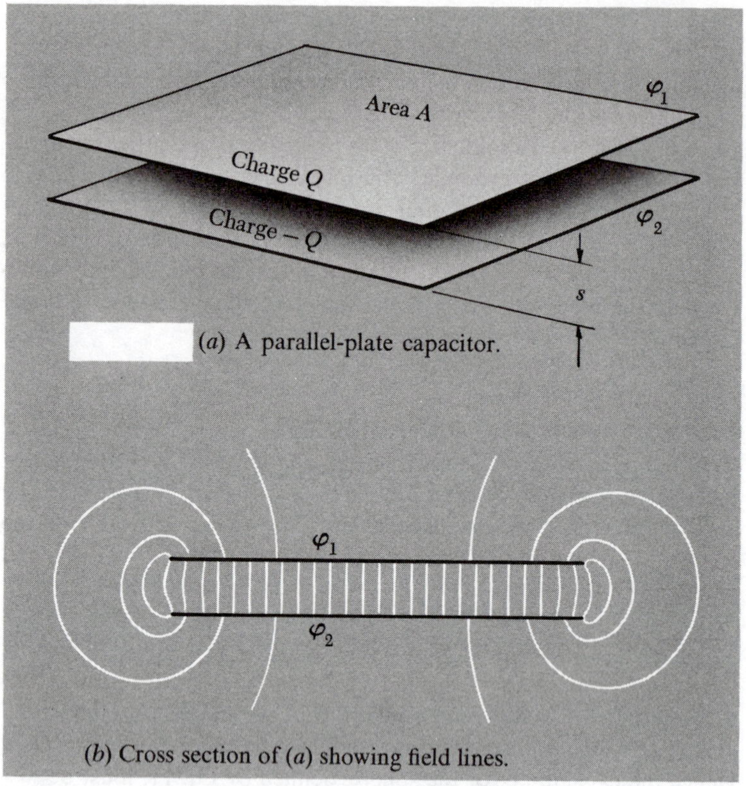

FIGURE 3.12

tively. Here the capacitance is defined as the ratio of the charge Q to the potential difference between the two conductors. The object itself, comprising the two conductors, insulating material to hold the conductors apart, and perhaps electrical terminals or leads, is called a *capacitor*. Most electronic circuits contain numerous capacitors. The parallel-plate capacitor is the simplest example.

Two similar flat conducting plates are arranged parallel to one another, separated by a distance s, as in Fig. 3.12*a*. Let the area of each plate be A and suppose that there is a charge Q on one plate and $-Q$ on the other. φ_1 and φ_2 are the values of the potential at each of the plates. Figure 3.12*b* shows in cross section the field lines in this system. Away from the edge, the field is very nearly uniform in the region between the plates. When it is treated as uniform, its magnitude must be $(\varphi_1 - \varphi_2)/s$. The corresponding density of the surface charge on the inner surface of one of the plates is

$$\sigma = \frac{E}{4\pi} = \frac{\varphi_1 - \varphi_2}{4\pi s} \tag{13}$$

If we may neglect the actual variation of E and therefore of σ which occurs principally near the edge of the plates, we can write a simple expression for the total charge on one plate:

$$Q = A\frac{\varphi_1 - \varphi_2}{4\pi s} \quad \text{(neglecting edge effects)} \qquad (14)$$

We should expect Eq. 14 to be more nearly accurate the smaller the ratio of the plate separation s to the lateral dimension of the plates. Of course, if we were to solve exactly the electrostatic problem, edge and all, for a particular shape of plate, we could replace Eq. 14 by an exact formula. To show how good an approximation Eq. 14 is, there are listed in Fig. 3.13 values of the correction factor f by which the charge Q given in Eq. 14 differs from the exact result, in the case of two conducting disks at various separations. The total charge is always a bit greater than Eq. 14 would predict. That seems reasonable as we look at Fig. 3.12*b*, for there is evidently an extra concentration of charge at the edge, and even some charge on the outer surfaces near the edge.

We are not concerned now with the details of such corrections but with the general properties of a two-conductor system, the *capacitor*. We are interested in the relation between the charge Q on one of the plates and the potential difference between the two plates. For the particular system to which Eq. 14 applies, the quotient $Q/(\varphi_1 - \varphi_2)$ is $A/4\pi s$. Even if this is only approximate, it is clear that the exact formula will depend only on the size and geometrical arrangement of the plates. That is, for a fixed pair of conductors, the ratio of charge to potential difference will be a constant. We call this constant the *capacitance* of the capacitor and denote it usually by C.

$$Q = C(\varphi_1 - \varphi_2) \qquad (15)$$

Thus the capacitance of the parallel-plate capacitor, with edge fields neglected, is given by

$$C = \frac{A \text{ (in cm}^2\text{)}}{4\pi s \text{ (in cm)}} \qquad (16)$$

Often, especially when we are concerned with electrical circuits, we shall want to measure charge in coulombs and potentials in volts. Then the capacitance, C in Eq. 14, will be measured in *farads*. If a capacitor has a capacitance of one *farad*, the charge Q is equal to one *coulomb* when the potential difference between the plates is one *volt*. Figure 3.14 summarizes the formulas for capacitance in both CGS and SI units. Refer to it when in doubt. As usual, the difference stems from a factor $4\pi\epsilon_0$ in any expression involving charge. The farad happens to be a gigantic unit; the capacitance of an isolated sphere the size of the earth is less than a tenth of a farad. But that causes no trouble. We deal on more familiar terms with the *microfarad* (μF),

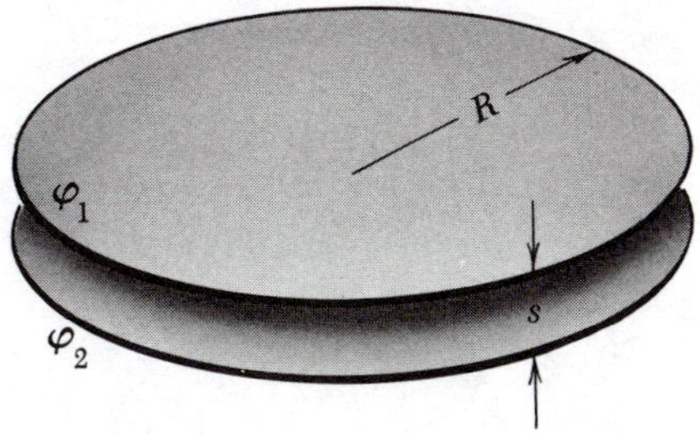

FIGURE 3.13
The true capacitance of parallel circular plates, compared to the prediction of Eq. 14, for various ratios of separation to plate radius. The effect of the edge correction can be represented by writing the charge Q as

$$Q = \frac{A(\phi_1 - \phi_2)}{4\pi s} f$$

For circular plates, the factor f depends on s/R as follows:

s/R	f
0.2	1.286
0.1	1.167
0.05	1.094
0.02	1.042
0.01	1.023

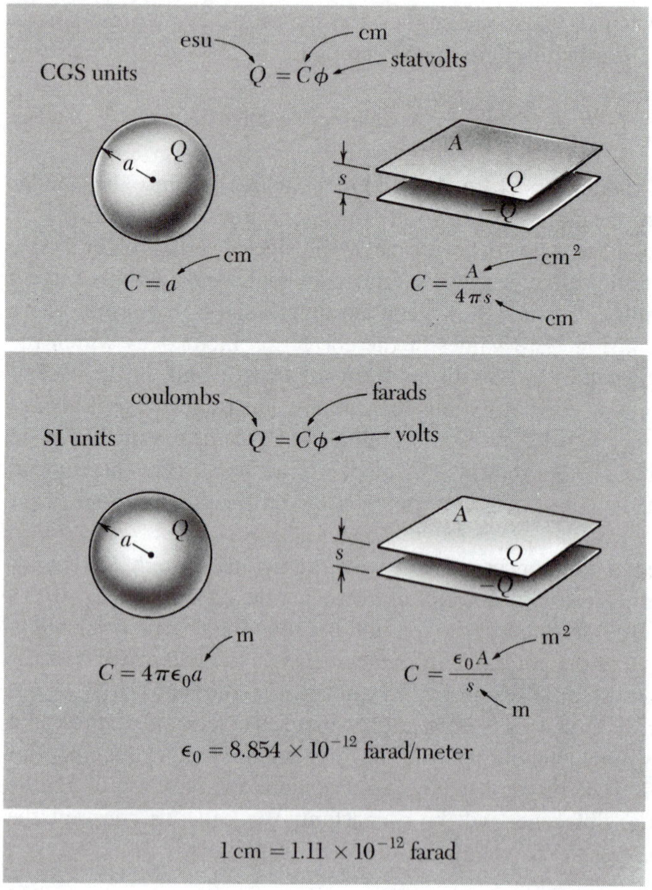

FIGURE 3.14
Summary of units associated with capacitance.

10^{-6} farad, and the *picofarad* (pF), 10^{-12} farad. One picofarad is roughly equivalent to 1 cm. With the farad defined as one coulomb per volt, the dimension of the constant ϵ_0 can be conveniently expressed as farads/meter.

Any pair of conductors, regardless of shape or arrangement, can be considered a capacitor. It just happens that the parallel-plate capacitor is a common arrangement and one for which an approximate calculation of the capacitance is very easy. Figure 3.15 shows two conductors, one inside the other. We can call this arrangement, too, a capacitor. As a practical matter, some mechanical support for the inner conductor would be needed, but that does not concern us. Also, to convey electric charge to or from the conductors we would need leads which are themselves conducting bodies. Since a wire leading out from the inner body, numbered 1, necessarily crosses the space between the conductors, it is bound to cause some perturbation of the electric field in that space. To minimize this we may suppose the lead wires to be extremely thin. Or we might imagine the leads removed before the potentials are determined.

In this system we can distinguish three charges: Q_1, the total charge on the inner conductor; $Q_2^{(i)}$, the amount of charge on the inner surface of the outer conductor; $Q_2^{(o)}$, the charge on the outer surface of the outer conductor. Observe first that $Q_2^{(i)}$ must equal $-Q_1$. We know this because a surface such as S in Fig. 3.15 encloses both these charges and no others and the flux through this surface is zero. The flux is zero because on the surface S, lying, as it does, in the interior of a conductor, the electric field is zero.

Evidently the value of Q_1 will uniquely determine the electric field within the region between the two conductors and thus will determine the difference between their potentials, $\varphi_1 - \varphi_2$. For that reason, if we are considering the two bodies as "plates" of a capacitor, it is only Q_1, or its counterpart $Q_2^{(i)}$, that is involved in determining the capacitance. The capacitance is:

$$C = \frac{Q_1}{\varphi_1 - \varphi_2} \tag{17}$$

$Q_2^{(o)}$, on which φ_2 itself depends, is here irrelevant. In fact, the complete enclosure of one conductor by the other makes the capacitance independent of everything outside.

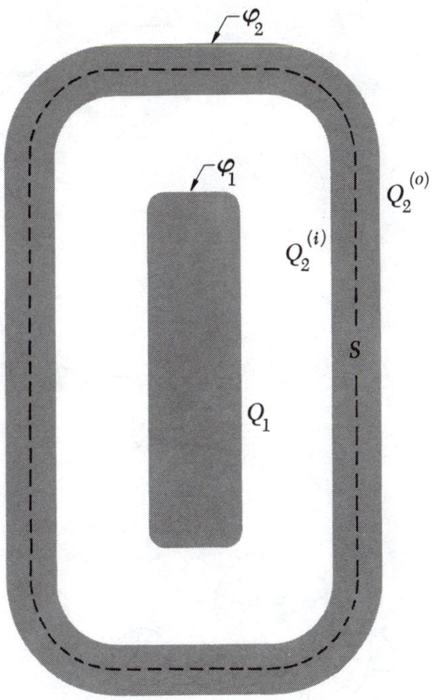

FIGURE 3.15
A capacitor in which one conductor is enclosed by the other.

POTENTIALS AND CHARGES ON SEVERAL CONDUCTORS

3.6 We have been skirting the edge of a more general problem, the relations among the charges and potentials of any number of conductors of some given configuration. The two-conductor capacitor is just a special case. It may surprise you that anything useful can be said about the general case. In tackling it, about all we can use is the uniqueness theorem and the superposition principle. To have something definite in mind, consider three separate conductors, all enclosed by a conducting shell, as in Fig. 3.16. The potential of this shell we may choose to be zero; with respect to this reference the potentials of the three conductors, for some particular state of the system, are φ_1, φ_2, and φ_3. The uniqueness theorem guarantees that, with φ_1, φ_2, and φ_3 given, the electric field is determined throughout the system. It follows that the charges Q_1, Q_2, and Q_3 on the individual conductors are likewise uniquely determined.

We need not keep account of the charge on the inner surface of the surrounding shell, since it will always be $-(Q_1 + Q_2 + Q_3)$. If you prefer, you can let "infinity" take over the role of this shell, imagining the shell to expand outward without limit. We have kept it in the picture because it makes the process of charge transfer easier to follow, for some people, if we have something to connect to.

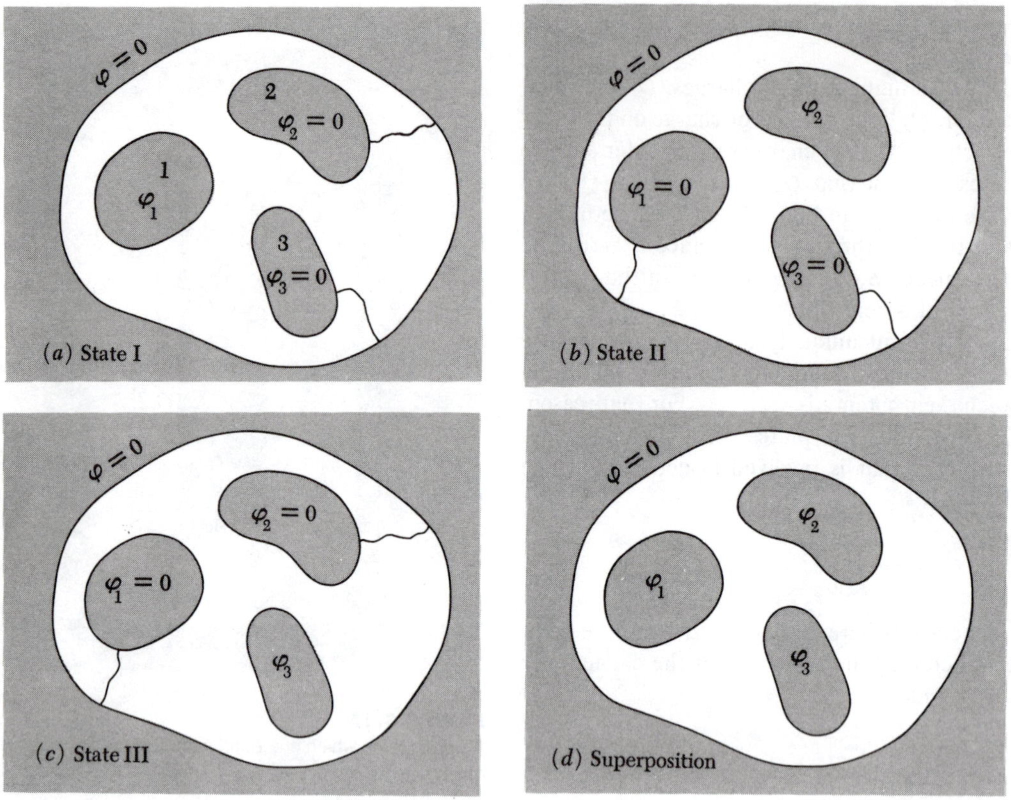

FIGURE 3.16
A general state of this system can be analyzed as the superposition (*d*) of three states (*a–c*) in each of which all conductors but one are at zero potential.

Among the possible states of this system are ones with φ_2 and φ_3 both zero. We could enforce this condition by connecting conductors 2 and 3 to the zero-potential shell, as indicated in Fig. 3.16*a*. As before, we may suppose the connecting wires are so thin that any charge residing on them is negligible. Of course, we really do not care how the specified condition is brought about. In such a state, which we shall call state I, the electric field in the whole system and the charge on every conductor is determined uniquely by the value of φ_1. Moreover, if φ_1 were doubled, that would imply a doubling of the field strength everywhere, and hence a doubling of each of the charges Q_1, Q_2, and Q_3. That is, with $\varphi_2 = \varphi_3 = 0$, each of the three charges must be proportional to φ_1. Stated mathematically:

$$\left.\begin{array}{l}\textit{State I}\\ \varphi_2 = \varphi_3 = 0\end{array}\right\} \quad Q_1 = C_{11}\varphi_1; \quad Q_2 = C_{21}\varphi_1; \quad Q_3 = C_{31}\varphi_1 \tag{18}$$

The three constants, C_{11}, C_{21}, and C_{31}, can depend only on the shape and arrangement of the conducting bodies.

In just the same way we could analyze states in which φ_1 and φ_3 are zero, calling such a condition state *II* (Fig. 3.16*b*). Again, we must first find a linear relation between the only nonzero potential, φ_2 in this case, and the various charges:

$$\left.\begin{matrix}\textit{State II}\\ \varphi_1 = \varphi_3 = 0\end{matrix}\right\} \quad Q_1 = C_{12}\varphi_2; \quad Q_2 = C_{22}\varphi_2; \quad Q_3 = C_{32}\varphi_2 \tag{19}$$

Finally, when φ_1 and φ_2 are held at zero, the field and the charges are proportional to φ_3:

$$\left.\begin{matrix}\textit{State III}\\ \varphi_1 = \varphi_2 = 0\end{matrix}\right\} \quad Q_1 = C_{13}\varphi_3; \quad Q_2 = C_{23}\varphi_3; \quad Q_3 = C_{33}\varphi_3 \tag{20}$$

Now the superposition of three states like *I*, *II*, and *III* is also a possible state. The electric field at any point is the vector sum of the electric fields at that point in the three cases, while the charge on a conductor is the sum of the charges it carried in the three cases. In this new state the potentials are φ_1, φ_2, and φ_3, none of them necessarily zero. In short, we have a completely general state. The relation connecting charges and potentials is obtained simply by adding Eqs. 18 through 20:

$$\begin{aligned} Q_1 &= C_{11}\varphi_1 + C_{12}\varphi_2 + C_{13}\varphi_3 \\ Q_2 &= C_{21}\varphi_1 + C_{22}\varphi_2 + C_{23}\varphi_3 \\ Q_3 &= C_{31}\varphi_1 + C_{32}\varphi_2 + C_{33}\varphi_3 \end{aligned} \tag{21}$$

It appears that the electrical behavior of this system is characterized by the nine constants $C_{11}, C_{12}, \ldots, C_{33}$. In fact only six constants are necessary, for it can be proved that in *any* system $C_{12} = C_{21}$, $C_{13} = C_{31}$, and $C_{23} = C_{32}$. Why this should be so is not obvious. Problem 3.27 will suggest a proof based on conservation of energy, but for that purpose you will need an idea developed in Section 3.7. The C's in Eqs. 21 are called the *coefficients of capacitance.* It is clear that our argument would extend to any number of conductors.

A set of equations like (21) can be solved for the φ's in terms of the Q's. That is, there is an equivalent set of linear relations of the form:

$$\begin{aligned} \varphi_1 &= P_{11}Q_1 + P_{12}Q_2 + P_{13}Q_3 \\ \varphi_2 &= P_{21}Q_1 + P_{22}Q_2 + P_{23}Q_3 \\ \varphi_3 &= P_{31}Q_1 + P_{32}Q_2 + P_{33}Q_3 \end{aligned} \tag{22}$$

The P's are called the *potential coefficients;* they could be computed from the C's, or vice versa.

We have here a simple example of the kind of relation we can expect to govern any *linear* physical system. Such relations turn up in the study of mechanical structures (connecting the strains with the loads), in the analysis of electrical circuits (connecting voltages and currents), and generally speaking, wherever the superposition principle can be applied.

ENERGY STORED IN A CAPACITOR

3.7 Consider a capacitor of capacitance C, with a potential difference φ_{12} between the plates. The charge Q is equal to $C\varphi_{12}$. There is a charge Q on one plate and $-Q$ on the other. Suppose we *increase* the charge from Q to $Q + dQ$ by transporting a positive charge dQ from the negative to the positive plate, working against the potential difference φ_{12}. The work that has to be done is $dW = \varphi_{12}\, dQ = Q\, dQ/C$. Therefore to charge the capacitor starting from the uncharged state to some final charge Q_f requires the work

$$W = \frac{1}{C}\int_{Q=0}^{Q_f} Q\, dQ = \frac{Q_f^2}{2C} \tag{23}$$

This is the energy U which is "stored" in the capacitor. It can also be expressed by

$$U = \tfrac{1}{2}C\varphi_{12}^2 \tag{24}$$

For the parallel-plate capacitor with plate area A and separation s we found the capacitance $C = A/4\pi s$ and the electric field $E = \varphi_{12}/s$. Hence Eq. 24 is also equivalent to

$$U = \frac{1}{2}\left(\frac{A}{4\pi s}\right)(Es)^2 = \frac{E^2}{8\pi}\cdot As = \frac{E^2}{8\pi}\cdot \text{volume} \tag{25}$$

This agrees with our general formula, Eq. 38 in Chapter 1 for the energy stored in an electric field.†

Equation 24 applies as well to the isolated charged conductor, which can be thought of as the inner plate of a capacitor, enclosed by an outer conductor of infinite size and potential zero. For the isolated sphere of radius a, we found $C = a$, so that $U = \tfrac{1}{2}a\phi^2$ or, equivalently, $U = \tfrac{1}{2}Q^2/a$, agreeing with our earlier calculation of the energy stored in the electric field of the charged sphere.

The oppositely charged plates of a capacitor will attract one another; some mechanical force will be required to hold them apart. This is obvious in the case of the parallel-plate capacitor, for which we could easily calculate the force on the surface charge. But we can make a more general statement based on Eq. 23, which related stored

†All this applies to the *vacuum capacitor* consisting of conductors with empty space in between. As you know from the laboratory, most capacitors used in electric circuits are filled with an insulator or "dielectric." We are going to study the effect that has in Chapter 10.

energy to charge Q and capacitance C. Suppose that C depends in some manner on a linear coordinate x which measures the displacement of one "plate" of a capacitor, which might be a conductor of any shape, with respect to the other. Let F be the magnitude of the force that must be applied to each plate to overcome their attraction and keep x constant. Now imagine the distance x is increased by an increment Δx with Q remaining constant and one plate fixed. The external force F on the other plate does work $F\,\Delta x$ and, if energy is to be conserved, this must appear as an increase in the stored energy $Q^2/2C$. That increase at constant Q is

$$\Delta U = \frac{dU}{dx}\Delta x = \frac{Q^2}{2}\frac{d}{dx}\left(\frac{1}{C}\right)\Delta x \tag{26}$$

Equating this to the work $F\,\Delta x$ we find

$$F = \frac{Q^2}{2}\frac{d}{dx}\left(\frac{1}{C}\right) \tag{27}$$

OTHER VIEWS OF THE BOUNDARY-VALUE PROBLEM

3.8 It would be wrong to leave the impression that there are no general methods for dealing with the Laplacian boundary-value problem. Although we cannot pursue this question much further, we shall mention some useful and interesting approaches which you are likely to meet in future study of physics or applied mathematics.

First, an elegant method of analysis, called conformal mapping, is based on the theory of functions of a complex variable. Unfortunately it applies only to two-dimensional systems. These are systems in which φ depends only on x and y, for example, all conducting boundaries being cylinders (in the general sense) with elements running parallel to z. Laplace's equation then reduces to

$$\frac{\partial^2\varphi}{\partial x^2} + \frac{\partial^2\varphi}{\partial y^2} = 0 \tag{28}$$

with boundary values specified on some lines or curves in the xy plane. Many systems of practical interest are like this or sufficiently like this to make the method useful, quite apart from its intrinsic mathematical interest. For instance, the exact solution for the potential around two long parallel strips is easily obtained by the method of conformal mapping. The field lines and equipotentials are shown in a cross-section plane in Fig. 3.17. This provides us with the edge field for any parallel-plate capacitor in which the edge is long compared with the gap. The field shown in Fig. 3.12*b* was copied from such a solution. You will be able to apply this method after you have studied in more advanced mathematics functions of a complex variable.

Second, we mention a numerical method for finding approxi-

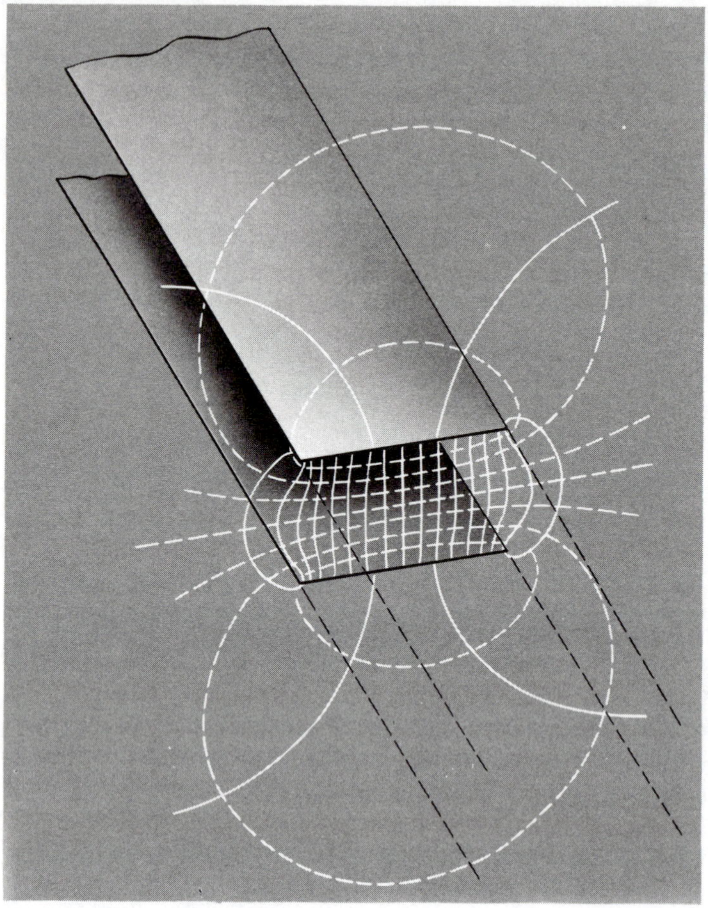

FIGURE 3.17
Field lines and equipotentials for two infinitely long conducting strips.

mate solutions of the electrostatic potential with given boundary values. Surprisingly simple and almost universally applicable, this method is based on that special property of harmonic functions with which we are already familiar: The value of the function at a point is equal to its average over the neighborhood of the point. In this method the potential function φ is represented by values at an array of discrete points only, including discrete points on the boundaries. The values at nonboundary points are then adjusted until each value is equal to the average of the neighboring values. In principle one could do this by solving a large number of simultaneous linear equations—as many as there are interior points. But an approximate solution can be obtained by the following procedure, called a *relaxation method*. Start with the boundary points of the array, or grid, set at the values prescribed. Assign starting values arbitrarily to the interior points. Now visit, in some order, all the interior points. At each point reset its value to the average of the values at the four (for a square grid) adjacent grid points. Repeat again and again, until all the changes made in the course of one sweep over the network of interior points are acceptably small. If you want to see how this method works, Problems 3.30 and 3.31 will provide an introduction. Whether convergence of the relaxation process can be ensured, or even hastened, and whether a relaxation method or direct solution of the simultaneous equations is the better strategy for a given problem are questions in applied mathematics that we cannot go into here. It is the high-speed computer, of course, that makes both methods feasible.

PROBLEMS

3.1 A spherical conductor A contains two spherical cavities. The total charge on the conductor itself is zero. However, there is a point charge q_b at the center of one cavity and q_c at the center of the other. A considerable distance r away is another charge q_d. What force acts on each of the four objects, A, q_b, q_c, q_d? Which answers, if any, are only approximate, and depend on r being relatively large?

3.2 What is wrong with the idea of a gravity screen, something that will "block" gravity the way a metal sheet seems to "block" the electric field. Think about the difference between the gravitational source and electrical sources. Note that the walls of the box in Fig. 3.6 do not block the field of the outside sources but merely allow the surface charges to set up a compensating field. Why can't something of this sort be contrived for gravity? What would you need to accomplish it?

3.3 In the field of the point charge over the plane (Fig. 3.9), if you follow a field line that starts out from the point charge in a horizontal direction, that is, parallel to the plane, where does it meet the surface of the conductor? (You'll need Gauss's law and a simple integration.)

3.4 A positive point charge Q is fixed 10 cm above a horizontal conducting plane. An equal negative charge $-Q$ is to be located somewhere along the perpendicular dropped from Q to the plane. Where can $-Q$ be placed so that the total force on it will be zero?

Ans. $y = 3.06$ cm.

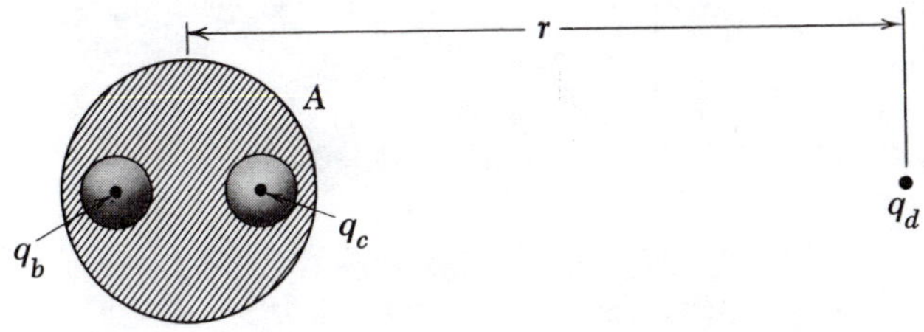

PROBLEM 3.1

3.5 A charge Q is located h cm above a conducting plane, just as in Fig. 3.8*a*. Asked to predict the amount of work that would have to be done to move this charge out to infinite distance from the plane, one student says that it is the same as the work required to separate to infinite distance two charges Q and $-Q$ which are initially $2h$ cm apart, hence $W = Q^2/2h$. Another student calculates the force that acts on the charge as it is being moved and integrates $F\,dx$, but gets a different answer. What did the second student get, and who is right?

3.6 By solving the problem of the point charge and the plane conductor we have, in effect, solved every problem that can be constructed from it by superposition. For instance, suppose we have a straight wire 200 meters long uniformly charged with 10^3 esu per centimeter of length, running parallel to the earth at a height of 5 meters. What is the field strength at the surface of the earth, immediately below the wire? (For steady fields the earth behaves like a good conductor.) What is the electrical force acting on the wire?

3.7 The two metal spheres in (a) are connected by a wire; the total charge is zero. In (b) two oppositely charged conducting spheres have been brought into the positions shown, inducing charges of opposite sign in A and in B. If now C and D are connected by a wire as in (c), it could be argued that something like the charge distribution in (b) ought to persist, each charge concentration being held in place by the attraction of the opposite charge nearby. What about that? Can you prove it won't happen?

3.8 Three conducting plates are placed parallel to one another as shown. The outer plates are connected by a wire. The inner plate is

PROBLEM 3.7

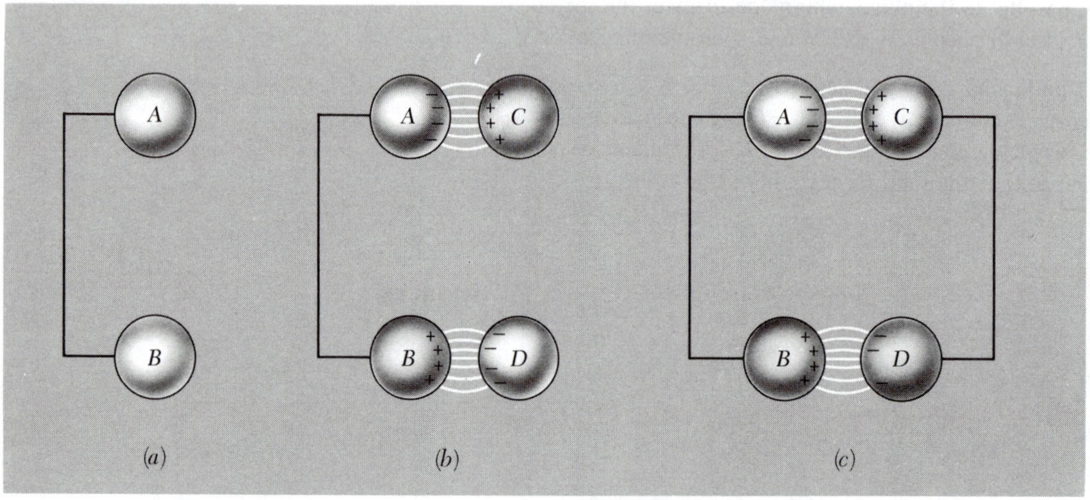

isolated and carries a charge amounting to 10 esu per square centimeter of plate. In what proportion must this charge divide itself into a surface charge σ_1 on one face of the inner plate and a surface charge σ_2 on the other side of the same plate?

PROBLEM 3.8

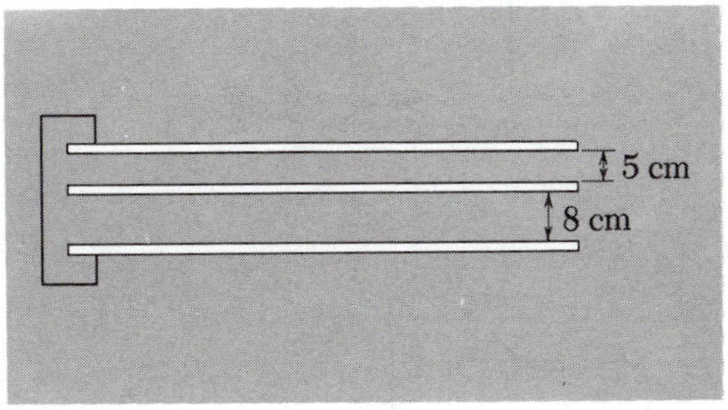

3.9 Locate two charges q each and two charges $-q$ each on the corners of a square, with like charges diagonally opposite one another. Show that there are two equipotential surfaces that are planes. In this way obtain, and sketch qualitatively, the field of a single point charge located symmetrically in the inside corner formed by bending a metal sheet through a right angle. Which configurations of conducting planes and point charges can be solved this way and which can't? How about a point charge located on the bisector of a 120° dihedral angle between two conducting planes?

3.10 What is the capacitance C of a capacitor that consists of two concentric spherical metal shells? The inner radius of the outer shell is a; the outer radius of the inner shell is b. Check your result by considering the limiting case with the gap between the conductors, $a - b$, much smaller than b. In that limit the formula for the capacitance of the flat parallel-plate capacitor ought to be applicable.

3.11 A 100-pF capacitor is charged to 100 volts. After the charging battery is disconnected, the capacitor is connected in parallel to another capacitor. If the final voltage is 30 volts, what is the capacitance of the second capacitor. How much energy was lost, and what happened to it?

3.12 Two aluminized optical flats 15 cm in diameter are separated by a gap of 0.04 mm, forming a capacitor. What is the capacitance in pF?

3.13 Make a rough estimate of the capacitance of an isolated human body. *Hint:* It must lie somewhere between that of an inscribed sphere and that of a circumscribed sphere. By shuffling over a nylon rug on a dry winter day you can easily charge yourself up to a couple of kilovolts—as shown by the length of the spark when your hand comes too close to a grounded conductor. How much energy would be dissipated in such a spark?

3.14 Given that the capacitance of an isolated conducting disk of radius a is $2a/\pi$, what is the energy stored in the electric field of such a disk when the net charge on the disk is Q? Compare this with the energy in the field of a nonconducting disk of the same radius which has an equal charge Q distributed with uniform density over its surface. (See Problem 2.27.) Which ought to be larger? Why?

3.15 Two coaxial aluminum tubes are 30 cm long. The outer diam-

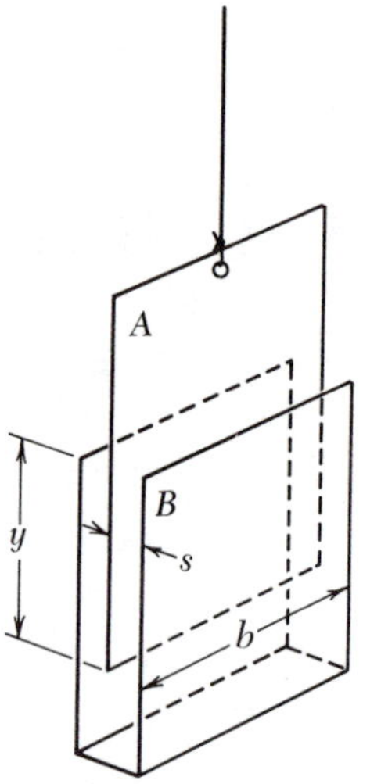

PROBLEM 3.18

eter of the inner tube is 3 cm, the inner diameter of the outer tube is 4 cm. When these are connected to a 45-volt battery, how much energy is stored in the electric field between the tubes?

3.16 Calculate the electrical force which acts on one plate of a parallel-plate capacitor. The potential difference between the plates is 10 statvolts, and the plates are squares 20 cm on a side with a separation of 3 cm. If the plates are insulated so the charge cannot change, how much external work could be done by letting the plates come together? Does this equal the energy that was initially stored in the electric field?

3.17 We want to design a spherical vacuum capacitor with a given radius a for the outer sphere, which will be able to store the greatest amount of electrical energy subject to the constraint that the electric field strength at the surface of the inner sphere may not exceed E_0. What radius b should be chosen for the inner spherical conductor, and how much energy can be stored?

Ans. $\frac{3}{4}a$; $(\frac{27}{512})a^3E_0^2$.

3.18 The aluminum sheet A is suspended by an insulating thread between the surfaces formed by the bent aluminum sheet B. The sheets are oppositely charged; the difference of potential, in statvolts, is V. This causes a force F, in addition to the weight of A, pulling A downward. If we can measure F and know the various dimensions, we should be able to infer V. As an application of Eq. 27, work out a formula giving V in terms of F and the relevant dimensions.

PROBLEM 3.19

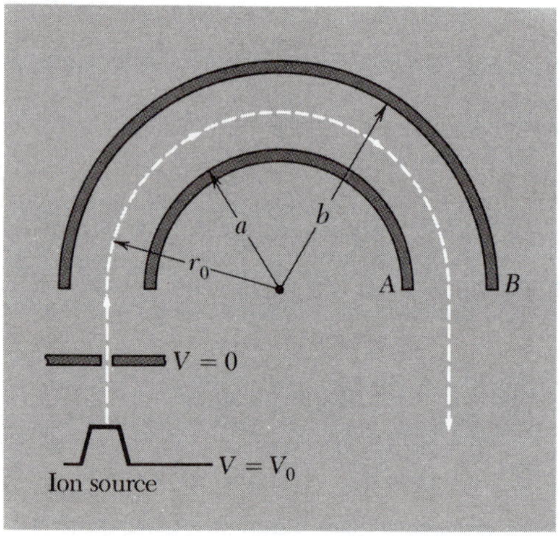

3.19 In the apparatus shown, ions are accelerated through a potential difference V_0 and then enter the space between the semicylindrical electrodes A and B. Show that an ion will follow the semicircular path of radius r_0 if the potentials of the outer and inner electrodes are maintained, respectively, at $2V_0 \ln(b/r_0)$ and $2V_0 \ln(a/r_0)$. (The cylindrical electrodes A and B are assumed to be long, in the direction perpendicular to the diagram, compared with the space between them.)

3.20 Here is the exact formula for the capacitance C of a conductor in the form of a prolate spheroid of length $2a$ and diameter $2b$.

$$C = \frac{2a\epsilon}{\ln\left(\dfrac{1+\epsilon}{1-\epsilon}\right)} \qquad \text{where } \epsilon = \sqrt{1 - \frac{b^2}{a^2}}$$

First verify that the formula reduces to the correct expression for the capacitance of a sphere if $b = a$. Now imagine that the spheroid is a charged water drop. If this drop is deformed at constant volume and constant charge Q from a sphere to a prolate spheroid, will the energy stored in the electric field increase or decrease? (The volume of the prolate spheroid is proportional to ab^2.)

3.21 Imagine the xy plane, the xz plane, and the yz plane all made of metal and soldered together at the intersections. A single point charge Q is located a distance d from each of the planes. Describe by a sketch the configuration of image charges you need to satisfy the boundary conditions. What is the direction and magnitude of the force that acts on the charge Q?

3.22 If you worked Problem 2.11, you should be able to derive from that result the formula given in Problem 3.20 for the capacitance of an isolated conductor of prolate spheroidal shape.

3.23 (*a*) Find the capacitance of a capacitor that consists of two coaxial cylinders, of radii a and b, and length L. Assume $L \gg b - a$, so that end corrections may be neglected. Check your results by showing that, if the gap between the cylinders, $b - a$, is very small compared with the radius, your formula reduces to one that could have been obtained by using the formula for the parallel-plate capacitor.

(*b*) A cylinder of 2.00-inch outer diameter hangs, with its axis vertical, from one arm of a beam balance. The lower portion of the hanging cylinder is surrounded by a stationary cylinder, coaxial, with inner diameter 3.00 inches. Calculate the magnitude of the force tending to pull the hanging cylinder further down when the potential difference between the two cylinders is 5 kilovolts.

3.24 Two parallel plates are connected by a wire so that they remain at the same potential. Let one plate coincide with the xz plane and the other with the plane $y = s$. The distance s between the plates is much smaller than the lateral dimensions of the plates. A point charge Q is located between the plates at $y = b$ (see figure). What is the magnitude of the total surface charge on the inner surface of each plate? The total surface charge on the inner surface of both plates must of course be $-Q$ (why?), and we can guess that a larger fraction of it will be found on the nearer plate. If the charge were very close to the left plate, $b \ll s$, the presence of the plate on the right couldn't make much difference. However, we want to know exactly how the charge divides. If you try to use an image method you will discover that you need an infinite chain of images, rather like the images you see in a barbershop with mirrors on both walls. It is not easy to calculate the resultant field at any point on one of the surfaces. Nevertheless, the question we asked can be answered by a very simple calculation based on superposition. (*Hint:* Adding another charge Q anywhere on the plane would just double the surface charge on each plate. In fact the total surface charge induced by any number of charges is independent of their position on the plane. If only we had a

PROBLEM 3.24

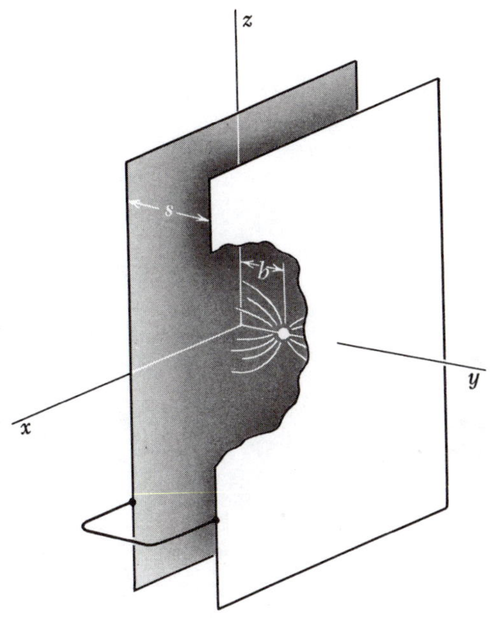

sheet of uniform charge on this plane the electric fields would be simple, and we could use Gauss's law. Take it from there.)

3.25 (*a*) Show that the square of a potential difference $(\phi_2 - \phi_1)^2$ has the same dimensions as force. This tells us that the electrostatic forces between bodies will largely be determined, as to order of magnitude, by the potential differences involved. Dimensions will enter only in ratios, and there may be some constants like 4π. What is the order of magnitude of force you expect with 1 statvolt potential difference between something and something else?

(*b*) Practically achievable potential differences are rather severely limited, for reasons having to do with the structure of matter. The highest man-made difference of electric potential is about 10^7 volts, achieved by a Van de Graaff electrostatic generator operating under high pressure. (Billion-volt accelerators do not involve potential differences that large.) How many pounds force are you likely to find associated with a "square megavolt"? These considerations may suggest why electrostatic motors have not found much application.

PROBLEM 3.26

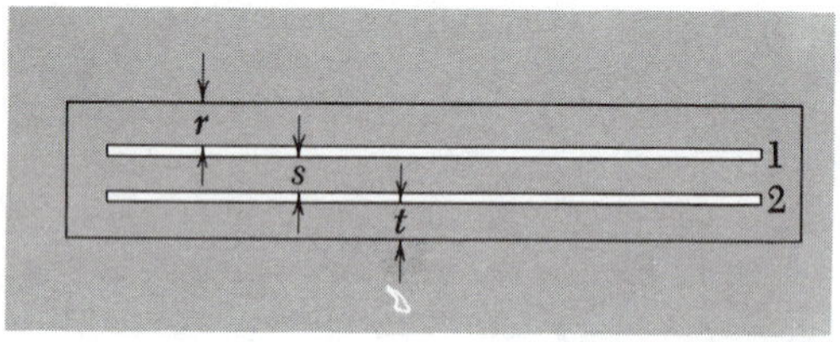

3.26 The figure shows in cross section a flat metal box in which there are two flat plates, 1 and 2, each of area A. The various distances separating the plates from each other and from the top and bottom of the box, labeled r, s, and t in the figure, are to be assumed small compared with the width and length of the plates, so that it will be a good approximation to neglect the edge fields in estimating the charges on the plates. In this approximation, work out the capacitance coefficients, C_{11}, C_{22}, and C_{12}. You might also work out C_{21} directly, to see that it comes out equal to C_{12} as asserted by the general theorem discussed in Problem 3.27.

3.27 Here are some suggestions which should enable you to construct a proof that C_{12} must always equal C_{21}. We know that, when an element of charge dQ is transferred from zero potential to a conductor at potential ϕ, some external agency has to supply an amount of energy $\phi\, dQ$. Consider a two-conductor system in which the two conductors have been charged so that their potentials are, respectively, ϕ_{1f} and ϕ_{2f} (f for "final"). This condition might have been brought about, starting from a state with all charges and potentials zero, in many different ways. Two possible ways are of particular interest:

(*a*) Keep ϕ_2 at zero while raising ϕ_1 gradually from zero to ϕ_{1f}; then raise ϕ_2 from zero to ϕ_{2f} while holding ϕ_1 constant at ϕ_{1f}.

(*b*) Carry out a similar program with the roles of 1 and 2 exchanged, that is, raise ϕ_2 from zero to ϕ_{2f} first, and so on.

Compute the total work done by external agencies, for each of the two charging programs. Then complete the argument.

3.28 A typical two-dimensional boundary-value problem is that of two parallel circular conducting cylinders, such as two metal pipes, of infinite length and at different potentials. These two-dimensional problems happen to be much more tractable than three-dimensional problems, mathematically. In fact, the key to all problems of the "two-pipe" class is given by the field around two parallel line charges of equal and opposite linear density. All equipotential surfaces in this field are circular cylinders! And all field lines are circular too. See if you can prove this. It is easiest to work with the potential, but you must note that one cannot set the potential zero at infinity in a two-dimensional system. Let zero potential be at the line midway between the two line charges, that is, at the origin in the cross-sectional diagram. The potential at any point is the sum of the potentials calculated for each line charge separately. This should lead you quickly to the discovery that the potential is simply proportional to ln (r_2/r_1) and is therefore constant on a curve traced by a point whose distances from two points are in a constant ratio. Make a sketch showing some of the equipotentials.

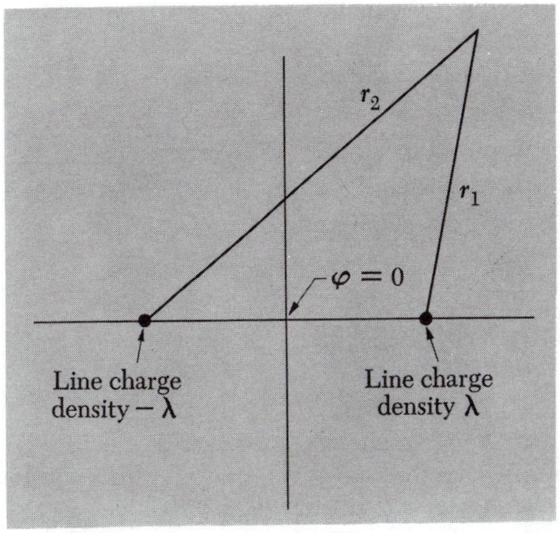

PROBLEM 3.28

3.29 Let $\varphi(x, y, z)$ be any function that can be expanded in a power series around a point (x_0, y_0, z_0). Write a Taylor series expansion for the value of φ at each of the six points $(x_0 + \delta, y_0, z_0)$, $(x_0 - \delta, y_0, z_0)$, $(x_0, y_0 + \delta, z_0)$, $(x_0, y_0 - \delta, z_0)$, $(x_0, y_0, z_0 + \delta)$, $(x_0, y_0, z_0 - \delta)$, which symmetrically surround the point (x_0, y_0, z_0) at a distance δ. Show that, if φ satisfies Laplace's equation, the average of these six values is equal to $\varphi(x_0, y_0, z_0)$ through terms of the third order in δ.

3.30 Here's how to solve Laplace's equation approximately, for given boundary values, using nothing but arithmetic. The method is the relaxation method mentioned in Section 3.8, and it is based on the result of Problem 3.29. For simplicity we take a two-dimensional example. In the figure there are two square equipotential boundaries, one inside the other. This might be a cross section through a capacitor made of two sizes of square metal tubing. The problem is to find, for an array of discrete points, numbers which will be a good approximation to the values at those points of the exact two-dimensional potential function $\phi(x, y)$. For this exercise, we'll make the array rather coarse, to keep the labor within bounds. Let us assign, arbitrarily, potential 100 to the inner boundary and zero to the outer. All points on these boundaries retain those values. You could start with any values at the interior points, but time will be saved by a little judicious guesswork. We know the correct values must lie between 0 and 100, and we expect that points closer to the inner boundary will have higher values than those closer to the outer boundary. Some reasonable starting values are suggested in the figure. Obviously, you should take advantage of the symmetry of the configuration: Only seven different interior values need to be computed. Now you simply go over these seven interior lattice points in some systematic manner,

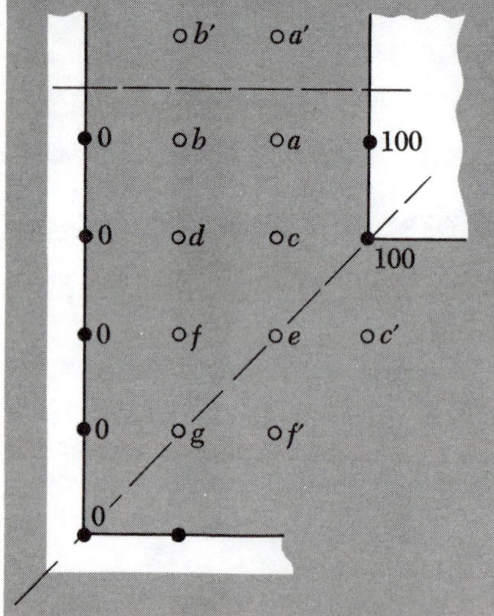

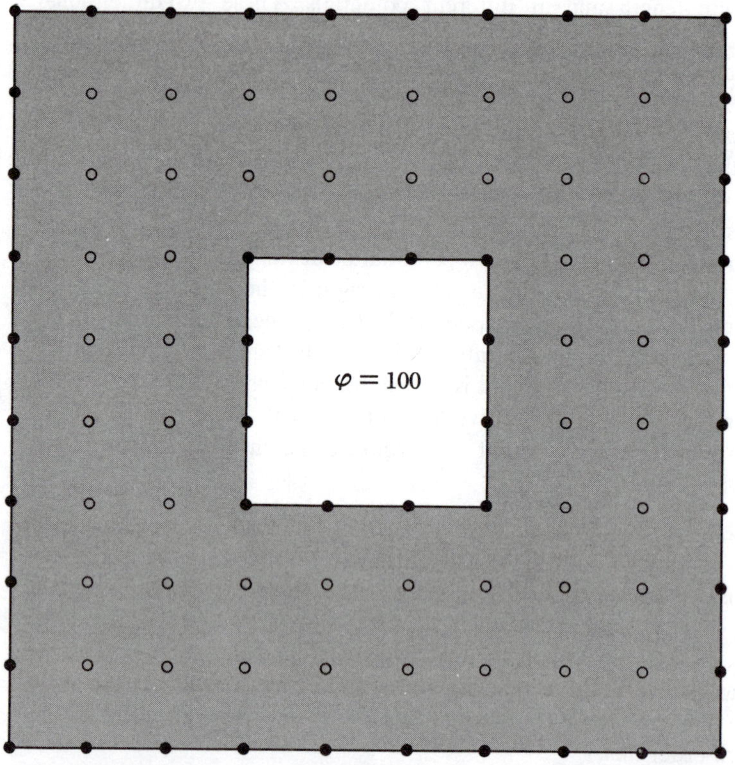

PROBLEM 3.30
Replace value at an interior point by ¼ × sum of its four neighbors: $c \rightarrow \frac{1}{4}(100 + a + d + e)$; keep $a' = a$, $b' = b$, $c' = c$, and $f' = f$. Suggested starting values:

$a = 50$	$e = 50$
$b = 25$	$f = 25$
$c = 50$	$g = 25$
$d = 25$	

replacing the value at each interior point by the average of its four neighbors. Repeat until all changes resulting from a sweep over the array are acceptably small. For this exercise, let us agree that it will be time to quit when no change larger in absolute magnitude than one unit occurs in the course of the sweep. The relaxation of the values toward an eventually unchanging distribution is closely related to the physical phenomenon of *diffusion.* If you start with much too high a value at one point, it will "spread" to its nearest neighbors, then to its next nearest neighbors, and so on, until the bump is smoothed out. Enter your final values on the array, and sketch the approximate course two equipotentials, for $\phi = 25$ and $\phi = 50$, would have in the actual continuous $\phi(x, y)$.

3.31 The relaxation method is clearly well adapted to machine computation. Write a program that will deal with the concentric square boundary problem on a finer mesh—say, a grid with four times as many points and half the spacing. It might be a good idea to utilize a coarse-mesh solution in assigning starting values for the relaxation on the finer mesh.

3.32 A capacitor consists of two concentric spherical shells. Call the inner shell, of radius a, conductor 1, and the outer shell, of radius b, conductor 2. For this two-conductor system, find C_{11}, C_{22}, and C_{12}.

Ans. $C_{11} = ab/(b - a)$; $C_{22} = b^2/(b - a)$; $C_{21} = -ab/(b - a)$.

4

ELECTRIC CURRENTS

ELECTRIC CURRENT AND CURRENT DENSITY

4.1 An electric current is charge in motion. The carriers of the charge can be physical particles like electrons or protons, which may or may not be attached to larger objects, atoms or molecules. Here we are not concerned with the nature of the charge carriers but only with the net transport of electric charge their motion causes. The electric current in a wire is the amount of charge passing a fixed mark on the wire in unit time. In CGS units current will be expressed in esu/sec. The SI unit is coulombs/sec, or *amperes* (amps). A current of 1 ampere is the same as a current of 2.998×10^9 esu/sec, which is equivalent to 6.24×10^{18} elementary electronic charges per second.

It is the net charge transport that counts, with due regard to sign. Negative charge moving east is equivalent to positive charge moving west. Water flowing through a hose could be said to involve the transport of an immense amount of charge—about 3×10^{23} electrons per gram of water! But since an equal number of protons move along with the electrons (every water molecule contains 10 of each), the electric current is zero. On the other hand, if you were to charge negatively a nylon thread and pull it steadily through a nonconducting tube, that would constitute an electric current, in the direction opposite the motion of the thread.

We have been considering current along a well-defined path, like a wire. If the current is *steady*—that is, unchanging in time—it must be the same at every point along the wire, just as with steady traffic the same number of cars must pass, per hour, different points along an unbranching road.

A more general kind of current, or charge transport, involves charge carriers moving around in three-dimensional space. To describe this we need the concept of *current density*. We have to consider average quantities, for charge carriers are discrete particles. We must suppose, as we did in defining the charge density ρ, that our scale of distances is such that any small region we wish to average over contains very many particles of any class we are concerned with.

Consider first a special situation in which there are n particles per cm^3, on the average, all moving with the same vector velocity $\mathbf{u}$ and carrying the same charge q. Imagine a small frame of area $\mathbf{a}$ fixed in some orientation, as in Fig. 4.1*a*. How many particles pass through the frame in a time interval Δt? If Δt begins the instant shown in Fig. 4.1*a* and *b*, the particles destined to pass through the frame in the next Δt sec will be just those now located within the oblique prism in Fig. 4.1*b*. This prism has the frame area as its base and an edge length $u \Delta t$, which is the distance any particle will travel in a time Δt. Particles outside this prism will either miss the window or fail to reach it. The volume of the prism is the product *base* $\times$ *altitude,* or $au\, \Delta t \cos\theta$,

FIGURE 4.1

(*a*) A swarm of charged particles all moving with the same velocity *u*. The frame has area *a*. The particles which will pass through the frame in the next Δt sec are those now contained in the oblique prism (*b*). The prism has base area $\mathbf{a}$ and altitude $u\,\Delta t \cos\theta$, hence its volume is $au\,\Delta t \cos\theta$ or $\mathbf{a} \cdot \mathbf{u}\,\Delta t$.

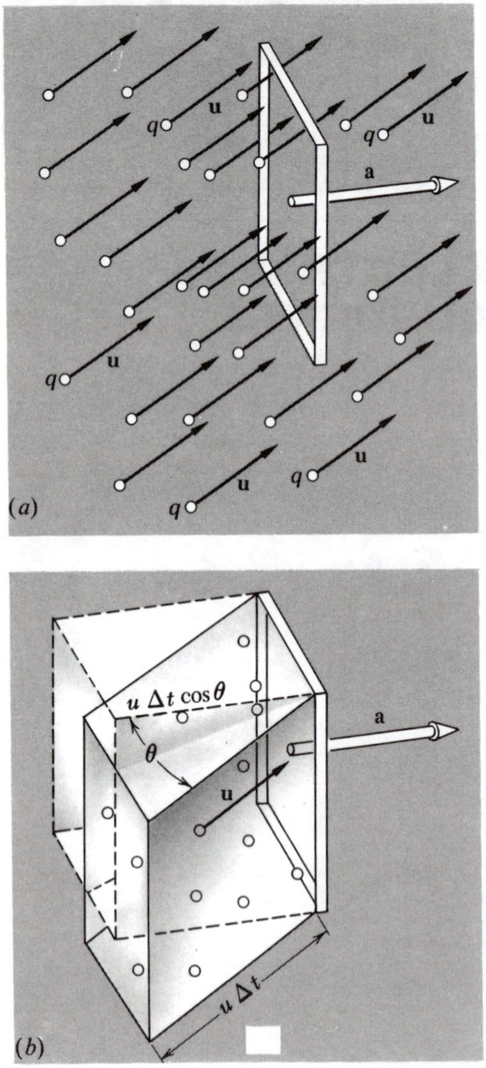

which can be written $\mathbf{a} \cdot \mathbf{u}\,\Delta t$. On the average, the number of particles found in such a volume will be $n\mathbf{a} \cdot \mathbf{u}\Delta t$. Hence the average *rate* at which charge is passing through the frame, that is, the current through the frame, which we shall call I_a, is

$$I_a = \frac{q(n\mathbf{a} \cdot \mathbf{u}\,\Delta t)}{\Delta t} = nq\mathbf{a} \cdot \mathbf{u} \tag{1}$$

Suppose we had many classes of particles in the swarm, differing in charge q, in velocity vector $\mathbf{u}$, or in both. Each would make its own contribution to the current. Let us tag each kind by a subscript k. The kth class has charge q_k on each particle, moves with velocity vector $\mathbf{u}_k$, and is present with an average population density of n_k such particles per cubic centimeter. The resulting current through the frame is then

$$I_a = n_1q_1\mathbf{a} \cdot \mathbf{u}_1 + n_2q_2\mathbf{a} \cdot \mathbf{u}_2 + \cdots = \mathbf{a} \cdot \sum_k n_kq_k\mathbf{u}_k \tag{2}$$

On the right is the scalar product of the vector $\mathbf{a}$ with a vector quantity that we shall call the current density $\mathbf{J}$:

$$\mathbf{J} = \sum_k n_kq_k\mathbf{u}_k \tag{3}$$

The magnitude of the current density J can be expressed in esu/sec-cm^2. In SI units current density is expressed in amperes per square meter (amp/m^2).†

Let's look at the contribution to the current density $\mathbf{J}$ from one variety of charge carriers, electrons say, which may be present with many different velocities. In a typical conductor, the electrons will have an almost random distribution of velocities, varying widely in direction and magnitude. Let N_e be the total number of electrons per unit volume, of all velocities. We can divide the electrons into many groups, each of which contains electrons with nearly the same speed and direction. The *average velocity* of all the electrons, like any average, would then be calculated by summing over the groups, weighting each velocity by the number in the group, and dividing by the total number. That is,

$$\bar{\mathbf{u}} = \frac{1}{N_e}\sum_k n_k\mathbf{u}_k \tag{4}$$

We use the bar over the top, as in $\bar{\mathbf{u}}$, to mean the average over a dis-

†Sometimes one encounters current density expressed in amps/cm^2. Nothing is wrong with that; the meaning is perfectly clear as long as the units are stated. (Long before SI was promulgated, two or three generations of electrical engineers made out quite well with amperes per square inch!)

tribution. Comparing Eq. 4 with Eq. 3, we see that the contribution of the electrons to the current density can be written simply in terms of the average electron velocity. Remembering that for the electron $q = -e$, and using subscript e to show that all quantities refer to this one type of charge carrier, we can write

$$\mathbf{J}_e = -eN_e\bar{\mathbf{u}}_e \tag{5}$$

This may seem rather obvious, but we have gone through it step by step to make clear that the current through the frame depends only on the average velocity of the carriers, which often is only a tiny fraction, in magnitude, of their random speeds.

STEADY CURRENTS AND CHARGE CONSERVATION

4.2 The current I flowing through any surface S is just the surface integral

$$I = \int_S \mathbf{J} \cdot d\mathbf{a} \tag{6}$$

We speak of a steady or stationary current system when the current density vector **J** remains constant in time everywhere. Steady currents have to obey the law of charge conservation. Consider some region of space completely enclosed by the balloonlike surface S. The surface integral of **J** over all of S gives the rate at which charge is leaving the volume enclosed. Now if charge forever pours out of, or into, a fixed volume, the charge density inside must grow infinite, unless some compensating charge is continually being created there. But charge creation is just what never happens. Therefore, for a truly time-independent current distribution, the surface integral of **J** over *any* closed surface must be zero. This is completely equivalent to the statement that, at every point in space:

$$\text{div}\,\mathbf{J} = 0 \tag{7}$$

To appreciate the equivalence, recall Gauss's theorem and our fundamental definition of divergence in terms of the surface integral over a small surface enclosing the location in question.

We can make a more general statement than Eq. 7. Suppose the current is not steady, **J** being a function of t as well as of x, y, and z. Then since $\int_S \mathbf{J} \cdot d\mathbf{a}$ is the instantaneous rate at which charge is *leaving* the enclosed volume, while $\int_V \rho\, dv$ is the total charge *inside* the volume at any instant, we have

$$\int_S \mathbf{J} \cdot d\mathbf{a} = -\frac{d}{dt}\int_V \rho\, dv \tag{8}$$

Letting the volume in question shrink down around any point (x, y, z), the relation expressed in Eq. 8 becomes:†

$$\text{div}\,\mathbf{J} = -\frac{\partial \rho}{\partial t} \qquad \text{(time-dependent charge distribution)} \qquad (9)$$

The time derivative of the charge density ρ is written as a partial derivative since ρ will usually be a function of spatial coordinates as well as time. Equations 8 and 9 express the *conservation of charge:* No charge can flow away from a place without diminishing the amount of charge that is there.

An instructive example of a stationary current distribution occurs in the plane diode, a two-electrode vacuum tube. One electrode, the cathode, is coated with a material that emits electrons copiously when heated. The other electrode, the anode, is simply a metal plate. By means of a battery the anode is maintained at a positive potential with respect to the cathode. Electrons emerge from this hot cathode with very low velocities and then, being negatively charged, are accelerated toward the positive anode by the electric field between cathode and anode. In the space between the cathode and anode the electric current consists of these moving electrons. The circuit is completed by the flow of electrons in external wires, possibly by the movement of ions in a battery, and so on, with which we are not here concerned. In this diode, ρ, the local density of charge in any region, is simply $-ne$, where n is the local density of electrons, in electrons per cubic centimeter. The local current density $\mathbf{J}$ is of course $\rho\mathbf{v}$, where $\mathbf{v}$ is the velocity of electrons in that region. In the plane-parallel diode we may assume $\mathbf{J}$ has no y or z components (Fig. 4.2). If conditions are steady, it follows then that J_x must be independent of x, for if div $\mathbf{J} = 0$ as Eq. 7 says, $\partial J_x/\partial x$ must be zero if $J_y = J_z = 0$. This is belaboring the obvious; if we have a steady stream of electrons moving in the x direction only, the same number per second have to cross any intermediate plane between cathode and anode. We conclude that ρv is constant. But observe that v is *not* constant; it varies with x because the electrons are accelerated by the field. Hence ρ is not constant either. Instead, the negative charge density is high near the cathode, low near the anode, just as the density of cars on an expressway is high near a traffic slowdown, low where traffic is moving at high speed.

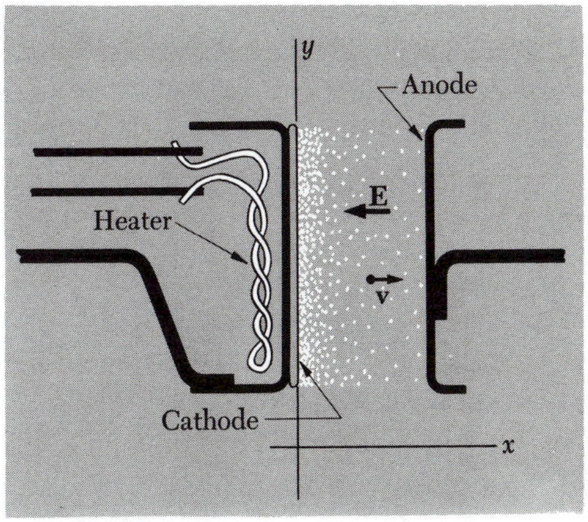

FIGURE 4.2
A vacuum diode with plane-parallel cathode and anode.

†If the step between Eqs. 8 and 9 is not obvious, look back at our fundamental definition of divergence in Chapter 2. As the volume shrinks, we can eventually take ρ outside the volume integral on the right. The volume integral is to be carried out at one instant of time. Its time derivative thus depends on the difference between the volume integral at t and at $t + dt$. The only difference is due to the change of ρ there, since the boundary of the volume remains in the same place.

ELECTRICAL CONDUCTIVITY AND OHM'S LAW

4.3 There are many ways of causing charge to move, including what we might call "bodily transport" of the charge carriers. In the Van de Graaff electrostatic generator (see Problem 4.3) an insulating belt is given a surface charge, which it conveys to another electrode for removal, much as an escalator conveys people. That constitutes a perfectly good current. In the atmosphere, charged water droplets falling because of their weight form a component of the electric current system of the earth. In this section we shall be interested in a more common agent of charge transport, the force exerted on a charge carrier by an electric field. An electric field **E** pushes positive charge carriers in one direction, negative charge carriers in the opposite direction. If either or both can move, the result is an electric current in the direction of **E.** In most substances, and over a wide range of electric field strengths, we find that the current density is proportional to the strength of the electric field that causes it. The linear relation between current density and field is expressed by

$$\mathbf{J} = \sigma \mathbf{E} \tag{10}$$

The factor σ is called the *conductivity* of the material. Its value depends on the material in question; it is very large for metallic conductors, extremely small for good insulators. It may depend too on the physical state of the material—on its temperature, for instance. But with such conditions given, it does not depend on the magnitude of **E.** If you double the field strength, holding everything else constant, you get twice the current density.

In Eq. 10 σ may be considered a scalar quantity, implying that the direction of **J** is always the same as the direction of **E.** That is surely what we would expect within a material whose structure has no "built-in" preferred direction. Materials do exist in which the electrical conductivity itself depends on the angle the applied field **E** makes with some intrinsic axis in the material. One example is a single crystal of graphite which has a layered structure on an atomic scale. For another example, see Problem 4.7. In such cases **J** may not have the direction of **E.** But there still are linear relations between the components of **J** and the components of **E,** relations expressed by Eq. 10 with σ a *tensor* quantity instead of a scalar.† From now on we'll consider only *isotropic* materials, those within which the electrical conductivity is the same in all directions.

†The most general linear relation between the two vectors **J** and **E** would be expressed as follows. In place of the three equations equivalent to Eq. 10, namely, $J_x = \sigma E_x$, $J_y = \sigma E_y$, and $J_z = \sigma E_z$, we would have $J_x = \sigma_{xx}E_x + \sigma_{xy}E_y + \sigma_{xz}E_z$, $J_y = \sigma_{yx}E_x + \sigma_{yy}E_y + \sigma_{yz}E_z$, and $J_z = \sigma_{zx}E_x + \sigma_{zy}E_y + \sigma_{zz}E_z$. The nine coefficients σ_{xx}, σ_{xy}, etc., make up a *tensor*. (In this case because of a symmetry requirement, it would turn out that $\sigma_{xy} = \sigma_{yx}$, $\sigma_{yz} = \sigma_{zy}$, $\sigma_{xz} = \sigma_{zx}$. Furthermore by a suitable orientation of the x, y, z axes, all the coefficients could be rendered zero except σ_{xx}, σ_{yy}, and σ_{zz}.)

Equation 10 is a statement of Ohm's law. It is an *empirical* law, a generalization derived from experiment, not a theorem that must be universally obeyed. In fact, Ohm's law is bound to fail in the case of any particular material if the electric field is too strong. And we shall meet some interesting and useful materials in which "nonohmic" behavior occurs in rather weak fields. Nevertheless, the remarkable fact is the enormous range over which, in the large majority of materials, current density is proportional to electric field. Later in this chapter we'll explain why this should be so. But now, taking Eq. 10 for granted, we want to work out its consequences. We are interested in the total current I flowing through a wire or a conductor of any other shape with well-defined ends, or terminals, and the difference in potential between those terminals, for which we'll use the symbol V (for *voltage*) rather than $\phi_1 - \phi_2$ or ϕ_{12}. If $\mathbf{J}$ is proportional to $\mathbf{E}$ everywhere inside the conductor, then I must surely be proportional to V. For $\mathbf{I}$ is the integral of $\mathbf{J}$ over a cross section of the conductor, while V is the line integral of $\mathbf{E}$ on a path through the conductor from one terminal to the other. The relation of V to I is therefore another expression of Ohm's law, which we'll write this way:

$$V = RI \tag{11}$$

The constant R is the *resistance* of the conductor between the two terminals. R depends on the size and shape of the conductor and the conductivity σ of the material. The simplest example is a solid rod of cross section area A and length L between its ends. A steady current I flows through this rod from one end to the other (Fig. 4.3). Of course there must be conductors to carry the current to and from the rod. We

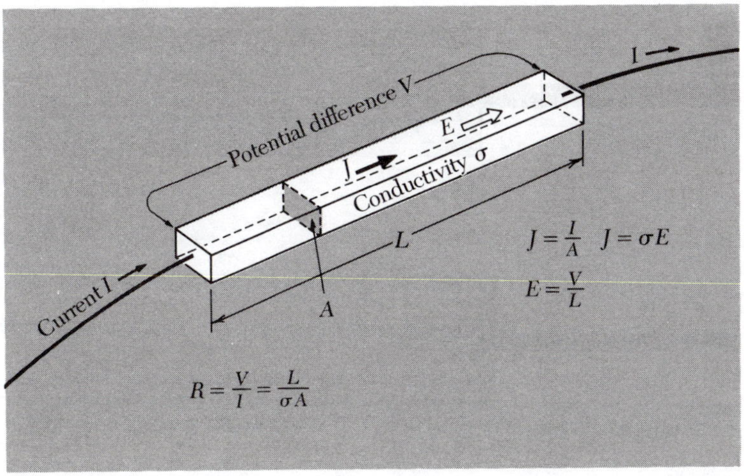

FIGURE 4.3
The resistance of a conductor of length L, uniform cross section area A, and conductivity σ.

consider the terminals of the rod to be the points where these conductors are attached. Inside the rod the current density is

$$J = \frac{I}{A} \tag{12}$$

and the electric field strength is

$$E = \frac{V}{L} \tag{13}$$

The resistance R in Eq. 11 is V/I. Using Eqs. 10, 12, and 13 we easily find that

$$R = \frac{V}{I} = \frac{LE}{AJ} = \frac{L}{A\sigma} \tag{14}$$

On the way to this simple formula we made some tacit assumptions. First, we assumed the current density is uniform over the cross section of the bar. To see why that must be so, imagine that J is actually greater along one side of the bar than on the other. Then E must also be greater along that side. But then the line integral of $\mathbf{E}$ from one terminal to the other would be greater for a path along one side than for a path along the other, and that cannot be true for an electrostatic field. A second assumption was that $\mathbf{J}$ kept its uniform magnitude and direction right out to the end of the bar. Whether that is true or not depends on the external conductors that carry current to and from the bar and how they are attached. Compare Fig. 4.4*a* with Fig. 4.4*b*. Suppose that the terminal in (*b*) is made of material with a conductivity much higher than that of the bar. That will make the plane of the end of the bar an equipotential surface, creating the current system to which Eq. 14 applies *exactly*. But all we can say in

FIGURE 4.4
Different ways in which the current I might be introduced into the conducting bar. In (*a*) it has to spread out before the current density **J** becomes uniform. In (*b*) if the external conductor has much higher conductivity than the bar, the end of the bar will be an equipotential and the current density will be uniform from the beginning. For long thin conductors like ordinary wires, the difference is negligible.

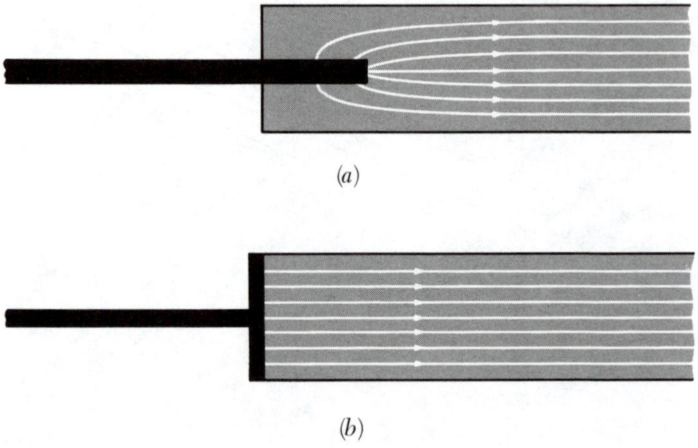

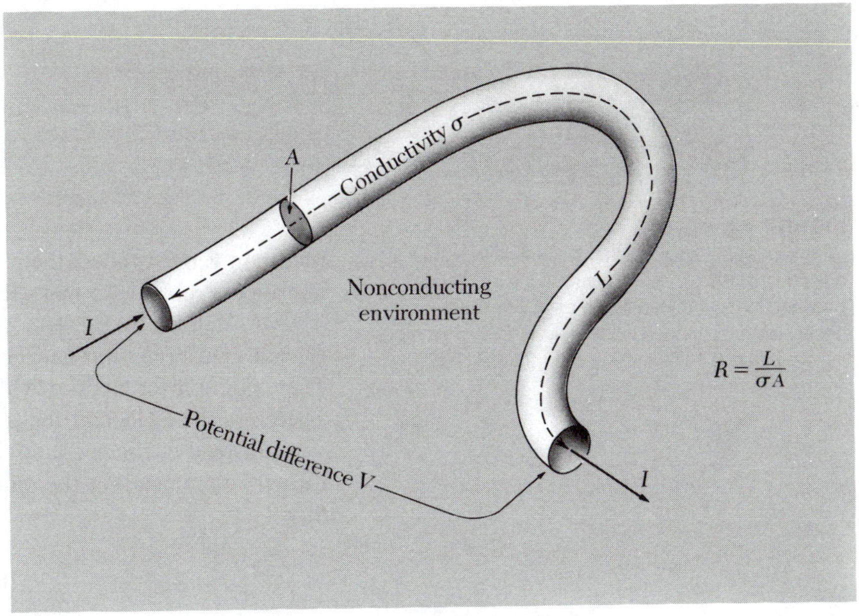

FIGURE 4.5
As long as our conductors are surrounded by a nonconducting medium (air, oil, vacuum, etc.) the resistance R between the terminals doesn't depend on the shape, only on the length of the conductor and its cross-sectional area.

general about such "end effects" is that Eq. 14 will give R to a good approximation if the width of the bar is small compared with its length.

A third assumption is that the bar is surrounded by an electrically nonconducting medium. Without that, we could not even *define* an isolated current path with terminals and talk about *the* current I and *the* resistance R. In other words, it is the enormous difference in conductivity between good insulators, including air, and conductors that makes *wires,* as we know them, possible. Imagine the conducting rod of Fig. 4.3 bent into some other shape, as in Fig. 4.5. Because it is embedded in a nonconducting medium into which current cannot leak, the problem presented in Fig. 4.5 is for all practical purposes the same as the one in Fig. 4.3 that we have already solved. Equation 14 applies to a bent wire as well as a straight rod, if we measured L along the wire.

In a region where the conductivity σ is constant, the steady current condition div $\mathbf{J} = 0$ (Eq. 7) together with Eq. 10 implies that div $\mathbf{E} = 0$ also. This tells us that the charge density is zero within that region. On the other hand, if σ varies from one place to another in the conducting medium, steady current flow may entail the presence of static charge within the conductor. Figure 4.6 shows a simple example, a bar made of two materials of different conductivity, σ_1 and σ_2. The current density $\mathbf{J}$ must be the same on the two sides of the interface; otherwise charge would continue to pile up there. It follows that the electric field $\mathbf{E}$ must be different in the two regions, with an abrupt jump in value at the interface. As Gauss's law tells us, such a discon-

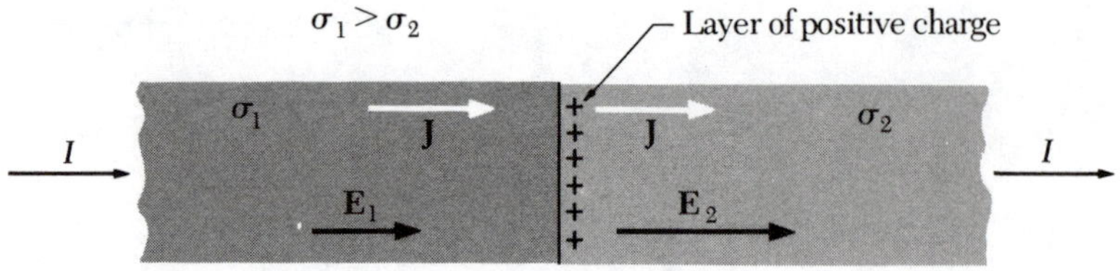

FIGURE 4.6
When current flows through this composite conductor, a layer of static charge appears at the interface between the two materials, so as to provide the necessary jump in the electric field **E.** In this example $\sigma_2 < \sigma_1$, hence E_2 must be greater than E_1.

tinuity in **E** must reflect the presence of a layer of static charge at the interface. Problem 4.5 looks further into this example.

As defined by Eq. 10, conductivity is current density divided by electric field strength. The CGS unit of current density is esu/sec-cm^2. The CGS unit for electric field strength can be expressed as esu/cm^2. Therefore the CGS unit for conductivity σ is just sec^{-1}.

Instead of the conductivity σ we could have used its reciprocal, *resistivity* ρ, in stating the relation between electric field and current density:

$$\mathbf{J} = \left(\frac{1}{\rho}\right)\mathbf{E} \tag{15}$$

It is customary to use ρ as the symbol for resistivity and σ as the symbol for conductivity in spite of their use in some of our other equations for volume charge density and surface charge density. In the rest of this chapter ρ will always denote resistivity and σ conductivity. Equation 14 written in terms of resistivity becomes

$$R = \frac{\rho L}{A} \tag{16}$$

The CGS unit for resistivity ρ is simply the *second.* This association of a resistivity with a time has a natural interpretation which will be explained in Section 4.11. The corresponding SI units are expressed by using a unit of resistance, the *ohm,* which is defined by Eq. 11 as one volt per ampere. If resistance R is in ohms, it is evident from Eq. 16 that ρ must have dimensions ohms $\times$ length. The official SI unit for ρ would therefore be the ohm-meter. But another unit of length can be used with perfectly clear meaning. In fact the unit most commonly used for resistivity, in both the physics and technology of electrical conduction, is the ohm-centimeter (ohm-cm). If one chooses to measure resistivity in ohm-cm, the corresponding unit for conductivity is written as ohm^{-1} cm^{-1}, or (ohm-cm)$^{-1}$, and called "reciprocal ohm-cm." It should be emphasized that Eqs. 10 through 16 are valid for any self-consistent choice of units.

In Table 4.1 the conductivity and resistivity of a few materials are given in different units for comparison. The key conversion factor is also given.

TABLE 4.1

Resistivity and its reciprocal, conductivity, for a few materials

Material	Resistivity ρ	Conductivity σ
Pure copper, 273 K	1.56×10^{-6} ohm-cm 1.73×10^{-18} sec	6.4×10^{5} (ohm-cm)$^{-1}$ 5.8×10^{17} sec^{-1}
Pure copper, 373 K	2.24×10^{-6} ohm-cm 2.47×10^{-18} sec	4.5×10^{5} (ohm-cm)$^{-1}$ 4.0×10^{17} sec^{-1}
Pure germanium, 273 K	200 ohm-cm 2.2×10^{-10} sec	0.005 (ohm-cm)$^{-1}$ 4.5×10^{9} sec^{-1}
Pure germanium, 500 K	0.12 ohm-cm 1.3×10^{-13} sec	8.3 (ohm-cm)$^{-1}$ 7.7×10^{12} sec^{-1}
Pure water, 291 K	2.5×10^{7} ohm-cm 2.8×10^{-5} sec	4.0×10^{-8} (ohm-cm)$^{-1}$ 3.6×10^{4} sec^{-1}
Seawater (varies with salinity)	25 ohm-cm 2.8×10^{-11} sec	0.04 (ohm-cm)$^{-1}$ 3.6×10^{10} sec^{-1}

Note: 1 ohm-meter = 100 ohm-cm = 1.11×10^{-10} sec.

THE PHYSICS OF ELECTRICAL CONDUCTION

4.4 To explain electrical conduction we have to talk first about atoms and molecules. Remember that a neutral atom, one that contains as many electrons as there are protons in its nucleus, is *precisely* neutral (Section 1.3). On such an object the net force exerted by an electric field is exactly zero. And even if the neutral atom were moved along by some other means, that would not be an electric current. The same holds for neutral molecules. Matter which consists only of neutral molecules ought to have zero electrical conductivity. Here one qualification is in order: We are concerned now with steady electric currents, that is, *direct* currents, not alternating currents. An alternating electric field could cause periodic deformation of a molecule, and that displacement of electric charge would be a true alternating electric current. We shall return to that subject in Chapter 10. For a steady current we need mobile charge carriers, or *ions*. These must be present in the material before the electric field is applied, for the electric fields we shall consider are not nearly strong enough to create ions by tearing electrons off molecules. Thus the physics of electrical conduction centers on two questions: How many ions are there in a unit volume of material, and how do these ions move in the presence of an electric field?

In pure water at room temperature approximately two H_2O molecules in a billion are, at any given moment, dissociated into negative ions, OH^-, and positive ions, H^+. (Actually the positive ion is better described as OH_3^+, that is, a proton attached to a water molecule.) This provides approximately 6×10^{13} negative ions and an equal number of positive ions in a cubic centimeter of water.† The motion of these ions in the applied electric field accounts for the conductivity of pure water given in Table 4.1. Adding a substance like sodium chloride whose molecules easily dissociate in water can increase enormously the number of ions. That is why seawater has electrical conductivity nearly a million times greater than that of pure water. It contains something like 10^{20} ions per cm^3, mostly Na^+ and Cl^-.

In a gas like nitrogen or oxygen at ordinary temperatures there would be no ions at all except for the action of some ionizing radiation such as ultraviolet light, x-rays, or nuclear radiation. For instance, ultraviolet light might eject an electron from a nitrogen molecule, leaving N_2^+, a molecular ion with a positive charge e. The electron thus freed is a negative ion. It may remain free or it may eventually stick to some molecule as an "extra" electron, thus forming a negative molecular ion. The oxygen molecule happens to have an especially high affinity for an extra electron; when air is ionized, N_2^+ and O_2^- are common ion types. In any case, the resulting conductivity of the gas depends on the number of ions present at any moment, which depends in turn on the intensity of the ionizing radiation and perhaps other circumstances as well. So we cannot find in a table *the* conductivity of a gas. Strictly speaking, the conductivity of pure nitrogen shielded from all ionizing radiation would be zero.‡

Given a certain concentration of positive and negative ions in a material, how is the resulting conductivity, σ in Eq. 10, determined? Let's consider first a slightly ionized gas. To be specific, suppose its density in molecules per cubic centimeter is like that of room air—about 10^{19} per cm^3. Here and there among these neutral molecules are positive and negative ions. Suppose there are N positive ions in unit volume, each of mass M_+ and carrying charge e, and an equal number of negative ions, each with mass M_- and charge $-e$. The number of ions in unit volume, $2N$, is very much smaller than the number of neutral molecules. When an ion collides with anything it is almost

†Students of chemistry may recall that the concentration of hydrogen ions in pure water corresponds to a pH value of 7.0, which means the concentration is $10^{-7.0}$ mole/liter. That is equivalent to $10^{-10.0}$ mole/cm^3. A mole of anything is 6.02×10^{23} things—hence the number 6×10^{13} given above.

‡But what about thermal energy? Won't that occasionally lead to the ionization of a molecule? In fact, the energy required to ionize, that is, to extract an electron from, a nitrogen molecule is several hundred times the mean thermal energy of a molecule at 300 K. You would not expect to find even one ion so produced in the entire earth's atmosphere!

always a neutral molecule rather than another ion. Occasionally a positive ion does encounter a negative ion and combine with it to form a neutral molecule. Such recombination† would steadily deplete the supply of ions if ions were not being continually created by some other process. But in any case the rate of change of N will be so slow that we can neglect it here.

Imagine now the scene, on a molecular scale, before an electric field is applied. The molecules, and the ions too, are flying about with random velocities appropriate to the temperature. The gas is mostly empty space, the mean distance between a molecule and its nearest neighbor being about 10 molecular diameters. The mean free path of a molecule, which is the average distance it travels before bumping into another molecule, is much larger, perhaps 10^{-5} cm, or several hundred molecular diameters. A molecule or an ion in this gas spends 99.9 percent of its time as a free particle. If we could look at a particular ion at a particular instant, say $t = 0$, we would find it moving through space with some velocity $\mathbf{u}$. What will happen next? The ion will move in a straight line at constant speed until, sooner or later, it chances to come close to a molecule, close enough for strong short-range forces to come into play. In this *collision* the total kinetic energy and the total momentum of the two bodies, molecule and ion, will be conserved, but the ion's velocity will be rather suddenly changed in both magnitude and direction to some new velocity $\mathbf{u}'$. It will then coast along freely with this new velocity until another collision changes its velocity to $\mathbf{u}''$, and so on. After at most a few such collisions the ion is as likely to be moving in any direction as in any other direction. The ion will have "forgotten" the direction it was moving at $t = 0$. To put it another way, if we picked 10,000 cases of ions moving horizontally south, and followed each of them for τ seconds, their final velocity directions would be distributed impartially over a sphere. It may take several collisions to wipe out most of the direction memory or only a few, depending on whether collisions involving large momentum changes or small momentum changes are the more common, and this depends on the nature of the interaction. An extreme case is the collision of hard elastic spheres, which turns out to produce a completely random new direction in just one collision. We need not worry about these differences. The point is that, whatever the nature of the collisions, there will be *some* time interval τ, characteristic of a given system, such that the lapse of τ seconds leads to substantial loss of

†In calling the process recombination we of course do not wish to imply that the two "recombining" ions were partners originally. Close encounters of a positive ion with a negative ion are made somewhat more likely by their electrostatic attraction. However, that effect is generally not important when the number of ions per unit volume is very much smaller than the number of neutral molecules.

correlation between the initial velocity direction and the final velocity direction of an ion in that system.† This characteristic time τ will depend on the ion and on the nature of its average environment; it will certainly be shorter the more frequent the collisions, since in our gas nothing happens to an ion between collisions.

Now we are ready to apply a uniform electric field $\mathbf{E}$ to the system. It will make the description easier if we imagine the loss of direction memory to occur completely at a single collision, as we have said it does in the case of hard spheres. Our main conclusion will actually be independent of this assumption. Immediately after a collision an ion starts off in some random direction. We will denote by $\mathbf{u}^c$ the velocity immediately after a collision. The electric force on the ion $\mathbf{E}e$ imparts momentum to the ion continuously. After time t it will have acquired from the field a momentum increment $\mathbf{E}et$, which simply adds vectorially to its original momentum $M\mathbf{u}^c$. Its momentum is now $M\mathbf{u}^c + \mathbf{E}et$. If the momentum increment is small relative to $M\mathbf{u}^c$, that implies that the velocity has not been affected much, so we can expect the next collision to occur about as soon as it would have in the absence of the electric field. In other words the average time between collisions, which we shall denote by $\bar{t}$, is independent of the field $\mathbf{E}$ if the field is not too strong.

The momentum acquired from the field is always a vector in the same direction. But it is lost, in effect, at every collision, since the direction of motion after a collision is random, regardless of the direction before.

What is the average momentum of all the positive ions at a given instant of time? This question is surprisingly easy to answer if we look at it this way: At the instant in question, suppose we stop the clock and ask each ion how long it has been since its last collision. Suppose we get the particular answer t_1 from positive ion 1. Then that ion must have momentum $e\mathbf{E}t_1$ *in addition* to the momentum $M\mathbf{u}_1^c$ with which it emerged from its last collision. The average momentum of all N positive ions is therefore

$$M\bar{\mathbf{u}}_+ = \frac{1}{N}\sum_j (M\mathbf{u}_j^c + e\mathbf{E}t_j) \tag{17}$$

Here $\mathbf{u}_j^c$ is the velocity the jth ion had just after its last collision. These velocities $\mathbf{u}_j^c$ are quite random in direction and therefore contribute zero to the average. The second part is simply $\mathbf{E}e$ times the *average of the* t_j, that is, times the *average of the time since the last collision.*

†It would be possible to define τ precisely for a general system by giving a quantitative measure of the correlation between initial and final directions. It is a statistical problem, like devising a measure of the correlation between the birth weights of rats and their weights at maturity. However, we shall not need a general quantitative definition to complete our analysis.

That must be the same as the average of the time until the *next* collision, and both are the same† as the average time between collisions, $\bar{t}$. We conclude that the average velocity of a positive ion, in the presence of the steady field **E**, is

$$\bar{\mathbf{u}}_+ = \frac{\mathbf{E}e\bar{t}_+}{M_+} \tag{18}$$

This shows that the average velocity of a charge carrier is proportional to the force applied to it. If we observe only the average velocity, it looks as if the medium were resisting the motion with a force proportional to the velocity. That is the kind of frictional drag you feel if you try to stir thick syrup with a spoon, a "viscous" drag. Whenever charge carriers behave like this, we can expect something like Ohm's law.

In Eq. 18 we have written $\bar{t}_+$ because the mean time between collisions may well be different for positive and negative ions. The negative ions acquire velocity in the opposite direction, but since they carry negative charge their contribution to the current density **J** adds to that of the positives. The equivalent of Eq. 4.5, with the two sorts of ions included is now

$$\mathbf{J} = Ne\left(\frac{e\mathbf{E}\bar{t}_+}{M_+}\right) - Ne\left(\frac{-e\mathbf{E}\bar{t}_-}{M_-}\right) = Ne^2\left(\frac{\bar{t}_+}{M_+} + \frac{\bar{t}_-}{M_-}\right)\mathbf{E} \tag{19}$$

Our theory predicts that the system will obey Ohm's law, for Eq. 19 expresses a linear relation between **J** and **E**, the other quantities being constants characteristic of the medium. Compare Eq. 19 with Eq. 10. The constant $Ne^2(\bar{t}_+/M_+ + \bar{t}_-/M_-)$ appears in the role of σ, the conductivity.

We made a number of rather special assumptions about this system, but looking back, we can see that they were not essential so far as the linear relation between **E** and **J** is concerned. Any system containing a constant density of free charge carriers, in which the motion of the carriers is frequently "rerandomized" by collisions or other interactions within the system, ought to obey Ohm's law if the field **E** is not too strong. The ratio of **J** to **E**, which is the conductivity σ of the medium, will be proportional to the number of charge carriers and to the characteristic time τ, the time for loss of directional correlation.

---

†You may think the average time between collisions would have to be equal to the *sum* of the *average time since the last collision* and the *average time to the next*. That would be true if collisions occurred at absolutely regular intervals, but they don't. They are independent random events, and for such the above statement, paradoxical as it may seem at first, is true. Think about it. The question does not affect our main conclusion, but if you unravel it you will have grown in statistical wisdom. (*Hint:* If one collision doesn't affect the probability of having another—that's what *independent* means—it can't matter whether you start the clock at some arbitrary time, or at the time of a collision.)

It is *only* through this last quantity that all the complicated details of the collisions enter the problem. The making of a detailed theory of the conductivity of any given system, assuming the number of charge carriers is known, amounts to making a theory for τ. In our particular example this quantity was replaced by $\bar{t}$, and a perfectly definite result was predicted for the conductivity σ. Introducing the more general quantity τ, and also allowing for the possibility of different numbers of positive and negative carriers, we can summarize our theory as follows:

$$\sigma \approx e^2 \left(\frac{N_+\tau_+}{M_+} + \frac{N_-\tau_-}{M_-} \right) \tag{20}$$

We use the sign $\approx$ to acknowledge that we did not give τ a precise definition. That could be done, however.

To emphasize the fact that electrical conduction ordinarily involves only a slight systematic drift superimposed on the random motion of the charge carriers, we have constructed Fig. 4.7 as an artificial microscopic view of the kind of system we have been talking about. Positive ions are represented by white dots, negative ions by circles. We assume the latter are electrons and hence, because of their small mass, so much more mobile than the positive ions that we may

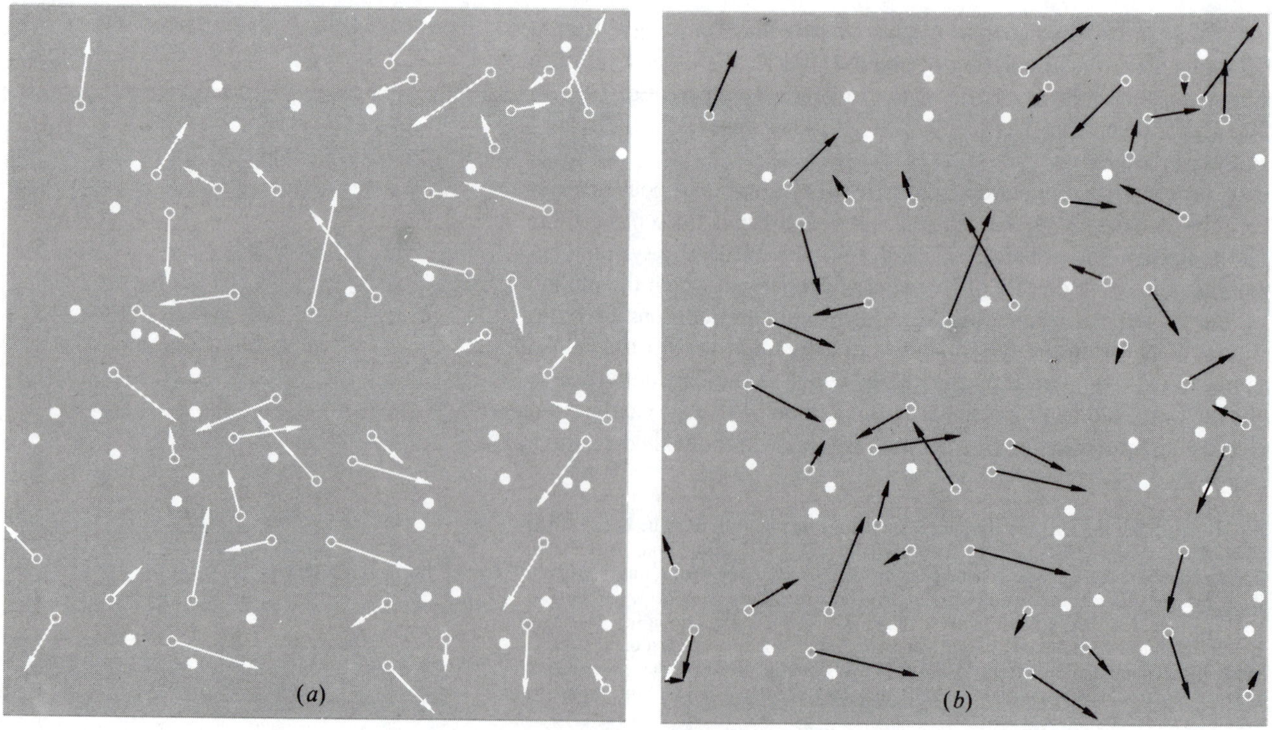

FIGURE 4.7
(*a*) A random distribution of electrons and positive ions with about equal numbers of each. Electron velocities are shown as vectors and in (*a*) are completely random. In (*b*) a drift toward the right, represented by the velocity vector →, has been introduced. This velocity was added to each of the original electron velocities, as shown in the case of the electron in the lower left corner.

neglect the motion of the positives altogether. In Fig. 4.7*a* we see a wholly random distribution of particles and of electron speeds. To make the diagram, the location and sign of a particle were determined by a random-number table. The electron velocity vectors were likewise drawn from a random distribution, one corresponding to the "maxwellian" distribution of molecular velocities in a gas. In Fig. 4.7*b* we have used the same positions, but now the velocities all have a small added increment to the right. That is, Fig. 4.7*b* is a view of an ionized material in which there is a net flow of negative charge to the right, equivalent to a positive current to the left. Figure 4.7*a* illustrates the situation with zero average current.

Obviously we should not expect the actual average of the velocities of the 46 electrons in Fig. 4.7*a* to be exactly zero, for they are statistically independent quantities. One electron doesn't affect the behavior of another. There will in fact be a randomly fluctuating electric current in the absence of any driving field, simply as a result of statistical fluctuations in the vector sum of the electron velocities. This spontaneously fluctuating current can be measured. It is a source of noise in all electric circuits, and often determines the ultimate limit of sensitivity of devices for detecting weak electric signals.

With these ideas in mind, consider the materials whose electrical conductivity is plotted, as a function of temperature, in Fig. 4.8. Glass at room temperature is a good insulator. Ions are not lacking in its internal structure, but they are practically immobile, locked in place. As a glass is heated, its structure becomes somewhat less rigid. An ion is able to move now and then, in the direction the electric field is pushing it. That happens in a sodium chloride crystal, too. The ions, in that case, Na^+ and Cl^-, move by infrequent short jumps.† Their average rate of progress is proportional to the electric field strength at any given temperature, so Ohm's law is obeyed. In both these materials, the main effect of raising the temperature is to increase the mobility of the charge carriers rather than their number.

Silicon and germanium are called *semiconductors*. Their conductivity, too, depends strongly on the temperature, but for a different reason. At zero absolute temperature, they would be perfect insulators, containing no ions at all, only neutral atoms. The effect of thermal energy is to create charge carriers by liberating electrons from some of the atoms. The steep rise in conductivity around room temperature and above reflects a great increase in the number of mobile electrons, not an increase in the mobility of an individual electron. We shall look more closely at semiconductors in Section 4.6.

The metals, exemplified by copper and lead in Fig. 4.8, are even better conductors. Their conductivity generally *decreases* with

---

†This involves some disruption of the perfectly orderly array of ions depicted in Fig. 1.7.

FIGURE 4.8
The electrical conductivity of some representative substances. Notice that logarithmic scales are used for both conductivity and absolute temperature.

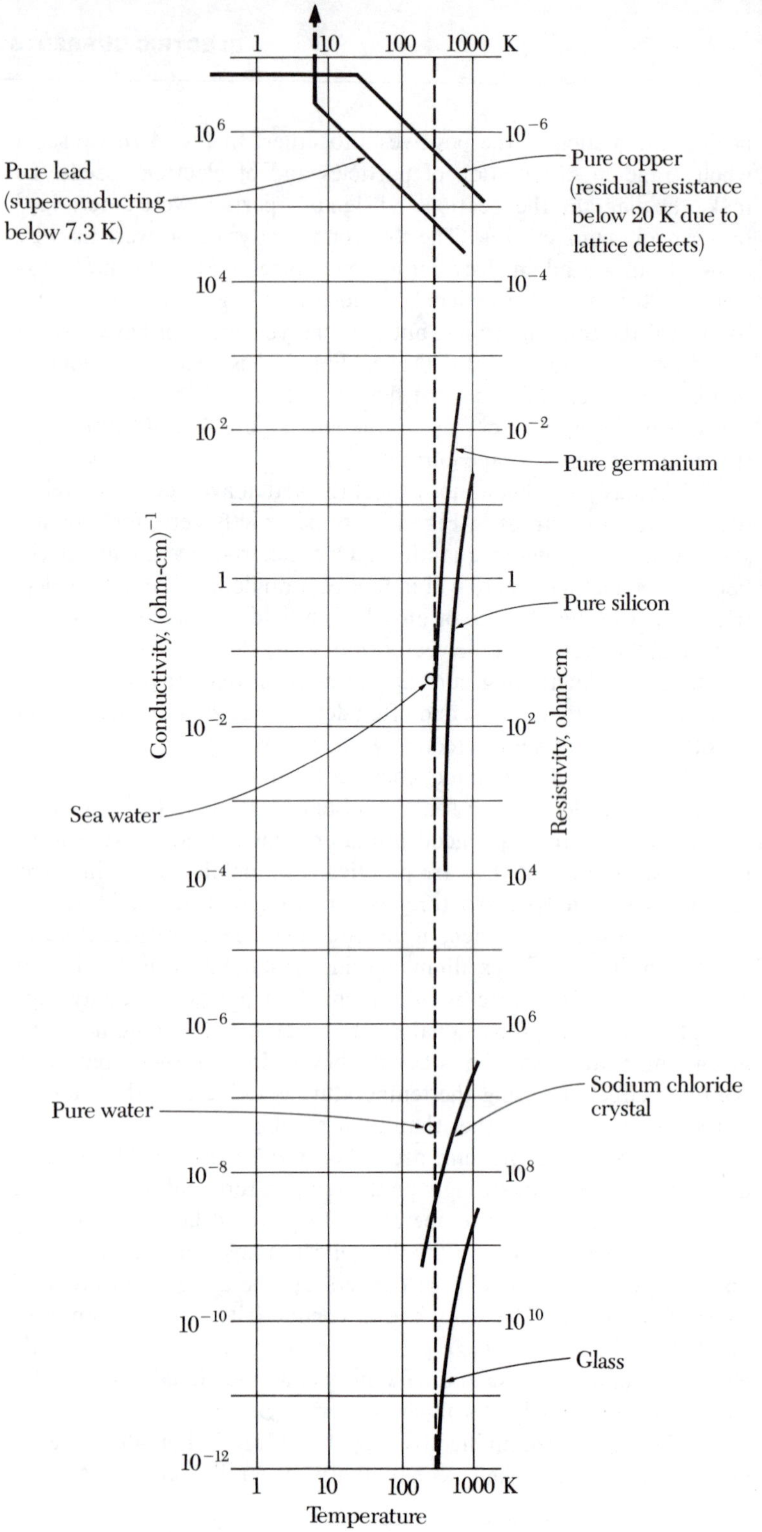

increasing temperature. In fact, over most of the range plotted, the conductivity of a pure metal like copper or lead is inversely proportional to the absolute temperature, as can be seen from the 45° slope of our logarithmic graph. Were that behavior to continue as copper and lead are cooled down toward absolute zero, we could expect an enormous increase in conductivity. At 0.001 K, a temperature now readily attainable in the laboratory, we should expect the conductivity of each metal to rise to 300,000 times its room temperature value. In the case of copper, we would be sadly disappointed. As we cool copper below about 20 K, its conductivity ceases to rise and remains constant from there on down. We'll try to explain that in the next section. In the case of lead, normally a somewhat poorer conductor than copper, something far more surprising happens. As a lead wire is cooled below 7.2 K, its resistance abruptly and completely *vanishes.* The metal becomes *superconducting.* This means, among other things, that an electric current, once started flowing in a circuit of lead wire, will continue to flow indefinitely (for years, even!) without any electric field to drive it. The conductivity may be said to be infinite, though the concept really loses its meaning in the superconducting state. Warmed above 7.2 K, the lead wire recovers its normal resistance as abruptly as it lost it. Many metals can become superconductors, including more than 20 elements and numerous metallic compounds. The temperature at which the transition from the normal to the superconducting state occurs depends on the material. The highest transition temperature yet observed is 21 K.

Our model of ions accelerated by the electric field, their progress being continually impeded by collisions, utterly fails us here. Somehow, in the superconducting state all impediment to the electrons' motion has vanished. Not only that, magnetic effects just as profound and mysterious are manifest in the superconductor. At this stage of our study we cannot fully describe, let alone explain, the phenomena of superconductivity. More will be said in Appendix C, which should be intelligible after our study of magnetism.

Superconductivity aside, all these materials obey Ohm's law. Doubling the electric field doubles the current if other conditions, including the temperature, are held constant. At least that is true if the field is not too strong. It is easy to see how Ohm's law could fail in the case of a partially ionized gas. Suppose the electric field is so strong that the additional velocity an electron acquires between collisions is comparable to its thermal velocity. Then the time between collisions will be shorter than it was before the field was applied, an effect not included in our theory and one that will cause the observed conductivity to depend on the field strength.

A more spectacular breakdown of Ohm's law occurs if the electric field is further increased until an electron gains so much energy between collisions that in striking a neutral atom it can knock another electron loose. The two electrons can now release still more electrons in the same way. Ionization increases explosively, quickly making a conducting path between the electrodes. This is a *spark*. It's what happens when a sparkplug fires, and when you touch a doorknob after walking over a rug on a dry day. There are always a few electrons in the air, liberated by cosmic rays if in no other way. Since one electron is enough to trigger a spark, this sets a practical limit to field strength that can be maintained in a gas. Air at atmospheric pressure will break down at roughly 30 kilovolts/cm or 100 statvolts/cm. In a gas at low pressure, where an electron's free path is quite long, as within the tube of an ordinary fluorescent lamp, a steady current can be maintained with a modest field, with ionization by electron impact occurring at a constant rate. The physics is fairly complex, and the behavior far from ohmic.

CONDUCTION IN METALS

4.5 The high conductivity of metals is due to electrons within the metal that are not attached to atoms but are free to move through the whole solid. Proof of this is the fact that electric current in a copper wire—unlike current in an ionic solution—transports no chemically identifiable substance. A current can flow steadily for years without causing the slightest change in the wire. It could only be electrons that are moving, entering the wire at one end and leaving it at the other.

We know from chemistry that atoms of the metallic elements rather easily lose their outermost electrons.† These would be bound to the atom if it were isolated, but become detached when many such atoms are packed close together in a solid. The atoms thus become positive ions, and these positive ions form the rigid lattice of the solid metal, usually in an orderly array. The detached electrons, which we shall call the conduction electrons, move through this three-dimensional lattice of positive ions.

The number of conduction electrons is large. The metal sodium, for instance, contains 2.5×10^{22} atoms in 1 cm^3, and each atom provides one conduction electron. No wonder sodium is a good conductor! But wait, there is a deep puzzle here. It is brought to light by applying our simple theory of conduction to this case. As we have seen, the mobility of a charge carrier is determined by the time τ during which it moves freely without bumping into anything. If we have 2.5×10^{22} electrons per cubic centimeter of mass m_e, we need only the experimentally measured conductivity of sodium to calculate an electron's

†This could even be taken as the property that defines a metallic element, making somewhat tautological the statement that metals are good conductors.

mean free time τ. The conductivity of sodium at room temperature, in CGS units, is 1.9×10^{17} sec^{-1}. Solving Eq. 20 for τ_-, with $N_+ = 0$ as there are no mobile positive carriers, we find

$$\tau_- = \frac{\sigma m_e}{Ne^2} = \frac{(1.9 \times 10^{17}) \times (9 \times 10^{-28})}{(2.5 \times 10^{22}) \times (23 \times 10^{-20})} = 3 \times 10^{-14} \text{ sec}$$

This seems a *surprisingly long* time for an electron to move through the lattice of sodium ions without suffering a collision. The thermal speed of an electron at room temperature ought to be about 10^7 cm/sec, according to kinetic theory, which in that time should carry it a distance of 3×10^{-7} cm. Now the ions in a crystal of sodium are practically touching one another. The centers of adjacent ions are only 3.8×10^{-8} cm apart, with strong electric fields and many bound electrons filling most of the intervening space. How could an electron travel nearly 10 lattice spaces through these obstacles without being deflected? Why is the lattice of ions so *easily penetrated* by the conduction electrons?

This puzzle baffled physicists until the *wave aspect* of the electrons' motion was recognized and explained by quantum mechanics. Here we can only hint at the nature of the explanation. It goes something like this. We should not now think of the electron as a tiny charged particle deflected by every electric field it encounters. It is *not localized* in that sense. It behaves more like a spread-out wave interacting, at any moment, with a larger region of the crystal. What interrupts the progress of this wave through the crystal is not the regular array of ions, dense though it is, but an *irregularity* in the array. (A light wave traveling through water can be scattered by a bubble or a suspended particle, but not by the water itself; the analogy has some validity.) In a geometrically perfect and flawless crystal the electron wave would never be scattered, which is to say that the electron would never be deflected; our time τ would be infinite. But real crystals are imperfect in at least two ways. For one thing, there is a random thermal vibration of the ions, which makes the lattice at any moment slightly irregular geometrically, and the more so the higher the temperature. It is this effect which makes the conductivity of a pure metal *decrease* as the temperature is raised. We see it in the sloping portions of the graph of σ for pure copper and pure lead in Fig. 4.8. A real crystal can have irregularities, too, in the form of foreign atoms, or impurities, and lattice defects—flaws in the stacking of the atomic array. Scattering by these irregularities limits the free time τ whatever the temperature. Such defects are responsible for the residual temperature-independent resistivity seen in the plot for copper in Fig. 4.8.

In metals Ohm's law is obeyed exceedingly accurately up to current densities far higher than any that can be long maintained. No deviation has ever been clearly demonstrated experimentally. According to one theoretical prediction, departures on the order of 1 percent might be expected at a current density of 10^9 amps/cm^2. That is more than a million times the current density typical of wires in ordinary circuits.

SEMICONDUCTORS

4.6 In a crystal of silicon each atom has four near neighbors. The three-dimensional arrangement of the atoms is shown in Fig. 4.9. Now silicon, like carbon which lies directly above it in the periodic table, has four valence electrons, just the number needed to make each bond between neighbors a shared electron pair—a covalent bond as it is called in chemistry. This neat arrangement makes a quite rigid structure. In fact, it is the way the carbon atoms are arranged in diamond, the hardest known substance. With its bonds all intact, the perfect silicon crystal is a perfect insulator; there are no mobile electrons. But imagine that we could extract an an electron from one of these bond pairs and move it a few hundred lattice spaces away in the crystal. This would leave a net positive charge at the site of the extraction and

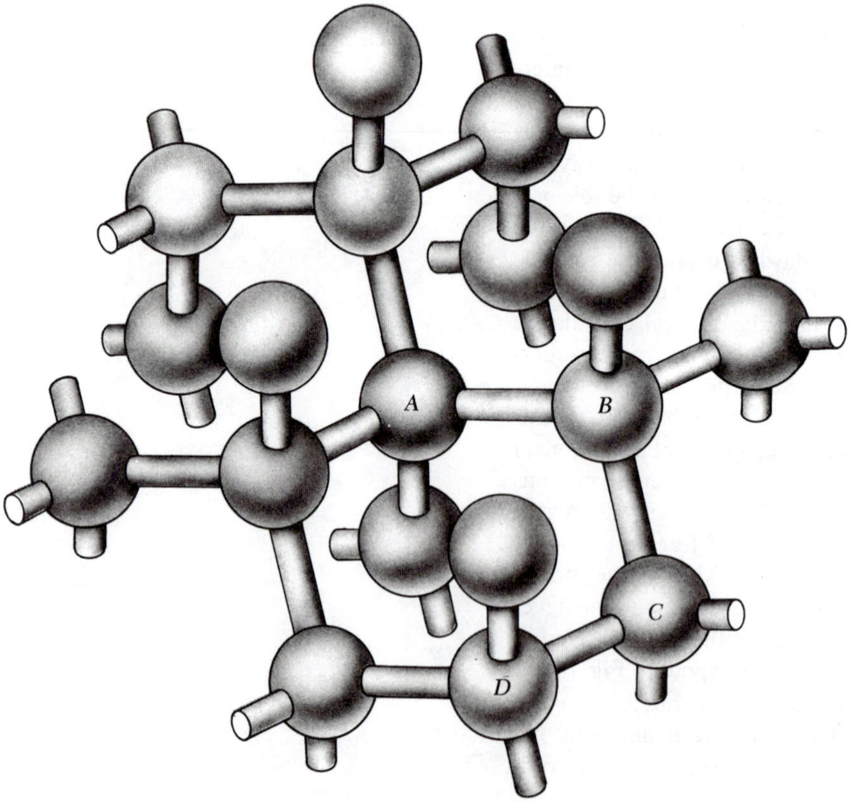

FIGURE 4.9
The structure of the silicon crystal. The balls are Si atoms. A rod represents a covalent bond between neighboring atoms, made by sharing a pair of electrons. This requires four valence electrons per atom. Diamond has this structure, and so does germanium.

would give us a loose electron. It would also cost a certain amount of energy. We'll take up the question of energy in a moment. But first let us note that we have created *two* mobile charges, not just one. The freed electron is mobile. It can move like a conduction electron in a metal, like which it is spread out, not sharply localized. The quantum state it occupies we call a state in the *conduction band.* The positive charge left behind is also mobile. If you think of it as an electron missing in the bond between atoms A and B in Fig. 4.9, you can see that this vacancy among the valence electrons could be transferred to the bond between B and C, thence to the bond between C and D and so on, just by shifting electrons from one bond to another. Actually, the motion of the hole, as we shall call it henceforth, is even freer than this would suggest. It sails through the lattice like a conduction electron. The difference is that it is a *positive* charge. An electric field **E** accelerates the hole in the direction of **E**, not the reverse. The hole acts as if it had a mass comparable with an electron's mass. This is really rather mysterious, for the hole's motion results from the collective motion of many valence electrons.† Nevertheless, and fortunately, it acts so much like a real positive particle that we may picture it as such from now on.

The minimum energy required to extract an electron from a valence state in silicon and leave it in the conduction band is 1.8×10^{-12} erg, or 1.12 electron-volts (ev). (One electron-volt is the work done in moving one electronic charge through a potential difference of one volt.) This is the *energy gap* between two bands of possible states, the valence band and the conduction band. States of intermediate energy for the electron simply do not exist. This energy ladder is represented in Fig. 4.10. Two electrons can never have the same quantum state—that is a fundamental law of physics. States ranging up the energy ladder must therefore be occupied even at absolute zero. As it happens, there are exactly enough states in the valence band to accommodate all the electrons. At $T = 0$, as shown in Fig. 4.10*a*, *all* of these valence states are occupied, and *none* of the conduction band states.

If the temperature is high enough, thermal energy can raise some electrons from the valence band to the conduction band. The effect of temperature on the probability that electron states will be occupied is expressed by the exponential factor $e^{-\Delta E/kT}$, called the Boltzmann factor. Suppose that two states labeled 1 and 2 are avail-

†This mystery is *not* explained by drawing an analogy, as is sometimes done, with a bubble in a liquid. In a centrifuge, bubbles in a liquid would go in toward the axis; the holes we are talking about would go out. A cryptic but true statement, which only quantum mechanics will make intelligible, is this: The hole behaves dynamically like a positive charge with positive mass because it is a vacancy in states with negative charge and negative mass.

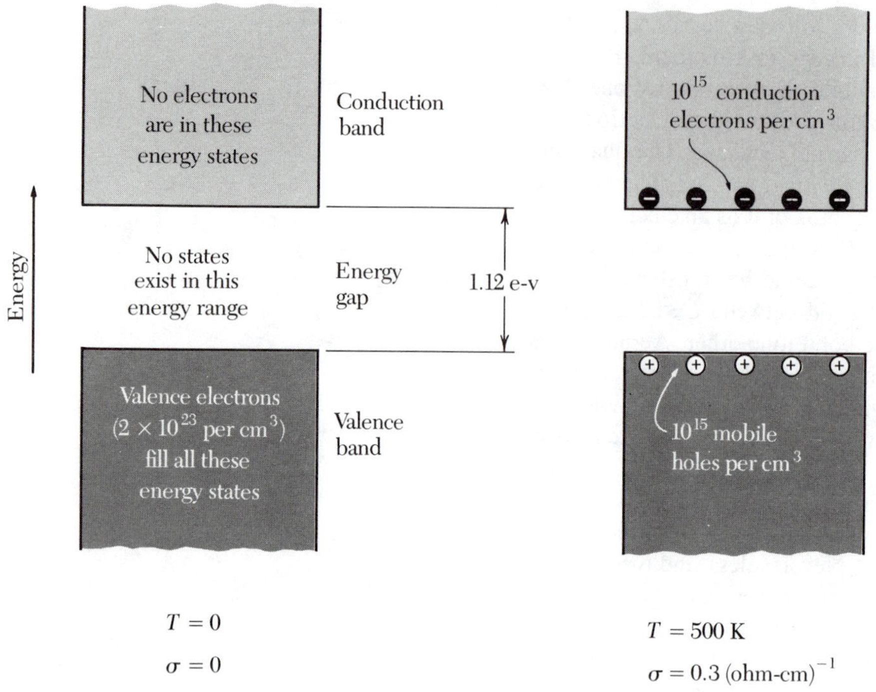

FIGURE 4.10
A schematic representation of the energy bands in silicon, which are all the possible states for the electrons, arranged in order of energy. Two electrons can't have the same state. At temperature zero the valence band is full; an electron occupies every available state. The conduction band is empty. At T = 500 K there are 10^{15} electrons in the lowest conduction band states, leaving 10^{15} holes in the valence band, in 1 cm^3 of the crystal.

able for occupation by an electron and that the electron's energy in state 1 would be E_1, while its energy in state 2 would be E_2. Let p_1 be the probability that the electron will be found occupying state 1, p_2 the probability that it will be found in state 2. In a system in thermal equilibrium at temperature T the ratio p_2/p_1 depends only on the energy *difference,* $\Delta E = E_2 - E_1$. It is given by

$$\frac{p_2}{p_1} = e^{-\Delta E/kT} \tag{21}$$

The constant k, Boltzmann's constant, has the value 1.38×10^{-16} erg/kelvin, or 1.38×10^{-23} joule/kelvin. This relation holds for any two states. It governs the population of available states on the energy ladder. To predict the resulting number of electrons in the conduction band at a given temperature we would have to know more about the number of states available. But this shows why the number of conduction electrons per unit volume depends so strongly on the temperature. For T = 300 K the energy kT is about 0.025 ev. The Boltzmann factor relating states 1 ev apart in energy would be e^{-40}, or

4×10^{-18}. In silicon at room temperature the number of electrons in the conduction band, per cubic centimeter is approximately 10^{10}. At 500 K one finds about 10^{15} electrons per cm^3 in the conduction band, and the same number of holes in the valence band (Fig. 4.10*b*). Both holes and electrons contribute to the conductivity, which is 0.3 $(\text{ohm-cm})^{-1}$ at that temperature. Germanium behaves like silicon, but the energy gap is somewhat smaller, 0.7 ev. At any given temperature it has more conduction electrons and holes than silicon, consequently higher conductivity, as is evident in Fig. 4.8. Diamond would be a semiconductor, too, if its energy gap weren't so large (5.5 ev) that there are no electrons in the conduction band at any attainable temperature.

With only 10^{10} conduction electrons and holes per cubic centimeter, the silicon crystal at room temperature is practically an insulator. But that can be changed dramatically by inserting foreign atoms into the pure silicon lattice. This is the basis for all the marvelous devices of semiconductor electronics. Suppose that some very small fraction of the silicon atoms—for example, 1 in 10^7—are replaced by phosphorus atoms. (This "doping" of the silicon can be accomplished in various ways.) The phosphorus atoms, of which there are now about 5×10^{15} per cm^3, occupy regular sites in the silicon lattice. A phosphorus atom has five valence electrons, one too many for the four-bond structure of the perfect silicon crystal. The extra electron easily comes loose. Only 0.044 ev of energy is needed to boost it to the conduction band. What is left behind in this case is not a mobile hole, but an immobile positive phosphorus ion. We now have nearly 5×10^{15} mobile electrons in the conduction band, and a conductivity of nearly 1 $(\text{ohm-cm})^{-1}$. There are a very few holes as well, the number that would be there in the pure crystal at room temperature. Because nearly all the charge carriers are *negative,* we call this "phosphorus-doped" crystal an *n-type semiconductor* (Fig. 4.11*a*).

Now let's dope a pure silicon crystal with aluminum atoms as the impurity. The aluminum atom has three valence electrons, one too few to construct four covalent bonds around its lattice site. That is cheaply remedied if one of the regular valence electrons joins the aluminum atom permanently, completing the bonds around it. The cost in energy is only 0.05 ev, much less than the 1.2 ev required to raise a valence electron up to the conduction band. This promotion creates a vacancy in the valence band, a mobile hole, and makes of the aluminum atom a fixed negative ion. Thanks to the holes thus created—at room temperature nearly equal in number to the aluminum atoms added—the crystal becomes a much better conductor. Of course there are also a few electrons in the conduction band, as there would be in the pure undoped silicon at the same temperature. But the overwhelming majority of the mobile charge carriers are positive, and we call this material a *p-type semiconductor* (Fig. 4.11*b*).

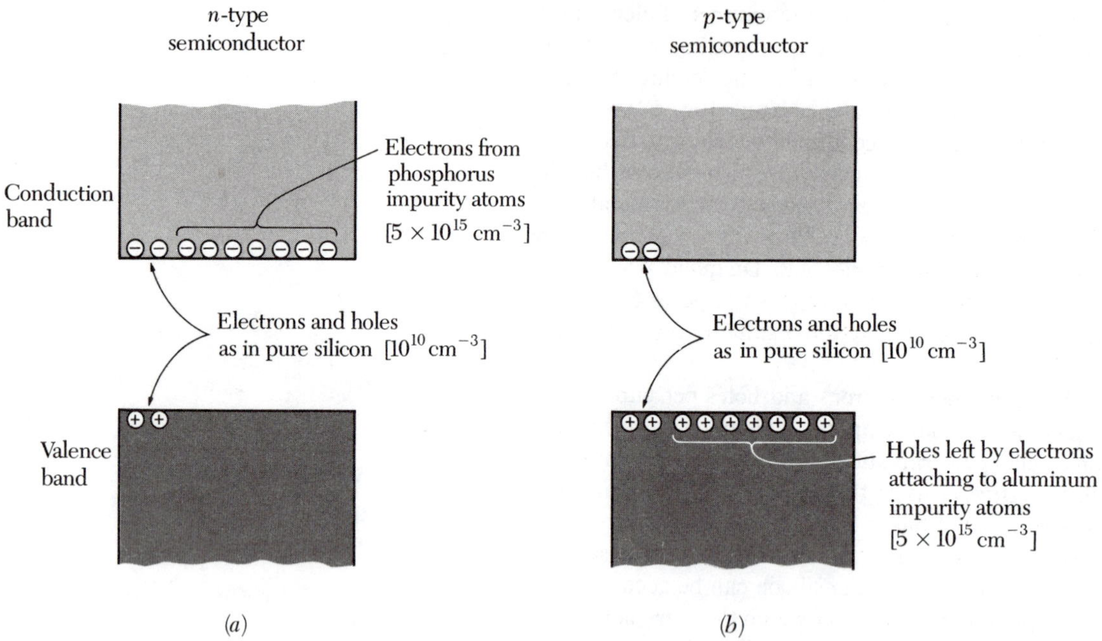

n-type semiconductor
p-type semiconductor
Conduction band
Electrons from phosphorus impurity atoms [5 × 10^15 cm^-3]
Electrons and holes as in pure silicon [10^10 cm^-3]
Electrons and holes as in pure silicon [10^10 cm^-3]
Valence band
Holes left by electrons attaching to aluminum impurity atoms [5 × 10^15 cm^-3]
(a)
(b)

FIGURE 4.11
In an n-type semiconductor most of the charge carriers are electrons released from pentavalent impurity atoms such as phosphorus. In the p-type semiconductor the majority of the charge carriers are holes. A hole is created when a trivalent impurity atom like aluminum grabs an electron to complete the covalent bonds to its four silicon neighbors. A few carriers of the opposite sign exist in each case, as they would in a pure silicon crystal at the same temperature. The number densities in brackets refer to our example of 5×10^{15} impurity atoms per cm^3, and room temperature. Under these conditions the number of majority charge carriers is practically equal to the number of impurity atoms, while the number of minority carriers is *very* much smaller.

Once the number of mobile charge carriers has been established, whether electrons or holes or both, the conductivity depends on their mobility, which is limited, as in metallic conduction, by scattering within the crystal. A single homogeneous semiconductor obeys Ohm's law. The spectacularly nonohmic behavior of semiconductor devices—as in a rectifier or a transistor—is achieved by combining n-type material with p-type material in various arrangements.

CIRCUITS AND CIRCUIT ELEMENTS

4.7 Electrical devices usually have well-defined terminals to which wires can be connected. Charge can flow into or out of the device over these paths. In particular, if two terminals, and only two, are connected by wires to something outside, and if the current flow is steady with constant potentials everywhere, then obviously the current must be equal and opposite at the two terminals.† In that case we can speak of *the* current I which flows through the device, and of *the* voltage V "between the terminals" or "across the terminals," which means their difference in electric potential. The ratio V/I for some given I is a

†It is perfectly possible to have 4 amps flowing into one terminal of a two-terminal object with 3 amps flowing out at the other terminal. But then the object is accumulating positive charge at the rate of 1 coulomb/sec. Its potential must be changing very rapidly—and that can't go on for long. Hence this cannot be a *steady,* or time-independent, current.

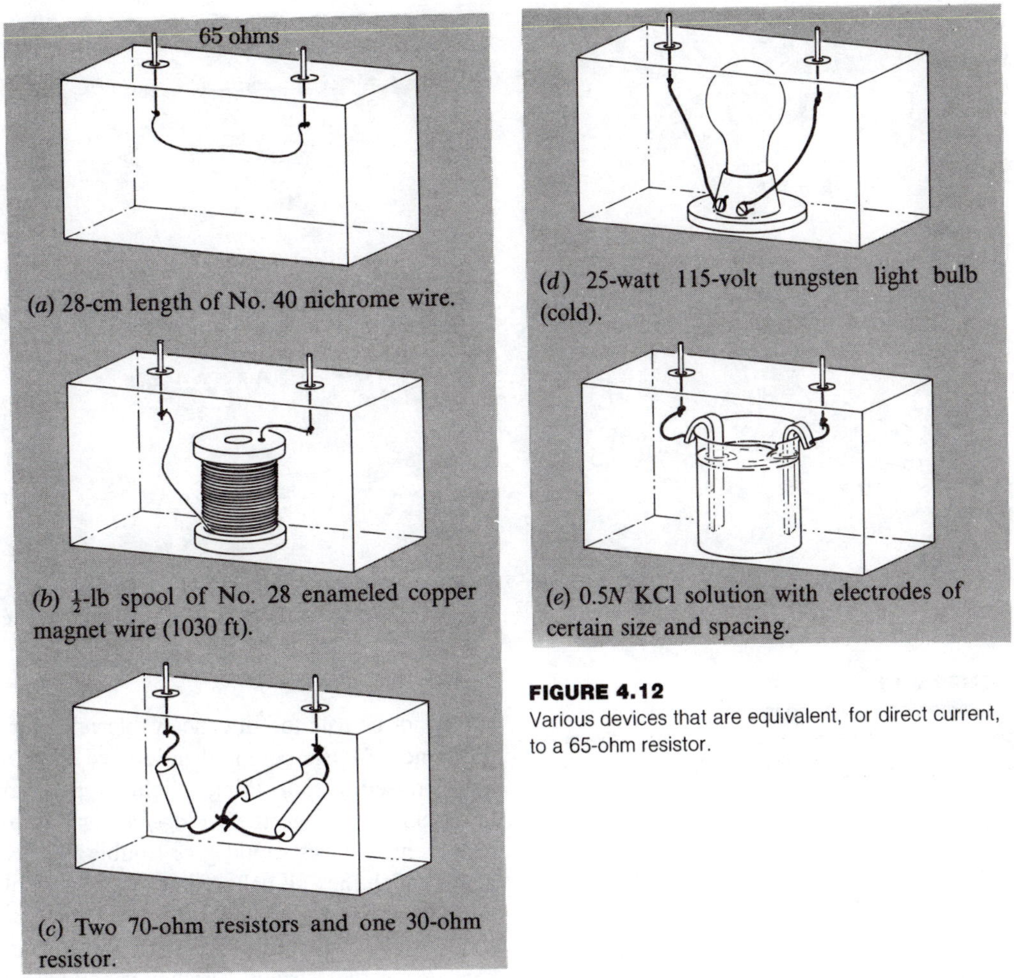

(*a*) 28-cm length of No. 40 nichrome wire.

(*b*) $\frac{1}{2}$-lb spool of No. 28 enameled copper magnet wire (1030 ft).

(*c*) Two 70-ohm resistors and one 30-ohm resistor.

(*d*) 25-watt 115-volt tungsten light bulb (cold).

(*e*) 0.5*N* KCl solution with electrodes of certain size and spacing.

FIGURE 4.12
Various devices that are equivalent, for direct current, to a 65-ohm resistor.

certain number of resistance units (ohms, if V is in volts and I in amp). If Ohm's law is obeyed in all parts of the object through which current flows, that number will be a constant, independent of the current. This one number completely describes the electrical behavior of the object, for steady current flow (dc) between the given terminals. With these rather obvious remarks we introduce a simple idea, the notion of a *circuit element*.

Look at the five boxes in Fig. 4.12. Each has two terminals, and inside each box there is some stuff, different in every box. If any one of these boxes is made part of an electrical circuit by connecting wires to the terminals, the ratio of the potential difference between the terminals to the current flowing in the wire that we have connected to the terminal will be found to be 65 ohms. We say the resistance between the terminals, in each box, is 65 ohms. This statement would

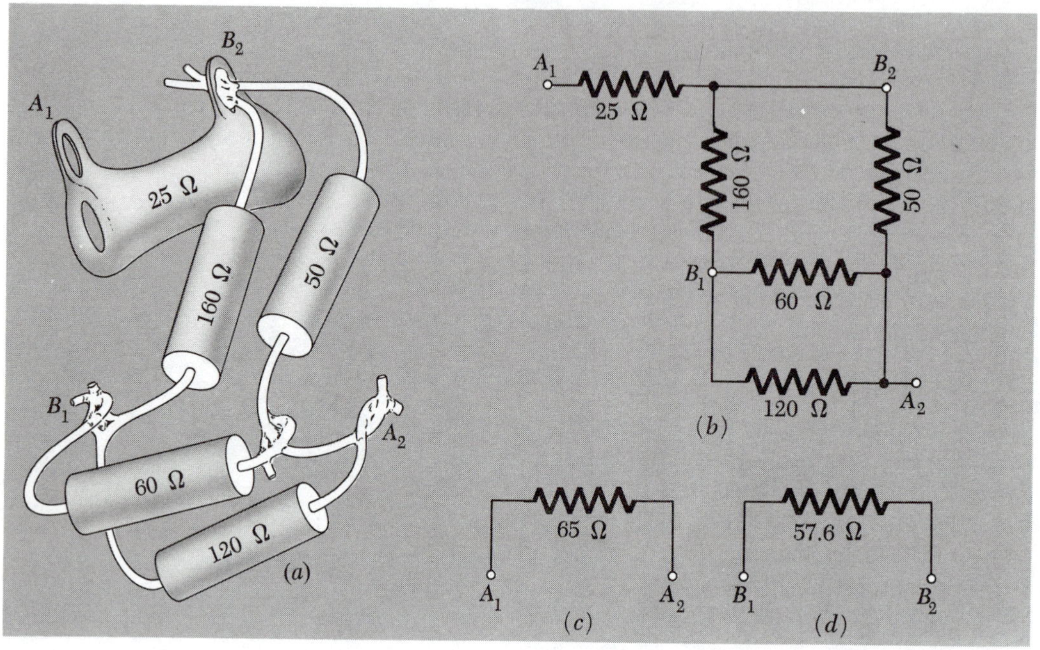

FIGURE 4.13

Some resistors connected together (*a*), the circuit diagram (*b*), and the equivalent resistance between certain pairs of terminals (*c*) and (*d*).

surely not be true for all conceivable values of the current or potential difference. As the potential difference or *voltage* between the terminals is raised, various things might happen, earlier in some boxes than in others, to change the *voltage/current* ratio. You might be able to guess which boxes would give trouble first. Still, there is *some* limit below which they all behave linearly, and within that range, for *steady* currents, the boxes are alike. They are alike in this sense: If any circuit contains one of these boxes, which box it is makes no difference in the behavior of that circuit. The box is equivalent to a 65-ohm resistor.† We represent it by the symbol -WWW- and in the description of the circuit of which the box is one component, we replace the box with this abstraction. An electrical circuit or network is then a collection of such circuit elements joined to one another by paths of negligible resistance.

Taking a network consisting of many elements connected together and selecting two points as terminals, we can regard the whole thing as equivalent, as far as these two terminals are concerned, to a single resistor. We say that the physical network of objects in Fig. 4.13*a* is represented by the diagram of Fig. 4.13*b* and for the termi-

†We use the term *resistor* for the actual object designed especially for that function. Thus a "200-ohm, 10-watt, wire-wound resistor" is a device consisting of a coil of wire on some insulating base, with terminals, intended to be used in such a way that the average power dissipated in it is not more than 10 watts.

nals A_1A_2 the equivalent circuit is Fig. 4.13*c*. The equivalent circuit for the terminals at B_1B_2 is given in Fig. 4.13*d*. If you put this assembly in a box with only that pair of terminals accessible, it will be indistinguishable from a resistor of 57.6 ohms resistance. There is one very important rule—only *direct-current* measurements are allowed! All that we have said depends on the current and electric fields being constant in time; if they are not, the behavior of a circuit element may not depend on its resistance alone. The concept of equivalent circuit can be extended from these dc networks to systems in which current and voltage vary with time. Indeed, that is where it is most valuable. We are not quite ready to explore that domain.

Little time will be spent here on methods for calculating the equivalent resistance of a network of circuit elements. The cases of series and parallel groups are easy. A combination like that in Fig. 4.14 is two resistors, of value R_1 and R_2, in series. The equivalent resistance is

$$R = R_1 + R_2 \tag{22}$$

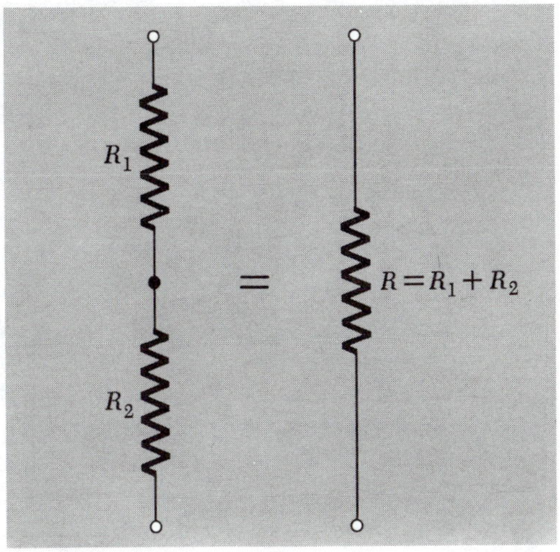

FIGURE 4.14
Resistances in series.

A combination like that in Fig. 4.15 is two resistors in parallel. By an argument that you should be able to give, the equivalent resistance R is found as follows:

$$\frac{1}{R} = \frac{1}{R_1} + \frac{1}{R_2} \qquad \text{or} \qquad R = \frac{R_1R_2}{R_1 + R_2} \tag{23}$$

That is all that is needed to handle a circuit like the one shown in Fig. 4.16, which, complicated as it looks, can be reduced, step by step, to series or parallel combinations. However, the simple network of Fig. 4.17 *cannot* be so reduced, so a more general method is required. Any conceivable network of resistors in which a constant current is flowing has to satisfy these conditions:

FIGURE 4.15
Resistances in parallel.

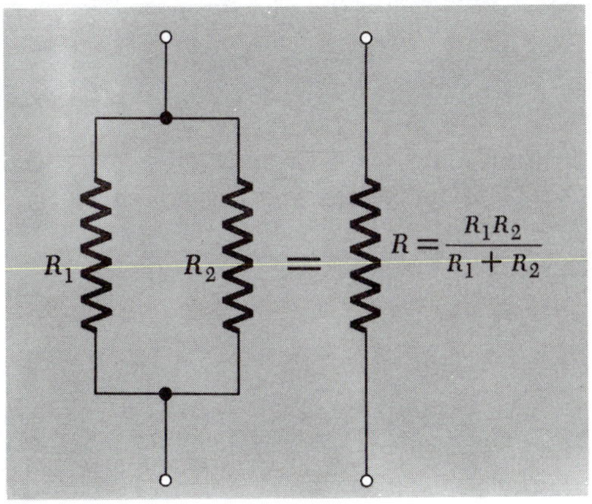

1. The current through each element must equal the voltage across that element divided by the resistance of the element.

2. At a *node* of the network, a point where three or more connecting wires meet, the algebraic sum of the currents into the node must be zero. (This is our old charge-conservation condition, Eq. 7, in circuit language.)

3. The sum of the potential differences taken in order around a *loop* of the network, a path beginning and ending at the same node, is zero. (This is network language for the general property of the static electric field: $\int \mathbf{E} \cdot d\mathbf{s} = 0$ for any closed path.)

The algebraic statement of these conditions for any network will provide exactly the number of independent linear equations needed to

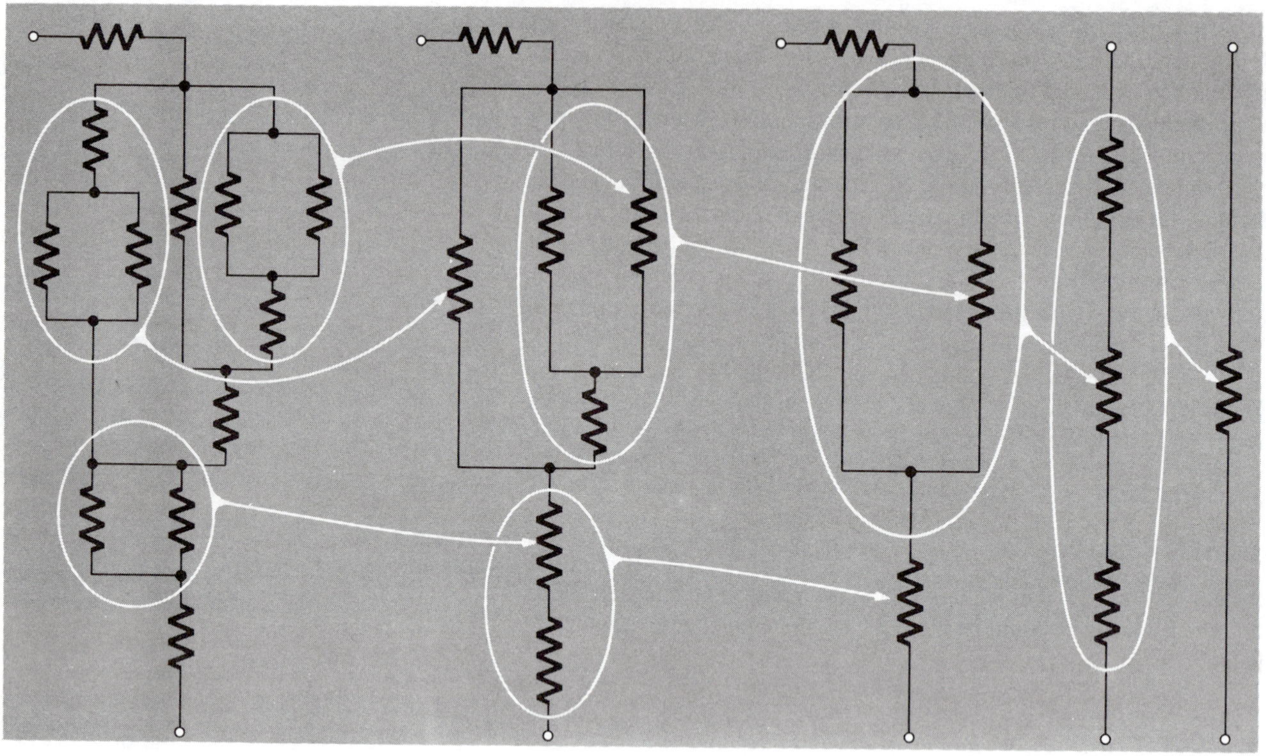

FIGURE 4.16
Reduction of a network that consists of series and parallel combinations only.

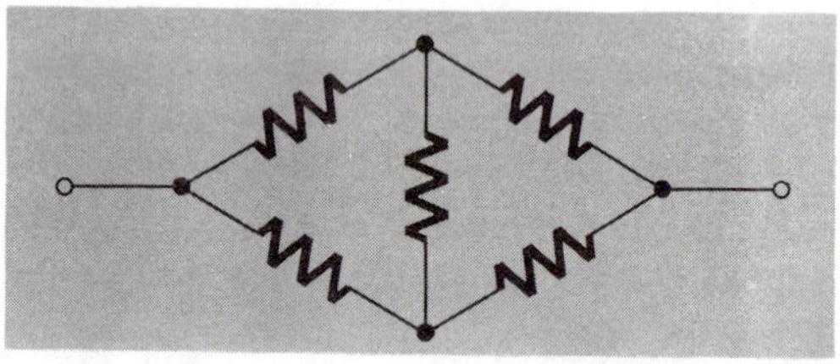

FIGURE 4.17
A simple bridge network. It can't be reduced in the manner of Fig. 4.16.

ensure that there is one and only one solution for the equivalent resistance between two selected nodes. We assert this without proving it. It is interesting to note that the structure of a dc network problem depends only on the *topology* of the network, that is, on those features of the diagram of connections that are independent of any distortion of the lines of the diagram.

A dc network of resistances is a linear system—the voltages and currents are governed by a set of linear equations, the statements of the conditions 1, 2, and 3. Therefore the superposition of different possible states of the network is also a possible state. Figure 4.18 shows a section of a network with certain currents, $I_1, I_2, \ldots$, flowing in the wires and certain potentials, $V_1, V_2, \ldots$, at the nodes. If some other set of currents and potentials, say $I_1', \ldots, V_1', \ldots$, is another possible state of affairs in this section of network, then so is the set $(I_1 + I_1')$, $\ldots, (V_1 + V_1'), \ldots$. These currents and voltages corresponding to the superposition will also satisfy the conditions 1, 2, and 3. Some general theorems about networks, interesting and useful to the electrical engineer, are based on this.

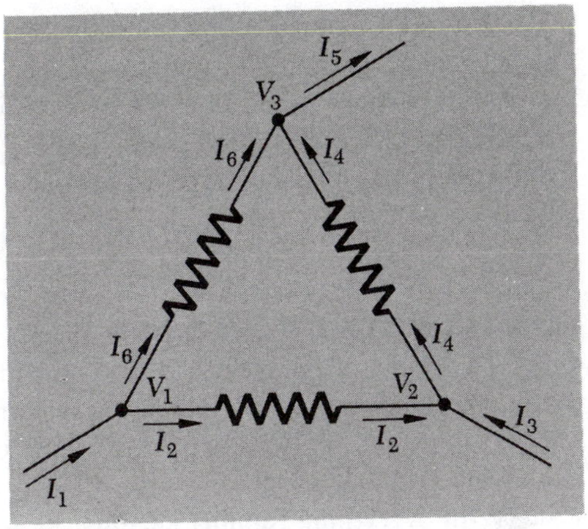

FIGURE 4.18
Currents and potentials at the nodes of a network.

ENERGY DISSIPATION IN CURRENT FLOW

4.8 The flow of current in a resistor involves the dissipation of energy. If it takes a force **F** to push a charge carrier along with average velocity **v**, any agency that accomplishes this must do work at the rate $\mathbf{F} \cdot \mathbf{v}$. If an electric field **E** is driving the ion of charge q, then $\mathbf{F} = q\mathbf{E}$, and the rate at which work is done is $q\mathbf{E} \cdot \mathbf{v}$. The energy thus expended shows up eventually as heat. In our model of ionic conduction the way this comes about is quite clear. The ion acquires some extra kinetic energy, as well as momentum, between collisions. A collision, or at most a few collisions, redirects its momentum at random but does not necessarily restore the kinetic energy to normal. For that to happen the ion has to transfer kinetic energy to the obstacle that deflects it. Suppose the charge carrier has a considerably smaller mass than the neutral atom it collides with. The average transfer of kinetic energy is small when a billiard ball collides with a bowling ball. Therefore the ion (billiard ball) will continue to accumulate extra energy until its average kinetic energy is so high that its average loss of energy in a collision equals the amount gained between collisions. In this way, by first "heating up" the charge carriers themselves, the work done by the electrical force driving the charge carriers is eventually passed on to the rest of the medium as random kinetic energy, or heat.

Suppose a steady current I, in amperes, flows through a resistor of R ohms. In every second, I coulombs of charge are transferred through a potential difference of V volts, where $V = IR$. Hence the work done in 1 sec is I^2R, in joules. (1 coulomb $\times$ 1 volt = 1 joule = 10^7 ergs.) The *watt*, or volt-ampere, is the corresponding unit of power P (rate of doing work) (1 watt = joule/sec).

$$P = I^2R \tag{24}$$

Naturally the steady flow of current in a dc circuit requires some source of energy capable of maintaining the electric field that drives the charge carriers. Until now we have avoided the question of the *electromotive force* by studying only parts of entire circuits; we kept the "battery" out of the picture. In Section 4.9 we shall discuss some sources of electromotive force.

ELECTROMOTIVE FORCE AND THE VOLTAIC CELL

4.9 The origin of the electromotive force in a direct-current circuit is some mechanism that transports charge carriers in a direction *opposite* that in which the electric field is trying to move them. A Van de Graaff electrostatic generator (Fig. 4.19) is an example on a large scale. With everything running steadily, we find current in the external resistance flowing in the direction of the electric field **E**, and energy being dissipated there (appearing as heat) at the rate IV_0, or I^2R. Inside the column of the machine, too, there is a downward-directed electric field. Here charge carriers can be moved against the field if they are stuck to a nonconducting belt. They are stuck so tightly that they can't slide backward along the belt in the generally downward electric field. (They can still be removed from the belt by a much stronger field localized at the brush in the terminal. We need not consider here the means for putting charge on and off the belt near the pulleys.) The energy needed to pull the belt is supplied from elsewhere—usually by an electric motor connected to a power line, but it could be a gasoline engine, or even a person turning a crank. This Van de Graaff generator is in effect a battery with an electromotive force, under these conditions, of V_0 volts.

FIGURE 4.19
In the Van de Graaff generator, charge carriers are mechanically transported in a direction opposite that in which the electric field would move them.

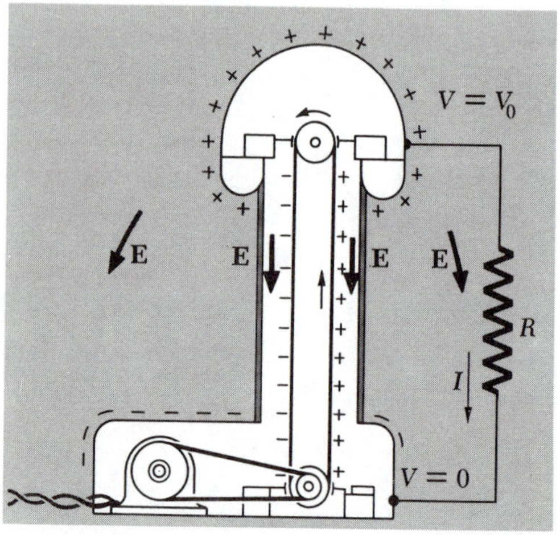

In ordinary batteries it is chemical energy that makes the charge carriers move through a region where the electric field opposes their motion. That is, a *positive* charge carrier may move to a place of *higher* electric potential if by so doing it can engage in a chemical reaction that will yield more energy than it costs to climb the electrical hill.

To see how this works, let us examine one particular voltaic cell. *Voltaic cell* is the generic name for a chemical source of electromotive force. In the experiments of Galvani around 1790 the famous twitching frogs' legs had signaled the chemical production of electric current. It was Volta who proved that the source was not "animal electricity," as Galvani maintained, but the contact of dissimilar metals in the circuit. Volta went on to construct the first battery, a stack of elementary cells, each of which consisted of a zinc disk and a silver disk separated by cardboard moistened with brine. The battery that powers your transistor radio comes in a tidier package, but the principle of operation is the same. Several kinds of voltaic cells are in use, differing in their chemistry but having common features: two electrodes of different material immersed in an ionized fluid, or electrolyte.

As an example, we'll describe the lead–sulfuric acid cell which is the basic element of the automobile battery. This cell has the important property that its operation is readily reversible. With a *storage battery* made of such cells, which can be charged and discharged repeatedly, energy can be stored and recovered electrically.

A fully charged lead–sulfuric acid cell has positive plates which hold lead dioxide, PbO_2, as a porous powder, and negative plates which hold pure lead of a spongy texture. The mechanical framework, or grid, is made of a lead alloy. All the positive plates are connected together and to the positive terminal of the cell. The negative plates, likewise connected, are interleaved with the positive plates, with a small separation. The schematic diagram in Fig. 4.20 shows only a small portion of a positive and a negative plate. The sulfuric acid electrolyte fills the cell, including the interstices of the active material, the porosity of which provides a large surface area for chemical reaction.

The cell will remain indefinitely in this condition if there is no external circuit connecting its terminals. The potential difference between its terminals will be close to 2.1 volts. This open-circuit potential difference is established "automatically" by the chemical interaction of the constituents. This is the *electromotive force* of the cell, for which the symbol $\mathcal{E}$ will be used. Its value depends on the concentration of sulfuric acid in the electrolyte, but not at all on the size, number, or separation of the plates.

Now connect the cell's terminals through an external circuit with resistance R. If R is not too small, the potential difference V between the cell terminals will drop only a little below its open-circuit value $\mathcal{E}$, and a current $I = V/R$ will flow around the circuit (Fig. 4.20*b*). Electrons flow *into* the positive terminal; other electrons flow *out* of the negative terminal. At each electrode chemical reactions are proceeding, the overall effect of which is to convert lead, lead dioxide, and sulfuric acid into lead sulfate and water. For every molecule of lead sulfate thus made, one charge e is passed around the circuit and an amount of energy $e\mathcal{E}$ is released. Of this energy the amount eV appears as heat in the external resistance R. The difference between $\mathcal{E}$ and V is caused by the resistance of the electrolyte itself, through which the current I must flow inside the cell. If we represent this internal resistance by R_i, the system can be quite well described by the equivalent circuit in Fig. 4.21.

As discharge goes on and the electrolyte becomes more diluted with water, the electromotive force $\mathcal{E}$ decreases somewhat. Normally, the cell is considered discharged when $\mathcal{E}$ has fallen below 1.75 volts. To recharge the cell, current must be forced around the circuit in the opposite direction by connecting a voltage source greater than $\mathcal{E}$ across the cell's terminals. The chemical reactions then run backward

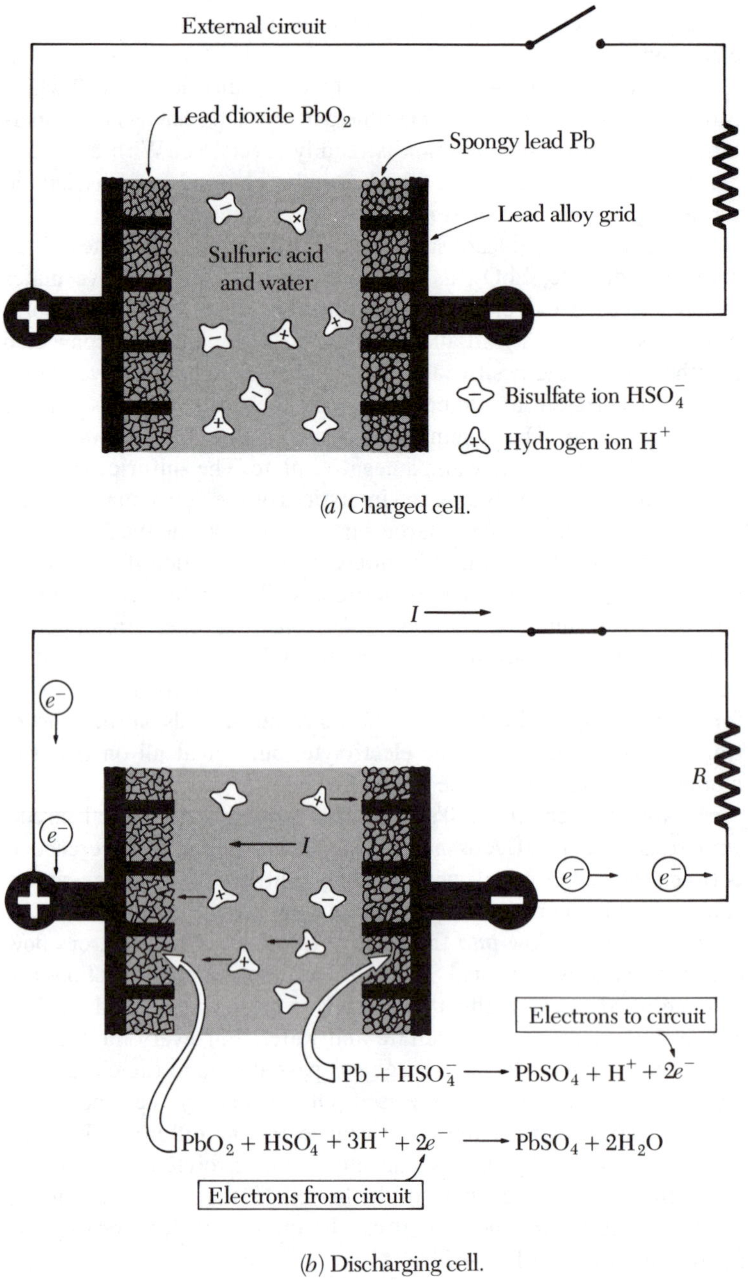
External circuit
Lead dioxide PbO_2
Spongy lead Pb
Lead alloy grid
Sulfuric acid and water
Bisulfate ion HSO_4^-
Hydrogen ion H^+
(a) Charged cell.
I
R
Electrons to circuit
$Pb + HSO_4^- \longrightarrow PbSO_4 + H^+ + 2e^-$
$PbO_2 + HSO_4^- + 3H^+ + 2e^- \longrightarrow PbSO_4 + 2H_2O$
Electrons from circuit
(b) Discharging cell.

FIGURE 4.20
A schematic diagram, not to scale, showing how the lead–sulfuric acid cell works. The electrolyte, sulfuric acid solution, permeates the lead dioxide granules in the positive plate and the spongy lead in the negative plate. The potential difference between the positive and negative terminals is 2.1 volts. With the external circuit closed, chemical reactions proceed at the solid-liquid interfaces in both plates, the depletion of sulfuric acid in the electrolyte, and the transfer of electrons through the external circuit from negative terminal to positive terminal, which constitutes the current *I*. To recharge the cell, replace the load *R* by a source with electromotive force greater than 2.1 volts, thus forcing current to flow through the cell in the opposite direction and reversing both reactions.

until all the lead sulfate is turned back into lead dioxide and lead. The investment of energy in charging the cell is somewhat more than the cell will yield on discharge, for the internal resistance R_i causes a power loss I^2R_i whichever way the current is flowing.

Notice in Fig. 4.20*b* that the current I in the electrolyte is produced by a net drift of positive ions toward the positive plate. Evidently the electric field in the electrolyte points toward, not away from, the positive plate. Nevertheless, the line integral of **E** around the whole circuit is zero, as it must be for any electrostatic field. The explanation is this: There are two very steep jumps in potential at the interface of the positive plate and the electrolyte and at the interface of the negative plate and the electrolyte. That is where the ions are moved *against* a strong electric field by forces arising in the chemical reactions. It is this region that corresponds to the belt in a Van de Graaff generator.

Every kind of voltaic cell has its characteristic electromotive force, falling generally in the range of 1 to 3 volts. The energy involved, per molecule, in any chemical reaction is essentially the gain or loss in the transfer of an outer electron from one atom to a different atom. That is never more than a few electron volts. We can be pretty sure that no one is going to invent a voltaic cell with a 12-volt electromotive force. The 12-volt automobile battery consists of six separate lead–sulfuric acid cells connected in series.

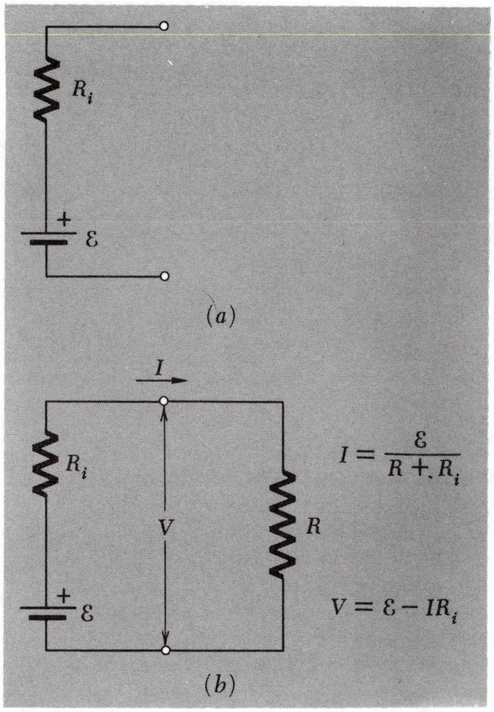

FIGURE 4.21
(*a*) The equivalent circuit for a voltaic cell is simply a resistance R_i in series with an electromotive force $\mathcal{E}$ of fixed value. (*b*) Calculation of the current in a circuit containing a voltaic cell.

NETWORKS WITH VOLTAGE SOURCES

4.10 A network of resistors could contain more than one electromotive force, or voltage source. The circuit in Fig. 4.22 contains two batteries with electromotive force $\mathcal{E}_1$ and $\mathcal{E}_2$, respectively. The positive terminal of each battery is indicated next to the conventional battery symbol. Assume that R_1 includes the internal resistance of one battery, R_2 that of the other. Supposing the resistances given, let us find the currents in this network. Having assigned directions arbitrarily to

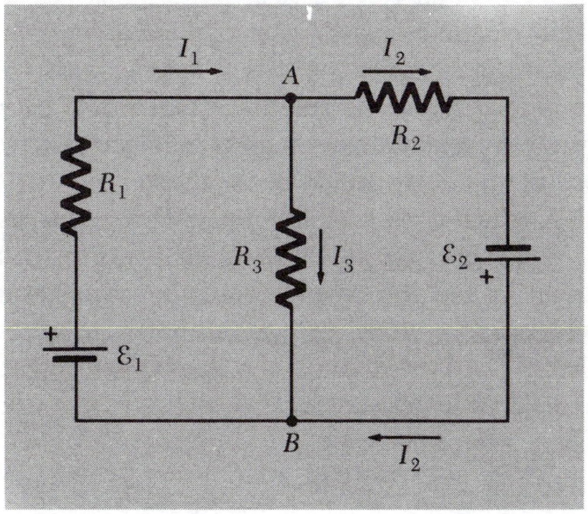

FIGURE 4.22
A network with two voltage sources.

the currents I_1, I_2, and I_3 in the branches, we impose the requirements stated in Section 4.7 and obtain three independent equations:

$$\begin{aligned} I_1 - I_2 - I_3 &= 0 \\ \mathcal{E}_1 - R_1 I_1 - R_3 I_3 &= 0 \\ \mathcal{E}_2 + R_3 I_3 - R_2 I_2 &= 0 \end{aligned} \tag{25}$$

To check the signs, note that in writing the two loop equations we have gone around each loop in the direction current would flow from the battery in that loop. The three equations can be solved for I_1, I_2, and I_3 with the result:

$$\begin{aligned} I_1 &= \frac{\mathcal{E}_1 R_2 + \mathcal{E}_1 R_3 + \mathcal{E}_2 R_3}{R_1 R_2 + R_2 R_3 + R_1 R_3} \\ I_2 &= \frac{\mathcal{E}_2 R_1 + \mathcal{E}_2 R_3 + \mathcal{E}_1 R_3}{R_1 R_2 + R_2 R_3 + R_1 R_3} \\ I_3 &= \frac{\mathcal{E}_1 R_2 - \mathcal{E}_2 R_1}{R_1 R_2 + R_2 R_3 + R_1 R_3} \end{aligned} \tag{26}$$

If in a particular case the value of I_3 turns out to be negative, it simply means that the current in that branch flows opposite to the direction we had assigned to positive current.

Suppose that a network such as this forms part of some larger system, to which it is connected at two of its nodes. For example, let us connect wires to the two nodes A and B and enclose the rest in a "black box" with these two wires as the only external terminals, as in Fig. 4.23*a*. A general theorem called Thévenin's theorem assures us that this two-terminal box is completely equivalent, in its behavior in any other circuit to which it may be connected, to a *single* voltage source $\mathcal{E}_{eq}$ with an internal resistance R_{eq}. This holds for any network of voltage sources and resistors, no matter how complicated. The values of $\mathcal{E}_{eq}$ and R_{eq} are easily determined. $\mathcal{E}_{eq}$ is the voltage between the two terminal wires when nothing is connected to them outside the box. In our example that is just $I_3 R_3$, with I_3 given by Eq. 26. The resistance R_{eq} is the resistance that would be measured between the two terminals with all the internal electromotive forces made zero. In our example that would be the resistance of R_1, R_2, and R_3 all in parallel, which is $R_1 R_2 R_3/(R_1 R_2 + R_2 R_3 + R_1 R_3)$.

What if we didn't know what was in the box? We could determine $\mathcal{E}_{eq}$ and R_{eq} experimentally by two measurements: Measure the *open-circuit voltage* with a voltmeter that draws negligible current; that is $\mathcal{E}_{eq}$. Now connect the terminals together through an ammeter of negligible resistance; this measures the *short-circuit current* I_{sc}. Then

$$R_{eq} = \frac{\mathcal{E}_{eq}}{I_{sc}} \tag{27}$$

FIGURE 4.23
Make R_{eq} equal to the resistance that would be measured between the terminals in (*a*) if all electromotive forces were zero. Make E_{eq} equal to the voltage observed between the terminals in (*a*) with the external circuit open. Then the circuit below is *equivalent* to the circuit above. You can't tell the difference by any direct-current measurement at those terminals.

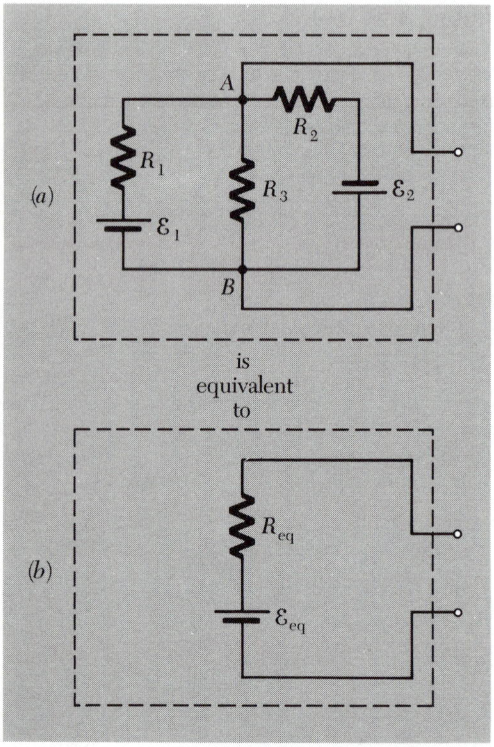

In analyzing a complicated circuit it sometimes helps to replace a two-terminal section by its equivalent $\mathcal{E}_{eq}$ and R_{eq}. Thévenin's theorem assumes the linearity of all circuit elements, including the reversibility of currents through batteries. If one of our batteries is a nonrechargeable dry cell with the current through it backward, caution is advisable!

VARIABLE CURRENTS IN CAPACITORS AND RESISTORS

4.11 Let a capacitor of capacitance C be charged to some potential V_0 and then discharged by suddenly connecting it across a resistance R. Figure 4.24 shows the capacitor indicated by the conventional symbol –||–, the resistor R, and a switch which we shall imagine to be closed at time $t = 0$. It is obvious that as current flows the capacitor will gradually lose its charge, the voltage across the capacitor will diminish, and this in turn will lessen the flow of current. To see exactly what happens we need only write down the conditions that govern the circuit. Let Q be the charge on the capacitor at any instant, V the potential difference between the plates which is also the voltage across the resistance R. Let I be the current, considered positive if it flows away from the positive side of the capacitor. These quantities, all functions of the time, must be related as follows:

$$Q = CV \qquad I = \frac{V}{R} \qquad -\frac{dQ}{dt} = I \tag{28}$$

Eliminating I and V, we obtain the equation which governs the time variation of Q:

$$\frac{dQ}{dt} = -\frac{Q}{RC} \tag{29}$$

Writing this in the form

$$\frac{dQ}{Q} = -\frac{dt}{RC} \tag{30}$$

we can integrate both sides, obtaining

$$\ln Q = \frac{-t}{RC} + \text{const} \tag{31}$$

The solution of our differential equation is therefore

$$Q = (\text{another constant}) \times e^{-t/RC} \tag{32}$$

We said that at $t = 0$, $V = V_0$, so that $Q = CV_0$ for $t = 0$.

FIGURE 4.24

Charge and current in an RC circuit. Charge decays by the factor $1/e$ in time RC.

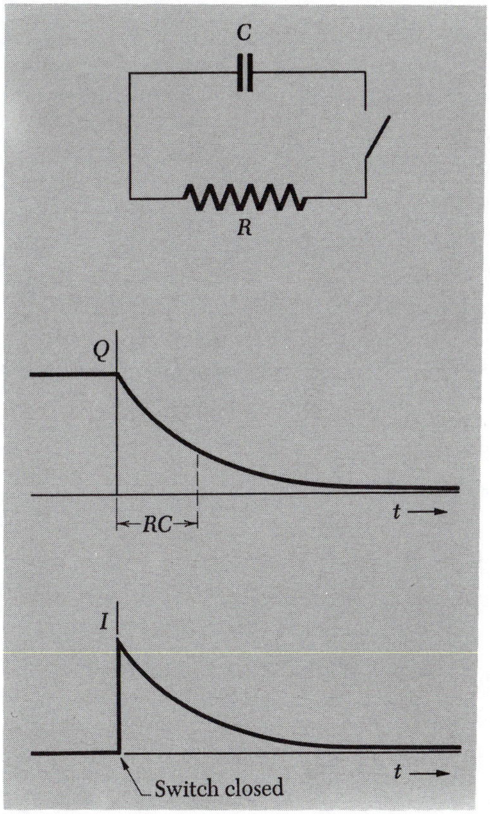

This determines the constant, and we now have the exact behavior of Q after the switch is closed:

$$Q = CV_0e^{-t/RC} \tag{33}$$

The behavior of the current I is found directly from this:

$$I = -\frac{dQ}{dt} = \frac{V_0}{R}e^{-t/RC} \tag{34}$$

At the closing of the switch the current rises at once to the value V_0/R and then decays exponentially to zero. The time that characterizes this decay is the constant RC. We should not be surprised to find that the product of resistance and capacitance has the dimensions of time, for we know that C has the dimensions of length, and we have already remarked that *resistance* $\times$ *length,* when it appears as ohm-cm, the unit of resistivity, has the dimensions of time. People often speak of the "RC time constant" associated with a circuit or part of a circuit.

In SI units the unit of capacitance is the farad. A capacitor of 1-farad capacitance has a charge of 1 coulomb for a potential difference of 1 volt. With R in ohms and C in farads, the product RC is a time in sec. Just to check this, note that ohm = volts/amp = volt-sec/coulomb, while farad = coulombs/volt. If we make the circuit of Fig. 4.24 out of a 0.05-microfarad capacitor and a 5-megohm resistor, both of which are reasonable objects to find around any laboratory, we would have $RC = 5 \times 10^6 \times 0.05 \times 10^{-6}$ or 0.25 sec.

Quite generally, in any electrical system made up of charged conductors and resistive current paths, one time scale—perhaps not the only one—for processes in the system is set by some resistance-capacitance product. This has a bearing on our earlier observation about the dimensions of resistivity. Imagine a capacitor with plates of area A and separation s. Its capacitance C is $A/4\pi s$. Now imagine the space between the plates suddenly filled with a conductive medium of resistivity ρ. To avoid any question of how this might affect the capacitance, let us suppose that the medium is a very slightly ionized gas; a substance of that density will hardly affect the capacitance at all. This new conductive path will discharge the capacitor as effectively as did the external resistor in Fig. 4.24. How quickly will this happen? The resistance of the path, R, is $\rho s/A$. Hence the time constant RC is just $(\rho s/A)(A/4\pi s) = \rho/4\pi$. For example, if our weakly ionized gas had a resistivity of 10^8 ohm-cm, the time constant for discharge of the capacitor would be about 10 microseconds. It does not depend on the size or shape of the capacitor.

What we have here is simply the time constant for the relaxation of an electric field in a conducting medium by redistribution of charge. We really don't need the capacitor plates to describe it. Imagine that we could suddenly imbed two sheets of charge, a negative sheet and a positive sheet, opposite one another in a conductor—for instance, in an n-type semiconductor (Fig. 4.25a). What will make these charges disappear? Do negative charge carriers move from the sheet on the left across the intervening space, neutralizing the positive charges when they arrive at the sheet on the right? Surely not—if that were the process, the time required would be proportional to the distance between the sheets. What happens instead is this. The *entire population* of negative charge carriers that fills the space between the sheets is caused to move by the electric field. Only a *very slight* displacement of this cloud of charge suffices to remove excess negative charge on the left, while providing on the right the extra negative charge needed to neutralize the positive sheet, as indicated in Fig. 4.25b. Within a conductor, in other words, neutrality is restored by a small readjustment of the entire charge distribution, not by a few charge carriers moving a long distance. That is why the relaxation time can be independent of the size of the system.

For a metal with resistivity typically 10^{-5} ohm-cm, $\rho/4\pi$ is about 10^{-18} sec, orders of magnitude shorter than the mean free time of a conduction electron in the metal. As a relaxation time this makes no sense. Our theory, at this stage, can tell us nothing about events on a time scale as short as that.

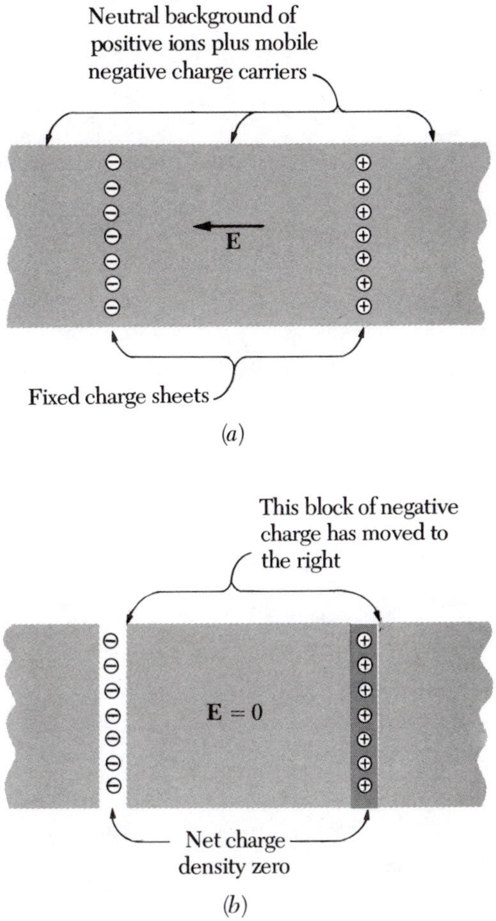

FIGURE 4.25
In a conducting medium, here represented by an n-type conductor, two fixed sheets of charge, one negative and one positive, can be neutralized by a slight motion of the entire block of mobile charge carriers lying between them. (a) Before the block of negative charge has moved. (b) After the net charge density has been reduced to zero at each sheet.

PROBLEMS

4.1 We have 5×10^{10} doubly charged positive ions per cm^3, all moving west with a speed of 10^7 cm/sec. In the same region there are 10^{11} electrons per cm^3 moving northeast with a speed of 10^8 cm/sec. (Don't ask how we managed it!) What is the direction of **J**? What is its magnitude in esu/sec-cm^2? In amps/meter^2?
Ans. 48.8° west of south; 5.14×10^9 esu/sec-cm^2; 1.71×10^4 amps/meter^2.

4.2 In a 6-gigaelectron-volt (GeV) electron synchrotron, electrons travel around the machine in an approximately circular path 240 meters long. It is normal to have about 10^{11} electrons circling on this path during a cycle of acceleration. The speed of the electrons is practically that of light. What is the current? We give this very simple problem to emphasize that nothing in our definition of current as rate of transport requires the velocities of the charge carriers to be nonrelativistic and that there is no rule against a given charged particle getting counted many times during a second as part of the current.
Ans. 0.020 amp.

4.3 In a Van de Graaff electrostatic generator, a rubberized belt 30 cm wide travels at a velocity of 20 meters/sec. The belt is given a surface charge at the lower roller, the surface charge density being high enough to cause a field of 40 statvolts/cm on each side of the belt. What is the current in milliamperes (milliamps)?

4.4 The first telegraphic messages crossed the Atlantic in 1858, by a cable 3000 km long laid between Newfoundland and Ireland. The conductor in this cable consisted of seven copper wires, each of diameter 0.73 mm, bundled together and surrounded by an insulating sheath.

(*a*) Calculate the resistance of the conductor. Use 3×10^{-6} ohm-cm for the resistivity of the copper, which was of somewhat dubious purity.

(*b*) A return path for the current was provided by the ocean itself. Given that the resistivity of seawater is about 25 ohm-cm, see if you can show that the resistance of the ocean return would have been much smaller than that of the cable.

4.5 Show that the total amount of charge at the junction of the two materials in Fig. 4.6 is $(I/4\pi)(1/\sigma_2 - 1/\sigma_1)$, where I is the current flowing through the junction, in esu/sec, and the conductivities σ_1 and σ_2 are expressed in CGS units of sec^{-1}.

4.6 A wire of pure tin is drawn through a die, reducing its diameter by 25 percent and increasing its length. By what factor will its resistance be increased? Then it is flattened into a ribbon by rolling, which results in a further increase in its length, which is now twice the original length. What has been the overall change in resistance? Assume the density and resistivity remain constant throughout.

4.7 A laminated conductor was made by depositing, alternately, layers of silver 100 angstroms thick and layers of tin 200 angstroms thick. The composite material, considered on a larger scale, may be considered a homogeneous but anisotropic material with an electrical conductivity $\sigma_\perp$ for currents perpendicular to the planes of the layers, and a different conductivity $\sigma_\parallel$ for currents parallel to that plane. Given that the conductivity of silver is 7.2 times that of tin, find the ratio $\sigma_\perp/\sigma_\parallel$.

Ans. 0.457.

4.8 A copper wire 1 km long is connected across a 6-volt battery. The resistivity of the copper is 1.7×10^{-6} ohm-cm; the number of conduction electrons per cubic centimeter is 8×10^{22}. What is the drift velocity of the conduction electrons under these circumstances? How long does it take an electron to drift once around the circuit?

4.9 Normally in the earth's atmosphere the greatest density of free electrons (liberated by ultraviolet sunlight) amounts to 10^6 per cm^3 and is found at an altitude of about 100 km where the density of air is so low that the mean free path of an electron is about 10 cm. At the temperature which prevails there an electron's mean speed is 10^7 cm/sec. What is the conductivity in sec^{-1} and in $(ohm\text{-}cm)^{-1}$?

Ans. $2 \times 10^8\ sec^{-1}$; $2 \times 10^{-4}\ (ohm\text{-}cm)^{-1}$

4.10 An ion in a liquid is so closely surrounded by neutral molecules that one can hardly speak of a "free time" between collisions. Still, it is interesting to see what value of τ is implied by Eq. 20 if we take the observed conductivity of pure water from Table 4.1 and the values given for N_+ and N_-, 10^{13} per cm^3. A typical thermal speed for a water molecule is 5×10^4 cm/sec. How far would it travel in that time τ?

Ans. 2.5×10^{-8} cm.

4.11 The resistivity of seawater is about 25 ohm-cm. The charge carriers are chiefly Na^+ and Cl^- ions, and of each there are about 3×10^{20} per cm^3. If we fill a plastic tube 2 meters long with seawater and connect a 12-volt battery to the electrodes at each end, what is the resulting average drift velocity of the ions, in cm/sec?

4.12 Use the figures given in Section 4.6 for the conductivity of pure silicon at 500 K and the density of conduction electrons and holes at that temperature to deduce the mean free time between collisions, assuming it is the same for electrons and holes.

4.13 In a silicon junction diode the region of the planar junction between *n*-type and *p*-type semiconductors can be approximately represented as two adjoining slabs of charge, one negative and one positive. Away from the junction, outside these charge layers, the potential is constant, its value being ϕ_n in the *n*-type material and ϕ_p in the *p*-type material. Given that the difference between ϕ_p and ϕ_n is 0.3 volt, and that the thickness of each of the two slabs of charge is 0.01 cm, find the charge density in each of the two slabs, and make a graph of the potential ϕ as a function of position x through the junction. What is the strength of the electric field at the midplane?

4.14 Refer to Eq. 20 and Fig. 4.10. Assume that $\tau_+ = \tau_-$ and $M_+ = M_- = m_e$, the electron mass. If a conductivity of 0.3 $(ohm\text{-}cm)^{-1}$ results from the presence of 10^{15} electrons per cm^3 in the conduction band and the same number of holes, what must be the value of the mean free time τ? The rms speed of an electron at 500 K is 1.5×10^7 cm/sec. Compare the mean free path with the distance between neighboring silicon atoms, which is 2.35×10^{-8} cm.

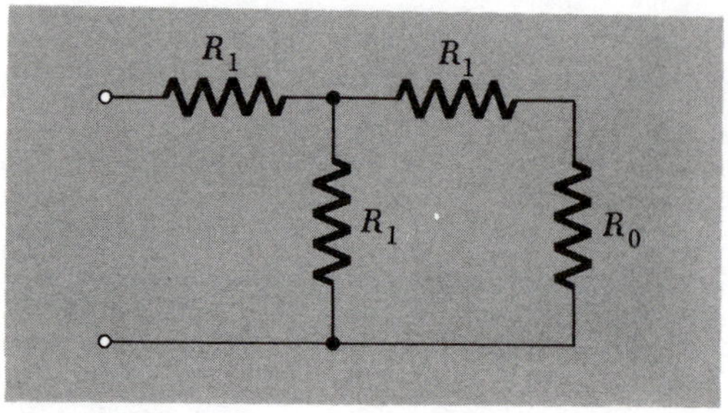

PROBLEM 4.16

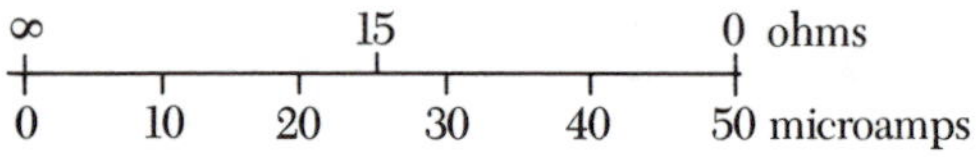

PROBLEM 4.19

4.15 Suppose that each of the resistors in the circuit at the left in Fig. 4.16 has the value 100 ohms. What is the resistance of the single equivalent resistor at the far right?

4.16 In the circuit, if R_0 is given, what value must R_1 have in order that the input resistance between the terminals shall be equal to R_0?

4.17 If the voltage at the terminals of an automobile battery drops from 12.3 to 9.8 volts when a 0.5-ohm resistor is connected across the battery, what is the internal resistance?

4.18 Show that, if a battery of fixed emf $\mathcal{E}$ and internal resistance R_i is connected to a variable external resistance R, the maximum power is delivered to the external resistor when $R = R_i$.

4.19 You have a microammeter which reads 50 microamps at full-scale deflection, and the coil in the meter movement has a resistance of 20 ohms. By adding two resistors, R_1 and R_2, and a 1.5-volt battery you can convert this into an ohmmeter. When the two outcoming leads of this ohmmeter are connected together, the meter is to register 0 ohms by giving exactly full-scale deflection. When the leads are connected across an unknown resistance R, the deflection will indicate the resistance value if the scale is appropriately marked. In particular, we want half-scale deflection to indicate 15 ohms. What values of R_1 and R_2 are required, how should the connections be made, and where on the ohm scale will the marks be (with reference to the old microammeter calibration) for 5 ohms and for 50 ohms?

4.20 A black box with three terminals, *a*, *b*, and *c*, contains nothing but three resistors and connecting wire. Measuring the resistance between pairs of terminals, we find R_{ab} = 30 ohms, R_{ac} = 60 ohms, and R_{bc} = 70 ohms. Show that the contents of the box could be either

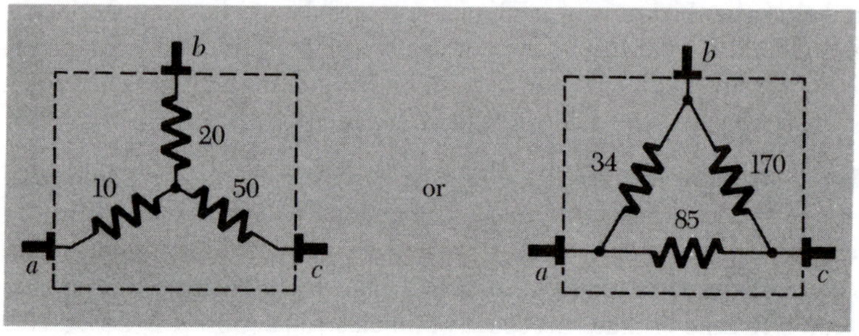

Is there any other possibility? Are the two boxes completely equivalent, or is there an external measurement that would distinguish between them?

4.21 In the circuit, all five resistors have the same value, 100 ohms, and each cell has an electromotive force of 1.5 volts. Find the open-circuit voltage and the short-circuit current for the terminals A, B. Then find $\mathcal{E}_0$ and R_0 for the Thévenin equivalent circuit.

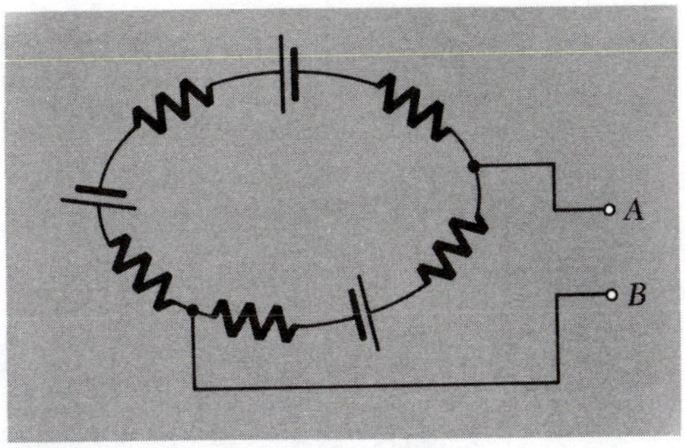

PROBLEM 4.21

4.22 A resistor R is to be connected across the terminals A, B, of the circuit on the right. For what value of R will the power dissipated in the resistor be greatest? To answer this, construct the Thévenin equivalent circuit and then invoke the result for Problem 4.18. How much power will be dissipated in R?

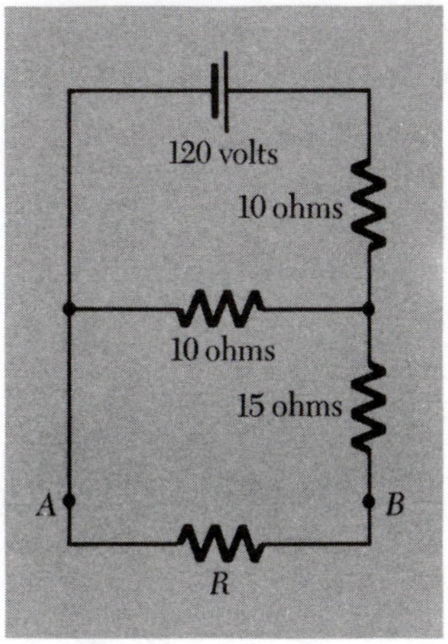

PROBLEM 4.22

4.23 Suppose the conducting medium in Fig. 4.25 is n-type silicon with 10^{15} electrons per cm^3 in the conduction band. Assume the initial density of charge on the sheets is such that the electric field strength is 1 statvolt/cm. By what distance must the intervening distribution of electrons be displaced to restore neutrality and reduce the electric field to zero?

4.24 As an illustration of the point made in the first footnote in Section 4.7 consider a black box which is approximately a 10-cm cube with two binding posts. Each of these terminals is connected by a wire to some external circuits. Otherwise, the box is well insulated from everything. A current of approximately 1 amp flows through this circuit element. Suppose now that the current in and the current out differ by one part in a million. About how long would it take, unless something else happens, for the box to rise in potential by 1000 volts?

4.25 Return to the example of the capacitor C discharging through the resistor R which was worked out in the text and show that the total energy dissipated in the resistor agrees with the energy originally stored in the capacitor. Suppose someone objects that the capacitor is never *really* discharged because Q only becomes zero for $t = \infty$. How would you counter this objection? You might find out how long it would take the charge to be reduced to one electron, with some reasonable assumptions.

4.26 Two graphite rods are of equal length. One is a cylinder of radius a. The other is conical, tapering linearly from radius a at one end to radius b at the other. Show that the end-to-end electrical resistance of the conical rod is a/b times that of the cylindrical rod. *Hint:* Consider the rod made up of thin, disklike slices, all in series.

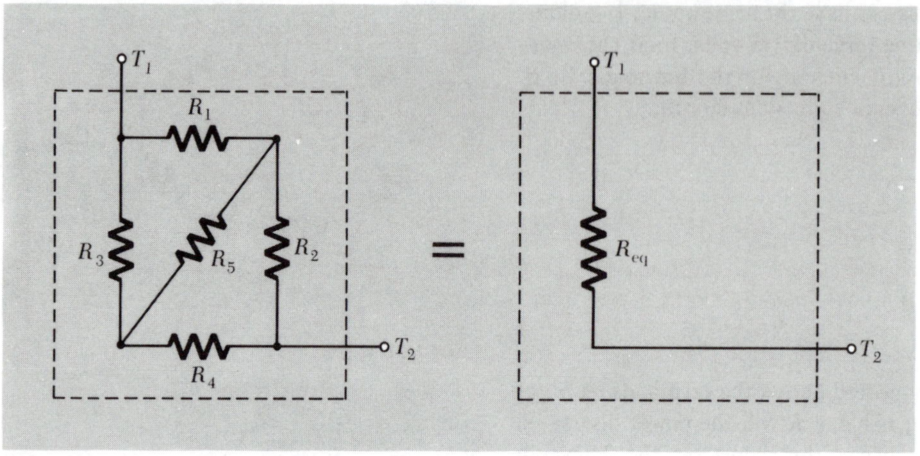

PROBLEM 4.27

4.27 This concerns the equivalent resistance R_{eq} between terminals T_1 and T_2 for the network of five resistors. One way to derive a formula for R_{eq} would be to solve the network for the current I that flows in at T_1 for a given voltage difference V between T_1 and T_2; then $R_{eq} = V/I$. The solution involves rather tedious algebra in which it is easy to make a mistake, so we'll tell you most of the answer:

$$R_{eq} = \frac{R_1R_2R_3 + R_1R_2R_4 + ? + R_2R_3R_4 + R_5(R_1R_3 + R_2R_3 + ? + R_2R_4)}{R_1R_2 + R_1R_4 + ? + R_3R_4 + R_5(R_1 + R_2 + R_3 + R_4)}$$

By considering the symmetry of the network you should be able to fill in the missing terms. Now check the formula by directly calculating R_{eq} in three special cases: (*a*) $R_5 = 0$, (*b*) $R_5 = \infty$, and (*c*) $R_1 = R_3 = 0$, and comparing your results with what the the formula gives.

4.28 A 12-volt lead-acid storage battery with a 20 ampere-hour capacity rating has a mass of 10 kg.

(*a*) How many kilograms of lead sulfate is formed when this battery is discharged. (Molecular weight of $PbSO_4$ is 303.)

(*b*) How many kilograms of batteries of this type would be required to store the energy derived from 1 kg of gasoline by an engine of 20 percent efficiency? (Heat of combustion of gasoline: 4.5×10^4 joules/gm.)

4.29 The common 1.5-volt dry cell used in flashlights and innumerable other devices releases its energy by oxidizing the zinc can which is its negative electrode, while reducing manganese dioxide, MnO_2, to Mn_2O_3 at the positive electrode. (It is called a carbon-zinc cell, but the carbon rod is just an inert conductor.) A cell of size D, weighing 90 gm, can supply 100 milliamps for about 30 hours.

(*a*) Compare its energy storage, in joules/kg, with that of the lead-acid battery described in Problem 4.28. Unfortunately the cell is not rechargeable.

(*b*) How high could you lift yourself with one D cell powering a 50 percent efficient winch?

4.30 The result for Problem 3.24 can help us to understand the flow of current in a circuit, part of which consists of charged particles moving through space between two electrodes. The question is, what is the nature of the current when only one particle traverses the space? (If we can work that out, we can easily describe any flow involving a larger number arriving on any schedule.) Consider the simple circuit in the figure, which consists of two electrodes in vacuum connected by a short wire. Suppose the electrodes are 2 mm apart. A rather slow alpha particle, of charge $2e$, is emitted by a radioactive nucleus in the left plate. It travels directly toward the right plate with a constant speed of 10^8 cm/sec and stops in this plate. Make a quantitative graph of the current in the connecting wire, plotting current against time. Do the same for an alpha particle that crosses the gap moving with the same speed but at an angle of 45° to the normal. (Actually for pulses as short as this the inductance of the connecting wire, here neglected, would affect the pulse shape.) Suppose we had a cylindrical arrangement of electrodes, with the alpha particles being emitted from a thin wire on the axis of a small cylindrical electrode. Would the current pulse have the same shape?

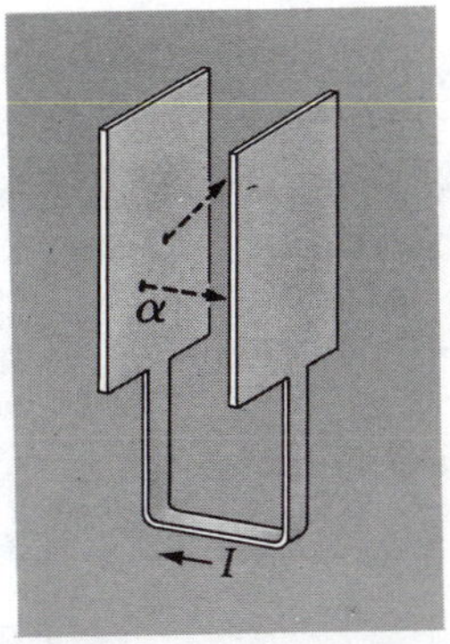

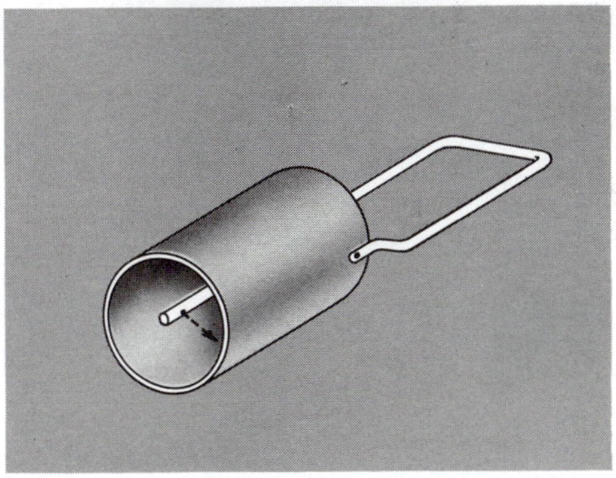

PROBLEM 4.30

4.31 All networks can be drawn flat if we adopt a conventional way of representing a "crossing without touching" such as ⤫ . Suppose a cube has a resistor along each edge. At each corner the leads from three resistors are soldered together. Flatten this network out into a circuit diagram. Find the equivalent resistance between two nodes that represent diagonally opposite corners of the cube, in the case where all resistors have the same value R_0. For this you do not need to solve a number of simultaneous equations; instead use symmetry arguments. Now find the equivalent resistance between two nodes that correspond to diagonally opposite corners of one face of the cube. Here again, considerations of symmetry will reduce the problem to a very simple one. For both these calculations, a sketch of the structure as a cube, rather than flattened out, will help you to spot the necessary symmetries in the currents.

4.32 Some important kinds of networks are infinite in extent. The figure shows a chain of series and parallel resistors stretching off endlessly to the right. The line at the bottom is the resistanceless return wire for all of them. This is sometimes called an attenuator chain, or a ladder network. The problem is to find the "input resistance," that is, the equivalent resistance between terminals A and B. Our interest

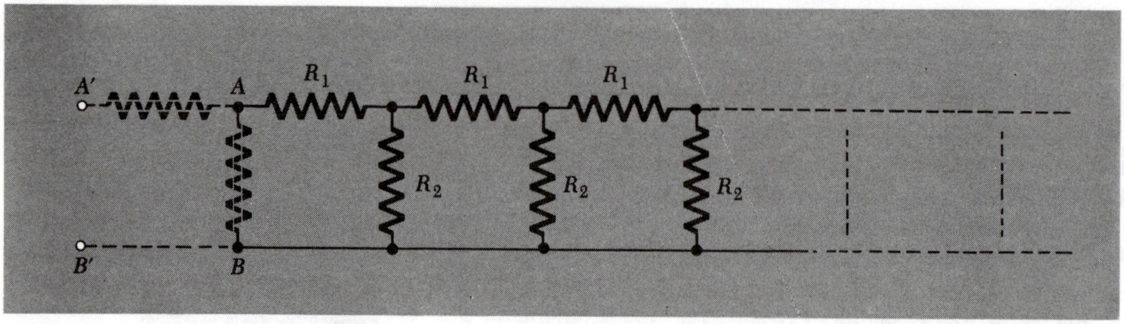

PROBLEM 4.32

in this problem mainly concerns the method of solution, which takes an odd twist and which can be used in other places in physics where we have an iteration of identical devices (even an infinite chain of lenses, in optics). The point is that the input resistance which we do not yet know—call it R—will not be changed by adding a new set of resistors to the front end of the chain to make it one unit longer. But now, adding this section, we see that this new input resistance is just R_1 in series with the parallel combination of R_2 and R. We get immediately an equation that can be solved for R. Show that, if voltage V_0 is applied at the input to such a chain, the voltage at successive nodes decreases in a geometric series. What ratio is required for the resistors to make the ladder an attenuator that halves the voltage at every step? Obviously a truly infinite ladder would not be practical. Can you suggest a way to terminate it after a few sections without introducing any error in its attenuation?

4.33 The figure shows two resistors in parallel, with values R_1 and R_2. The current I_0 divides somehow between them. Show that the condition that $I_1 + I_2 = I_0$, together with the requirement of *minimum power dissipation,* leads to the same current values that we would calculate with ordinary circuit formulas. This illustrates a general variational principle that holds for direct current networks: The distribution of currents within the network, for given input current I_0, is always that which gives the *least* total power dissipation.

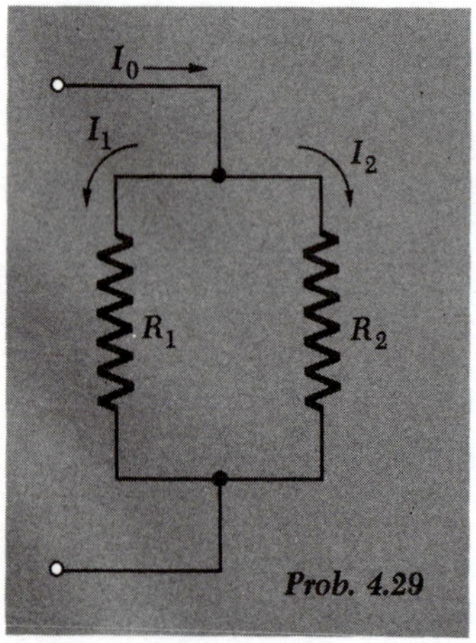

PROBLEM 4.33

5

THE FIELDS OF MOVING CHARGES

FROM OERSTED TO EINSTEIN

5.1 In the winter of 1819–1820 Hans Christian Oersted was lecturing on electricity, galvanism, and magnetism to advanced students at the University of Copenhagen. *Electricity* meant electrostatics; *galvanism* referred to the effects produced by continuous currents from batteries, a subject opened up by Galvani's chance discovery and the subsequent experiments of Volta; *magnetism* dealt with the already ancient lore of lodestones, compass needles, and the terrestrial magnetic field. It seemed clear to some that there must be a relation between galvanic currents and electric charge, although there was little more direct evidence than the fact that both could cause shocks. On the other hand, magnetism and electricity appeared to have nothing whatever to do with one another. Still Oersted had a notion, vague perhaps, but tenaciously pursued, that magnetism like the galvanic current might be a sort of "hidden form" of electricity. Groping for some manifestation of this, he tried before his class the experiment of passing a galvanic current through a wire which ran above and at right angles to a compass needle. It had no effect. After the lecture, something impelled him to try the experiment with a wire running parallel to the compass needle. The needle swung wide—and when the galvanic current was reversed it swung the other way!

The scientific world was more than ready for this revelation. A ferment of experimentation and discovery followed as soon as the word reached other laboratories. Before long Ampère, Faraday, and others had worked out an essentially complete and exact description of the magnetic action of electric currents. Faraday's crowning discovery of electromagnetic induction came less than 12 years after Oersted's experiment. In the previous two centuries since the publication in 1600 of William Gilbert's great work *De Magnete,* man's understanding of magnetism had advanced not at all. Out of these experimental discoveries there grew the complete classical theory of electromagnetism. Formulated mathematically by Maxwell, it was triumphantly corroborated by Hertz's demonstration of electromagnetic waves in 1888.

Special relativity has its historical roots in electromagnetism. Lorentz, exploring the electrodynamics of moving charges, was led very close to the final formulation of Einstein. And Einstein's great paper of 1905 was entitled not "Theory of Relativity," but rather "On the Electrodynamics of Moving Bodies." Today we see in the postulates of relativity and their implications a wide framework, one that embraces all physical laws and not solely those of electromagnetism. We expect any complete physical theory to be relativistically invariant. It ought to tell the same story in all inertial frames of reference. As it happened, physics already *had* one relativistically invariant theory—Maxwell's electromagnetic theory—long before the significance of relativistic invariance was recognized. Whether the ideas of special relativity could have evolved in the absence of a complete theory of the electromagnetic field is a question for the historian of science to speculate about; probably it can't be answered. We can only say that the actual history shows rather plainly a path running from Oersted's compass needle to Einstein's postulates.

Still, relativity is not a branch of electromagnetism, nor a consequence of the existence of light. The central postulate of special relativity, which no observation has yet contradicted, is the equivalence of reference frames moving with constant velocity with respect to one another. Indeed, it is possible, without even mentioning light, to derive the formulas of special relativity from nothing more than that postulate and the assumption that all spatial directions are equivalent.* The universal constant c then appears in these formulas as a limiting velocity, approached by an energetic particle but never exceeded. Its value can be ascertained by an experiment that does not involve light or anything else (such as neutrinos) which are believed to travel at precisely that speed. In other words, we would have special relativity even if electromagnetic waves could not exist.

Later in this chapter we are going to follow the historical path from Oersted to Einstein almost in reverse. We'll take special relativity as given, and ask how an electostatic system of charges and fields looks in another reference frame. In this way we shall find the forces that act on electric charges in motion, including the force that acts between electric currents. Magnetism, seen from this viewpoint, is a relativistic aspect of electricity.† But first, let's review some of the phenomena we shall be trying to explain.

MAGNETIC FORCES

5.2 Two wires running parallel to one another and carrying currents in the same direction are drawn together. The force on one of the wires, per unit length of wire, is inversely proportional to the dis-

*See N. David Mermin, "Relativity Without Light," *American Journal of Physics,* **52:**119 (1984), in which it is shown that the most general law for the addition of velocities which is consistent with the equivalence of inertial frames must have the form $v = (v_1 + v_2)/(1 + v_1v_2/c^2)$, identical to our Eq. 6 in Appendix A. To discover the value of the constant c in our universe we need only measure with adequate accuracy three lower speeds v, v_1, and v_2. For references to other articles on the same theme see also N.D. Mermin, *American Journal of Physics,* **52,** 967 (1984).

†The earliest exposition of this approach, to my knowledge, is the article by L. Page, A Derivation of the Fundamental Relations of Electrodynamics from Those of Electrostatics, *American Journal of Science,* **XXXIV**: 57 (1912). It was natural for Page, writing only 7 years after Einstein's revolutionary paper, to consider relativity more in need of confirmation than electrodynamics. His concluding sentence reads: "Viewed from another standpoint, the fact that we have been able, by means of the principle of relativity, to deduce the fundamental relations of electrodynamics from those of electrostatics, may be considered as some confirmation of the principle of relativity."

tance between the wires (Fig. 5.1*a*). Reversing the direction of one of the currents changes the force to one of repulsion. Thus the two sections of wire in Fig. 5.1*b*, which are part of the same circuit, tend to fly apart. There is some sort of "action at a distance" between the two filaments of steady electric current. It seems to have nothing to do with any static electric charge on the surface of the wire. There may be some such charge and the wires may be at different potentials, but the force we are concerned with depends only on the charge *movement* in the wires, that is, on the two currents. You can put a sheet of metal between the two wires without affecting this force at all (Fig. 5.1*c*). These new forces that come into play when charges are moving are called *magnetic*.

Oersted's compass needle (Fig. 5.2*a*) doesn't look much like a direct-current circuit. We now know, however, as Ampère was the first to suspect, that magnetized iron is full of perpetually moving charges—electric currents on an atomic scale. A slender coil of wire with a battery to drive current through it (Fig. 5.2*b*) behaves just like the compass needle under the influence of a nearby current.

Observing the motion of a free charged particle, instead of a wire carrying current, we find the same thing happening. In a cathode ray tube, electrons that would otherwise follow a straight path are deflected toward or away from an external current-carrying wire, depending on the relative direction of the current in that wire (Fig. 5.3). You are already familiar with this from the laboratory, and you know that this interaction of currents and other moving charges can

FIGURE 5.1
(*a*) Parallel wires carrying currents in the same direction are pulled together. (*b*) Parallel wires carrying currents in opposite directions are pushed apart. (*c*) These forces are not affected by putting a metal plate between the wires.

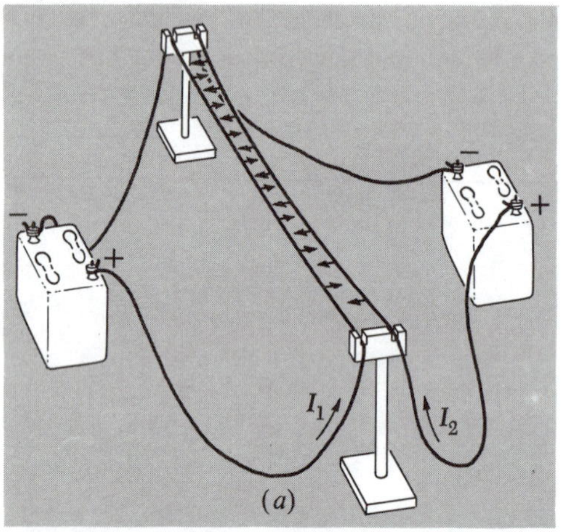

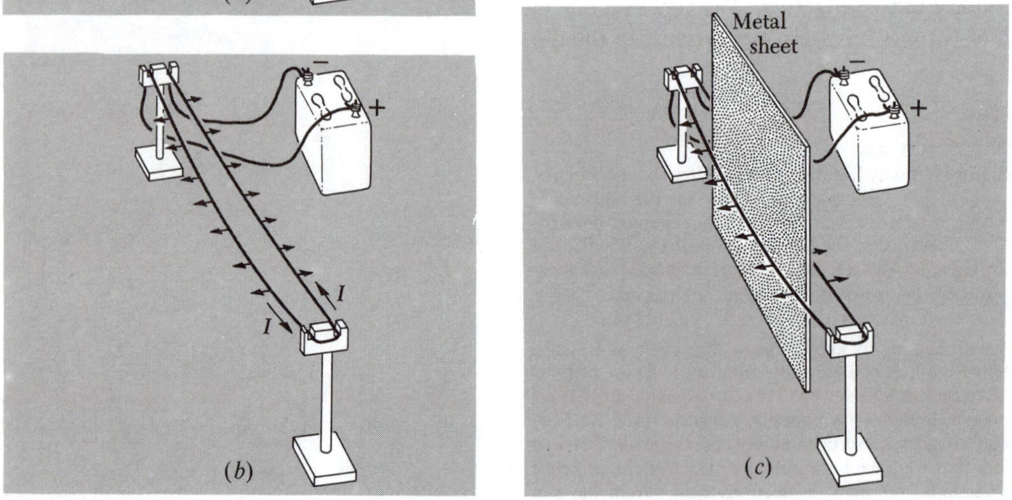

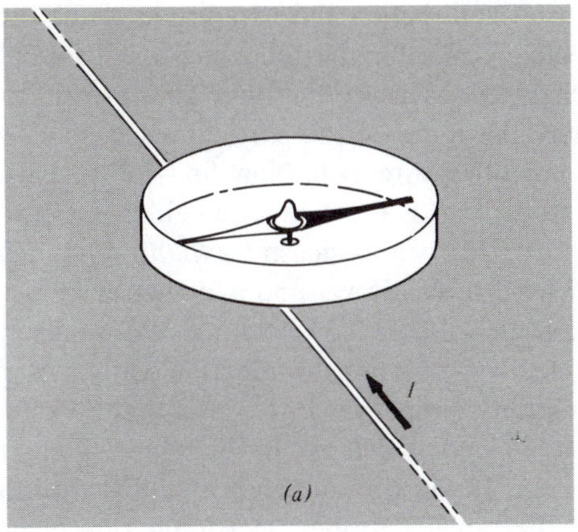

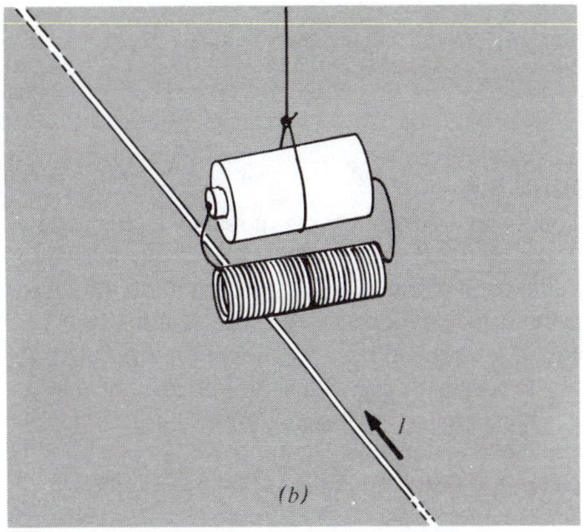

FIGURE 5.2
A compass needle (*a*) and a coil of wire carrying current (*b*) are similarly influenced by current in a nearby conductor. The direction of the current *I* is understood to be that in which positive ions would be moving if they were the carriers of the current. In the earth's magnetic field the black end of the compass needle would point north.

be described by introducing a *magnetic field*. (The electric field, remember, was simply a way of describing the action at a distance between stationary charges that is expressed in Coulomb's law.) We say that an electric current has associated with it a magnetic field which pervades the surrounding space. Some other current, or any moving charged particle which finds itself in this field, experiences a force proportional to the strength of the magnetic field in that locality. The force is always perpendicular to the velocity, for a charged particle. The entire force on a particle carrying charge q is given by

$$\mathbf{F} = q\mathbf{E} + \frac{q}{c}\mathbf{v} \times \mathbf{B} \qquad (1)$$

where $\mathbf{B}$ is the magnetic field.†

We shall take Eq. 1 as the definition of **B**. The inclusion of a factor $1/c$ in the second term appears, at this stage, quite arbitrary. We are free to include it since we have not previously specified the units for **B**. We shall deal with the question of units at the beginning of the next chapter. All that concerns us now is that the magnetic field strength is a vector which determines the velocity-proportional part of the force on a moving charge. In other words, the command, "Mea-

†Here for the first time we make use of the vector product, or *cross product*, of two vectors. A reminder: The vector $\mathbf{v} \times \mathbf{B}$ is a vector perpendicular to both $\mathbf{v}$ and $\mathbf{B}$ and of magnitude $vB \sin\theta$, where θ is the angle between the directions of $\mathbf{v}$ and $\mathbf{B}$. A right-hand rule determines the sense of the direction of the vector $\mathbf{v} \times \mathbf{B}$. In our Cartesian coordinates $\hat{\mathbf{x}} \times \hat{\mathbf{y}} = \hat{\mathbf{z}}$ and $\mathbf{v} \times \mathbf{B} = \hat{\mathbf{x}}(v_yB_z - v_zB_y) + \hat{\mathbf{y}}(v_zB_x - v_xB_z) + \hat{\mathbf{z}}(v_xB_y - v_yB_x)$.

FIGURE 5.3
An example of the attraction of currents in the same direction. Compare with Fig. 5.1*a*. We can also describe it as the deflection of an electron beam by a magnetic field.

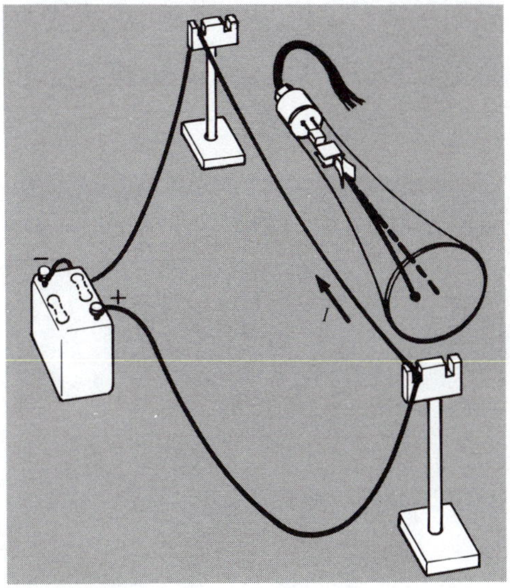

FIGURE 5.4

(*a*) The magnitude of a charge at rest is determined by the force on a test charge at rest and Coulomb's law. (*b*) In the case of a moving charge, the force, for all we know now, may depend on the position of the test charge. If so, we can't use procedure (*a*). (*c*) At the instant Q passes through the center of the spherical array of test charges, measure the radial force component on each, and use the average value of F_r to determine Q. This is equivalent to measuring the surface integral of **E**.

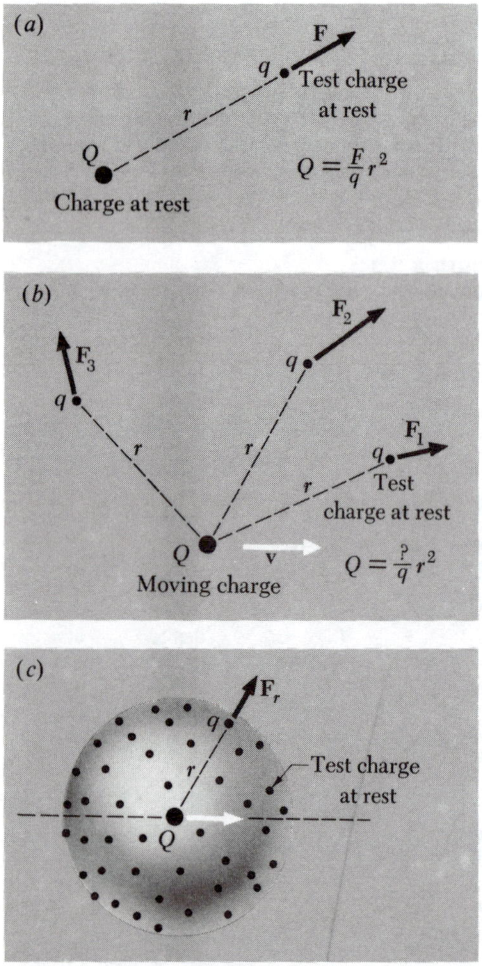

sure the direction and magnitude of the vector **B** at such and such a place," calls for the following operations: Take a particle of known charge q. Measure the force on q at rest, to determine **E.** Then measure the force on the particle when its velocity is **v**; repeat with **v** in some other direction. Now find a **B** that will make Eq. 1 fit all these results; that is the magnetic field at the place in question.

Clearly this doesn't *explain* anything. Why does Eq. 1 work? Why can we always find a **B** that is consistent with this simple relation, for all possible velocities? We want to understand why there is a velocity-proportional force. It is really most remarkable that this force is strictly proportional to v, and that the effect of the electric field does not depend on v at all! In the following pages we'll see how this comes about. It will turn out that a field **B** with these properties *must* exist if the forces between electric charges obey the postulates of special relativity. Seen from this point of view, magnetic forces are a relativistic aspect of charge in motion.

A review of the essential ideas and formulas of special relativity is provided in Appendix A. This would be a good time to read through it.

MEASUREMENT OF CHARGE IN MOTION

5.3 How are we going to measure the quantity of electric charge on a moving particle? Until this question is settled, it is pointless to ask what effect motion has on charge itself. A charge can only be measured by the effects it produces. A point charge Q which is at rest can be measured by determining the force that acts on a test charge q a certain distance away (Fig. 5.4*a*). That is based on Coulomb's law. But if the charge we want to measure is moving, we are on uncertain ground. There is now a special direction in space, the instantaneous direction of motion. It could be that the force on the test charge q depends on the *direction* from Q to q, as well as on the distance between the two charges. For different positions of the test charge, as in Fig. 5.4*b*, we would observe different forces. Putting these into Coulomb's law would lead to different values for the same quantity Q. Also we have as yet no assurance that the force will always be in the direction of the radius vector **r.**

To allow for this possibility, let's agree to define Q by averaging over all directions. Imagine a large number of infinitesimal test charges distributed evenly over a sphere (Fig. 5.4*c*). At the instant the moving charge passes the center of the sphere, the radial component of force on each test charge is measured, and the average of these force magnitudes is used to compute Q. Now this is just the operation that would be needed to determine the surface integral of the electric field over that sphere, at time t. The test charges here are all at rest, remember; the force on q per unit charge gives, by definition, the electric field at that point. This suggests that Gauss's law, rather than Coulomb's law, offers the natural way† to define quantity of charge for a moving charged particle, or for a collection of moving charges. We can frame such a definition as follows.

The amount of electric charge in a region is defined by the surface integral of the electric field **E** over a surface S enclosing the region. This surface S is fixed in some coordinate frame F. The field **E** is measured, at any point (x, y, z) and at time t in F, by the force on a test charge *at rest in F*, at that time and place. The surface integral is to be determined for a particular time t. That is, the field values used are those measured simultaneously by observers deployed all over S. (This presents no difficulty, for S is stationary in the frame F.) Let us denote such a surface integral, over S at time t, by

$$\int_{S(t)} \mathbf{E} \cdot d\mathbf{a} \tag{2}$$

We define the amount of charge inside S as $1/4\pi$ times this integral:

$$Q = \frac{1}{4\pi} \int_{S(t)} \mathbf{E} \cdot d\mathbf{a} \tag{3}$$

It would be embarrassing if the value of Q so determined depended on the size and shape of the surface S. For a stationary charge it doesn't—that is Gauss's law. But how do we know that Gauss's law holds when charges are moving? Fortunately it does. We can take that as an experimental fact. This fundamental property of the electric field of moving charges permits us to define quantity of charge by Eq. 3. From now on we can speak of the amount of charge in a region or on a particle, and that will have a perfectly definite meaning even if the charge is in motion.

Figure 5.5 summarizes these points in an example. Two protons and two electrons are shown in motion, at a particular instant of time. It is a fact that the surface integral of the electric field **E** over the surface S_1 is precisely equal to the surface integral over S_2 evaluated at the same instant, and we may use this integral, as we always have used Gauss's law in electrostatics, to determine the total charge enclosed. Figure 5.6 raises a new question. What if the same particles had some other velocities? For instance, suppose the two protons and two electrons combine to form a hydrogen molecule. Will the total charge appear exactly the same as before?

†It is not the only *possible* way. You could, for instance, adopt the arbitrary rule that the test charge must always be placed directly ahead (in the direction of motion) of the charge to be measured. Charge so defined would *not* have the simple properties we are about to discuss, and your theory would prove clumsy and complicated.

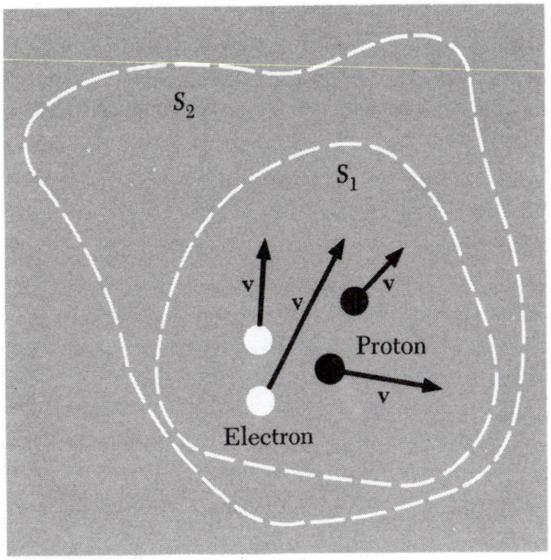

FIGURE 5.5
Gauss' law remains valid for the field of moving charges. The flux of **E** through S_2 is equal to the flux of **E** through S_1, evaluated at the same instant of time.

FIGURE 5.6
Does the flux of **E** through S depend on the state of motion of the charged particles? Is the surface integral of **E** over S the same as in Fig. 5.5? Here the particles are bound together as a hydrogen molecule.

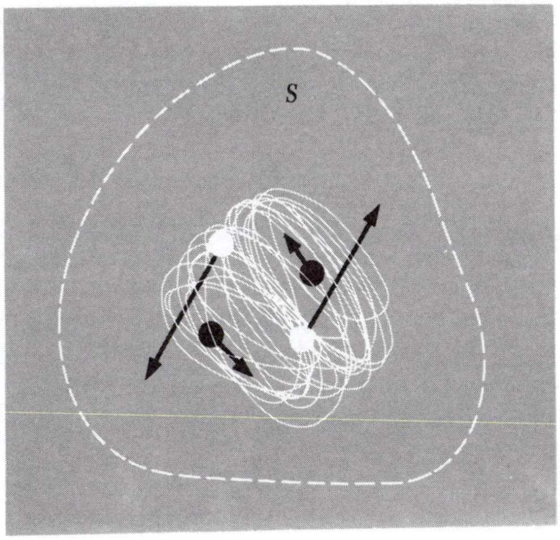

INVARIANCE OF CHARGE

5.4 There is conclusive experimental evidence that the total charge in a system is not changed by the motion of the charge carriers. We are so accustomed to taking this for granted that we seldom pause to think how remarkable and fundamental a fact it is. For proof, we can point to the exact electrical neutrality of atoms and molecules. We have already described in Chapter 1 the experimental test of the neutrality of the hydrogen molecule, which proved that the electron and proton carry charges equal in magnitude to better than 1 part in 10^{20} A similar experiment was performed with helium atoms. Now the helium atom contains two protons and two electrons, the same charged particles that make up the hydrogen molecule. In the helium atom their motion is very different. The protons, in particular, instead of revolving slowly 0.7 angstrom apart, are tightly bound into the helium nucleus where they move with kinetic energies in the range of 1 million ev. If *motion* had any effect on the amount of charge, we could not have exact cancellation of nuclear and electronic charge in *both* the hydrogen molecule and the helium atom. In fact, the helium atom was shown to be neutral with nearly the same experimental accuracy.

Another line of evidence comes from the optical spectra of isotopes of the same element, atoms with different nuclear masses but, nominally at least, the same nuclear charge. Here again we find a marked difference in the motion of the protons within the nucleus, but comparison of the spectral lines of the two species shows no discrepancy that could be attributed to even a slight difference in total nuclear charge.

Mass is *not* invariant in the same way. We know that the mass of a particle is changed by its motion, by the factor $1/(1 - v^2/c^2)^{1/2}$. To emphasize the difference, we show in Fig. 5.7 an imaginary experiment. In the box on the right the two massive charged particles, which are fastened to the end of a pivoted rod, have been set revolving with speed v. The entire mass on the right is *greater* than the mass on the left, as demonstrated by weighing the box on a spring balance or by measuring the force required to accelerate it.† The total electric charge, however, is unchanged. A real experiment equivalent to this can be carried out with a mass spectrograph, which can reveal quite plainly a mass difference between an ionized deuterium molecule (two protons, two neutrons, one electron) and an ionized helium atom (also two protons, two neutrons, and one electron). These are two very different structures, within which the component particles are whirling

†The difference in mass depends not only on the kinetic energy of the particles, but also on any change in potential energy, as in the elastic strain in the rod that holds the particles. If the rod is fairly stiff, this contribution will be small compared with the v^2/c^2 term. See if you can show why.

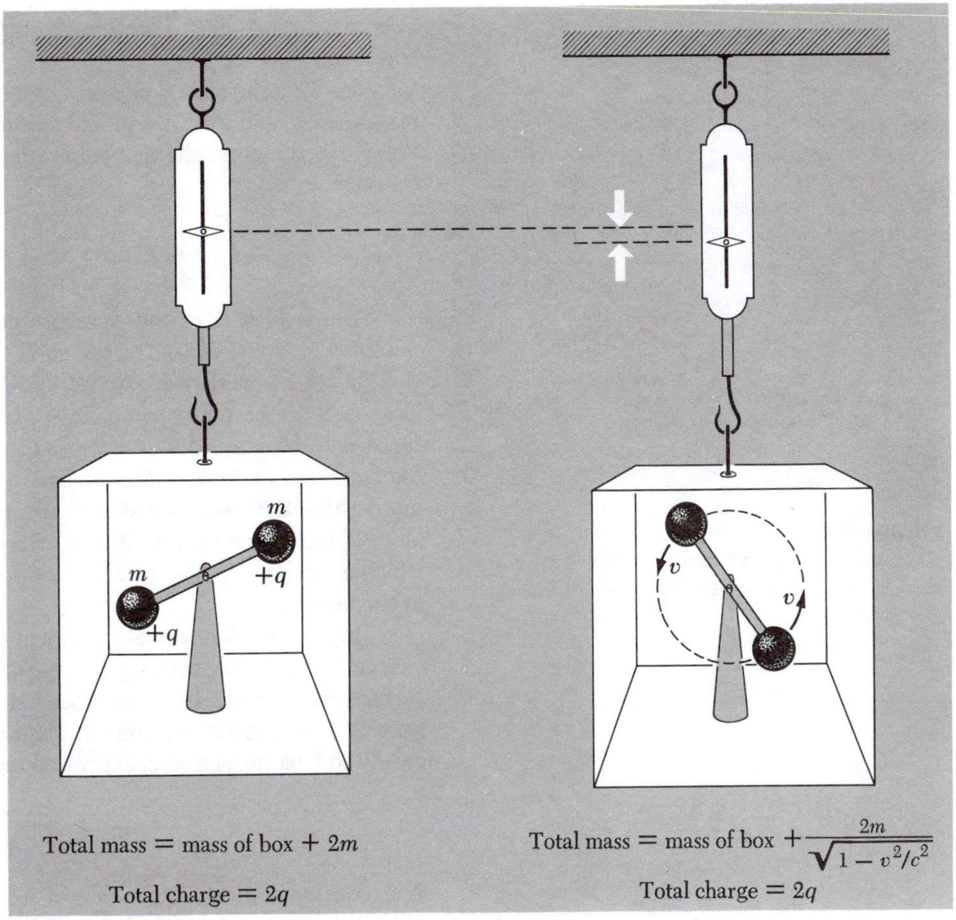
m
m
+q
+q
v
v
Total mass = mass of box + 2m
Total charge = 2q
Total mass = mass of box + $\frac{2m}{\sqrt{1 - v^2/c^2}}$
Total charge = 2q

FIGURE 5.7
An imaginary experiment to show the invariance of charge. The charge in the box is to be measured by measuring the electric field all around the box or, equivalently, by measuring the force on a distant test charge.

around with very different speeds. The difference in energy shows up as a measurable difference in mass. There is no detectable difference, to very high precision, in the electric charge of the two ions.

This invariance of charge lends a special significance to the fact of charge quantization. We emphasized in Chapter 1 the importance—and the mystery—of the fact that every elementary charged particle has a charge equal in magnitude to that of every other such particle. We now observe that this precise equality holds not only for two particles at rest with respect to one another, but for *any* state of relative motion.

The experiments we have described, and many others, show that the value of our Gauss's law surface integral $\int_S \mathbf{E} \cdot d\mathbf{a}$ *depends only on the number and variety of charged particles inside S, and not on*

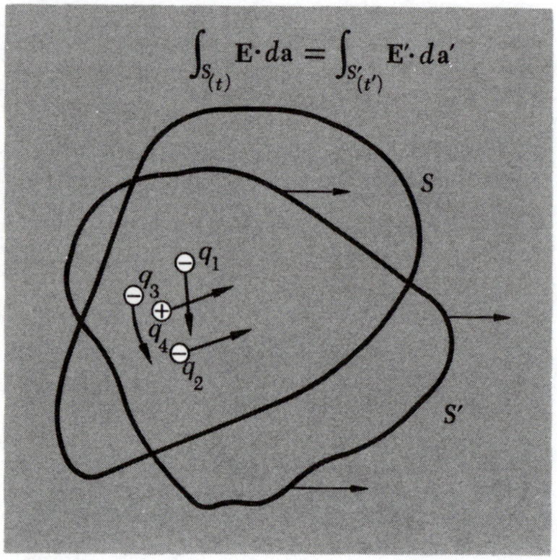

FIGURE 5.8
The surface integral of **E** over S is equal to the integral of **E**′ over S'. The charge is the same in all frames of reference.

how they are moving. According to the postulate of relativity, such a statement must be true for *any* inertial frame of reference if it is true for one. Therefore if F' is some *other* inertial frame, moving with respect to F, and if S' is a closed surface in *that* frame which at time t' encloses the same charged bodies that were enclosed by S at time t, we must have

$$\int_{S(t)} \mathbf{E} \cdot d\mathbf{a} = \int_{S'(t')} \mathbf{E}' \cdot d\mathbf{a}' \tag{4}$$

The field $\mathbf{E}'$ is of course measured in F', that is, it is defined by the force on a test charge at rest in F'. The distinction between t and t' must not be overlooked. As we know, events that are simultaneous in F need not be simultaneous in F'. Each of the surface integrals in Eq. 4 is to be evaluated at one instant in *its* frame. If charges lie on the boundary of S, or of S', one has to be rather careful about ascertaining that the charges within S at t are the same as those within S' at t'. If the charges are well away from the boundary, as in Fig. 5.8 which is intended to illustrate the relation in Eq. 4, there is no problem in this respect.

Equation 4 is a formal statement of the relativistic invariance of charge. We can choose our gaussian surface in *any* inertial frame; the surface integral will give a number independent of the frame. Invariance of charge is not the same as charge conservation, which was discussed in Chapter 4 and is expressed mathematically in the equation

$$\text{div}\,\mathbf{J} = \frac{-\partial\rho}{\partial t}$$

Charge *conservation* implies that, if we take a closed surface fixed in some coordinate system and containing some charged matter, and if no particles cross the boundary, then the total charge inside that surface remains constant. Charge *invariance* implies that, if we look at this collection of stuff from any other frame of reference, we will measure exactly the same amount of charge. Energy is conserved, but energy *is not* a relativistic invariant. Charge is conserved, and charge *is* a relativistic invariant. In the language of relativity theory, energy is one component of a four-vector, while charge is a scalar, an invariant number, with respect to the Lorentz transformation. This is an observed fact with far-reaching implications. It completely determines the nature of the field of moving charges.

ELECTRIC FIELD MEASURED IN DIFFERENT FRAMES OF REFERENCE

5.5 If charge is to be invariant under a Lorentz transformation, the electric field **E** has to transform in a particular way. "Transforming

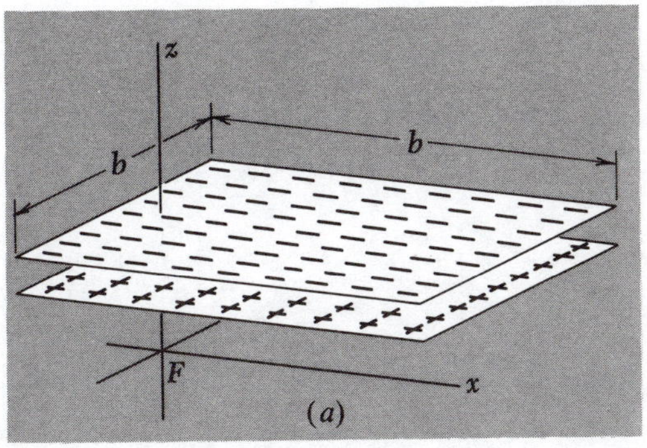

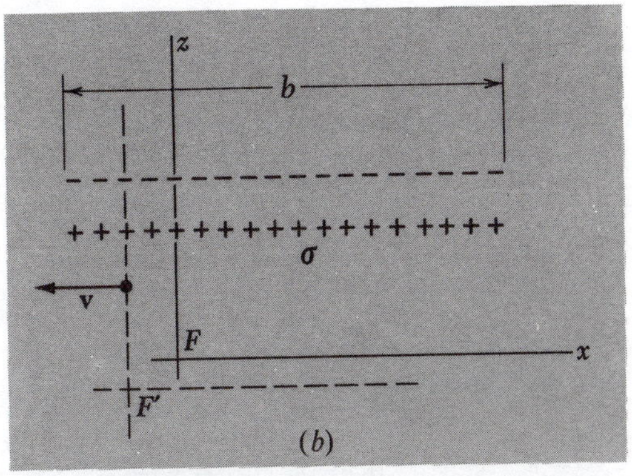

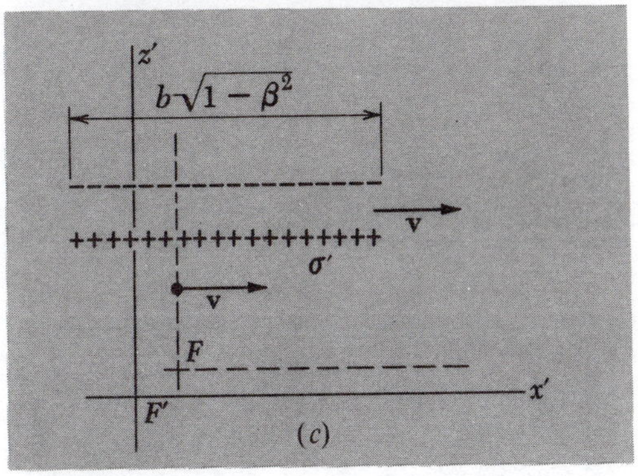

FIGURE 5.9

(*a*) Two square sheets of surface density $+\sigma$ and $-\sigma$, stationary in an inertial frame *F*. (*b*) A cross-section view in the *F* frame. *F*′ is a different frame moving in the $-\hat{\mathbf{x}}$ direction with respect to *F*. (*c*) Cross section of the charge sheets as seen in frame *F*′. Same charge is on shorter sheet, so charge density is greater: $\sigma' = \gamma\sigma$.

E" means answering a question like this: If an observer in a certain inertial frame *F* measures an electric field **E** as so-and-so-many statvolts/cm, at a given point in space and time, what field will be measured at the same space-time point by an observer in a different inertial frame *F*′? For a certain class of fields, we can answer this question by applying Gauss's law to some simple systems.

In the frame *F* (Fig. 5.9*a*) there are two stationary sheets of charge of uniform density σ and $-\sigma$ esu/cm^2, respectively. They are squares *b* cm on a side lying parallel to the *xy* plane, and their separation is supposed to be so small compared with their extent that the field between them can be treated as uniform. The magnitude of this field, as measured by an observer in *F*, is of course just $4\pi\sigma$. Now consider an inertial frame *F*′ which moves toward the left, with respect to *F*, with velocity **v**. To an observer in *F*′, the charged "squares" are no longer square. Their x' dimension is contracted from b to $b\sqrt{1-\beta^2}$, where β stands for v/c, as usual. But total charge is invariant, that is, independent of reference frame, so the charge *density* measured in *F*′ must be *greater* than σ in the ratio γ, that is, $1/\sqrt{1-\beta^2}$. Figure 5.9 shows the system in cross section, (*b*) as seen in *F* and (*c*) as seen in *F*′. What can we say about the electric field in *F*′ if all we know about the electric field of moving charges is contained in Eq. 4?

For one thing, we can be sure that the electric field is zero outside the sandwich, and uniform between the sheets, at least in the limit as the extent of the sheets becomes infinite. The field of an infinite uniform sheet could not depend on the distance from the sheet, nor on position along the sheet. There is nothing in the system to fix a position along the sheet. But for all we know at this point, the field of a single moving sheet of positive charge *might* look like Fig. 5.10*a*. However, even if it did, the field of a sheet of negative charge moving with the same velocity would have to look like Fig. 5.10*b*, and the superposition of the two fields would still give zero field outside our two charged sheets and a uniform perpendicular field between them, as in Fig.

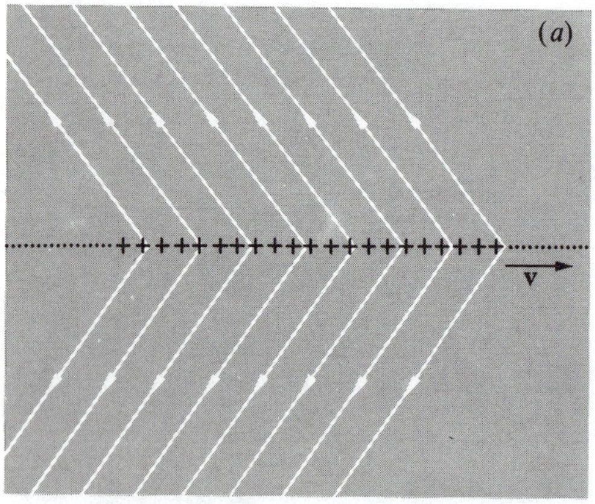

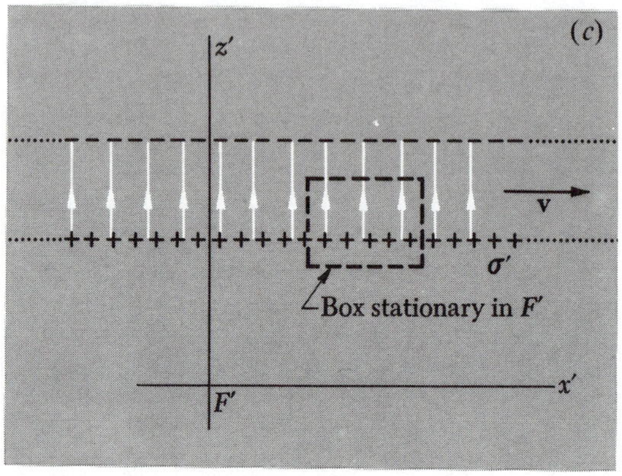

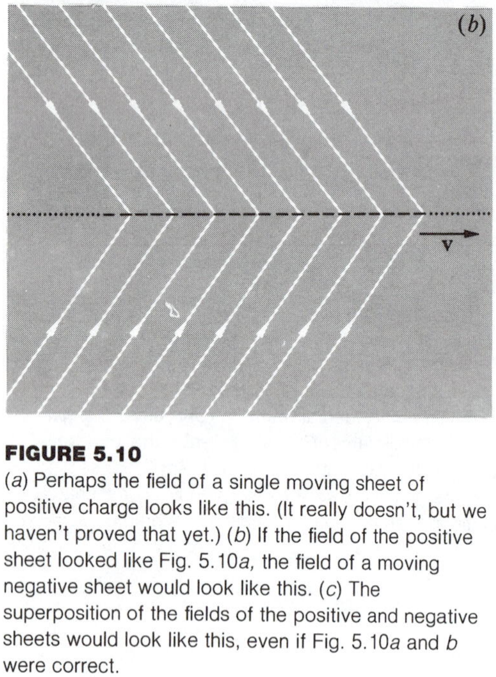

FIGURE 5.10

(*a*) Perhaps the field of a single moving sheet of positive charge looks like this. (It really doesn't, but we haven't proved that yet.) (*b*) If the field of the positive sheet looked like Fig. 5.10*a*, the field of a moving negative sheet would look like this. (*c*) The superposition of the fields of the positive and negative sheets would look like this, even if Fig. 5.10*a* and *b* were correct.

5.10*c*. (Actually, as we shall prove before long, the field of a single sheet of charge moving in its own plane is perpendicular to the sheet, unlike the hypothetical fields pictured in Fig. 5.10*a* and *b*.)

We can apply Gauss's law to a box stationary in frame F', the box shown in cross section in Fig. 5.10*c*. The charge content is determined by σ', and the field is zero outside the sandwich. Gauss's law tells us that the magnitude of E'_z, which is the only field component inside, must be $4\pi\sigma'$, or $4\pi\sigma/\sqrt{1-\beta^2}$.

$$E'_z = \frac{E_z}{\sqrt{1-\beta^2}} = \gamma E_z \tag{5}$$

Now imagine a different situation with the stationary charged sheets in the frame F oriented perpendicular to the x axis, as in Fig. 5.11. The observer in F now reports a field in the x direction of magnitude $E_x = 4\pi\sigma$. In this case, the surface charge density observed in the frame F' is the *same* as that observed in F. The sheets are not contracted; only the distance between them is contracted, but that doesn't enter into the determination of the field. This time we find by applying Gauss's law to the box stationary in F':

$$E'_x = 4\pi\sigma' = 4\pi\sigma = E_x \tag{6}$$

That is all very well for the particularly simple arrangement of charges here pictured; do our conclusions have more general validity? This question takes us to the heart of the meaning of *field*. If the electric field **E** at a point in space-time is to have a unique meaning, then the way **E** appears in other frames of reference, in the same space-time neighborhood, cannot depend on the nature of the sources, wherever they may be, that produced **E.** In other words, the observer in F, having measured the field in his neighborhood at some time, ought to be able to predict *from these measurements alone* what observers in other frames of reference would measure at the same space-time point. Were this not true, *field* would be a useless concept. The evidence that it *is* true is the eventual agreement of our field theory with experiment.

Seen in this light, the relations expressed in Eqs. 5 and 6 take on a significance beyond the special case of charges on parallel sheets. Consider *any* charge distribution, all parts of which are at rest with respect to the frame F. If an observer in F measures a field E_z in the z direction, then an observer in the frame F' will report, for the same space-time point, a field $E'_z = \gamma E_z$. That is, he will get a number, as the result of his E'_z measurement, which is larger by the factor γ than the number the F observer got in his E_z measurement. On the other hand, if the observer in F measures a field E_x in the x direction, the direction of the velocity of F' with respect to F, then the observer in F' reports a field E'_x *equal* to E_x. Obviously the y and the z directions are equivalent, both being transverse to the velocity $\mathbf{v}$. Anything we have said about E'_z applies to E'_y too. Whatever the direction of $\mathbf{E}$ in the frame F, we can treat it as a superposition of fields in the x, the y, and the z directions, and from the transformation of each of these predict the vector field $\mathbf{E}'$ at that point in F'. Let's summarize this in words appropriate to relative motion in any direction: Charges at rest in frame F are the source of a field $\mathbf{E}$. Let frame F' move with speed $\mathbf{v}$ relative to F. At any point in F, resolve $\mathbf{E}$ into a longitudinal component $E_\parallel$ parallel to $\mathbf{v}$ and a transverse component $E_\perp$ perpendicular to the direction of $\mathbf{v}$. At the same space-time point in F', the field $\mathbf{E}'$ is to be resolved into $E'_\parallel$ and $E'_\perp$, $E'_\parallel$ being parallel to $\mathbf{v}$ and $E'_\perp$ perpendicular thereto. We have now learned that

$$\boxed{\begin{aligned} E'_\parallel &= E_\parallel \\ E'_\perp &= \gamma E_\perp \end{aligned}} \tag{7}$$

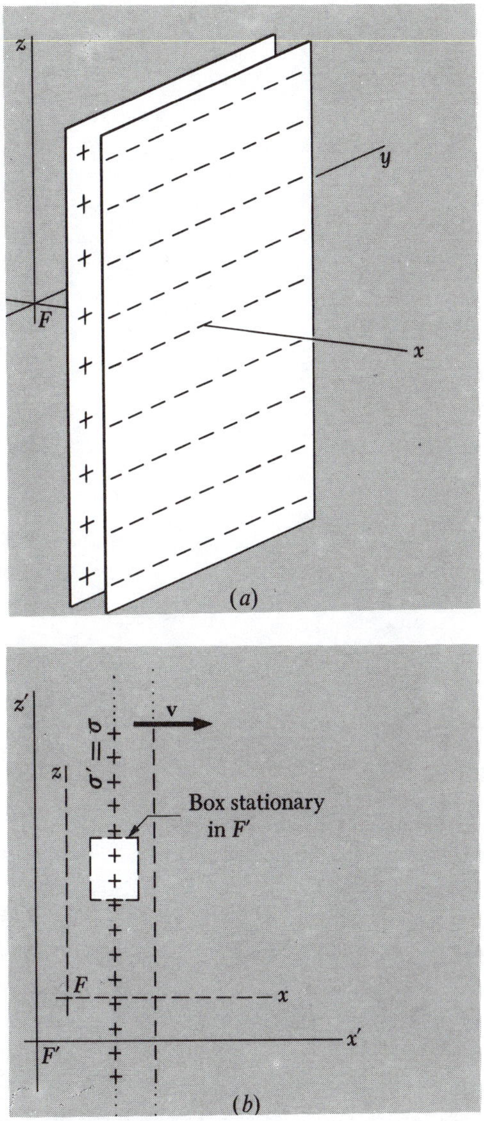

FIGURE 5.11
The electric field in another frame of reference (relative velocity parallel to field direction). (*a*) In reference frame F. (*b*) Cross-sectional view in reference frame F'.

Our conclusion holds only for fields that arise from charges stationary in F. As we shall see presently, if charges in F are moving, the prediction of the electric field in F' involves knowledge of *two* fields in F, the electric and the magnetic. But we already have a useful result, one that suffices whenever we can find any inertial frame of reference in which all the charges remain at rest. We shall use it now to study the electric field of a point charge moving with constant velocity.

FIELD OF A POINT CHARGE MOVING WITH CONSTANT VELOCITY

5.6 In the frame F the point charge Q remains at rest at the origin (Fig. 5.12*a*). At every point the electric field $\mathbf{E}$ has the magnitude Q/r^2 and is directed radially outward. In the xz plane its components at any point (x, z) are

$$E_x = \frac{Q}{r^2}\cos\theta = \frac{Qx}{(x^2+z^2)^{3/2}}$$
$$E_z = \frac{Q}{r^2}\sin\theta = \frac{Qz}{(x^2+z^2)^{3/2}} \tag{8}$$

Consider another frame F' which is moving in the negative x direction, with speed v, with respect to frame F. We need the relation between the coordinates of an event in the two frames, for which we turn to the Lorentz transformation given in Eq. 2 of Appendix A. It simplifies the description to assume, as we are free to do, that the origins of the two frames coincide at time zero according to observers in both frames. In other words that event, the coincidence of the origins, can be the event A referred to by Eq. 2, with coordinates $x_A = 0$, $y_A = 0$, $z_A = 0$, $t_A = 0$ in frame F and $x'_A = 0$, $y'_A = 0$, $z'_A = 0$, $t'_A = 0$ in frame F'. Then event B is the space-time point we are trying to locate. We can omit the tag B and call its coordinates in F just x, y, z, t, and its coordinates in F' just x', y', z', t'. Then Eq. 2 of Appendix A would become

$$x' = \gamma x - \gamma\beta ct \qquad y' = y \qquad z' = z \qquad t' = \gamma t - \frac{\gamma\beta x}{c}$$

However, *that* transformation was for an F' frame moving in the positive x direction with respect to F, as one can quickly verify by noting that, with increasing time t, x' gets smaller. To construct the Lorentz transformation for our problem, in which the F' frame moves in the opposite direction, we must either reverse the sign of β or switch the

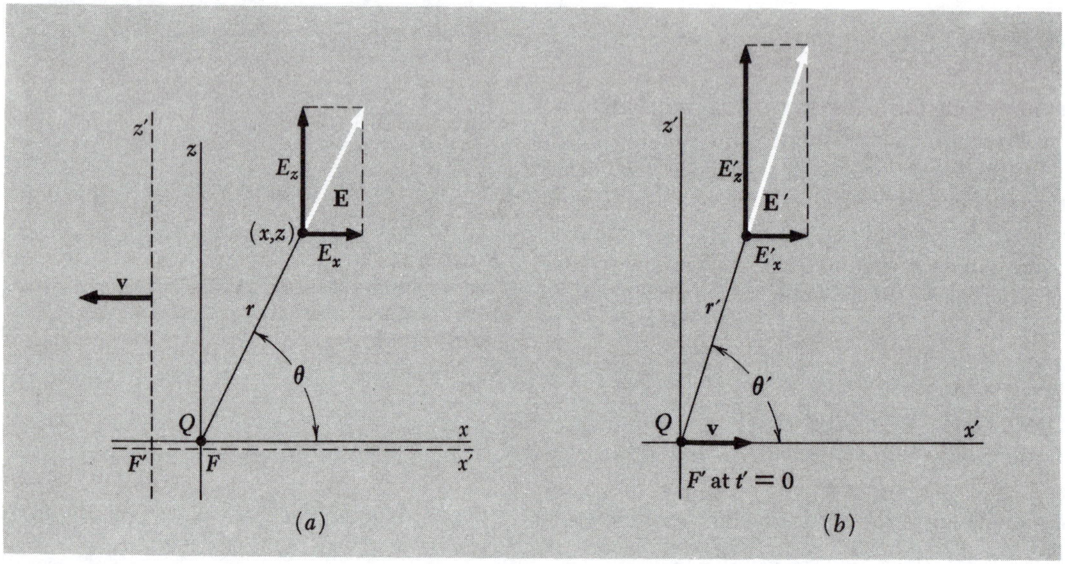

FIGURE 5.12
The electric field of a point charge (*a*) in a frame in which the charge is at rest, and (*b*) in a frame in which the charge moves with constant velocity.

primes. We'll choose to do the latter because we want to express x and z in terms of x' and z'. The Lorentz transformation we need is therefore

$$x = \gamma x' - \gamma\beta ct' \qquad y = y' \qquad z = z' \qquad t = \gamma t' - \frac{\gamma\beta x'}{c} \tag{9}$$

According to Eqs. 5 and 6, $E'_z = \gamma E_z$ and $E'_x = E_x$. Using Eqs. 8 and 9, we can express the field components E'_z and E'_x in terms of the coordinates in F'. For the instant $t' = 0$, when Q passes the origin in F', we have

$$\begin{aligned} E'_x = E_x &= \frac{\gamma Q x'}{[(\gamma x')^2 + z'^2]^{3/2}} \\ E'_z = \gamma E_z &= \frac{\gamma Q z'}{[(\gamma x')^2 + z'^2]^{3/2}} \end{aligned} \tag{10}$$

Note first that $E'_z/E'_x = z'/x'$. This tells us that the vector $\mathbf{E}'$ makes the same angle with the x' axis as does the radius vector $\mathbf{r}'$. Hence $\mathbf{E}'$ points radially outward along a line drawn from the instantaneous position of Q, as in Fig. 5.12*b*. Pause a moment to let this conclusion sink in! It means that, if Q passed the origin of the primed system at precisely 12:00 noon, "prime time," an observer stationed anywhere in the primed system will report that the electric field in his vicinity was pointing, at 12:00 noon, exactly radially from the origin. This sounds at first like instantaneous transmission of information! How can an observer a mile away know where the particle is at the same instant? He can't. That wasn't implied. This particle, remember, has been moving at constant speed forever, on a "flight plan" that calls for it to pass the origin at noon. That information has been available for a long time. It is the *past history* of the particle that determined the field observed, if you want to talk about cause and effect. We'll inquire presently into what happens when there is an unscheduled change in the flight plan.

To find the strength of the field, we compute $E'^2_x + E'^2_z$, which is the square of the magnitude of the field, E'^2.

$$\begin{aligned} E'^2 = E'^2_x + E'^2_z &= \frac{\gamma^2 Q^2(x'^2 + z'^2)}{[(\gamma x')^2 + z'^2]^3} = \frac{Q^2(x'^2 + z'^2)}{\gamma^4[x'^2 + z'^2 - \beta^2 z'^2]^3} \\ &= \frac{Q^2(1-\beta^2)^2}{(x'^2 + z'^2)^2\left(1 - \dfrac{\beta^2 z'^2}{x'^2 + z'^2}\right)^3} \end{aligned} \tag{11}$$

(Here, for once, it was neater with β worked back into the expression.) Let r' denote the distance from the charge Q, which is momentarily at the origin, to the point (x',z') where the field is measured: $r' = (x'^2 + z'^2)^{1/2}$. Let θ' denote the angle between this radius vector and

FIGURE 5.13
The intensity in various directions of the field of a moving charge. At this instant, the charge is passing the origin of the $x'y'z'$ frame. The numbers give the field strength relative to Q/r'^2.

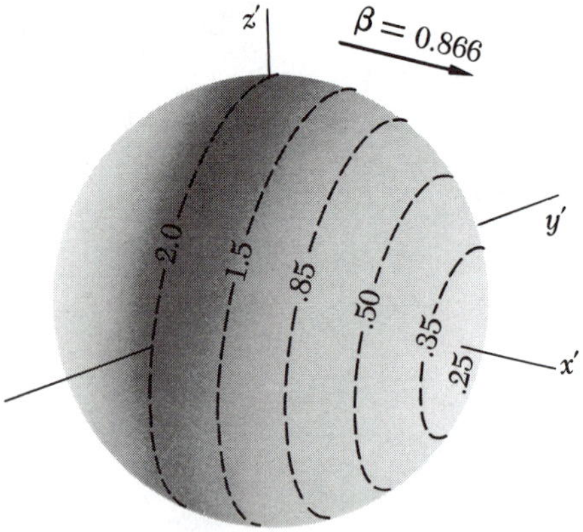

the velocity of the charge Q, which is moving in the positive x' direction in the frame F'. Then since $z' = r' \sin \theta'$, the magnitude of the field can be written as

$$E' = \frac{Q}{r'^2} \frac{1 - \beta^2}{(1 - \beta^2 \sin^2 \theta')^{3/2}} \tag{12}$$

There is nothing special about the origin of coordinates, nor about the $x'z'$ plane as compared with any other plane through the x' axis. Therefore we can say quite generally that the electric field of a charge which has been in uniform motion is at a given instant of time directed radially from the instantaneous position of the charge, while its magnitude is given by Eq. 12 with θ' the angle between the direction of motion of the charge and the radius vector from the instantaneous position of the charge to the point of observation.

For low speeds the field reduces simply to $E' \approx Q/r'^2$, and is practically the same, at any instant, as the field of a point charge stationary in F' at the instantaneous location of Q. But if β^2 is not negligible, the field is stronger at right angles to the motion than in the direction of the motion, at the same distance from the charge. If we were to indicate the intensity of the field by the density of field lines, as is often done, the lines tend to concentrate in a pancake perpendicular to the direction of motion. Figure 5.13 shows the density of lines as they pass through a unit sphere, from a charge moving in the x' direction with a speed $v/c = 0.866$. A simpler representation of the field is shown in Fig. 5.14, a cross section through the field with some field lines in the $x'z'$ plane indicated.†

This is a remarkable electric field. It is not spherically symmetrical, which is not surprising because in this frame there is a preferred direction, the direction of motion of the charge. However, the field is symmetrical about a plane perpendicular to the direction of motion of the charge. That, by the way, is sufficient to prove that the field of a uniform sheet of charge moving in its own plane must be perpendicular to the sheet. Think of that field as the sum of the fields of charge elements spread uniformly over the sheet. Since each of these individual fields has the fore-and-aft symmetry of Fig. 5.14 with respect to the direction of motion, their sum could only be perpendicular to the sheet. It could not look like Fig. 5.10*a*.

FIGURE 5.14
Another representation of the field of a uniformly moving charge.

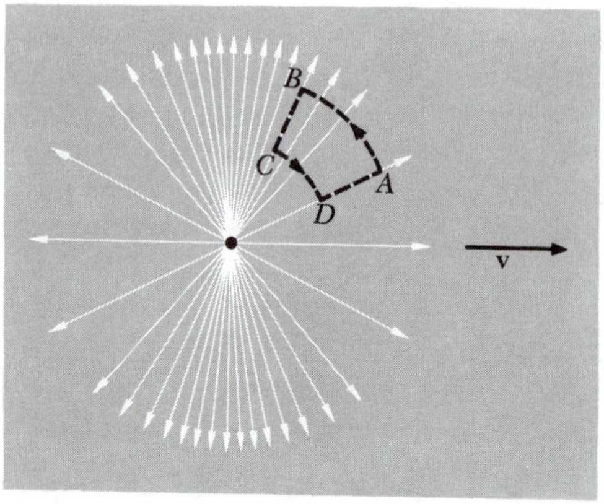

The field in Fig. 5.14 is a field that *no stationary charge distribution,* whatever its form, could produce. For in this field the line integral of $\mathbf{E}'$ is *not zero* around every closed path. Consider, for example,

†A *two-dimensional* diagram like Fig. 5.14 cannot faithfully represent the field intensity by the density of field lines. Unless we arbitrarily break off some of the lines, the density of lines in the picture will fall off as $1/r'$, whereas the intensity of the field we are trying to represent falls off as $1/r'^2$. So Fig. 5.14 gives only a qualitative indication of the variation of E' with r' and θ'.

the closed path *ABCD* in Fig. 5.14. The circular arcs contribute nothing to the line integral, being perpendicular to the field; on the radial sections, the field is *stronger* along *BC* than along *DA*, so the *circulation* of **E**′ on this path is not zero. But remember, this is not an electro*static* field. In the course of time the electric field **E**′ at any point in the frame *F*′ changes as the source charge moves.

Figure 5.15 shows the electric field at certain instants of time observed in a frame of reference through which an electron is moving at constant velocity in the *x* direction.† In Figure 5.15, the speed of the electron is 0.33*c*, its kinetic energy therefore about 30,000 ev [30 kiloelectron-volts (kev)]. The value of β^2 is ⅑, and the electric field does not differ greatly from that of a charge at rest. In Fig. 5.16, the speed is 0.8*c*, corresponding to a kinetic energy of 335 kev. If the time unit for each diagram is taken as 1.0×10^{-10} sec, the distance scale is life-size, as drawn. Of course, the diagram holds equally well for *any* charged particle moving at the specified fraction of the speed of light. We mention the equivalent energies for an electron merely to remind the reader that relativistic speeds are nothing out of the ordinary in the laboratory.

FIELD OF A CHARGE THAT STARTS OR STOPS

5.7 It must be clearly understood that *uniform velocity,* as we have been using the term, implies a motion at constant speed in a straight line that has been going on forever. What if our electron had *not* been traveling in the distant past along the negative *x* axis until it came into view in our diagram at $t = 0$? Suppose it had been sitting quietly at rest at the origin, waiting for the clock to read $t = 0$. Just prior to $t = 0$, something gives the electron a sudden large acceleration, up to the speed v, and it moves away along the positive *x* axis at this speed. Its motion *from then on* precisely duplicates the motion of the electron for which Fig. 5.16 was drawn. But Fig. 5.16 does *not* correctly represent the field of the electron whose history was just described. To see that it cannot do so, consider the field at the point marked *P*, at time $t = 2$, which means 2×10^{-10} sec. In 2×10^{-10} sec a light signal travels 6 cm. Since this point lies more than 6 cm distant from the origin, it could not have received the news that the electron had started to move at $t = 0$! Unless there is a gross violation of relativity—and we are taking the postulates of relativity as basis for this whole discussion—the field at the point *P* at time $t = 2$, and indeed at all points outside the sphere of radius 6 cm centered on the origin, *must be the field of a charge at rest at the origin.*

On the other hand, close to the moving charge itself, what hap-

†Previously we had the charge at rest in the unprimed frame, moving in the primed frame. Here we adopt *xyz* for the frame in which the charge is moving, to avoid cluttering the subsequent discussion with primes.

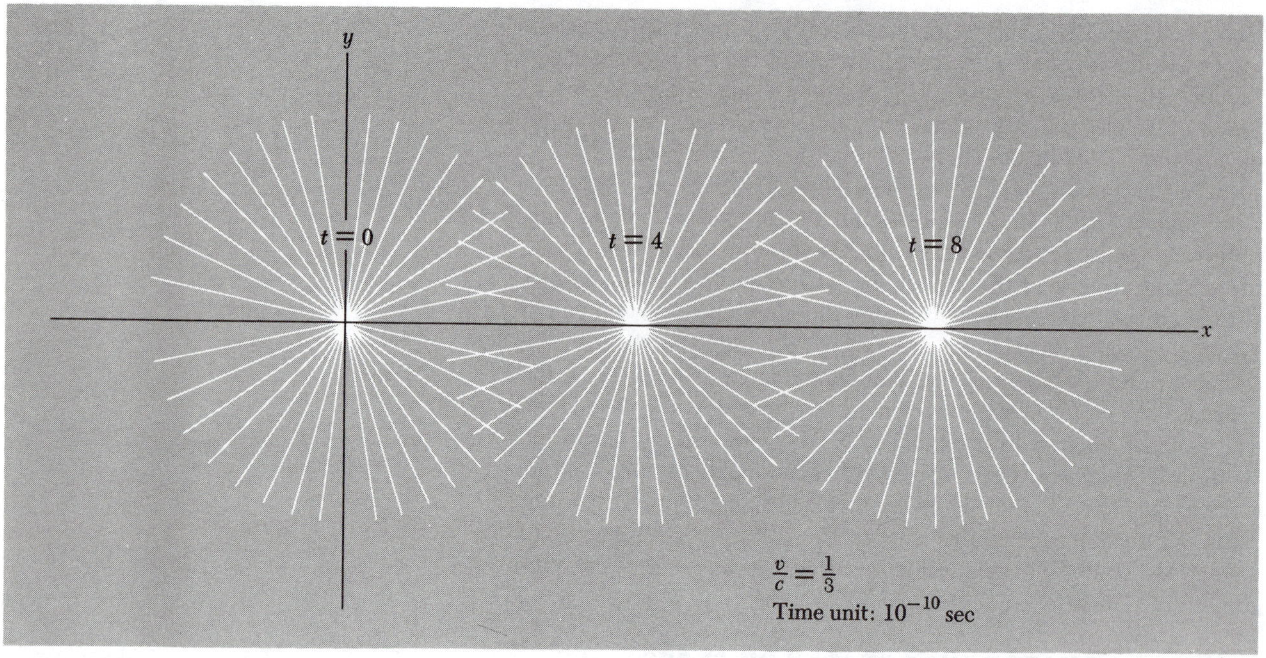
y
t = 0
t = 4
t = 8
x
v/c = 1/3
Time unit: 10^−10 sec

FIGURE 5.15
The electric field of a moving charge, shown for three instants of time; $v/c = 1/3$.

pened in the remote past can't make any difference. The field must somehow change, as we consider regions farther and farther from the charge, at the given instant $t = 2$, from the field shown in the second diagram of Fig. 5.16 to the field of a charge at the origin. We can't deduce more than this without knowing how fast the news *does* travel. Suppose—just suppose—it travels as fast as it can without conflicting with the relativity postulates. Then if the period of acceleration is neglected, we should expect the field within the entire 6-cm-radius sphere, at $t = 2$, to be the field of a uniformly moving point charge. If that is so, the field of the electron which starts from rest, suddenly acquiring the speed v at $t = 0$, must look something like Fig. 5.17. There is a thin spherical shell (whose thickness in an actual case will depend on the duration of the interval required for acceleration) within which the transition from one type of field to the other takes place. This shell simply expands with speed c, its center remaining at $x = 0$. The arrowheads on the field lines indicate the direction of the field when the source is a negative charge, as we have been assuming.

Figure 5.18 shows the field of an electron which had been moving with uniform velocity *until* $t = 0$, at which time it reached $x = 0$ where it was abruptly *stopped*. Now the news that it was stopped cannot reach, by time t, any point farther than ct from the origin. The field outside the sphere of radius $R = ct$ must be that which would have prevailed if the electron had kept on moving at its original speed. That is why we see the "brush" of field lines on the right in Fig. 5.18

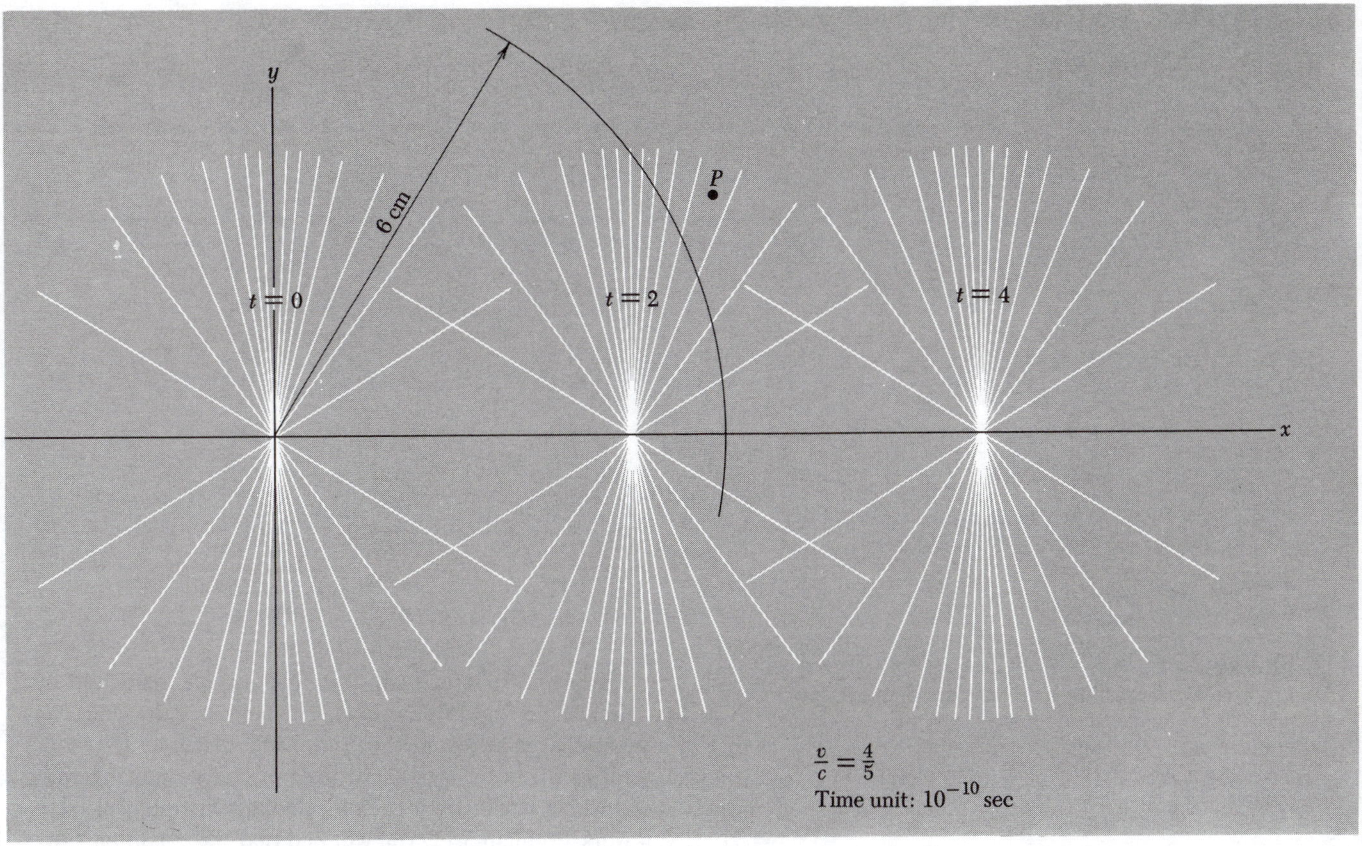
y
6 cm
P
t = 0
t = 2
t = 4
x
v/c = 4/5
Time unit: 10^-10 sec

FIGURE 5.16
The electric field of a moving charge, shown for three instants of time; $v/c = 4/5$.

pointing precisely down to the position where the electron would be if it hadn't stopped. (Note that this last conclusion does not depend on the assumption we introduced in the previous paragraph, that the news travels as fast as it can.) The field almost seems to have a life of its own!

It is a relatively simple matter to connect the inner and outer field lines. There is only one way it can be done that is consistent with Gauss's law. Taking Fig. 5.18 as an example, from some point such as A on the radial field line making angle θ_0 with the x axis, follow the field line wherever it may lead until you emerge in the outer field on some line making an angle that we may call φ_0 with the x axis. (This line of course is radial from the extrapolated position of the charge, the apparent source of the outer field.) Connect A and D to the x axis by circular arcs, arc AE centered on the source of the inner field, arc DF centered on the apparent source of the outer field. Rotate the curve $EABCDF$ around the x axis to generate a surface of revolution. As the surface encloses no charge, the surface integral of $\mathbf{E}$ over the entire surface must be zero. The only contributions to the

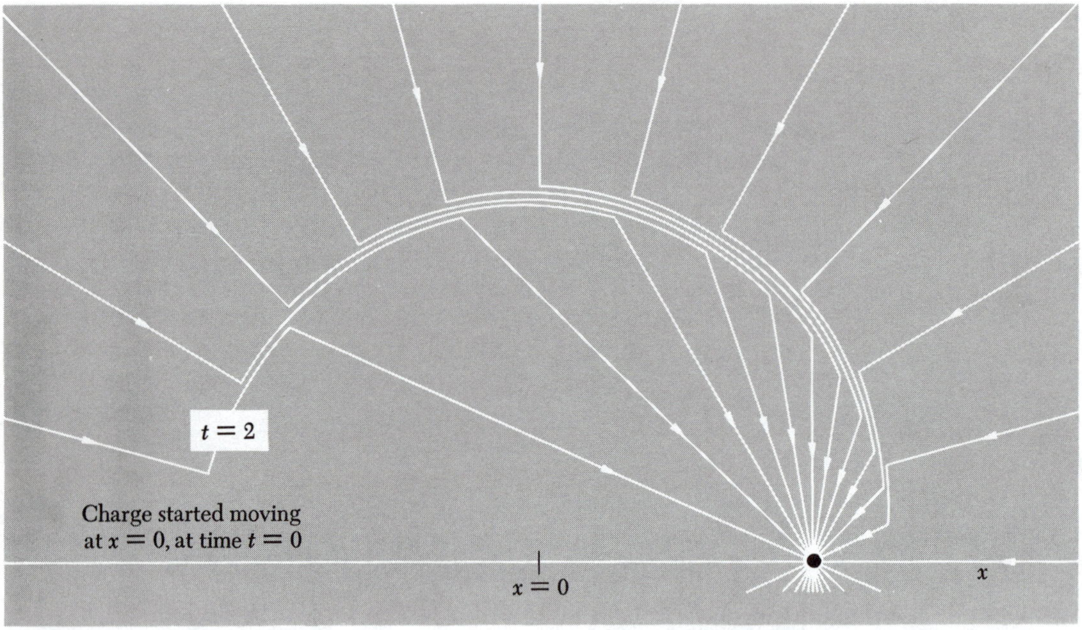
t = 2
Charge started moving at x = 0, at time t = 0
x = 0
x

FIGURE 5.17
An electron initially at rest in the laboratory frame is suddenly accelerated at $t = 0$ and moves with constant velocity thereafter. This is how the electric field looks at the instant $t = 2$ all over the laboratory frame.

integral come from the spherical caps, for the surface generated by *ABCD* is parallel to the field by definition. The field over the inner cap is that of a point charge at rest at the origin. The field over the outer cap is the field, as given by Eq. 12, of a point charge moving with constant speed which would have been located, at this moment, at $x = 2v$. If you work Problem 5.11, you will find that the condition "flux in through one cap equals flux out through the other" requires

$$\tan \phi_0 = \gamma \tan \theta_0 \tag{13}$$

The presence of γ in this formula is not surprising. We had already noticed the "relativistic compression" of the field pattern of a rapidly moving charge, illustrated in Fig. 5.14. The important new feature in Fig. 5.18 is the zigzag in the field line *ABCD*. The cause of this is not the γ in Eq. 13, but the fact that the apparent source of the outer field is displaced from the source of the inner field. If *AB* and *CD* belong to the same field line, the connecting segment *BC* has to run *nearly perpendicular* to a radial vector. We have a *transverse* electric field there, and one that, to judge by the crowding of the field lines, is relatively intense compared with the radial field. As time goes on, the zigzag in the field lines will move radially outward with speed c. But the thickness of the shell of transverse field will not increase, for that was determined by the duration of the deceleration process.

The ever-expanding shell of transverse electric field would keep on going *even if* at some later time—at $t = 3$, say—we suddenly accelerated the electron back to its original velocity. That would only

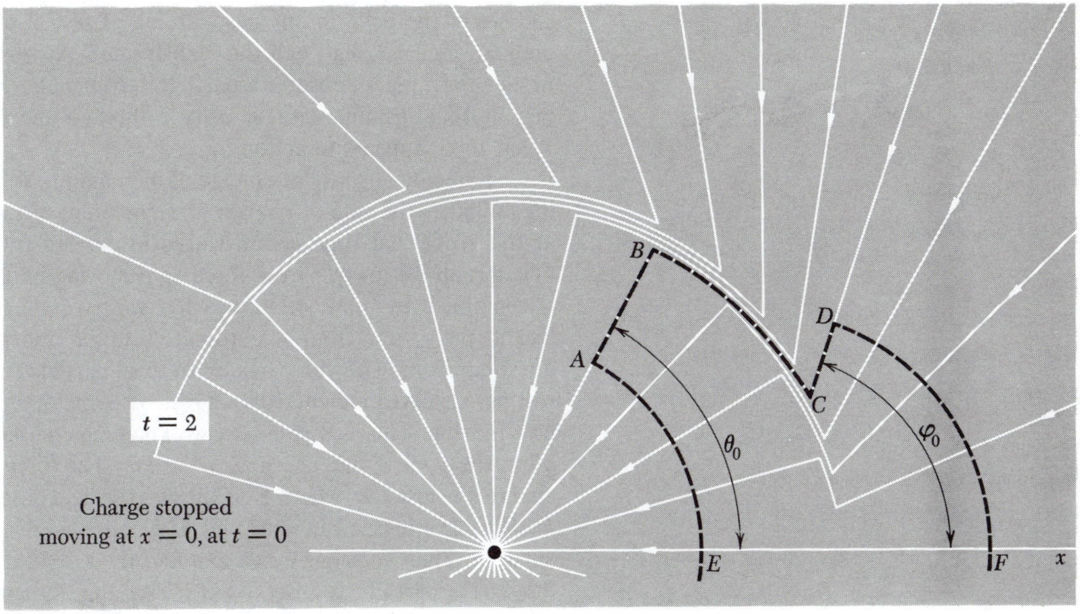
t = 2
Charge stopped moving at x = 0, at t = 0
A
B
C
D
E
F
x
θ0
φ0

FIGURE 5.18
An electron that has been moving with constant velocity reaches the origin at $t = 0$, is abruptly stopped, and remains at rest thereafter. This is how the field looks in the laboratory frame at the instant $t = 2$. The dashed outline follows a field line from A to D. Rotating the whole outline $EABCDF$ around the x axis generates a closed surface, the total flux through which must be zero. The flux in through the spherical cap FD must equal the flux out through the spherical cap EA. This condition suffices to determine the relation between θ_0 and ϕ_0.

launch a new outgoing shell, this one looking very much like the field in Fig. 5.17. The field *does* have a life of its own! What has been created here before our eyes is an *electromagnetic wave*. The magnetic field that is also part of it was not revealed in this view. Later, in Chapter 9, we shall learn how the electric and magnetic fields work together in propagating an electrical disturbance through empty space. What we have discovered here is that such waves *must* exist if nature conforms to the postulates of special relativity and if electric charge is a relativistic invariant.

More can be done with our "zigzag-in-the-field-line" analysis. Appendix B shows how to derive, rather simply, an accurate and simple formula for the rate of radiation of energy by an accelerated electric charge. We must return now to the uniformly moving charge, which has more surprises in store.

FORCE ON A MOVING CHARGE

5.8 Equation 12 tells us the force experienced by a stationary charge in the field of another charge that is moving at constant velocity. We now ask a different question: What is the force that acts on a moving charge, one that moves in the field of some other charges?

We shall look first into the case of a charge moving through the field produced by stationary charges. It might be an electron moving between the charged plates of an oscilloscope, or an alpha particle moving through the Coulomb field around an atomic nucleus. The sources of the field, in any case, are all at rest in some frame of reference which we shall call the "lab frame." At some place and time in the lab frame we observe a particle carrying charge q which is moving, at that instant, with velocity $\mathbf{v}$ through the electrostatic field. What force appears to act on q?

Force means rate of change of momentum, so we are really asking, What is the rate of change of momentum of the particle, $d\mathbf{p}/dt$, at this place and time, as measured in our lab frame of reference? (That is all we mean by the force on a moving particle.) The answer is contained, by implication, in what we have already learned. Let's look at the system from a coordinate frame F' moving, at the time in question, along with the particle. In this "particle frame" the particle will be, at least momentarily, at rest. It is the other charges that are now moving. This is a situation we know how to handle. The charge q has the same value; charge is invariant. The force on the stationary charge q is just $\mathbf{E}'q$, where $\mathbf{E}'$ is the electric field observed in the frame F'. We have learned how to find $\mathbf{E}'$ when $\mathbf{E}$ is given; Eq. 7 provides our rule. Thus knowing $\mathbf{E}$, we can find the rate of change of momentum of the particle as observed in F'. All that remains is to transform this quantity back to F. So our problem hinges on the question, How does force, that is, rate of change of momentum, transform from one inertial frame to another?

The answer to that question is worked out later and is expressed in Eqs. 12 and 13 of Appendix A. The force component *parallel* to the relative motion of the two frames has the *same* value in the moving frame as it does in the rest frame of the particle. A force component *perpendicular* to the relative frame velocity is always *smaller*, by $1/\gamma$, than its value in the particle's rest frame. Let us summarize this in Eq. 14 using subscripts $\|$ and $\perp$ to label momentum components, respectively, parallel to and perpendicular to the relative velocity of F' and F, as we did in Eq. 7.

$$\frac{dp_{\|}}{dt} = \frac{dp'_{\|}}{dt'} \qquad \frac{dp_{\perp}}{dt} = \frac{1}{\gamma}\frac{dp'_{\perp}}{dt'} \tag{14}$$

Note that this is not a symmetrical relation between the primed and unprimed quantities. The rest frame of the particle, which we have chosen to call F' in this case, is special. In it the magnitude of the transverse force component is greater than in any other frame.

Equipped with the force transformation law, Eq. 14, and the transformation law for electric field components, Eq. 7, we return now to our charged particle moving through the field $\mathbf{E}$, and we discover an astonishingly simple fact. Consider first $E_{\|}$, the component of $\mathbf{E}$ parallel to the instantaneous direction of motion of our charged par-

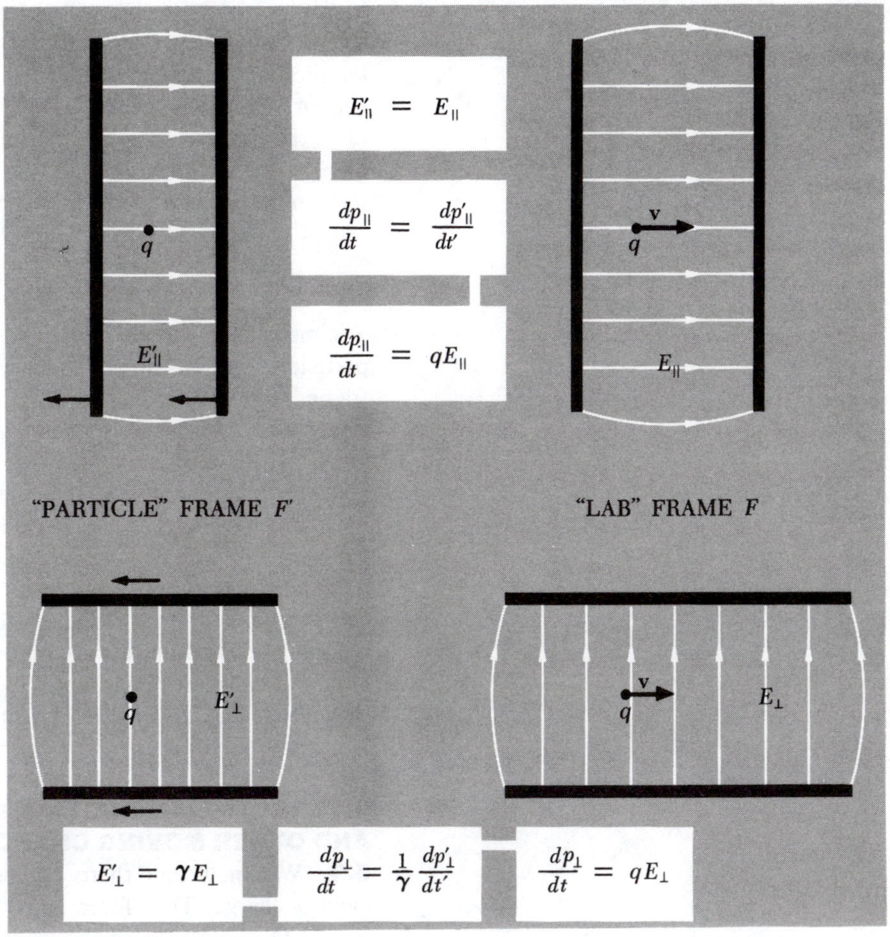

FIGURE 5.19
In a frame in which the charges producing the field **E** are at rest, the force on a charge q moving with any velocity is simply $q\mathbf{E}$.

ticle. Transform to a frame F' moving, at that instant, with the particle. In that frame the longitudinal electric field is $E'_{\parallel}$, and according to Eq. 7, $E'_{\parallel} = E_{\parallel}$. So the force $dp'_{\parallel}/dt'$ is

$$\frac{dp'_{\parallel}}{dt'} = qE'_{\parallel} = qE_{\parallel} \tag{15}$$

Back in frame F, observers are measuring the longitudinal force, that is, the rate of change of the longitudinal momentum component, $dp_{\parallel}/dt$. According to Eq. 14, $dp_{\parallel}/dt = dp'_{\parallel}/dt'$, so in frame F the longitudinal force component they find is

$$\frac{dp_{\parallel}}{dt} = \frac{dp'_{\parallel}}{dt'} = qE_{\parallel} \tag{16}$$

Of course the particle does not *remain* at rest in F' as time goes on. It will be accelerated by the field $\mathbf{E}'$, and $\mathbf{v}'$, the velocity of the particle in the inertial frame F', will gradually increase from zero. However, as we are concerned with the instantaneous acceleration, only infinitesimal values of v' are involved anyway, and the restriction on Eq. 14 is rigorously fulfilled. For $E_\perp$, the transverse field component in F, the transformation is $E'_\perp = \gamma E_\perp$, so that $dp'_\perp/dt' = qE'_\perp = q\gamma E_\perp$. But on transforming the force back to frame F we have $dp_\perp/dt = (1/\gamma)(dp'_\perp/dt')$, so the γ drops out after all:

$$\frac{dp_\perp}{dt} = \frac{1}{\gamma}(\gamma E_\perp q) = qE_\perp \tag{17}$$

The message of Eqs. 16 and 17 is simply this: The force on a charged particle in motion through F is q times the electric field $\mathbf{E}$ in that frame, *strictly independent* of the velocity of the particle. Figure 5.19 is a reminder of this fact, and of the way we discovered it.

You have already used this result earlier in the course, where you were simply told that the contribution of the electric field to the force on a moving charge is $q\mathbf{E}$. Because this is familiar and so simple, you may think it is obvious and we have been wasting our time proving it. Now we could have taken it as an empirical fact. It has been verified over an enormous range, up to velocities so close to the speed of light, in the case of electrons, that the factor γ is 10^4. From that point of view it is a most remarkable law. Our discussion in this chapter has shown that this fact is also a direct consequence of charge invariance.

INTERACTION BETWEEN A MOVING CHARGE AND OTHER MOVING CHARGES

5.9 We know that there can be a velocity-dependent force on a moving charge. That force is associated with a *magnetic field,* the sources of which are electric currents, that is, other charges in motion. Oersted's experiment showed that electric currents could influence magnets, but at that time the nature of a magnet was totally mysterious. Soon Ampère and others unraveled the interaction of electric currents with each other, as in the attraction observed between two parallel wires carrying current in the same direction. This led Ampère to the hypothesis that a magnetic substance contains permanently circulating electric currents. If so, Oersted's experiment could be understood as the interaction of the galvanic current in the wire with the permanent microscopic currents which gave the compass needle its special properties. Ampère gave a complete and elegant mathematical formulation of the interaction of steady currents, and of the equivalence of magnetized matter to systems of permanent currents. His brilliant conjecture about the actual nature of magnetism in iron had to wait a century, more or less, for its ultimate confirmation.

Whether the magnetic manifestations of electric currents arose from anything *more* than the simple transport of charge was not clear to Ampère and his contemporaries. Would the motion of an electrostatically charged object cause effects like those produced by a continuous galvanic current? Later in the century Maxwell's theoretical work suggested the answer should be *yes*. The first direct evidence was obtained by Henry Rowland, to whose experiment we shall return at the end of Chapter 6.

From our present vantage point, the magnetic interaction of electric currents can be recognized as an inevitable corollary to Coulomb's law. If the postulates of relativity are valid, if electric charge is invariant, and if Coulomb's law holds, then, as we shall now show, the effects we commonly call "magnetic" are bound to occur. They will emerge as soon as we examine the electric interaction between a moving charge and other moving charges. A very simple system will illustrate this.

In the lab frame of Fig. 5.20*a*, with spatial coordinates x, y, z, there is a line of positive charges, at rest and extending to infinity in both directions. We shall call them ions for short. Indeed, they might represent the copper ions that constitute the solid substance of a copper wire. There is also a line of negative charges that we shall call electrons. These are all moving to the right with speed v_0. In a real wire the electrons would be intermingled with the ions; we've separated them in the diagram for clarity. The linear density of positive charge is λ_0 in esu/cm. It happens that the linear density of negative charge along the line of electrons is exactly equal in magnitude. That is, any given length of "wire" contains at a given instant the same number of electrons and protons.† The net charge on the wire is zero. Gauss' law tells us there can be no flux from a cylinder that contains no charge, so the electric field must be zero everywhere outside the wire. A test charge q at rest near this wire experiences no force whatever.

Suppose the test charge is not at rest in the lab frame but is moving with speed v in the x direction. Transform to a frame moving with the test charge, the x', y' frame in Fig. 5.20*b*. The test charge q is here at rest, but something else has changed: The wire appears to be charged! There are two reasons for that: The positive ions are closer together, and the electrons are farther apart. Because the lab frame in which the positive ions are at rest is moving with speed v, the distance between positive ions as seen in the test charge frame is contracted by $\sqrt{1 - v^2/c^2}$, or $1/\gamma$. The linear density of positive charge in this frame is correspondingly greater; it must be $\gamma\lambda_0$. The density of negative charge takes a little longer to calculate, for the electrons were already moving with speed v_0 in the lab frame. Hence their linear density in the lab frame, which was $-\lambda_0$, had already been increased

†It doesn't have to, but that equality can always be established, if we choose, by adjusting the number of electrons per unit length. We assume that has been done.

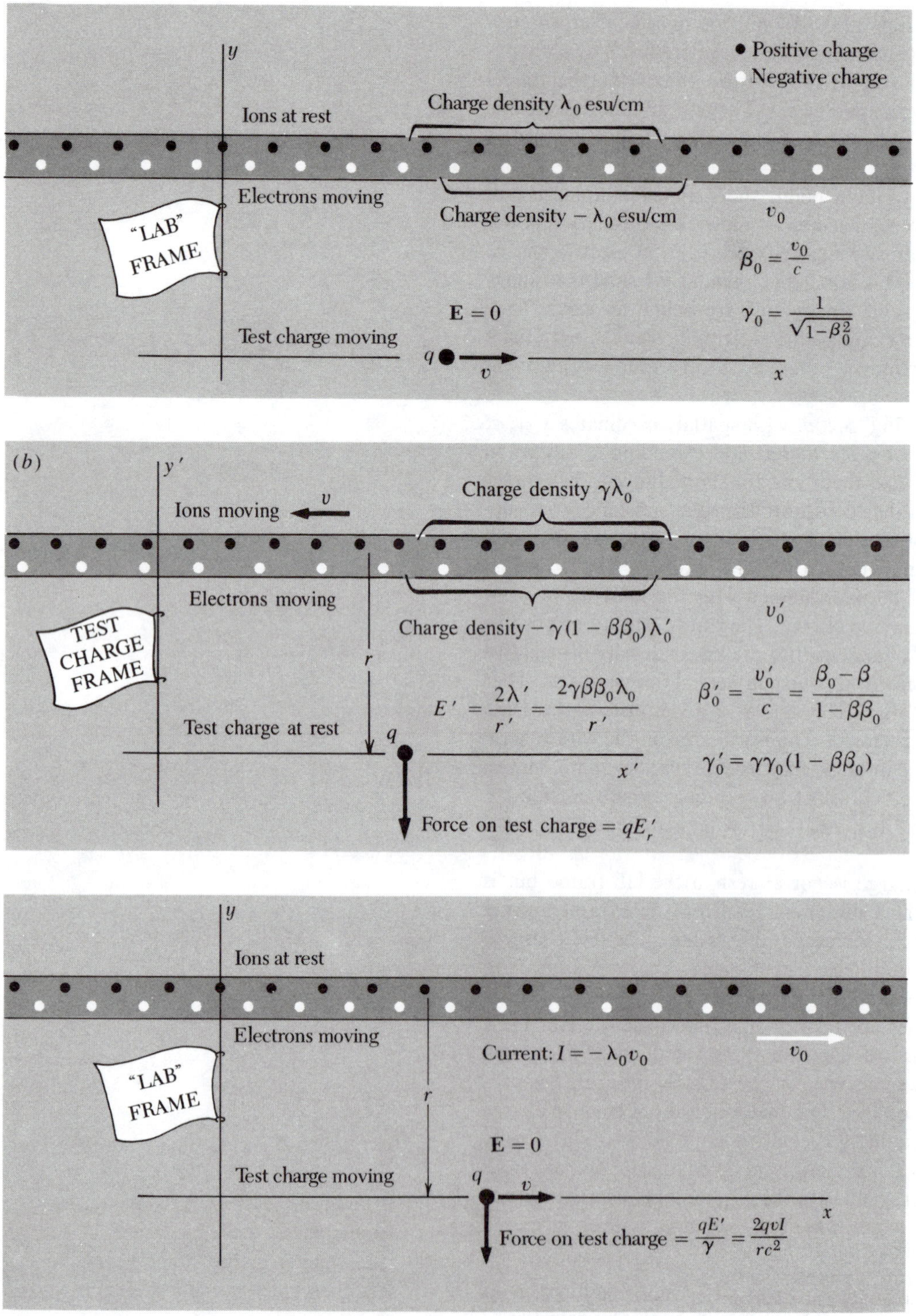

Positive charge
Negative charge
y
Ions at rest
Charge density λ_0 esu/cm
Electrons moving
Charge density $-\lambda_0$ esu/cm
"LAB" FRAME
v_0
$\beta_0 = \frac{v_0}{c}$
$\mathbf{E} = 0$
$\gamma_0 = \frac{1}{\sqrt{1-\beta_0^2}}$
Test charge moving
q
v
x
(b)
y′
Ions moving
v
Charge density $\gamma\lambda_0'$
Electrons moving
v_0'
TEST CHARGE FRAME
Charge density $-\gamma(1-\beta\beta_0)\lambda_0'$
r
$E' = \frac{2\lambda'}{r'} = \frac{2\gamma\beta\beta_0\lambda_0}{r'}$
$\beta_0' = \frac{v_0'}{c} = \frac{\beta_0-\beta}{1-\beta\beta_0}$
Test charge at rest
q
x′
$\gamma_0' = \gamma\gamma_0(1-\beta\beta_0)$
Force on test charge = qE_r'
y
Ions at rest
Electrons moving
Current: $I = -\lambda_0 v_0$
v_0
"LAB" FRAME
r
$\mathbf{E} = 0$
Test charge moving
q
v
x
Force on test charge = $\frac{qE'}{\gamma} = \frac{2qvI}{rc^2}$

by a Lorentz contraction. In the electrons' own rest frame the negative charge density must have been $-\lambda_0/\gamma_0$, where γ_0 is the *Lorentz factor* that goes with v_0.

Now we need the speed of the electrons in the test charge frame in order to calculate their density there. To find that velocity (v_0' in Fig. 5.20*b*) we must add the velocity $-v$ to the velocity v_0, remembering to use the relativistic formula for the addition of velocities (Eq. 6 in Appendix A). Let $\beta_0' = v_0'/c$, $\beta_0 = v_0/c$, and $\beta = v/c$. Then

$$\beta_0' = \frac{\beta_0 - \beta}{1 - \beta\beta_0} \tag{18}$$

The corresponding Lorentz factor γ_0', obtained from Eq. 18 with a little algebra, is

$$\gamma_0' = (1 - \beta_0'^2)^{-1/2} = \gamma\gamma_0(1 - \beta\beta_0) \tag{19}$$

This is the factor by which the linear density of negative charge in the electrons' own rest frame is enhanced when it is measured in the test charge frame. The total linear density of charge in the wire in the test charge frame, λ', can now be calculated:

$$\lambda' = \gamma\lambda_0 - \frac{\lambda_0}{\gamma_0}\gamma\gamma_0(1 - \beta\beta_0) = \gamma\beta\beta_0\lambda_0 \tag{20}$$

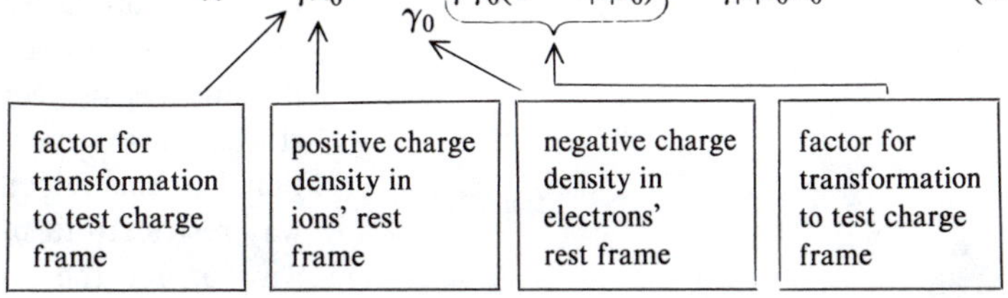
factor for transformation to test charge frame
positive charge density in ions' rest frame
negative charge density in electrons' rest frame
factor for transformation to test charge frame

The wire is positively charged. Gauss's law guarantees the existence of a radial electric field E_r' given by our familiar formula for the field of any infinite line charge:

$$E_r' = \frac{2\lambda'}{r'} = \frac{2\gamma\beta\beta_0\lambda_0}{r'} \tag{21}$$

At the location of the test charge q this field is in the $-y'$ direction. The test charge will experience a force

$$F_y' = qE_y' = -\frac{2q\gamma\beta\beta_0\lambda_0}{r'} \tag{22}$$

Now let's return to the lab frame, pictured again in Fig. 5.20*c*. What is the magnitude of the force on the charge q as measured there? If its value is qE_y' in the rest frame of the test charge, observers in the lab frame will report a force smaller by the factor $1/\gamma$. Since $r = r'$, the force on our moving test charge, measured in the lab frame, is

FIGURE 5.20
A test charge q moving parallel to a current in a wire. (*a*) In the lab frame the wire, in which the positive charges are fixed, is at rest. The current consists of electrons moving to the right with speed v_0. The net charge on the wire is zero. There is no electric field outside the wire. (*b*) In a frame in which the test charge is at rest the positive ions are moving to the left with speed v and the electrons are moving to the right with speed v_0'. The linear density of a positive charge is greater than the linear density of negative charge. The wire appears positively charged, with an external field E_r' which causes a force qE_r' on the stationary test charge q. (*c*) That force transformed back to the lab frame has the magnitude qE_r'/γ, which is proportional to the product of the speed v of the test charge and the current in the wire, $-\lambda_0 v_0$.

$$F_y = \frac{F'_y}{\gamma} = -\frac{2q\beta\beta_0\lambda_0}{r} \tag{23}$$

Now $-\lambda_0 v_0$ or $-\lambda_0\beta_0 c$ is just the total current I in the wire, in the lab frame, for it is the amount of charge flowing past a given point per second. We'll call current positive if it is equivalent to positive charge flowing in the positive x direction. Our current in this example is negative. Our result can be written this way:

$$F_y = \frac{2I}{rc^2} q v_x \tag{24}$$

We have found that in the lab frame the moving test charge experiences a force in the y direction which is proportional to the current in the wire, and to the velocity of the test charge in the x direction.

It is a remarkable fact that the force on the moving test charge does not depend separately on the velocity or density of the charge carriers but only on the product, $\beta_0\lambda_0$ in our example, that determines the charge transport. If we have a certain current I, say 10^7 esu/sec which is the same as 3.3 milliamps, it does not matter whether this current is composed of high-energy electrons moving with 99 percent of the speed of light, of electrons in a metal executing nearly random thermal motions with a slight drift in one direction, or of charged ions in solution with positive ions moving one way, negatives the other. Or it could be any combination of these, as Problem 5.18 will demonstrate. Furthermore, the force on the test charge is strictly proportional to the velocity of the test charge v. Our derivation was in no way restricted to small velocities, either for the charge carriers in the wire or for the moving charge q. Equation 24 is exact, with no restrictions.

Let's see how this explains the mutual repulsion of conductors carrying currents in opposite directions, as shown in Fig. 5.1*b* at the beginning of this chapter. Two such wires are represented in the lab frame in Fig. 5.21*a*. Assume the wires are uncharged in the lab frame. Then there is no electrical force from the opposite wire on the positive ions which are stationary in the lab frame. Transferring to a frame in which one set of electrons is at rest (Fig. 5.21*b*), we find that in the other wire the electron distribution is Lorentz-contracted more than the positive ion distribution. Because of that the electrons at rest in this frame will be repelled by the other wire. But when we transfer to the frame in which those other electrons are at rest (Fig. 5.21*c*), we find the same situation. They too will be repelled. These repulsive forces will be observed in the lab frame as well, modified only by the factor γ. We conclude that the two streams of electrons will repel one another in the lab frame. The stationary positive ions, although they feel no direct electrical force from the other wire, will be the indirect bearers of this repulsive force if the electrons remain confined within the wire. So the wires will be pushed apart, as in Fig. 5.1*b*, until some external force balances the repulsion.

Moving parallel to a current-carrying conductor, the charged particle experienced a force perpendicular to its direction of motion. What if it moves, instead, at right angles to the conductor? A velocity perpendicular to the wire will give rise to a force parallel to the wire—again, a force perpendicular to the particle's direction of motion. To see how this comes about, let us return to the lab frame of that system and give the test charge a velocity v in the y direction, as in Fig. 5.22*a*. Transferring to the rest frame of the test charge (Fig. 5.22*b*), we find the positive ions moving vertically downward. Certainly they cannot cause a horizontal field at the test-charge position. The x' component of the field from an ion on the left will be exactly cancelled by the x' component of the field of a symmetrically positioned ion on the right. The effect we are looking for is caused by the electrons. They are all moving obliquely in this frame, downward and toward the right. Consider the two symmetrically located electrons e_1 and e_2. Their electric fields, relativistically compressed in the direction of the electrons' motion, have been represented by a brush of field lines in the manner of Fig. 5.14. You can see that, although e_1 and e_2 are equally far away

FIGURE 5.21
(*a*) Lab frame with two wires carrying current in opposite directions. As in metal wire, current is due to motion of negative ions (electrons) only. (*b*) Rest frame of electrons in wire 1. Note that in wire 2 positive ions are compressed, but electron distribution is contracted even more. (*c*) Rest frame of electrons in wire 2. Just as in (*b*), the *other* wire appears to these electrons at rest to be negatively charged.

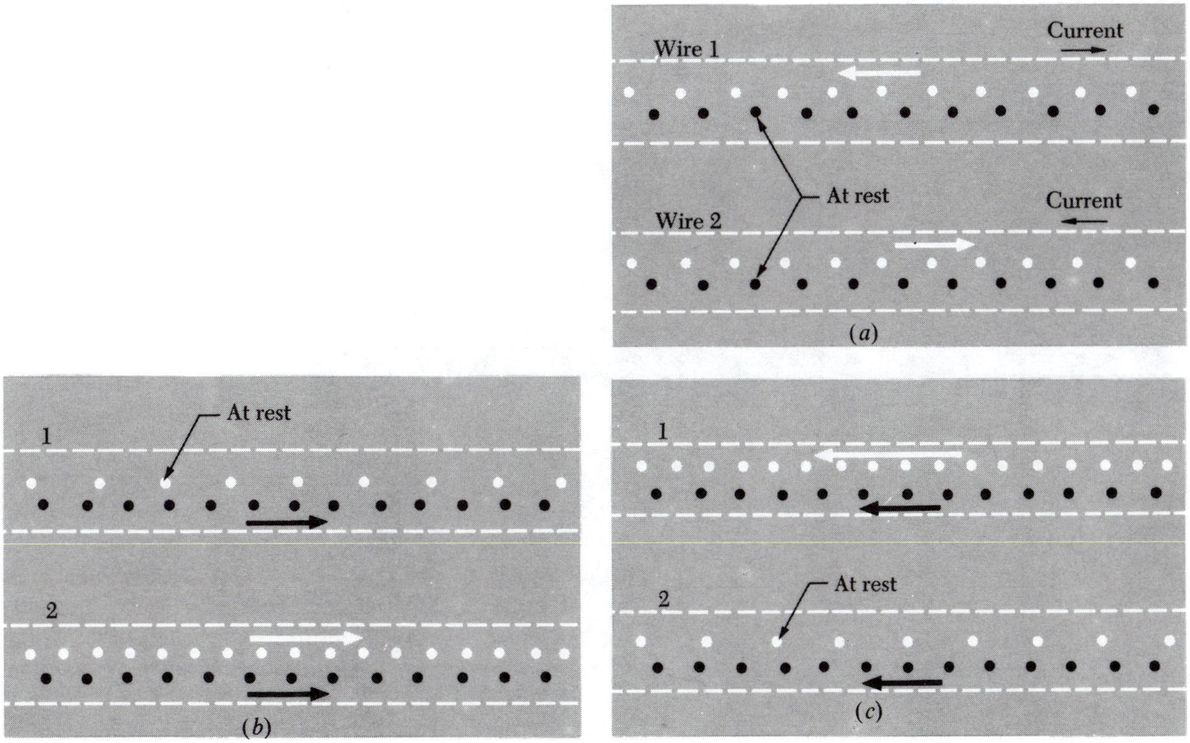

from the test charge, the field of electron e_2 will be *stronger* than the field of electron e_1 at that location. That is because the line from e_2 to the test charge is more nearly perpendicular to the direction of motion of e_2. In other words, the angle θ' that appears in the denominator of Eq. 12 is here different for e_1 and e_2, so that $\sin^2 \theta'_2 > \sin^2 \theta'_1$. That will be true for any symmetrically located pair of electrons on the line, as you can verify with the aid of Fig. 5.23. The electron on the right always wins. Summing over all the electrons is therefore bound to yield a resultant field E' in the $\hat{\mathbf{x}}$ direction. The y' component of the electrons' field will be exactly cancelled by the field of the ions. That E'_y is zero is guaranteed by Gauss's law, for the number of charges per unit length of wire is the same as it was in the lab frame. The wire is uncharged in both frames.

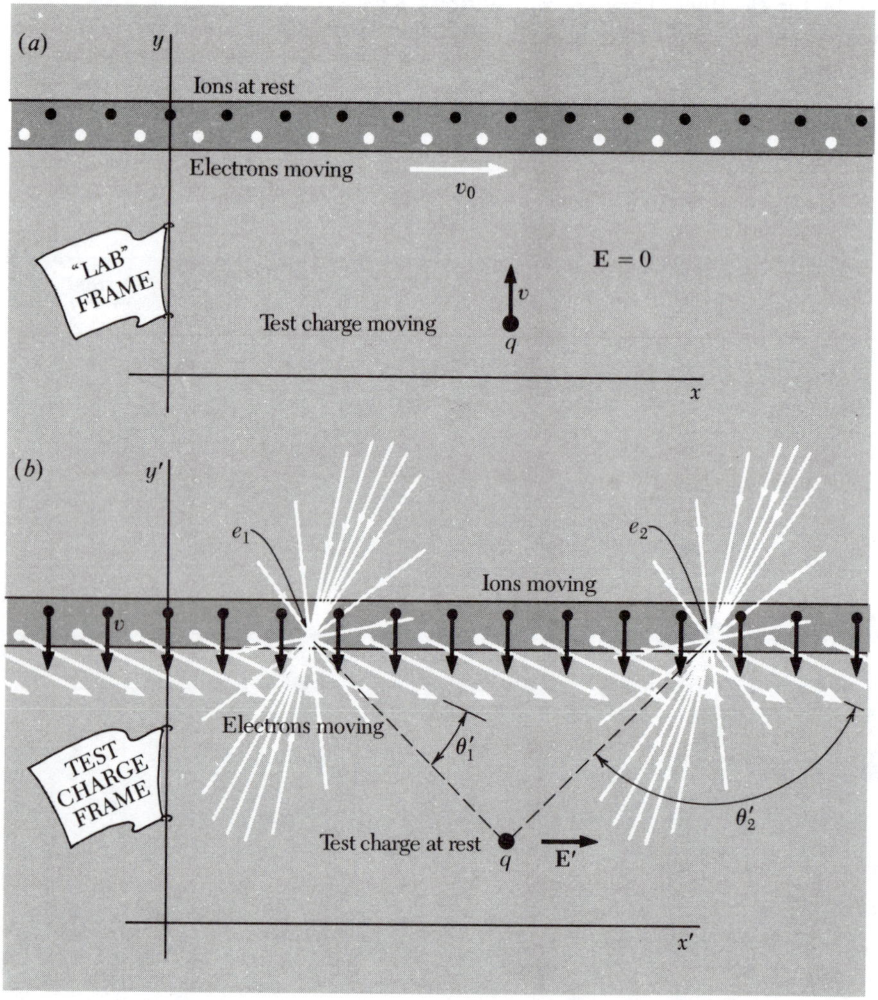

FIGURE 5.22
(*a*) The "wire" with its current of moving negative charges, or "electrons," is the same as in Fig. 5.20, but now the test charge is moving toward the wire. (*b*) In the rest frame of the test charge the positive charges, or "ions," are moving in the $-\hat{\mathbf{y}}$ direction. The electrons are moving obliquely. Because the field of a moving charge is stronger in directions more nearly perpendicular to its velocity, an electron on the right, such as e_2, causes a stronger field at the position of the test charge than does a symmetrically located electron on the left. Therefore the vector sum of the fields has in this frame a component in the $\hat{\mathbf{x}}'$ direction.

The force on our test charge, qE'_x, when transformed back into the lab frame, will be a force proportional to v in the $\hat{\mathbf{x}}$ direction, which is the direction of $\mathbf{v} \times \mathbf{B}$ if $\mathbf{B}$ is a vector in the $\hat{\mathbf{z}}$ direction, pointing at us out of the diagram. We could show that the magnitude of this velocity-dependent force is given here also by Eq. 24: $F = 2qvI/rc^2$. The physics needed is all in Eq. 12, but the integration is somewhat laborious and will not be undertaken here.

In this chapter we have seen how the fact of charge invariance implies forces between electric currents. That does not oblige us to look on one fact as the cause of the other. These are simply two aspects of electromagnetism whose relationship beautifully illustrates the more general law: Physics is the same in all inertial frames of reference.

If we had to analyze every system of moving charges by transforming back and forth among various coordinate systems, our task would grow both tedious and confusing. There is a better way. The overall effect of one current on another, or of a current on a moving charge, can be described completely and concisely by introducing a new field, the magnetic field.

FIGURE 5.23
A closer look at the geometry of Fig. 5.22*b*, showing that, for *any* pair of electrons equidistant from the test charge, the one on the right will have a larger value of $\sin^2 \theta'$. Hence, according to Eq. 5.12, it will produce the stronger field at the test charge.

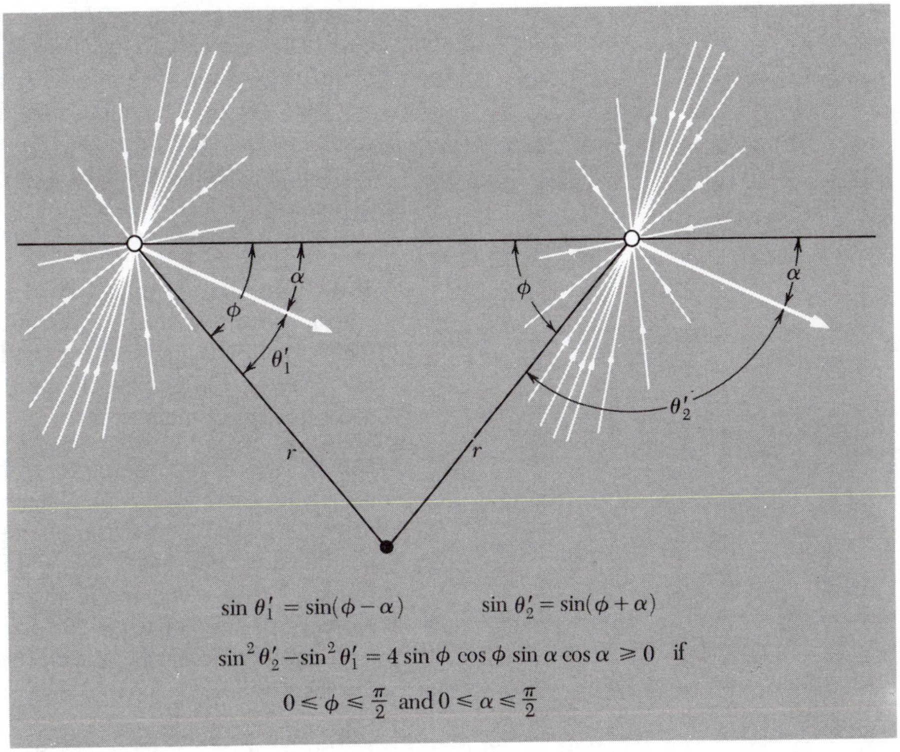

α
ϕ
θ'_1
θ'_2
r
$\sin \theta'_1 = \sin(\phi - \alpha)$
$\sin \theta'_2 = \sin(\phi + \alpha)$
$\sin^2 \theta'_2 - \sin^2 \theta'_1 = 4 \sin \phi \cos \phi \sin \alpha \cos \alpha \geqslant 0$ if
$0 \leqslant \phi \leqslant \frac{\pi}{2}$ and $0 \leqslant \alpha \leqslant \frac{\pi}{2}$

PROBLEMS

5.1 A capacitor consists of two parallel rectangular plates with a vertical separation of 2 cm. The east-west dimension of the plates is 20 cm, the north-south dimension is 10 cm. The capacitor has been charged by connecting it temporarily to a battery of 300 volts (1 statvolt). How many excess electrons are on the negative plate? What is the electric field strength between the plates? Now give the following quantities as they would be measured in a frame of reference which is moving eastward, relative to the laboratory in which the plates are at rest, with speed $0.6c$: the three dimensions of the capacitor; the number of excess electrons on the negative plate; the electric field strength between the plates. Answer the same questions for a frame of reference which is moving upward with speed $0.6c$.

5.2 On a nylon filament 0.01 cm in diameter and 4 cm long there are 5.0×10^8 extra electrons distributed uniformly over the surface. What is the electric field strength at the surface of the filament:

(*a*) In the rest frame of the filament?

(*b*) In a frame in which the filament is moving at a speed $0.9c$ in a direction parallel to its length?

5.3 A beam of 9.5-megaelectron-volt (Mev) electrons ($\gamma = 20$) amounting as current to 0.05 microamperes, is traveling through vacuum. The transverse dimensions of the beam are less than 1 mm, and there are no positive charges in or near it.

(*a*) In the lab frame, what is approximately the electric field strength 1 cm away from the beam, and what is the average distance between an electron and the next one ahead of it, measured parallel to the beam?

(*b*) Answer the same questions for the electron rest frame.

5.4 Consider the trajectory of a charged particle which is moving with a speed $0.8c$ in the x direction when it enters a large region in which there is a uniform electric field in the y direction. Show that the x velocity of the particle must actually *decrease*. What about the x component of momentum?

5.5 Fixed in the frame F is a sheet of charge, of uniform surface density σ, which bisects the dihedral angle formed by the xy and the yz planes. The electric field of this stationary sheet is of course perpendicular to the sheet. How will this be described by observers in a frame F' that is moving in the x direction with velocity $0.6c$ with respect to F? What is the surface charge density σ' and what is the direction and strength of the electric field in F'? Is it perpendicular to the sheet?

5.6 In a colliding beam storage ring an antiproton going east passed a proton going west, the distance of closest approach being 10^{-8} cm.

The kinetic energy of each particle in the lab frame was 93 Gev, corresponding to $\gamma = 100$. In the rest frame of the proton, what was the maximum intensity of the electric field at the proton due to the charge on the antiproton? For about how long, approximately, did the field exceed half its maximum intensity?

5.7 The most extremely relativistic charged particles we know about are cosmic rays which arrive from outer space. Occasionally one of these particles has so much kinetic energy that it can initiate in the atmosphere a "giant shower" of secondary particles, dissipating, in total, as much as 10^{19} ev of energy (more than 1 joule!). The primary particle, probably a proton, must have had $\gamma \approx 10^{10}$. How far away from such a proton would the field rise to 1 volt/meter as it passes? Roughly how thick is the "pancake" of field lines at that distance?

Ans. 4 meters; 4×10^{-10} meter.

5.8 In the laboratory frame a proton is at rest at the origin at $t = 0$. At that instant a negative pion which has been traveling in along the x axis at a speed of $0.6c$ has reached the point $x = 0.01$ cm. There are no other charges around. What is the magnitude of the force on the pion? What is the magnitude of the force on the proton? What about Newton's third law?

5.9 The deflection plates in a high-voltage cathode ray oscilloscope are two rectangular plates, 4 cm long and 1.5 cm wide, spaced 0.8 cm apart. There is a difference in potential of 6000 volts between the plates. An electron which has been accelerated through a potential difference of 250 kilovolts enters this deflector from the left, moving parallel to the plates and halfway between them, initially. We want to find the position of the electron and its direction of motion when it leaves the deflecting field at the other end of the plates. We shall neglect the fringing field and assume the electric field between the plates is uniform right up to the end. The rest mass of the electron may be taken as 500 kev. First carry out the analysis in the lab frame by answering these questions: $\gamma = ?$; $\beta = ?$; p_x, in units of mc, $= ?$; time spent between the plates = ? (Neglect the change in horizontal velocity discussed in Problem 5.4); transverse momentum component acquired, in units of mc, = ? ; transverse velocity at exit = ? ; vertical position at exit = ? ; direction of flight at exit? Now describe this whole process as it would appear in an inertial frame which moved with the electron at the moment it entered the deflecting region: What do the plates look like? What is the field between them? What happens to the electron in this coordinate system? Your main object in this exercise is to convince yourself that the two descriptions are completely consistent.

5.10 In the rest frame of a particle with charge q_1 another particle with charge q_2 is approaching, moving with velocity v not small compared with c. If it continues to move in a straight line, it will pass a distance d from the position of the first particle. It is so massive that its displacement from the straight path during the encounter is small compared with d. Likewise, the first particle is so massive that its displacement from its initial position while the other particle is nearby is also small compared with d.

(a) Show that the increment in momentum acquired by each particle as a result of the encounter is perpendicular to $\mathbf{v}$ and in magnitude $2q_1q_2/vd$. (Gauss's law can be useful here.)

(b) Expressed in terms of the other quantities, how large must the masses of the particles be to justify our assumptions?

5.11 Derive Eq. 13 by performing the integration to find the flux of E through each of the spherical caps described in the legend of Fig. 5.18. On the inner cap the field strength is constant, and the element of surface area may be taken as $2\pi r^2 \sin\theta\, d\theta$. On the outer cap the field is described by Eq. 12 with the appropriate changes in symbols, and the element of surface area is $2\pi r^2 \sin\phi\, d\phi$. The integral you will need is

$$\int \frac{dx}{(a^2 + x^2)^{3/2}} = \frac{x}{a^2(a^2 + x^2)^{1/2}}$$

5.12 In the field of the moving charge Q, given by Eq. 12, we want to find an angle δ such that half of the total flux from Q is contained between the two conical surfaces $\theta' = \pi/2 + \delta$ and $\theta' = \pi/2 - \delta$. If you have done Problem 5.11 you have already done most of the work. You should find that, for $\gamma \gg 1$, the angle between the two cones is roughly $1/\gamma$.

5.13 In the figure you see an electron at time $t = 0.0$ and the associated electric field at that instant. Distances in centimeters are given.

PROBLEM 5.13

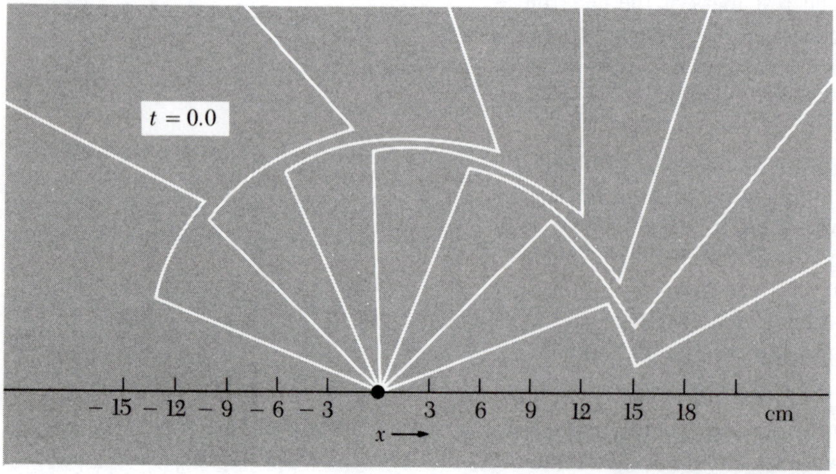

in the diagram.

(*a*) Describe what *has been* going on. Make your description as complete and quantitative as you can.

(*b*) Where was the electron at the time $t = -7.5 \times 10^{-10}$ sec?

(*c*) What was the strength of the electric field at the origin at that instant?

5.14 The figure shows a highly relativistic positive particle approaching the origin from the left and a negative particle approaching with equal speed from the right. They collide at the origin at $t = 0$, find some way to dispose of their kinetic energy, and remain there as a neutral entity. What do you think the electric field looks like at some time $t > 0$? Sketch the field lines. How does the field change as time goes on?

5.15 In Fig. 5.20 the relative spacing of the black and white dots was designed to be consistent with $\gamma = 1.2$ and $\beta_0 = 0.8$. Calculate β_0'. Find the value, as a fraction of λ_0, of the net charge density λ' in the test-charge frame.

5.16 Suppose that the velocity of the test charge in Fig. 5.20 is made equal to that of the electrons, v_0. What would then be the linear densities of positive charge, and of negative charge, in the test-charge frame?

5.17 Two protons are moving parallel to one another a distance r apart, with the same velocity βc in the lab frame. According to Eq. 12, at the instantaneous position of one of the protons the electric field E caused by the other, as measured in the lab frame, is $\gamma e/r^2$. But the force on the proton measured in the lab frame is *not* $\gamma e^2/r^2$. Verify that by finding the force in the proton rest frame and transforming

PROBLEM 5.14

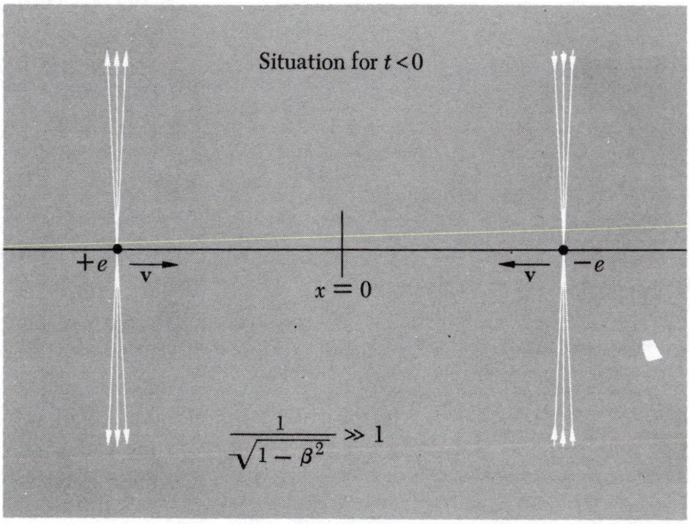

that force back to the lab frame. Show that the discrepancy can be accounted for if there is a magnetic field β times as strong as the electric field, accompanying this proton as it travels through the lab frame.

5.18 Consider a composite line charge consisting of several kinds of carriers, each with its own velocity. For one kind, k, the linear density of charge measured in frame F is λ_k and the velocity is $\beta_k c$ parallel to the line. The contribution of these carriers to the current in F is then $I_k = \lambda_k \beta_k c$. How much do these k-type carriers contribute to the charge and current in a frame F' which is moving parallel to the line at velocity $-\beta c$ with respect to F? By following the steps we took in the transformations in Fig. 5.20, you should be able to show that

$$\lambda'_k = \gamma\left(\lambda_k + \frac{\beta I_k}{c}\right) \qquad I'_k = \gamma(I_k + \beta c \lambda_k)$$

If each component of the linear charge density and current transforms in this way, then so must the total λ and I:

$$\lambda' = \gamma\left(\lambda + \frac{\beta I}{c}\right) \qquad I' = \gamma(I + \beta c \lambda)$$

You have now derived the Lorentz transformation to a parallel-moving frame for *any* line charge and current, whatever its composition.

5.19 A proton moves in along the x axis toward the origin at a velocity $v_x = -c/2$. At the origin it collides with a massive nucleus, rebounds elastically and moves outward on the x axis with nearly the same speed. Make a sketch showing approximately how the electric field of which the proton is the source looks at an instant 10^{-10} sec after the proton reached the origin.

5.20 A stationary proton is located on the z axis at $z = a$. A negative muon is moving with speed $0.8c$ along the x axis. Consider the total electric field of these two particles, in this frame, at the time when the muon passes through the origin. What are the values at that instant of E_x and E_z at the point $(a, 0, 0)$ on the x axis?

Ans. $E_x = -0.00645\ e/a^2$; $E_z = -0.354\ e/a^2$.

5.21 In a high-voltage oscilloscope the source of electrons is a cathode at potential -125 kilovolts with respect to the anode and the enclosed region beyond the anode aperture. Within this region there is a pair of parallel plates 5 cm long in the x-direction (the direction of the electron beam) and 8 mm apart in the y-direction. An electron leaves the cathode with negligible velocity, is accelerated toward the anode, and subsequently passes between these deflecting plates at a time when the potential of the lower plate is -120 volts, that of the upper plate $+120$ volts.

Fill in the blanks. Use rounded-off constants: electron rest mass

$= 5 \times 10^5$ ev, etc. When the electron arrives at the anode, its kinetic energy is ________ ev, its mass has increased by a factor of ______, and its velocity is ____________ c. Its momentum is ____________ gm-cm/sec in the x direction. Beyond the anode the electron passes between parallel metal plates. The field between the plates is ______ statvolts/cm; the force on the electron is ____________ dynes upward. The electron spends ____________ sec between the plates and emerges, having acquired y momentum of magnitude p_y = ________ gm cm/sec. Its trajectory now slants upward at an angle θ = ______ radians.

A fast neutron which just happened to be moving along with the electron when it passed through the anode reported subsequent events as follows: "We were sitting there when this capacitor came flying at us at ____________ cm/sec. It was ____________ cm long, so it surrounded us for ____________ sec. That didn't bother me, but the electric field of ____________ statvolts/cm accelerated the electron so that after the capacitor left us the electron was moving away from me at ____________ cm/sec."

6

THE MAGNETIC FIELD

DEFINITION OF THE MAGNETIC FIELD

6.1 A charge which is moving parallel to a current of other charges experiences a force perpendicular to its own velocity. We can see it happening in the deflection of the electron beam in Fig. 5.3. We discovered in Section 5.9 that this is consistent with—indeed, is required by—Coulomb's law with charge invariance and special relativity. And we found that a force perpendicular to the charged particle's velocity also arises in motion at right angles to the current-carrying wire. For a given current the magnitude of the force, which we calculated for the particular case in Fig. 5.20*a,* is proportional to the product of the particle's charge q and its speed v in our frame. Just as we defined the electric field **E** as the vector force on unit charge at rest, so we can define another field **B** by the *velocity-dependent* part of the force that acts on a charge in motion. The defining relation was introduced at the beginning of Chapter 5. Let us state it again more carefully.

At some instant t a particle of charge q passes the point (x, y, z) in our frame, moving with velocity **v**. At that moment the force on the particle (its rate of change of momentum) is **F**. The electric field at that time and place is known to be **E**. Then the magnetic field at that time and place is defined as the vector **B** which satisfies the vector equation

$$\mathbf{F} = q\mathbf{E} + \frac{q}{c}\mathbf{v} \times \mathbf{B} \tag{1}$$

Of course, **F** here includes only the charge-dependent force and not, for instance, the weight of the particle carrying the charge. A vector **B** satisfying Eq. 1 always exists. Given the values of **E** and **B** in some region, we can with Eq. 1 predict the force on any other particle moving through that region with any other velocity. For fields that vary in time and space Eq. 1 is to be understood as a local relation among the instantaneous values of **F**, **E**, **v**, and **B**. Of course, all four of these quantities must be measured in the same inertial frame.

In the case of our "test charge" in the lab frame of Fig. 5.20*a,* the electric field **E** was zero. With the charge q moving in the positive x direction, $\mathbf{v} = \hat{\mathbf{x}}v$, we found that the force on it was in the negative y direction, with magnitude $2Iqv/rc^2$:

$$\mathbf{F} = -\hat{\mathbf{y}}\,\frac{2Iqv}{rc^2} \tag{2}$$

In this case the magnetic field must be

$$\mathbf{B} = \hat{\mathbf{z}}\,\frac{2I}{rc} \tag{3}$$

for then Eq. 1 becomes

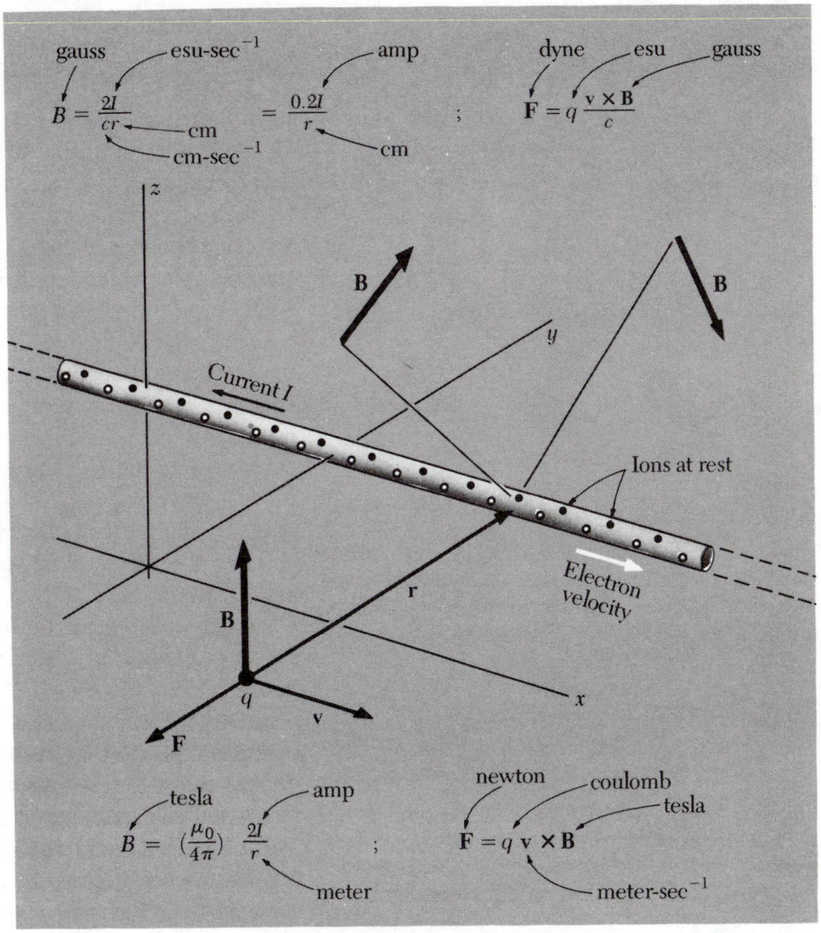

FIGURE 6.1
The magnetic field of a current in a long straight wire and the force on a charged particle moving through that field.

$$\mathbf{F} = \frac{q}{c}\mathbf{v} \times \mathbf{B} = (\hat{\mathbf{x}} \times \hat{\mathbf{z}})\left(\frac{qv}{c}\right)\left(\frac{2I}{rc}\right) = -\hat{\mathbf{y}}\,\frac{2Iqv}{rc^2} \qquad (4)$$

in agreement with Eq. 2.

The relation of $\mathbf{B}$ to $\mathbf{v}$ and to the current I is shown in Fig. 6.1. Three mutually perpendicular directions are involved: the direction of $\mathbf{B}$ at the point of interest, the direction of a vector $\mathbf{r}$ from that point to the wire, and the direction of current flow in the wire. Here questions of *handedness* arise for the first time in our study. Having adopted Eq. 1 as the definition of $\mathbf{B}$ and agreed on the conventional rule for the vector product, that is, $\hat{\mathbf{x}} \times \hat{\mathbf{y}} = \hat{\mathbf{z}}$, etc., in coordinates like those of Fig. 6.1, we have determined the direction of $\mathbf{B}$. That relation has a handedness, as you can see by imagining a particle that

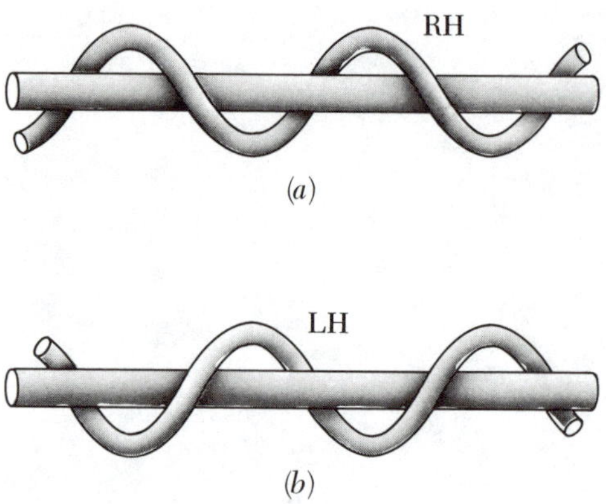

FIGURE 6.2
A reminder. The helix in (*a*) is called a right-handed helix, that in (*b*) a left-handed helix.

moves along the wire in the direction of the current while circling around the wire in the direction of **B**. Its trail, no matter how you look at it, would form a right-hand helix, like that in Fig. 6.2*a,* not a left-hand helix like that in Fig. 6.2*b*.

Consider an experiment like Oersted's, as pictured in Fig. 5.2*a*. The direction of the current was settled when the wire was connected to the battery. Which way the compass needle points can be stated if we color one end of the needle and call it the head of the arrow. By tradition long antedating Oersted the "north-seeking" end of the needle is so designated, and that is the black end of the needle in Fig. 5.2*a*.† If you compare that picture with Fig. 6.1 you will see that we have defined **B** so that it points in the direction of "local magnetic north." Or to put it another way, the current arrow and the compass needle in Fig. 5.2*a* define a right-handed helix (see Fig. 6.2), as do the current direction and the vector **B** in Fig. 6.1. This is not to say that there is anything intrinsically right-handed about electromagnetism. It is only the self-consistency of our rules and definitions that concerns us here. Let us note, however, that a question of handedness *could never arise* in electrostatics. In this sense the vector **B** differs in character from the vector **E**. In the same way, a vector representing an angular velocity, in mechanics, differs from a vector representing a linear velocity.

As for the units in which magnetic field strength is to be expressed, notice that our defining equation, Eq. 1, gives **B** the same dimensions as **E**, the factor $\mathbf{v}/c$ being dimensionless. With force F in dynes and charge q in esu, unit magnetic field strength is 1 dyne/esu. This unit has a name, the *gauss*. There is no special name for the unit dyne/esu when it is used as a unit of electric field strength. It is the same as 1 statvolt/cm, which is the term we shall usually use for unit electric field strength in our CGS system. When we use Eq. 3 to calculate the strength of the magnetic field at distance r caused by a current I in the straight wire, **B** will be in gauss (or dynes/esu) if I is in esu/sec, r in cm, and c in cm/sec.

In SI units the equations look a bit different because the force equation equivalent to our Eq. 1 defining **B** is written like this:

$$\mathbf{F} = q\mathbf{E} + q\mathbf{v} \times \mathbf{B} \qquad (1')$$

F is in newtons, q in coulombs, **E** in volts/meter, and **v** in meters/sec. Notice that c does not appear. In a magnetic field of unit strength a charge of one coulomb moving with a velocity of one meter/sec perpendicular to the field experiences a force of one newton. The unit of **B** so defined is called the *tesla*. One tesla is equivalent to precisely 10^4

†We now know that the earth's magnetic field has reversed many times in geologic history. See page 296 and the reference there given.

gauss. The relation between field and current, the equivalent of our Eq. 3, now takes the following form in SI units:

$$\mathbf{B} = \hat{\mathbf{z}}\,\frac{\mu_0 I}{2\pi r} \tag{3'}$$

where **B** is in teslas, I is in amps, and r is in meters. The constant μ_0, like the constant ϵ_0 we met in electrostatics, is a fundamental constant in the SI unit system. Its value is exactly $4\pi \times 10^{-7}$.

Let us use Eqs. 1 and 3 to calculate the magnetic force between parallel wires carrying current. Let r be the distance between the wires, and let I_1 and I_2 be the currents which we'll assume are flowing in the same direction, as shown in Fig. 6.3. The wires are assumed to be infinitely long—a fair assumption in a practical case if they are very long compared with the distance r between them. We want to predict the force that acts on some finite length l of one of the wires. The current in wire 1 causes a magnetic field of strength

$$B_1 = \frac{2I_1}{cr} \tag{5}$$

at the location of wire 2. Within wire 2 there are n_2 moving charges per centimeter length of wire, each with charge q_2 and speed v_2. They constitute the current I_2:

$$I_2 = n_2 q_2 v_2 \tag{6}$$

According to Eq. 1 the force on each charge is $q_2 v_2 B_1/c$.† The force on each centimeter length of wire is therefore $n_2 q_2 v_2 B_1/c$, or simply $I_2 B_1/c$, in dynes/cm. In view of Eq. 4, the force on a length l of wire 2 is

$$F = \frac{2I_1 I_2 l}{c^2 r} \tag{7}$$

Obviously the force on an equal length of wire 1 caused by the field of wire 2 must be given by the same formula. We have not bothered to keep track of signs because we knew already that currents in the same direction attract one another.

The same exercise carried out in SI units, with Eqs. 1′ and 3′, will lead to

$$F = \frac{\mu_0}{2\pi}\,\frac{I_1 I_2 l}{d} \tag{7'}$$

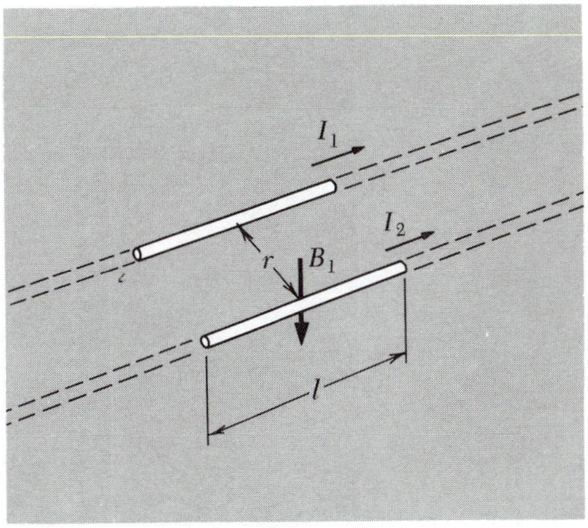

FIGURE 6.3
Current I_1 produced magnetic field B_1 at conductor 2. The force on a length l of conductor 2 is given by Eq. 7 or 7′.

†B_1 is the field *inside* wire 2, caused by the current in wire 1. When we study magnetic fields inside matter in Chapter 11, we'll find that most conductors, including copper and aluminum, but *not* including iron, have very little influence on a magnetic field. For the present, let us agree to avoid things like iron, ferromagnetic materials. Then we can safely assume that the magnetic field inside the wire is practically what it would be in vacuum with the same currents flowing.

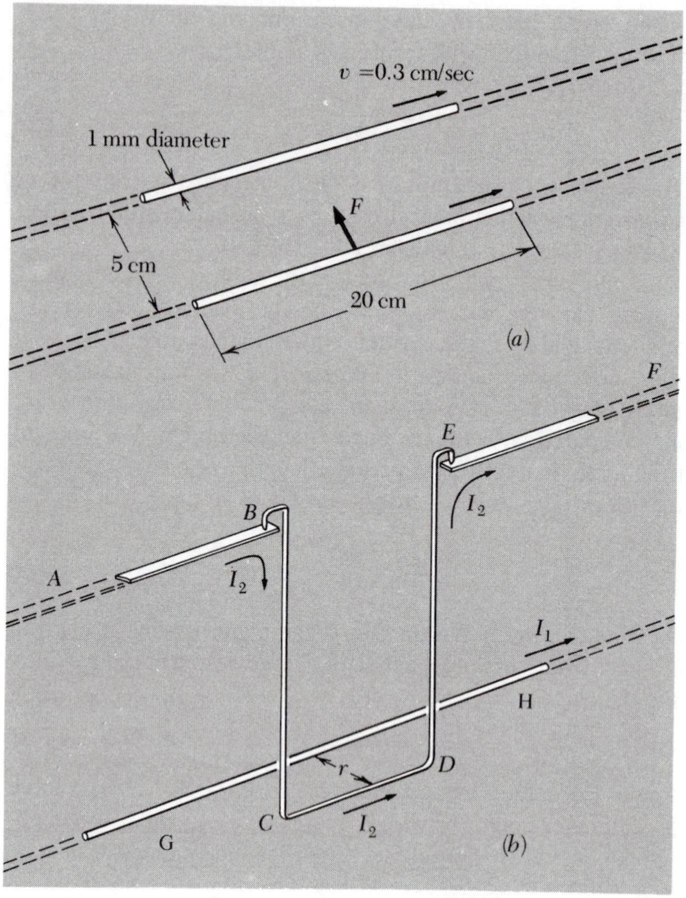

FIGURE 6.4
(*a*) The current in each copper wire is 9.5 × 10^{10} esu/sec, and the force F on the 20-cm length of conductor is 80 dynes. (*b*) One way to measure the force on a length of conductor. The section *BCDE* swings like a pendulum below the conducting pivots. The force on the length *CD* due to the field of the straight conductor *GH* is the only force deflecting the pendulum from the vertical.

Here F is in newtons, while I_1 and I_2 are in amps. As the factor l/d which appears in both Eqs. 7 and 7′ is dimensionless, l and d could be in any units.†

Let's apply Eq. 6 to the pair of wires in Fig. 6.4*a*. They are copper wires 1 mm in diameter and 5 cm apart. In copper the number of conduction electrons per cubic centimeter, already mentioned in Chapter 4, is 8.45×10^{22}, so there are $(\pi/4)(0.01)(8.45 \times 10^{22})$ or 6.6×10^{20} conduction electrons in a 1-cm length of this wire. Suppose

†Equation 7′ has usually been regarded as the primary definition of the ampere in the SI system, μ_0 being *assigned* the value $4\pi \times 10^{-7}$. That is to say, one ampere is that current which, flowing in each of two infinitely long parallel wires a distance d apart, will cause a force of exactly 2×10^{-7} newton on a length d of one of the wires. The other SI electrical units are then defined in terms of the ampere. Thus a coulomb is one ampere-second, a volt is one joule/coulomb, and an ohm is one volt/ampere. See Appendix E.

their mean drift velocity $\bar{v}$ is 0.3 cm/sec. (Of course their random speeds are vastly greater.) The current in each wire is then

$$\begin{aligned} I = nq\bar{v} &= (6.6 \times 10^{20}\ \text{cm}^{-1})(4.80 \times 10^{-10}\ \text{esu})(0.3\ \text{cm/sec}) \\ &= 9.5 \times 10^{10}\ \text{esu/sec} \end{aligned} \tag{8}$$

The attractive force on a 20-cm length of wire is

$$F = \frac{2I^2 l}{c^2 d} = \frac{2(9.5 \times 10^{10})^2 \times 20}{(3 \times 10^{10})^2 \times 5} = 80\ \text{dynes} \tag{9}$$

Now 80 dynes is not an enormous force, but it is easily measurable. Figure 6.4*b* shows how the force on a given length of conductor could be observed. The c^2 in the denominator of Eq. 9 reminds us that, as we discovered in the last chapter, this is a relativistic effect, strictly proportional to v^2/c^2 and traceable to a Lorentz contraction. And here with v less than the speed of a healthy ant, it is causing a quite respectable force! The explanation is the immense amount of negative charge the conduction electrons represent, charge which ordinarily is so precisely neutralized by positive charge that we hardly notice it. To appreciate that, consider the force with which our wires in Fig. 6.3 would repel one another if the charge of the 6.6×10^{20} electrons per cm were *not* neutralized. You will find that the force is just c^2/v^2 times the force we calculated above, or roughly 40 trillion tons per centimeter of wire. So full of electricity is all matter! If the electrons in one raindrop were removed from the earth, the whole earth's potential would rise by several million volts.

Matter in bulk, from raindrops to planets, is almost exactly neutral. You will find that any piece of it much larger than a molecule contains nearly the same number of electrons as protons. If it didn't, the resulting electric field would be so strong that the excess charge would be irresistibly blown away. That would happen to electrons in our copper wire even if the excess of negative charge were no more than 10^{-10} of the total. A magnetic field, on the other hand, cannot destroy itself in this way. No matter how strong it may be, it exerts no force on a stationary charge. That is why forces that arise from the *motion* of electric charges can dominate the scene. The second term on the right in Eq. 1 can be much larger than the first. Thanks to that second term, an electric motor can start your car. In the atomic domain, however, where the coulomb force between pairs of charged particles comes into play, magnetic forces do take second place relative to electrical forces. They are weaker, generally speaking, by just the factor we should expect, the square of the ratio of the particle speed to the speed of light.

Inside atoms we find magnetic fields as large as 10^5 gauss. The strongest large-scale fields easily produced in the laboratory are on that order of magnitude too, although fields up to several million gauss have been created for short times. In ordinary electrical machinery, electric motors for instance, 10^4 gauss (or 1 tesla†) would be more typical. The strength of the earth's magnetic field is a few tenths of a gauss at the earth's surface, and presumably many times stronger down in the earth's metallic core where the currents that cause the field are flowing. We see a spectacular display of magnetic fields on and around the sun. A *sunspot* is an eruption of magnetic field with local intensity of a few thousand gauss. Some other stars have stronger magnetic fields. Strongest of all is the magnetic field at the surface of a neutron star, or pulsar, where the intensity is believed to reach the hardly conceivable range of 10^{12} gauss. On a vaster scale, our galaxy is pervaded by magnetic fields which extend over thousands of light years of interstellar space. The field strength can be deduced from observations in radioastronomy. It is a few microgauss—enough to make the magnetic field a significant factor in the dynamics of the interstellar medium.

SOME PROPERTIES OF THE MAGNETIC FIELD

6.2 The magnetic field, like the electric field, is a device for describing how charged particles interact with one another. If we say that the magnetic field at the point (4.5, 3.2, 6.0) at 12:00 noon points horizontally in the negative y direction and has a magnitude of 5 gauss, we are making a statement about the acceleration a moving charged particle at that point in space-time would exhibit. The remarkable thing is that a statement of this form, giving simply a vector quantity **B**, says all there is to say. With it one can predict uniquely the velocity-dependent part of the force on *any* charged particle moving with *any* velocity. It makes unnecessary any further description of the other charged particles which are the sources of the field. In other words, if two quite different systems of moving charges happen to produce the same **E** and **B** at a particular point, the behavior of any test particle at the point would be exactly the same in the two systems. It is for this reason that the concept of field, as an intermediary in the interaction of particles, is useful. And it is for this reason that we think of the field as an independent entity.

Is the field more, or less, real than the particles whose interaction, as seen from our present point of view, it was invented to describe? That is a deep question which we would do well to set aside

†Nikola Tesla (1856–1943) inventor and electrical engineer for whom the SI unit was named, invented the alternating-current induction motor and other useful electromagnetic devices. Gauss's work in magnetism was concerned mainly with the earth's magnetic field. Perhaps this will help you to remember which is the larger unit.

for the time being. People to whom the electric and magnetic fields were vividly real—Faraday and Maxwell, to name two—were led thereby to new insights and great discoveries. Let's view the magnetic field as concretely as they did and learn some of its properties.

So far we have studied only the magnetic field of a straight wire or filament of steady current. The field direction, we found, is everywhere perpendicular to a plane containing the filament and the point where the field is observed. The magnitude of the field is proportional to $1/r$. The field lines are circles surrounding the filament, as shown in Fig. 6.5. The sense of direction of **B** is determined by our previously adopted convention about the vector cross-product, by the (arbitrary) decision to write the second term in Eq. 1 as $(q/c)\mathbf{v} \times \mathbf{B}$, and by the *physical fact* that a positive charge moving in the direction of a positive current is attracted to it rather than repelled. These are all consistent if we relate the direction of **B** to the direction of the current that is its source in the manner shown in Fig. 6.5. Looking in the direction of positive current, we see the **B** lines curling clockwise. Or you may prefer to remember it as a right-hand-thread relation.

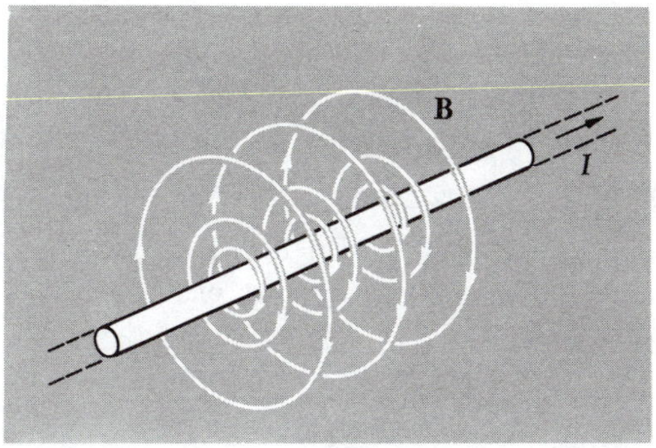

FIGURE 6.5
Magnetic field lines around a straight wire carrying current.

Let's look at the line integral of **B** around a closed path in this field. (Remember that a similar inquiry in the case of the electric field of a point charge led us to a simple and fundamental property of all electrostatic fields.) Consider first the path *ABCD* in Fig. 6.6*a*. This lies in a plane perpendicular to the wire; in fact, we need only work in this plane, for **B** has no component parallel to the wire. The line integral of **B** around the path shown is zero, for the following reason. Paths *BC* and *DA* are perpendicular to **B** and contribute nothing. Along *AB*, **B** is stronger in the ratio r_2/r_1 than it is along *CD*; but *CD* is longer than *AB* by the same factor, for these two arcs subtend the same angle at the wire. So the two arcs give equal and opposite contributions, and the whole integral is zero.

It follows that the line integral is also zero on any path that can be constructed out of radial segments and arcs, such as the path in Fig. 6.6*b*. From this it is a short step to conclude that the line integral is zero around *any* path that does not enclose the wire. To smooth out the corners we would only need to show that the integral around a little triangular path vanishes. The same step was involved in the case of the electric field.

A path which does not enclose the wire is one like the path in Fig. 6.6*c*, which, if it were made of string, could be pulled free. The line integral around any such path is zero.

Now consider a circular path that encloses the wire, as in Fig. 6.6*d*. Here the circumference is $2\pi r$ and the field is $2I/cr$ and everywhere parallel to the path, so the value of the line integral around this particular path is $(2\pi r)(2I/rc)$, or $4\pi I/c$. We now claim that *any* path looping once around the wire must give the same value. Consider, for instance, the crooked path C in Fig. 6.6*e*. Let us construct the path C'

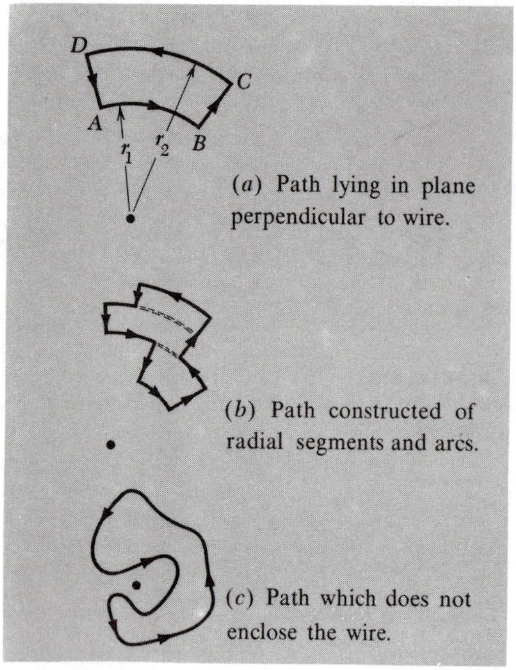

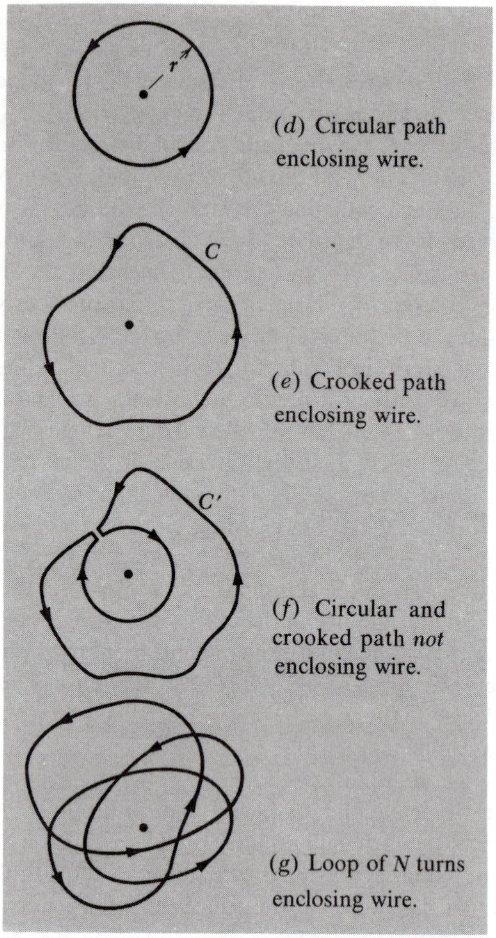

FIGURE 6.6
The line integral of the magnetic field **B** over any closed path depends only on the current enclosed.

in Fig. 6.6*f* made of a path like *C* and a circular path, but *not* enclosing the wire. The line integral around *C*′ must be zero, and therefore the integral around *C* must be the negative of the integral around the circle, which we have already evaluated as $4\pi I/c$ in magnitude. The sign will depend in an obvious way on the sense of traversal of the path. Our general conclusion is:

$$\int \mathbf{B} \cdot d\mathbf{s} = \frac{4\pi}{c} \times \text{current enclosed by path} \tag{10}$$

Equation 10 holds when the path loops the current filament once. Obviously a path which loops it *N* times, like the one in Fig. 6.6*g*, will give just *N* times as big a result for the line integral.

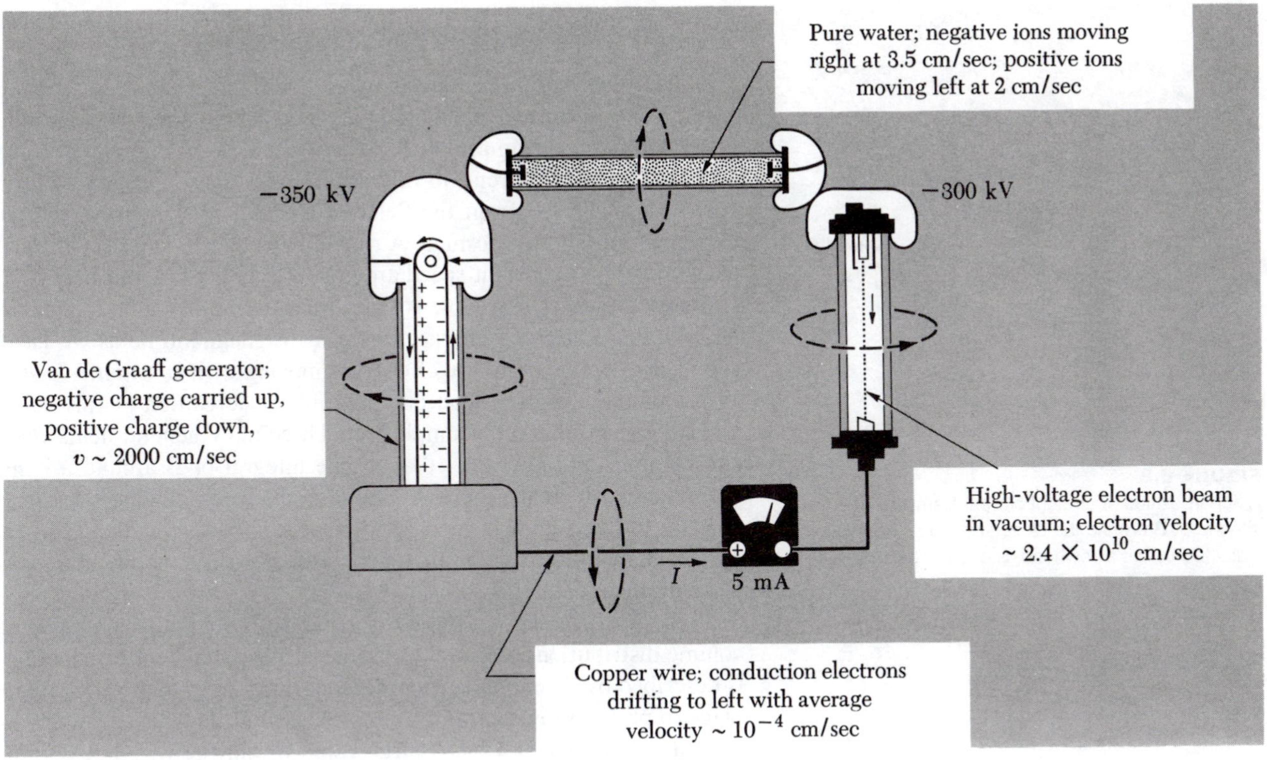
Pure water; negative ions moving right at 3.5 cm/sec; positive ions moving left at 2 cm/sec
−350 kV
−300 kV
Van de Graaff generator; negative charge carried up, positive charge down, $v \sim 2000$ cm/sec
High-voltage electron beam in vacuum; electron velocity $\sim 2.4 \times 10^{10}$ cm/sec
I
5 mA
Copper wire; conduction electrons drifting to left with average velocity $\sim 10^{-4}$ cm/sec

FIGURE 6.7

The line integral of **B** has precisely the same value around every part of this circuit, although the velocity of the charge carriers is quite different in different parts.

The magnetic field, as we have emphasized before, depends only on the rate of charge transport, the number of units of charge passing a given point in the circuit, per second. Figure 6.7 shows a circuit with a current of 5 milliamperes, equivalent to 15×10^6 esu/sec. The average velocity of the charge carriers ranges from 10^{-4} cm/sec in one part of the circuit to 0.8 times the speed of light in another. The line integral of **B** over a closed path has the same value around every part of this circuit, namely:

$$\int \mathbf{B} \cdot d\mathbf{s} = \frac{4\pi I}{c} = \frac{4\pi \times (15 \times 10^6 \text{ esu/sec})}{3 \times 10^{10} \text{ cm/sec}} \qquad (11)$$
$$= 0.00628 \text{ gauss-cm}$$

What we have proved for the case of a long straight filament of current clearly holds, by superposition, for the field of any system of straight filaments. In Fig. 6.8 several wires are carrying currents in different directions. If Eq. 10 holds for the magnetic field of one of these wires, it must hold for the total field which is the vector sum, at every point, of the fields of the individual wires. That is a pretty complicated field. Nevertheless, we can predict the value of the line inte-

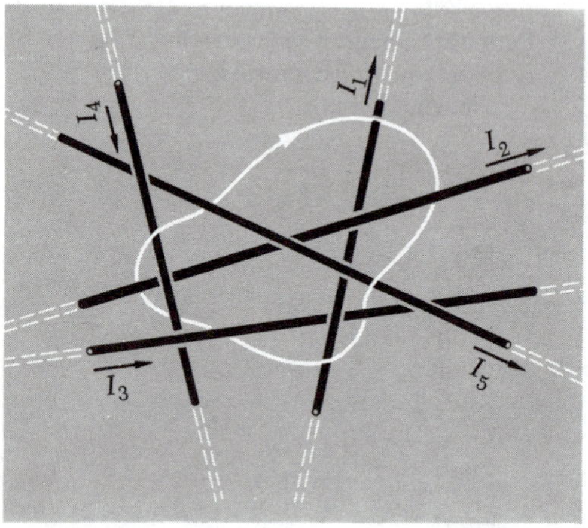

FIGURE 6.8
A superposition of straight current filaments. The line integral of **B** around the closed path, in the direction indicated by the arrowhead, is equal to $(4\pi/c)(-I_4 + I_5)$.

gral of **B** around the closed path in Fig. 6.8 merely by noting which currents the path encircles, and in which sense.

However, we are interested in other things than long straight wires. We want to understand the magnetic field of any sort of current distribution—for example, that of a current flowing in a closed loop, a circular ring of current, to take the simplest case. Perhaps we can derive this field too from the fields of the individual moving charge carriers, properly transformed. A ring of current could be a set of electrons moving at constant speed around a circular path. But here that strategy fails us. The trouble is that an electron moving on a circular path is an *accelerated* charge, whereas the magnetic fields we have rigorously derived are those of charges moving with *constant velocity.* We shall therefore abandon our program of derivation at this point and state the remarkably simple fact: These more general fields *obey exactly the same law,* Eq. 10. The line integral of **B** around a bent wire is equal to that around a long straight wire carrying the same current. As this goes beyond anything we have so far deduced, we must look on it here as a postulate confirmed by the experimental tests of its implications.

To state the law in the most general way, we must talk about volume distributions of current. A general steady current distribution is described by a volume current density $\mathbf{J}(x, y, z)$ which varies from place to place but is constant in time. A current in a wire is merely a special case in which **J** has a large value within the wire but is zero elsewhere. We discussed volume distribution of current in Chapter 4, where we noted that, for time-independent currents, **J** has to satisfy the continuity equation, or conservation-of-charge condition,

$$\operatorname{div} \mathbf{J} = 0 \tag{12}$$

FIGURE 6.9
J is the local current density. The surface integral of **J** over S is the current enclosed by the curve C.

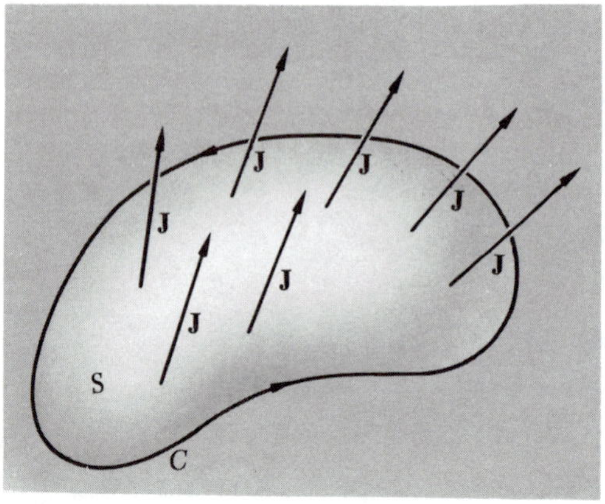

Take any closed curve C in a region where currents are flowing. The total current enclosed by C is the flux of **J** through the surface spanning C, that is, the surface integral $\int_S \mathbf{J} \cdot d\mathbf{a}$ over this surface S (see Fig. 6.9). A general statement of the relation in Eq. 10 is therefore

$$\int_C \mathbf{B} \cdot d\mathbf{s} = \frac{4\pi}{c} \int_S \mathbf{J} \cdot d\mathbf{a} \tag{13}$$

Let us compare this with Stokes' theorem, which we developed in Chapter 2:

$$\int_C \mathbf{F} \cdot d\mathbf{s} = \int_S (\operatorname{curl} \mathbf{F}) \cdot d\mathbf{a} \tag{14}$$

We see that a statement equivalent to Eq. 13 is this:

$$\boxed{\text{curl } \mathbf{B} = \frac{4\pi \mathbf{J}}{c}} \tag{15}$$

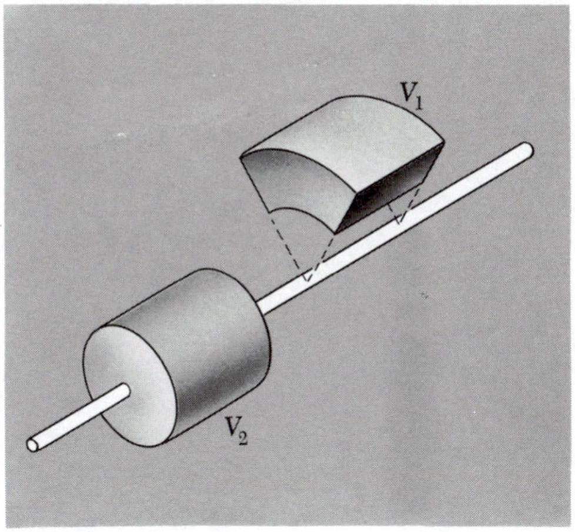

FIGURE 6.10
There is zero net flux of **B** from either box.

This is the simplest and most general statement of the relation between the magnetic field and the moving charges which are its source.

However, Eq. 15 is not enough to determine $\mathbf{B}(x, y, z)$, given $\mathbf{J}(x, y, z)$, for many different vector fields could have the same curl. We need to complete it with another condition. We had better think about the divergence of **B**. Going back to the magnetic field of a single straight wire, we observe that the divergence of that field is zero. You can't draw a little box anywhere, even one enclosing the wire, which will have a net outward or inward flux. It is enough to note that the boxes V_1 and V_2 in Fig. 6.10 have no net flux and can shrink to zero without developing any. For this field then, div $\mathbf{B} = 0$, and hence also for all superpositions of such fields. Again we postulate that the principle can be extended to the field of any distribution of currents, so that a companion to Eq. 12 is the condition

$$\boxed{\text{div } \mathbf{B} = 0} \tag{16}$$

In SI units the relation of magnetic field **B** in teslas to current density **J** in amps/m^2 is

$$\text{curl } \mathbf{B} = \mu_0 \mathbf{J} \tag{15$'$}$$

We are concerned with fields whose sources lie within some finite region. We won't consider sources that are infinitely remote and infinitely strong. With that proviso, Eqs. 15 and 16 together determine **B** uniquely if **J** is given. For suppose both equations are satisfied by two different fields $\mathbf{B}_1$ and $\mathbf{B}_2$. Their difference, the vector field $\mathbf{D} = \mathbf{B}_1 - \mathbf{B}_2$, is a field with zero divergence and zero curl everywhere. What could it be like? Having zero curl, it must be the gradient of some potential function $f(x, y, z)$: $\mathbf{D} = \nabla f$. But $\nabla \cdot \mathbf{D} = 0$, too, so $\nabla \cdot \nabla f$ or $\nabla^2 f = 0$ everywhere. Over a sufficiently remote enclosing boundary f must take on some constant value f_0. Since f satisfies Laplace's equation everywhere inside that boundary, it cannot have a maximum or a minimum anywhere in that region (Section 2.11) and so it must have the value f_0 everywhere. Hence $\mathbf{D} = \nabla f = 0$, and $\mathbf{B}_1 = \mathbf{B}_2$.

In the case of the electrostatic field the counterpart of Eqs. 15 and 16 was

$$\text{div}\,\mathbf{E} = 4\pi\rho \qquad \text{and} \qquad \text{curl}\,\mathbf{E} = 0 \tag{17}$$

In the case of the electric field, however, we could begin with Coulomb's law, which expressed directly the contribution of each charge to the electric field at any point. Here we shall have to work our way back to some relation of that type.† We shall do so by means of a *potential function.*

VECTOR POTENTIAL

6.3 We found that the scalar potential function $\varphi(x, y, z)$ gave us a simple way to calculate the electrostatic field of a charge distribution. If there is some charge distribution $\rho(x, y, z)$, the potential at any point (x_1, y_1, z_1) is given by the volume integral

$$\varphi(x_1, y_1, z_1) = \int \frac{\rho(x_2, y_2, z_2)\, dv_2}{r_{12}} \tag{18}$$

The integration is extended over the whole charge distribution, and r_{12} is the magnitude of the distance from (x_2, y_2, z_2) to (x_1, y_1, z_1). The electric field $\mathbf{E}$ is obtained as the negative of the gradient of φ:

$$\mathbf{E} = -\text{grad}\,\varphi \tag{19}$$

The same trick won't work here, because of the essentially different character of $\mathbf{B}$. The curl of $\mathbf{B}$ is *not* necessarily zero, so $\mathbf{B}$ can't, in general, be the gradient of a scalar potential. However, we know another kind of vector derivative, the curl. It turns out that we can usefully represent $\mathbf{B}$, not as the gradient of a scalar function but as the curl of a *vector* function, like this:

$$\boxed{\mathbf{B} = \text{curl}\,\mathbf{A}} \tag{20}$$

By obvious analogy, we call $\mathbf{A}$ the *vector potential.* It is *not* obvious, at this point, why this tactic is helpful. That will have to emerge as we proceed. It is encouraging that Eq. 16 is automatically

†The student may wonder why we couldn't have started from some equivalent of Coulomb's law for the interaction of currents. The answer is that a piece of a current filament, unlike an electric charge, is not an independent object that can be physically isolated. You cannot perform an experiment to determine the field from *part* of a circuit; if the rest of the circuit isn't there, the current can't be steady without violating the continuity condition.

satisfied, since div curl **A** = 0, for any **A**.† Or to put it another way, the fact that div **B** = 0 presents us with the opportunity to represent **B** as the curl of another vector function. Our job now is to discover how to calculate **A**, when the current distribution **J** is given, so that Eq. 20 will indeed yield the correct magnetic field. In view of Eq. 15, the relation between **J** and **A** is

$$\text{curl}\,(\text{curl}\,\mathbf{A}) = \frac{4\pi\mathbf{J}}{c} \tag{21}$$

Equation 21, being a vector equation, is really three equations. We shall work out one of them, say the x-component equation. The x component of curl **B** is $\partial B_z/\partial y - \partial B_y/\partial z$. The z and y components of **B** are, respectively,

$$B_z = \frac{\partial A_y}{\partial x} - \frac{\partial A_x}{\partial y} \qquad B_y = \frac{\partial A_x}{\partial z} - \frac{\partial A_z}{\partial x} \tag{22}$$

Thus the x component part of Eq. 21 reads

$$\frac{\partial}{\partial y}\left(\frac{\partial A_y}{\partial x} - \frac{\partial A_x}{\partial y}\right) - \frac{\partial}{\partial z}\left(\frac{\partial A_x}{\partial z} - \frac{\partial A_z}{\partial x}\right) = \frac{4\pi J_x}{c} \tag{23}$$

We assume our functions are such that the order of partial differentiation can be interchanged. Taking advantage of that and rearranging a little, we can write Eq. 23 in this way:

$$-\frac{\partial^2 A_x}{\partial y^2} - \frac{\partial^2 A_x}{\partial z^2} + \frac{\partial}{\partial x}\left(\frac{\partial A_y}{\partial y}\right) + \frac{\partial}{\partial x}\left(\frac{\partial A_z}{\partial z}\right) = \frac{4\pi J_x}{c} \tag{24}$$

To make the thing more symmetrical, let's add and subtract the same term, $\partial^2 A_x/\partial x^2$, on the left:

$$-\frac{\partial^2 A_x}{\partial x^2} - \frac{\partial^2 A_x}{\partial y^2} - \frac{\partial^2 A_x}{\partial z^2} + \frac{\partial}{\partial x}\left(\frac{\partial A_x}{\partial x} + \frac{\partial A_y}{\partial y} + \frac{\partial A_z}{\partial z}\right) = \frac{4\pi J_x}{c} \tag{25}$$

We can now recognize the first three terms as the negative of the Laplacian of A_x. The quantity in parentheses is the divergence of **A**. Now we have a certain latitude in the construction of **A**. All we care about is its curl; its divergence can be anything we like. Let us *require* that‡

$$\text{div}\,\mathbf{A} = 0 \tag{26}$$

In other words, among the various functions which might satisfy our

†If you are not familiar with this fact, refer back to Problem 2.16.

‡To see why we are free to do this, suppose we had an **A** such that curl **A** = **B**, but div **A** = $f(x, y, z) \neq 0$. Treating f like the charge density ρ in an electrostatic field, we obviously can find a field **F**, the analog of the electrostatic **E**, such that div **F** = f. But we know that the curl of such a field will be zero. Hence we could add −**F** to **A**, making a new field with the correct curl and zero divergence.

requirement that curl $\mathbf{A} = \mathbf{B}$, let us consider as candidates only those which also have zero divergence. Then that part of Eq. 25 drops away and we are left simply with

$$\frac{\partial^2 A_x}{\partial x^2} + \frac{\partial^2 A_x}{\partial y^2} + \frac{\partial^2 A_x}{\partial z^2} = -\frac{4\pi J_x}{c} \tag{27}$$

J_x is a known scalar function of x, y, z. Let us compare Eq. 27 with Poisson's equation, Eq. 2.54, which reads

$$\frac{\partial^2 \varphi}{\partial x^2} + \frac{\partial^2 \varphi}{\partial y^2} + \frac{\partial^2 \varphi}{\partial z^2} = -4\pi\rho \tag{28}$$

The two equations are identical in form. We already *know* how to find a solution to Eq. 28. The volume integral in Eq. 18 is the prescription. Therefore a solution to Eq. 27 must be given by

$$A_x(x_1, y_1, z_1) = \frac{1}{c}\int \frac{J_x(x_2, y_2, z_2)\, dv_2}{r_{12}} \tag{29}$$

The other components must satisfy similar formulas. They can all be combined neatly in one vector formula:

$$\boxed{\mathbf{A}(x_1, y_1, z_1) = \frac{1}{c}\int \frac{\mathbf{J}(x_2, y_2, z_2)\, dv_2}{r_{12}}} \tag{30}$$

There is only one snag. We stipulated that div $\mathbf{A} = 0$, in order to get Eq. 27. How do we know the $\mathbf{A}$ given by Eq. 30 will have this special property? Fortunately, it can be shown that it does.

As an example of a vector potential, consider a long straight wire carrying a current I. In Fig. 6.11 we see the current coming toward us out of the page, flowing along the positive z axis. We know what the magnetic field of the straight wire looks like. The field lines are circles, as sketched already in Fig. 6.5. A few are shown in Fig. 6.11. The magnitude of $\mathbf{B}$ is $2I/cr$. Using a unit vector $\hat{\varphi}$ in the "circumferential" direction we can write the vector $\mathbf{B}$ as

$$\mathbf{B} = \frac{2I\hat{\varphi}}{cr} \tag{31}$$

Noting that the unit vector $\hat{\varphi}$ is $-\sin\varphi\hat{\mathbf{x}} + \cos\varphi\hat{\mathbf{y}}$, we can write this in terms of x and y as follows:

$$\mathbf{B} = \frac{2I(-\sin\varphi\hat{\mathbf{x}} + \cos\varphi\hat{\mathbf{y}})}{c\sqrt{x^2+y^2}} = \frac{2I}{c}\left(\frac{-y\hat{\mathbf{x}} + x\hat{\mathbf{y}}}{x^2+y^2}\right) \tag{32}$$

FIGURE 6.11
Some field lines around a current filament. Current flows toward you (out of the plane of the paper).

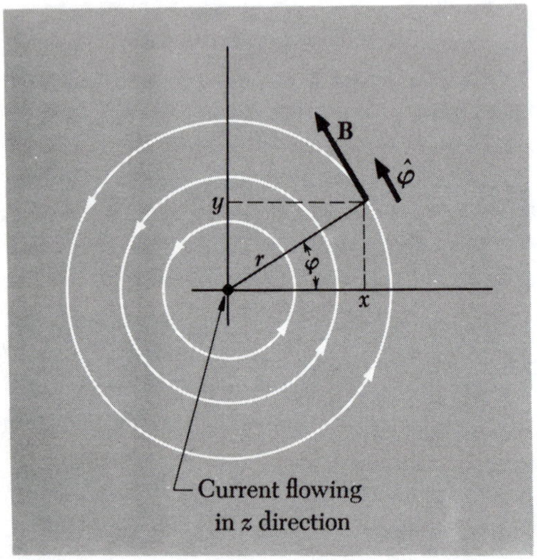

One vector function $\mathbf{A}(x, y, z)$ that will satisfy $\nabla \times \mathbf{A} = \mathbf{B}$ is the following:

$$\mathbf{A} = -\hat{\mathbf{z}}\frac{I}{c}\ln(x^2 + y^2) \tag{33}$$

To verify this, we calculate the components of $\nabla \times \mathbf{A}$:

$$(\nabla \times \mathbf{A})_x = \frac{\partial A_z}{\partial y} - \frac{\partial A_y}{\partial z} = \frac{-2Iy}{c(x^2 + y^2)} \qquad (= B_x)$$

$$(\nabla \times \mathbf{A})_y = \frac{\partial A_x}{\partial z} - \frac{\partial A_z}{\partial x} = \frac{2Ix}{c(x^2 + y^2)} \qquad (= B_y) \tag{34}$$

$$(\nabla \times \mathbf{A})_z = \frac{\partial A_y}{\partial x} - \frac{\partial A_x}{\partial y} = 0 \qquad (= B_z)$$

Of course, this is not the only function that could serve as the vector potential for this particular **B**. To the **A** of Eq. 33 could be added any vector function with zero curl. This all holds for the space outside the wire. Inside the wire, **B** is different, so **A** must be different also. It is not hard to find the appropriate vector potential function for the interior of a solid round wire—see Problem 6.26.

Incidentally, the **A** for our particular example above could not have been obtained by Eq. 30. The integral would diverge owing to the infinite extent of the wire. This may remind you of the difficulty we encountered in Chapter 2 in setting up a scalar potential for the electric field of a charged wire. Indeed the two problems are very closely related, as we should expect from their identical geometry and the similarity of Eqs. 30 and 18. We found (Eq. 17 of Chapter 2) that a suitable scalar potential for the line charge problem is $-\lambda \ln(x^2 + y^2) +$ *arbitrary constant*. This assigns zero potential to some arbitrary point which is neither on the wire nor an infinite distance away. Both that scalar potential and the vector potential of Eq. 33 are singular at the origin and at infinity.

FIELD OF ANY CURRENT-CARRYING WIRE

6.4 Figure 6.12 shows a loop of wire carrying current I. The vector potential **A** at the point (x_1, y_1, z_1) is given according to Eq. 30 by the integral over the loop. For current confined to a thin wire we may take as the volume element dv_2 a short section of the wire of length dl. The current density J is I/a, where a is the cross-section area, and $dv_2 = a\,dl$. Hence $\mathbf{J}\,dv_2 = I\,d\mathbf{l}$, and if we make the vector $d\mathbf{l}$ point in the direction of positive current, we can simply replace $\mathbf{J}\,dv_2$ by $I\,d\mathbf{l}$. Thus

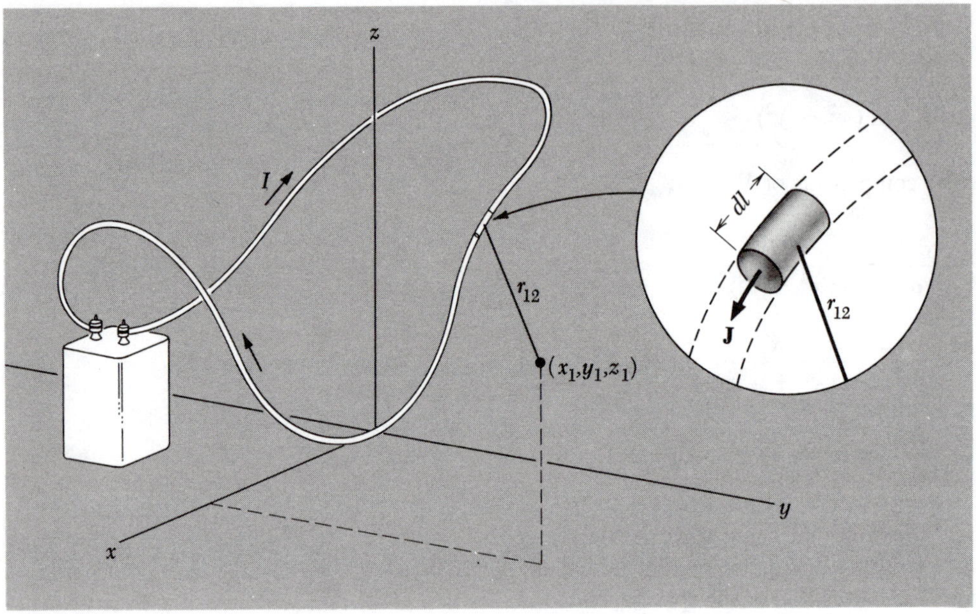

FIGURE 6.12
Each element of the current loop contributes to the vector potential **A** at the point (x_1, y_1, z_1).

for a thin wire or filament, we can write Eq. 30 as a line integral over the circuit:

$$\mathbf{A} = \frac{I}{c} \int \frac{d\mathbf{l}}{r_{12}} \tag{35}$$

To calculate **A** everywhere and then find **B** by taking the curl of **A** might be a long job. It will be more useful to isolate one contribution to the line integral for **A**, the contribution from the segment of wire at the origin, where the current happens to be flowing in the x direction (Fig. 6.13). We shall denote the length of this segment by dl. Let $d\mathbf{A}$ be the contribution of this part of the integral to **A**. Then at the point $(x, y, 0)$ in the xy plane, $d\mathbf{A}$, which points in the positive x direction, is

$$d\mathbf{A} = \hat{\mathbf{x}} \frac{(I/c)\, dl}{\sqrt{x^2 + y^2}} \tag{36}$$

It is clear from symmetry that the contribution of this part of **A** to curl **A** must be perpendicular to the xy plane. Denoting the corresponding part of **B** by $d\mathbf{B}$ we have

$$\begin{aligned} d\mathbf{B} &= \text{curl}\,(d\mathbf{A}) = \hat{\mathbf{z}}\left(-\frac{\partial A_x}{\partial y}\right) \\ &= \hat{\mathbf{z}} \frac{(I/c)\, dl y}{(x^2 + y^2)^{3/2}} = \hat{\mathbf{z}} \frac{(I/c)\, dl \sin\varphi}{r^2} \end{aligned} \tag{37}$$

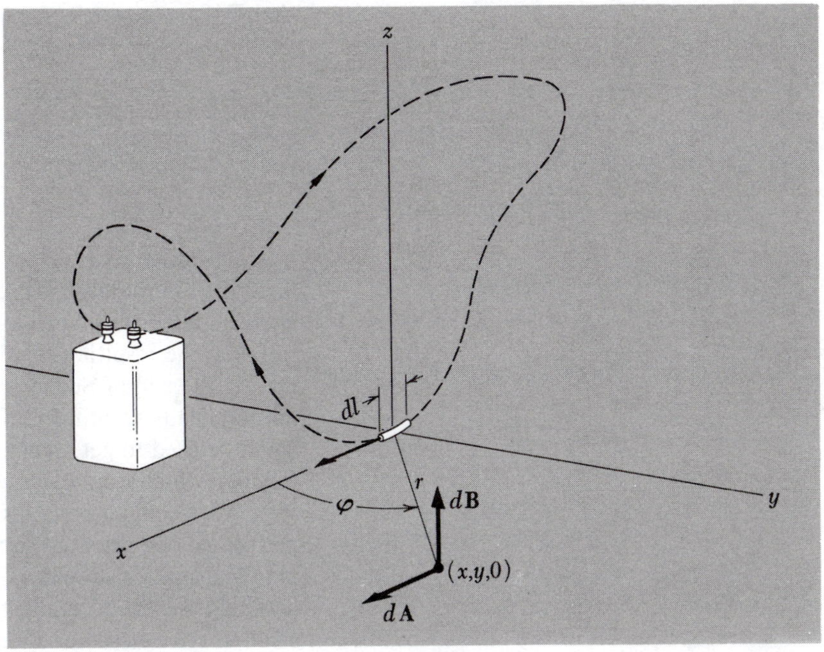

FIGURE 6.13
If we find $d\mathbf{A}$, the contribution to **A** of the particular element shown, its contribution to **B** can be calculated using **B** = curl **A**.

With this result we can free ourselves at once from a particular coordinate system. Obviously all that matters is the relative orientation of the element dl and the radius vector $\mathbf{r}$ from that element to the point where the field $\mathbf{B}$ is to be found. The contribution to $\mathbf{B}$ from any short segment of wire dl can be taken to be a vector perpendicular to the plane containing dl and $\mathbf{r}$, of magnitude $I\,dl \sin\varphi/r^2c$, where φ is the angle between dl and $\mathbf{r}$. This can be written compactly using the cross-product and is illustrated in Fig. 6.14.

$$\boxed{d\mathbf{B} = \frac{I\,d\mathbf{l} \times \hat{\mathbf{r}}}{cr^2}} \tag{38}$$

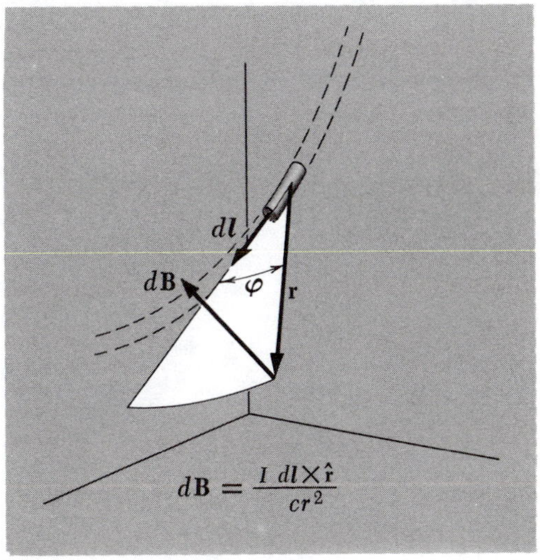

FIGURE 6.14
The field of any circuit can be calculated by using this relation for the contribution of each circuit element.

If you are familiar with the rules of the vector calculus, you can take a short cut from Eq. 35 to Eq. 38. Writing $d\mathbf{B} = \nabla \times d\mathbf{A}$, with $d\mathbf{A} = I\,d\mathbf{l}/cr$, we treat ∇ as a vector, reversing the order of the cross-product and changing the sign. Here $d\mathbf{l}$ is a constant, so that ∇ operates only on $1/r$, otherwise we couldn't get away with this! We recall that $\nabla(1/r) = -\hat{\mathbf{r}}/r^2$ (as in going from the Coulomb potential to the Coulomb field). Thus:

$$d\mathbf{B} = \nabla \times \frac{I\,d\mathbf{l}}{cr} = -\frac{I}{c}\,d\mathbf{l} \times \nabla\left(\frac{1}{r}\right)$$
$$= -\frac{I}{c}\,d\mathbf{l} \times \left(-\frac{\hat{\mathbf{r}}}{r^2}\right) = \frac{I\,d\mathbf{l} \times \hat{\mathbf{r}}}{cr^2} \tag{39}$$

Historically, Eq. 38 is known as the Biot-Savart law. The meaning of Eq. 38 is that, if **B** is computed by integrating over the *complete circuit,* taking the contribution from each element to be given by this formula, the resulting **B** will be correct. As we remarked in the footnote at the end of Section 6.2, the contribution of part of a circuit is not physically identifiable. In fact, Eq. 38 is not the only formula that could be used to get a correct result for **B**—to it could be added any function which would give zero when integrated around a closed path.

We seem to have discarded the vector potential as soon as it performed one essential service for us. Indeed, it is often easier, as a practical matter, to calculate the field of a current system directly, now that we have Eq. 38, than to find the vector potential first. We shall practice on some examples in the next section. However, the vector potential is important for deeper reasons. For one thing, it has revealed to us a striking parallel between the relation of the electrostatic field **E** to its sources, electric charges, and the relation of the magnetic field **B** to steady currents. Its greatest usefulness lies ahead, in the study of time-varying fields, and electromagnetic radiation.

FIELDS OF RINGS AND COILS

6.5 A current filament in the form of a circular ring of radius b is shown in Fig. 6.15*a*. We could predict without any calculation that the magnetic field of this source must look something like Fig. 6.15*b*, where we have sketched some field lines in a plane through the axis of symmetry. The field as a whole must be rotationally symmetrical about this axis, the z axis in Fig. 6.15*a*, and the field lines themselves must be symmetrical with respect to the plane of the loop, the xy plane. Very close to the filament the field will resemble that near a long straight wire, since the distant parts of the ring are there relatively unimportant.

It is easy to calculate the field on the axis, using Eq. 38. Each element of the ring of length dl contributes a $d\mathbf{B}$ perpendicular to $\mathbf{r}$. We need only include the z component of $d\mathbf{B}$, for we know the total field on the axis must point in the z direction,

$$dB_z = \frac{I\,dl}{cr^2}\cos\theta = \frac{I\,dl}{cr^2}\frac{b}{r} \tag{40}$$

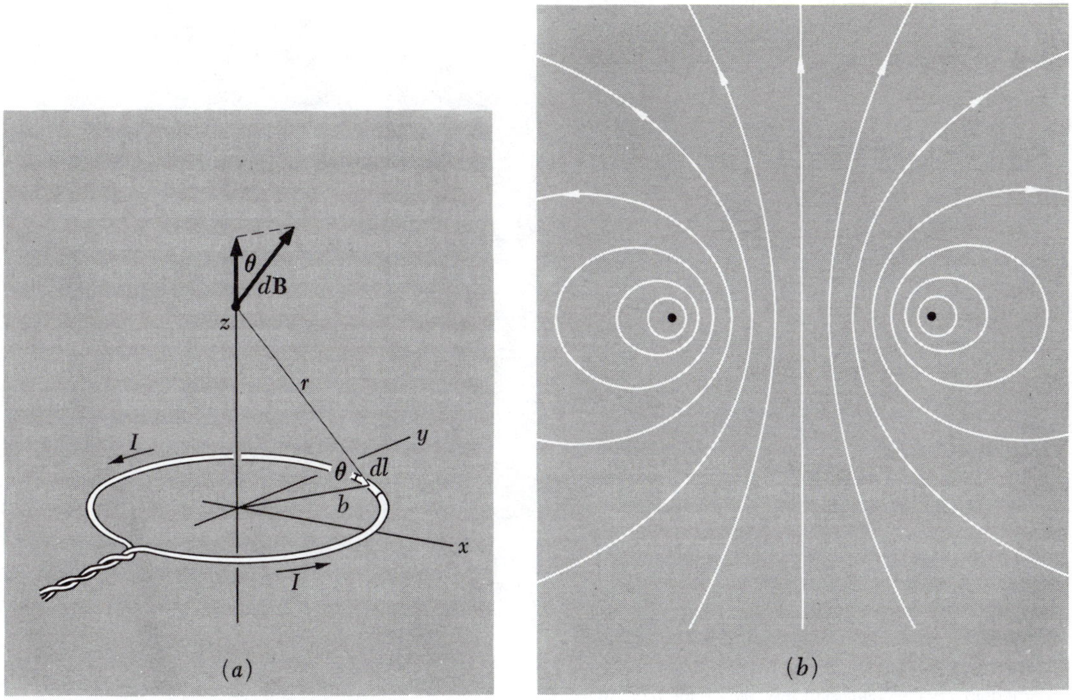

FIGURE 6.15
The magnetic field of a ring of current. (*a*) Calculation of field on the axis. (*b*) Some field lines.

Integrating over the whole ring, we have simply $\int dl = 2\pi b$, so the field on the axis at any point z is

$$B_z = \frac{2\pi b^2 I}{cr^3} = \frac{2\pi b^2 I}{c(b^2 + z^2)^{3/2}} \qquad \text{(field on axis)} \qquad (41)$$

At the center of the ring, $z = 0$, the magnitude of the field is

$$B_z = \frac{2\pi I}{cb} \qquad \text{(field at center)} \qquad (42)$$

The cylindrical coil of wire shown in Fig. 6.16*a* is usually called a solenoid. We assume the wire is closely and evenly spaced so that the number of turns in the winding, per centimeter length along the cylinder, is a constant, n. Now the current path is actually helical, but if the turns are many and closely spaced, we can ignore this and regard the whole solenoid as equivalent to a stack of current rings. Then we can use Eq. 41 as a basis for calculating the field at any point, such as the point z, on the axis of the coil. Take first the contribution from the current ring included between radii from the point z making angles θ and $\theta + d\theta$ with the axis. The length of this segment of the solenoid, shaded in Fig. 6.16*b*, is $r\, d\theta/\sin\theta$, and it is therefore equiv-

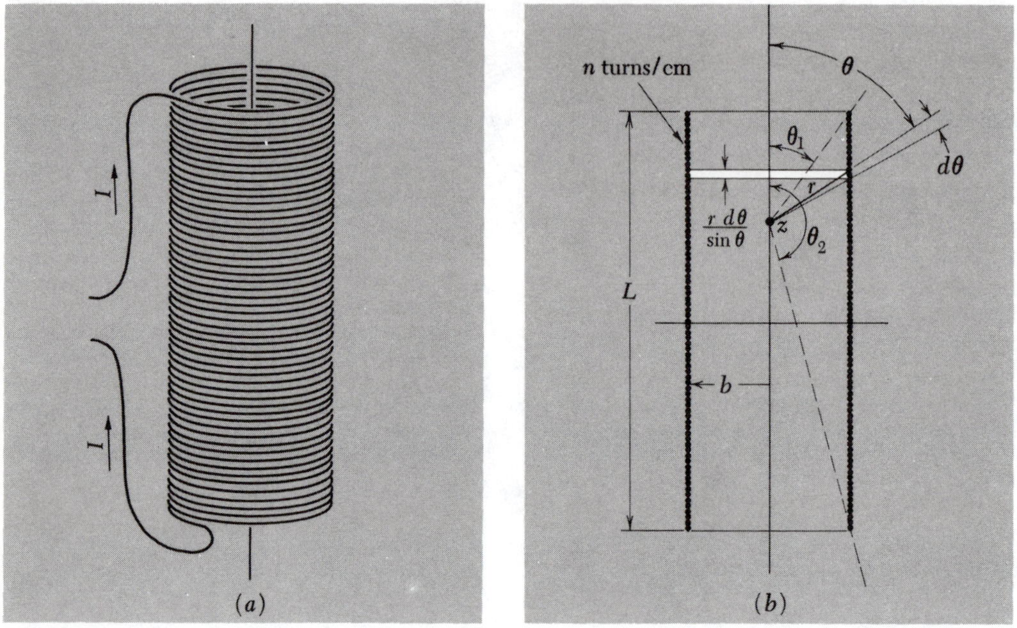

FIGURE 6.16
(*a*) Solenoid. (*b*) Calculation of the field on the axis of a solenoid.

alent to a ring carrying a current $Inr\,d\theta/\sin\theta$. Since $r = b/\sin\theta$, we have, for the contribution of this ring to the axial field:

$$dB_z = \frac{2\pi b^2}{cr^3}\frac{Inr\,d\theta}{\sin\theta} = \frac{2\pi In}{c}\sin\theta\,d\theta \tag{43}$$

Carrying out the integration between the limits θ_1 and θ_2 gives

$$B_z = \frac{2\pi In}{c}\int_{\theta_1}^{\theta_2}\sin\theta\,d\theta = \frac{2\pi In}{c}(\cos\theta_1 - \cos\theta_2) \tag{44}$$

We have used Eq. 44 to make a graph, in Fig. 6.17, of the field strength on the axis of a coil the length of which is four times its diameter. The ordinate is the field strength B_z relative to the field strength in a coil of infinite length with the same number of turns per centimeter and the same currents in each turn. For the infinite coil, $\theta_1 = 0$ and $\theta_2 = \pi$, so

$$B_z = \frac{4\pi In}{c} \qquad \text{(infinitely long solenoid)} \tag{45}$$

At the center of the "four-to-one" coil the field is very nearly as large as this, and it stays pretty nearly constant until we approach one of the ends.

Figure 6.18 shows the magnetic field lines in and around a coil of these proportions. Note that some field lines actually penetrate the

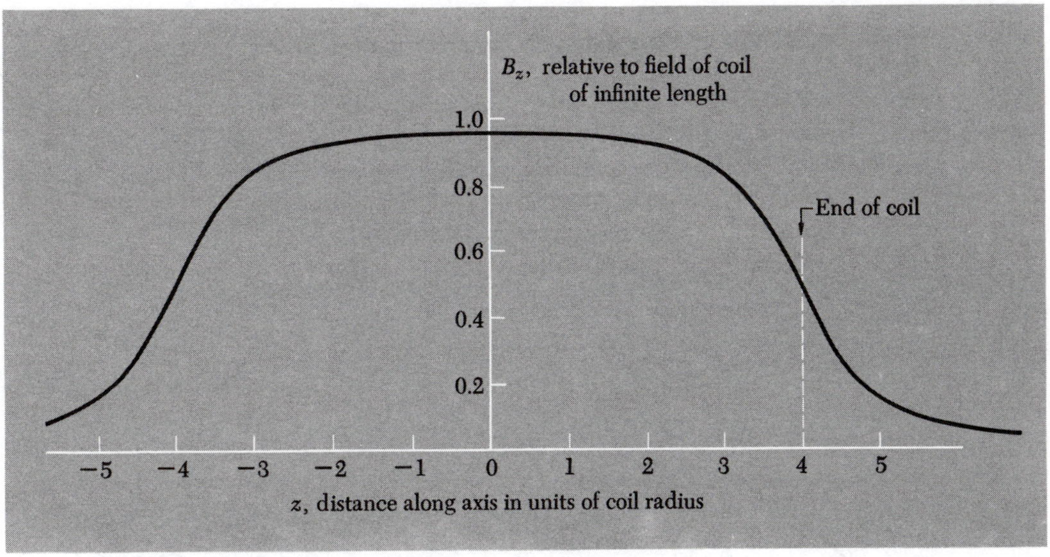

FIGURE 6.17
Field strength B_z on the axis, for the solenoid shown in Fig. 6.18.

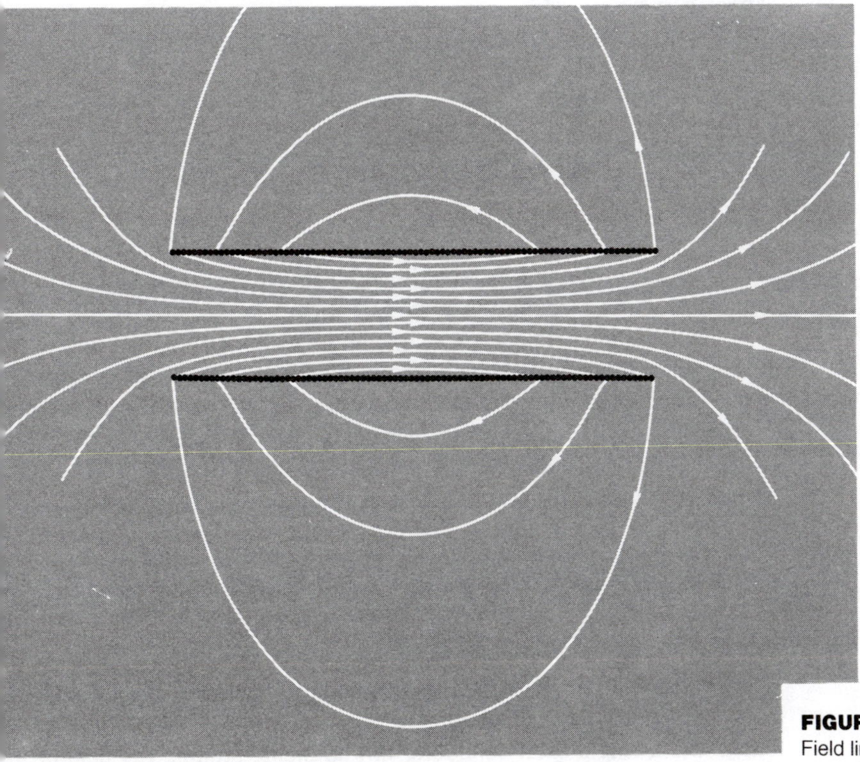

FIGURE 6.18
Field lines in and around a solenoid.

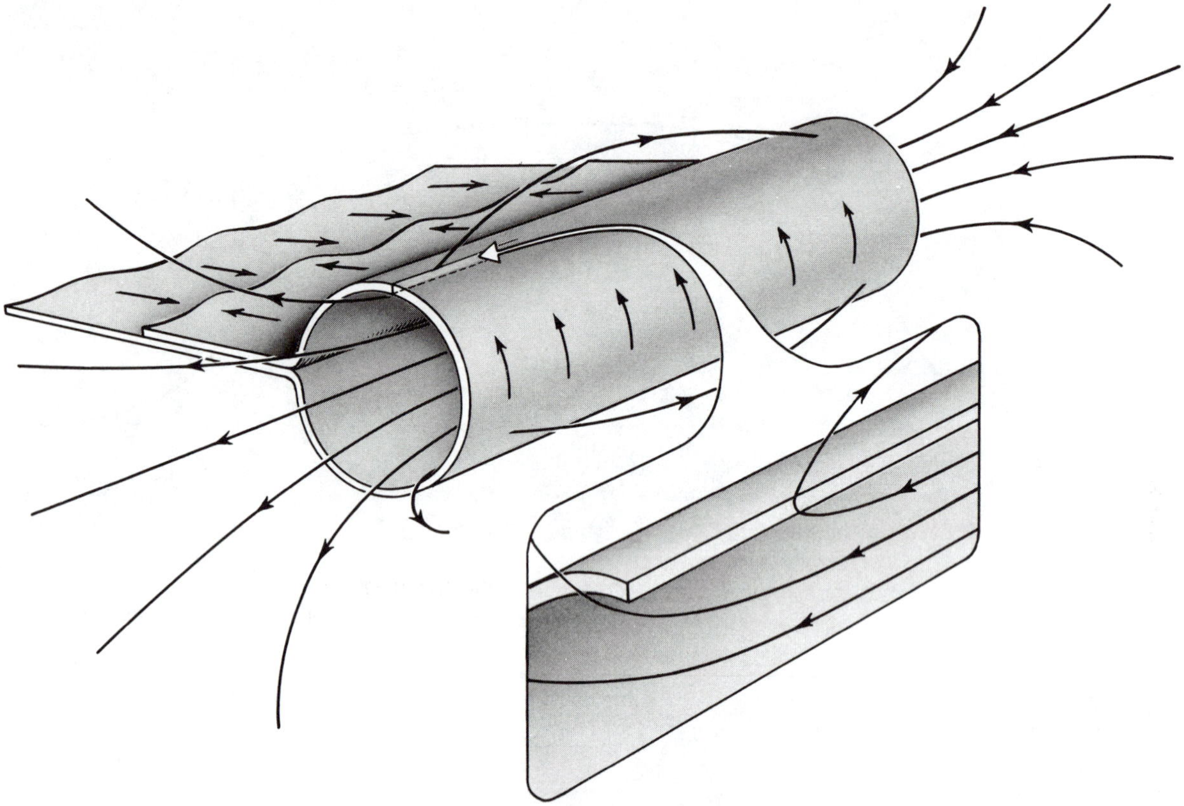

FIGURE 6.19

A solenoid formed by a single cylindrical conducting sheet. Inset shows how the field lines change direction inside the current-carrying conductor.

winding. The cylindrical sheath of current is a surface of discontinuity for the magnetic field. Of course, if we were to examine the field very closely in the neighborhood of the wires, we would not find any infinitely abrupt kinks, but we would find a very complicated, ripply pattern around and through the individual wires.

It is quite possible to make a long solenoid with a *single* turn of a thin wide ribbonlike conductor, as in Fig. 6.19. To this our calculation and the diagram in Fig. 6.18 apply exactly, the quantity nI being merely replaced by the current per centimeter flowing in the sheet. Now the change in direction of a field line that penetrates the wall occurs entirely within the thickness of the sheet, as suggested in the inset in Fig. 6.19.

In calculating the field of the solenoid in Fig. 6.16 we treated it as a stack of rings, ignoring the longitudinal current which must exist in any coil in which the current enters at one end and leaves at the other. Let us see how the field is modified if that is taken into account. The helical coil in Fig. 6.20*c* is equivalent, so far as its external field is concerned, to the superposition of the stack of current rings in Fig. 6.20*a* and a single axial conductor in Fig. 6.20*b*. Adding the field of

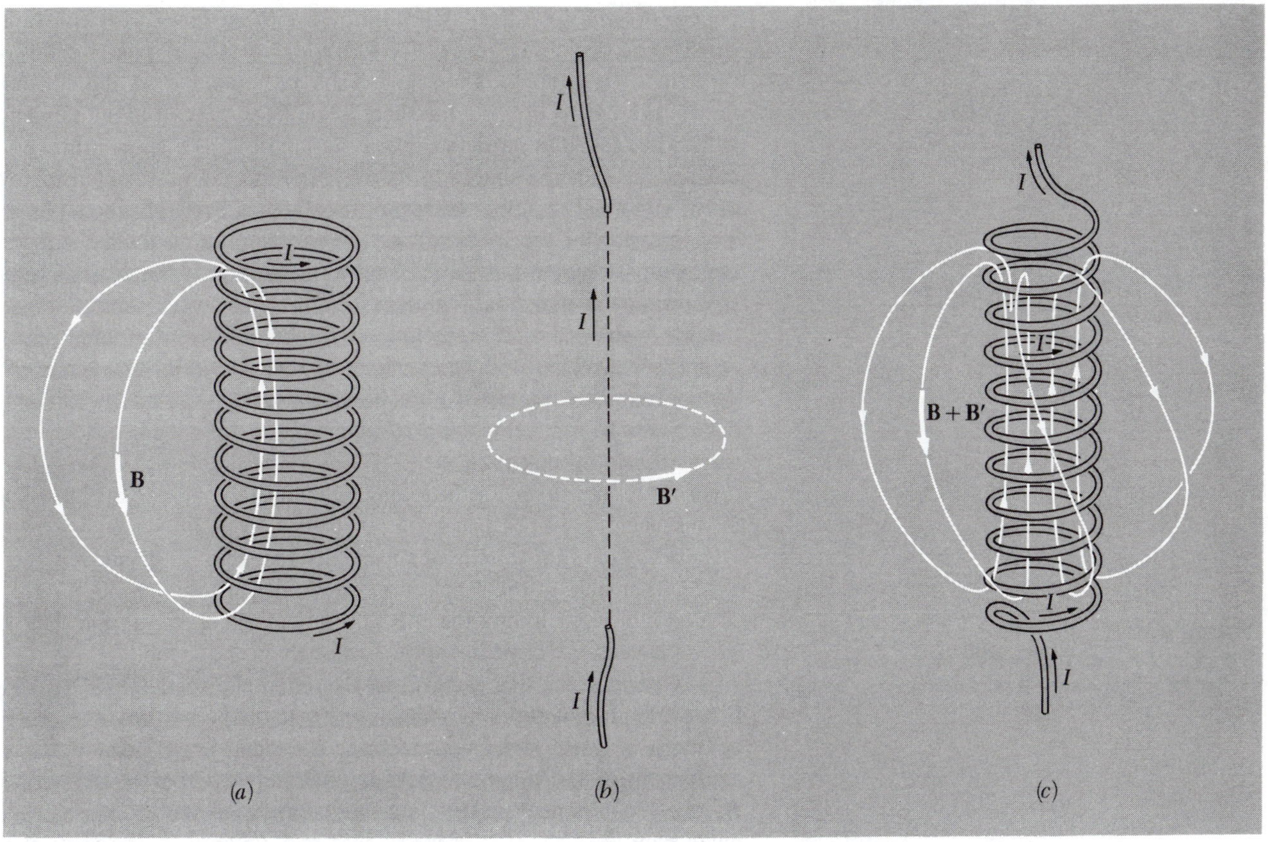

FIGURE 6.20
The helical coil (*c*) is equivalent to a stack of circular rings, each carrying current *I* and shown in (*a*), plus a current *I* parallel to the axis of the coil. A path around the coil encloses the current *I*, the field of which, **B′**, must be added to the field **B** of the rings to form the external field of the helical coil.

the latter, **B′**, to the field **B** of the former, we get the external field of the coil. It has a helical twist. Some field lines have been sketched in Fig. 6.20*c*. As for the field inside the solenoid, the longitudinal current *I* flows, in effect, on the cylinder itself. Such a current distribution, a uniform hollow tube of current, produces zero field inside the cylinder—leaving unmodified the interior field we calculated before. If you follow a looping field line from inside to outside to inside again, you will discover that it does *not* close on itself. Field lines generally don't. You might find it interesting to figure out how this picture would be changed if the wire that leads the current *I* away from the coil were brought down along the axis of the coil to emerge at the bottom.

CHANGE IN B AT A CURRENT SHEET

6.6 In the example of Fig. 6.19 we had a solenoid constructed from a single curved sheet of current. Let's look at something even simpler, a flat, unbounded current sheet. You may think of this as a sheet of

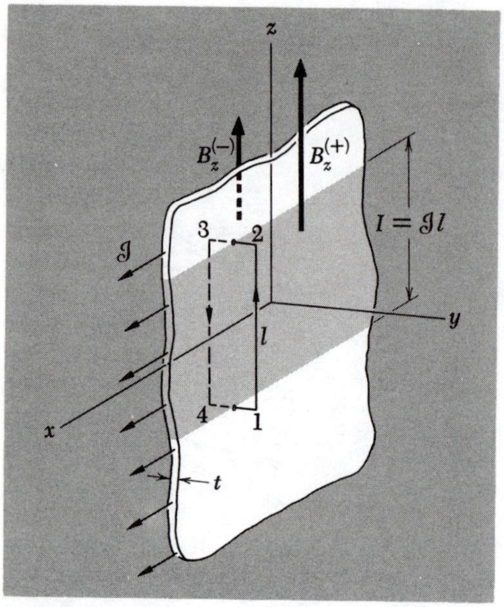

FIGURE 6.21
At a sheet of surface current there must be a change in the parallel component of **B** from one side to the other.

copper of uniform thickness in which a current flows with constant density and direction everywhere within the metal. In order to refer to directions, let us locate the sheet in the xz plane and let the current flow in the x direction. As the sheet is supposed to be of infinite extent with no edges, it is hard to draw a picture of it! We show a broken-out fragment of the sheet in Fig. 6.21, in order to have something to draw; you must imagine the rest of it extending over the whole plane. The thickness of the sheet will not be very important, finally, but we may suppose that it has some definite thickness t. If the current density inside the metal is J in esu/sec-cm^2, then every centimeter of height, in the z direction, includes a ribbon of current amounting to Jt esu/sec. We call this the *surface current density* or *sheet current density* and use the symbol $\mathcal{J}$ to distinguish it from the volume current density **J**. The units of $\mathcal{J}$ are esu/sec-cm. If we are not concerned with what goes on inside the sheet itself, $\mathcal{J}$ is a useful quantity. It is $\mathcal{J}$ that determines the *change* in the magnetic field from one side of the sheet to the other, as we shall see.

The field in Fig. 6.21 is not merely that due to the sheet alone. Some other field in the z direction was present, from another source. The total field, including the effect of the current sheet, is represented by the **B** vectors drawn in front of and behind the sheet.

Consider the line integral of **B** around the rectangle 12341 in Fig. 6.21. One of the long sides is in front of the surface, the other behind it, with the short sides piercing the sheet. Let B_z^+ denote the z component of the magnetic field immediately in front of the sheet, B_z^- the z component of the field immediately behind the sheet. We mean here the field of *all* sources that may be around, including the sheet itself. The line integral of **B** around the long rectangle is simply $l(B_z^+ - B_z^-)$. (Even if there were some other source which caused a field component parallel to the short legs of the rectangle, these legs themselves can be kept much shorter than the long sides, since we assume the sheet is thin, in any case, compared with the scale of any field variation.) The current enclosed by the rectangle is just $l\mathcal{J}$. Hence we have the relation $l(B_z^+ - B_z^-) = 4\pi\mathcal{J}l/c$, or

$$B_z^+ - B_z^- = \frac{4\pi\mathcal{J}}{c} \tag{46}$$

A current sheet of density $\mathcal{J}$ gives rise to a jump in that component of **B** which is parallel to the surface and perpendicular to $\mathcal{J}$. This may remind you of the change in electric field at a sheet of charge. There, the *perpendicular* component of **E** is discontinuous, the magnitude of the jump depending on the density of surface charge.

If the sheet is the only current source we have, then of course the field is symmetrical about the sheet. B_z^+ is $2\pi\mathcal{J}/c$, and B_z^- is $-2\pi\mathcal{J}/c$. This is shown in Fig. 6.22*a*. Some other situations, in which the effect of the current sheet is superposed on a field already present

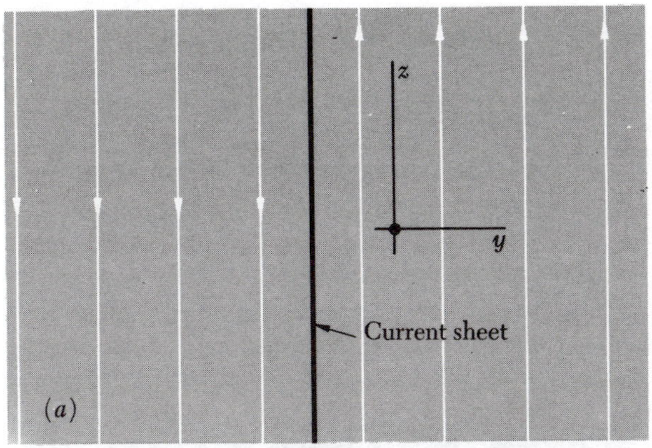
z
y
Current sheet
(a)

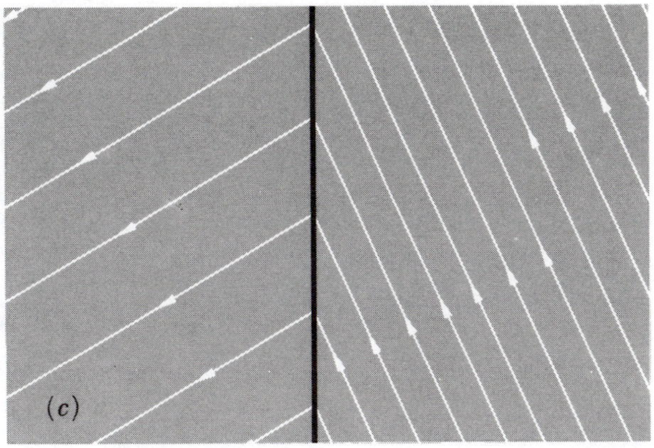
(c)

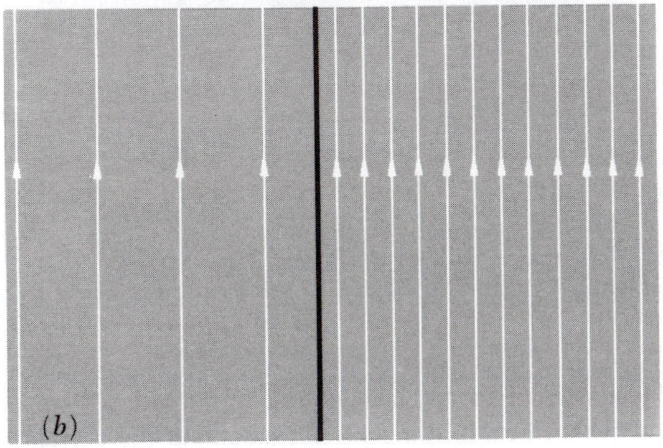
(b)

FIGURE 6.22
Some possible forms of the total magnetic field near a current sheet. Current flows in the $\hat{\mathbf{x}}$ direction (out of the page). (*a*) The field of the sheet alone. (*b*) Superposed on a uniform field in the *z* direction (this is like the situation in Fig. 6.21). (*c*) Superposed on a uniform field in another direction. In every case the component B_z changes by $4\pi\mathcal{J}/c$, on passing through the sheet, with no change in B_y.

FIGURE 6.23
The magnetic field between plane-parallel current sheets.

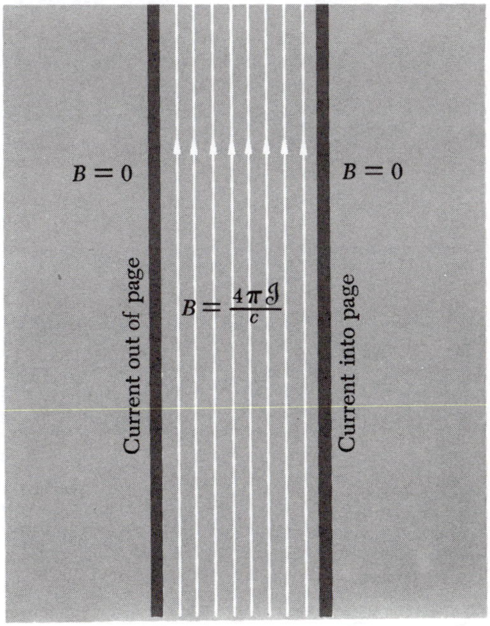
B = 0
B = 0
Current out of page
$B = \frac{4\pi\mathcal{J}}{c}$
Current into page

from another source, are shown in Fig. 6.22*b* and *c*. Suppose there are two sheets carrying equal and opposite surface currents, as shown in cross section in Fig. 6.23, with no other sources around. The direction of current flow is perpendicular to the plane of the paper, out on the left and in on the right. The field between the sheets is $4\pi\mathcal{J}/c$, and there is no field at all outside. Something like this is found when current is carried by two parallel ribbons or slabs, close together compared with their width, as sketched in Fig. 6.24. Often *bus bars* for distributing heavy currents in power stations are of this form.

The change in **B** takes place within the sheet, as we already remarked in connection with Fig. 6.19. For the same $\mathcal{J}$, the thinner the sheet, the more abrupt the transition. We looked at a situation very much like this in Chapters 1 and 2 when we examined the discontinuity in the perpendicular component of **E** that occurs at a sheet of surface charge. It was instructive then to ask about the force on the surface charge, and we shall ask a similar question here.

Consider a square portion of the sheet, 1 cm on a side. The cur-

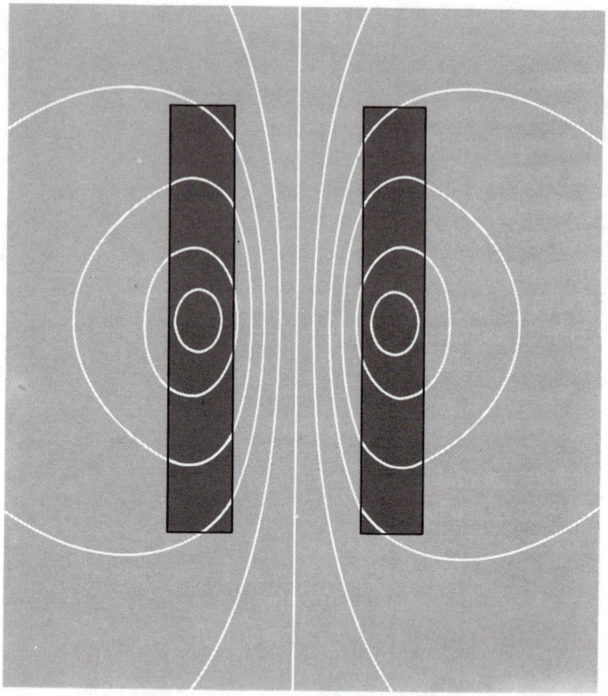

FIGURE 6.24
The magnetic field of a pair of copper bus bars, shown in cross section, carrying current in opposite directions.

rent included is equal to $\mathcal{J}$, the length of current path is 1 cm, and the *average* field that acts on this current, assuming the current is uniformly distributed through the thickness of the sheet, is $\frac{1}{2}(B_z^+ + B_z^-)$. Therefore the force on this portion of the current distribution is

$$\text{Force on 1 cm}^2 \text{ of sheet} = \frac{1}{2}(B_z^+ + B_z^-)\frac{\mathcal{J}}{c} \tag{47}$$

In view of Eq. 46, we can substitute $(B_z^+ - B_z^-)/4\pi$ for $\mathcal{J}/c$, so that the force per square centimeter can be expressed in this way:

$$\begin{aligned}\text{Force per cm}^2 &= \left(\frac{B_z^+ + B_z^-}{2}\right)\left(\frac{B_z^+ - B_z^-}{4\pi}\right) \\ &= \frac{1}{8\pi}\left[(B_z^+)^2 - (B_z^-)^2\right]\end{aligned} \tag{48}$$

The force is perpendicular to the surface and proportional to the area, like the stress caused by hydrostatic pressure. To make sure of the sign, we can figure out the direction of the force in a particular case, such as that in Fig. 6.23. The force is *outward* on each conductor. It is as if the high-field region were the region of high pressure. The repulsion of any two conductors carrying current in opposite directions, as in Fig. 6.24, can be seen as an example of that.

We have been considering an infinite flat sheet, but things are much the same in the immediate neighborhood of any surface where there is a change in **B.** Wherever the component of **B** parallel to the surface changes from B_1 to B_2, from one side of the surface to the other, we may conclude not only that there is a sheet of current flowing in the surface, but that the surface must be under a perpendicular stress of $(B_1^2 - B_2^2)/8\pi$, measured in dynes/cm^2. This is one of the controlling principles in *magnetohydrodynamics,* the study of electrically conducting fluids, a subject of interest both to electrical engineers and to astrophysicists.

HOW THE FIELDS TRANSFORM

6.7 A sheet of surface charge, if it is moving parallel to itself, constitutes a surface current. If we have a uniform charge density of σ on the surface, with the surface itself sliding along at speed v, the surface current density is just $\mathcal{J} = \sigma v$. This simple idea will help us to see how the electric and magnetic field quantities must change when we transform from one inertial frame of reference to another.

Let's imagine two plane sheets of surface charge, parallel to the xz plane as in Fig. 6.25. Again, we show fragments of surfaces only in the sketch; the surfaces are really infinite in extent. In the inertial frame F, with coordinates x, y, and z, the density of surface charge is σ on one sheet and $-\sigma$ on the other. Here σ means the amount of charge within unit area when area is measured by observers stationary in F. (It is not the density of charge in the rest frame of the charges themselves, which would be smaller by $1/\gamma$.) In the frame F the uni-

FIGURE 6.25
As observed in the frame F on the left, the surface charge density is σ and the surface current density is σv_0. Frame F' on the right moves in the x direction with speed v as seen from F. In F' the surface charge density is σ' and the current density is $\sigma' v_0'$.

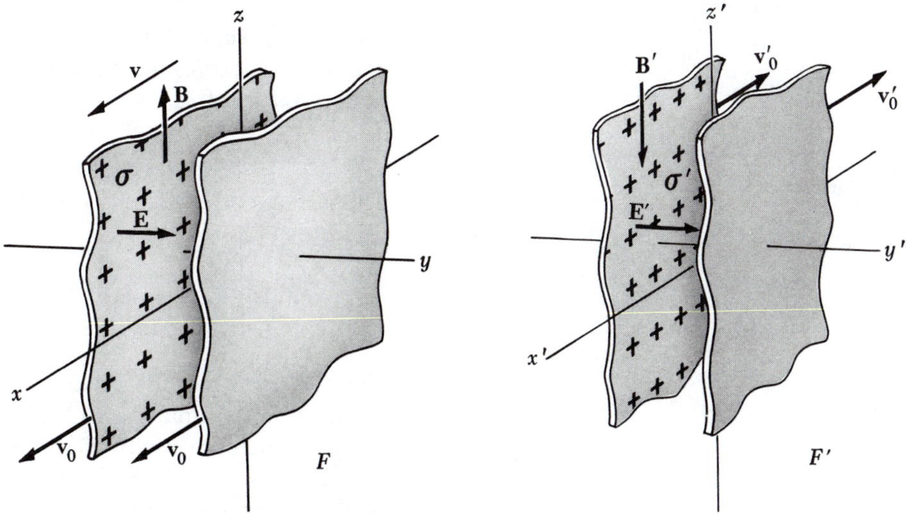

form electric field **E** points in the positive y direction, and Gauss' law assures us, as usual, that its strength is

$$E_y = 4\pi\sigma \tag{49}$$

In this frame F the sheets are both moving in the positive x direction with speed v_0, so that we have a pair of current sheets. The density of surface current is $\mathcal{J}_x = \sigma v_0$ in one sheet, the negative of that in the other. As in the arrangement in Fig. 6.23, the field between two such current sheets is

$$B_z = \frac{4\pi \mathcal{J}_x}{c} = \frac{4\pi\sigma v_0}{c} \tag{50}$$

The inertial frame F' is one that moves, as seen from F, with a speed v in the positive x direction. *What fields will an observer in F' measure?* To answer this we need only find out what the sources look like in F'.

In F' the x' velocity of the charge-bearing sheets is v_0', given by the velocity addition formula

$$v_0' = \frac{v_0 - v}{1 - v_0 v/c^2} = c\,\frac{\beta_0 - \beta}{1 - \beta_0\beta} \tag{51}$$

There is a different Lorentz contraction of the charge density in this frame, exactly as in our earlier example of the moving line charge in Section 5.9. We can repeat the argument we used then: The density in the rest frame of the charges themselves is $\sigma(1 - v_0^2/c^2)^{1/2}$, or σ/γ_0, and therefore the density of surface charge in the frame F' is

$$\sigma' = \sigma\,\frac{\gamma_0'}{\gamma_0} \tag{52}$$

As usual, γ_0' stands for $(1 - v_0'^2/c^2)^{-1/2}$. By means of Eq. 51 we can eliminate γ_0', expressing it in terms of β_0 and β, or γ_0 and γ. When we do this, the result is

$$\sigma' = \sigma\gamma\,(1 - \beta_0\,\beta) \tag{53}$$

The surface current density in the frame F' is *charge density* $\times$ *charge velocity:*

$$\mathcal{J}' = \sigma' v_0' = \sigma\gamma(1 - \beta_0\beta)c\,\frac{(\beta_0 - \beta)}{1 - \beta_0\beta} = \sigma\gamma\,(v_0 - v) \tag{54}$$

We now know how the sources appear in frame F', so we know what the fields in that frame must be. In saying this, we are again invoking the postulate of relativity. The laws of physics must be the same in all inertial frames, and that includes the formulas connecting electric field with surface charge density, and magnetic field with surface current density. It follows then that

$$E'_y = 4\pi\sigma' = \gamma\left[4\pi\sigma - \left(\frac{4\pi\sigma v_0}{c}\right)\left(\frac{v}{c}\right)\right] \tag{55}$$

$$B'_z = \frac{4\pi}{c}\mathcal{J}' = \gamma\left[\frac{4\pi\sigma v_0}{c} - 4\pi\sigma\left(\frac{v}{c}\right)\right] \tag{56}$$

If we look back at the values of E_y and B_z in Eqs. 49 and 50, we see that our result can be written as follows:

$$E'_y = \gamma(E_y - \beta B_z) \qquad B'_z = \gamma(B_z - \beta E_y) \tag{57}$$

If the sandwich of current sheets had been oriented parallel to the xy plane, instead of the xz plane, we would have obtained relations connecting E'_z with E_z and B_y, and B'_y with B_y and E_z. Of course they would have the same form as the relations above, but if you trace the directions through, you will find that there are differences in sign, following from the rules for the direction of **B.**

Now we must learn how the field components in the direction of motion change. We discovered in Section 5.5 that a longitudinal component of **E** has the same magnitude in the two frames. That this is true also of a longitudinal component of **B** can be seen as follows. Suppose a longitudinal component of **B**, a B_x component in the arrangement in Fig. 6.25, is produced by a solenoid around the x axis in frame F. The field strength inside a solenoid, as we know, depends only on the current in the wire, I, which is charge per second, and n, the number of turns of wire per centimeter of axial length. In the frame F' the solenoid will be Lorentz-contracted, so the number of turns per centimeter in that frame will be greater. But the current, as reckoned by the observer in F', will be reduced, since from his point of view, the F observer who measured the current by counting the number of electrons passing a point on the wire, per second, was using a slow-running watch. The time dilation just cancels the length contraction in the product nI. Indeed any quantity of the dimensions (longitudinal length)$^{-1}$ $\times$ (time)$^{-1}$ is unchanged in a Lorentz transformation. So $B'_x = B_x$.

Remember the point made early in Chapter 5, in the discussion following Eq. 5.6: The transformation properties of the field are *local* properties. The values of **E** and **B** at some space-time point in one frame must uniquely determine the field components observed in any other frame at that same space-time point. Therefore the fact that we have used an especially simple kind of source (the parallel uniformly charged sheets) in our derivation in no way compromises the generality of our result. We have in fact arrived at the general laws for the transformation of all components of the electric and magnetic field, of whatever origin or configuration.

We give below the full list of transformations. All primed quantities are measured in the frame F', which is moving in the positive x direction with speed v as seen from F. Unprimed quantities are the numbers which are the results of measurement in F. As usual, β stands for v/c and γ for $(1 - \beta^2)^{-1/2}$.

$$\boxed{\begin{array}{lll} E'_x = E_x & E'_y = \gamma(E_y - \beta B_z) & E'_z = \gamma(E_z + \beta B_y) \\ B'_x = B_x & B'_y = \gamma(B_y + \beta E_z) & B'_z = \gamma(B_z - \beta E_y) \end{array}} \quad (58)$$

The equations in the box confront us with an astonishing fact, their symmetry with respect to **E** and **B**. If the printer had mistakenly interchanged E's with B's, and y's with z's, the equations would come out exactly the same! Yet our previous view was that magnetism is a kind of "second-order" effect arising from relativistic changes in the electric fields of moving charges. Certainly magnetic phenomena as we find them in Nature are distinctly different from electrical phenomena. The world around us is by no means symmetrical with respect to electricity and magnetism. Nevertheless, with the sources out of the picture, we find that the fields themselves, **E** and **B**, are connected to one another in a highly symmetrical way.

It appears too that the electric and magnetic fields are in some sense aspects, or components, of a single entity. We can speak of the *electromagnetic* field, and we may think of E_x, E_y, E_z, B_x, B_y, and B_z as six components of the electromagnetic field. The *same* field viewed in different inertial frames will be represented by different sets of values for these components, somewhat as a vector is represented by different components in different coordinate systems rotated with respect to one another. However, the electromagnetic field so conceived is not a vector, mathematically speaking, but rather something called a *tensor*. The totality of the equations in the box forms the prescription for transforming the components of such a tensor when we shift from one inertial frame to another. We are not going to develop that mathematical language here. In fact, we shall return now to our old way of talking about the electric field as a vector field, and the magnetic field as another vector field coupled to the first in a manner to be explored further in Chapter 7. To follow up on this brief hint of the unity of the electromagnetic field as represented in four-dimensional spacetime, you will have to wait for a more advanced course.

We can express the transformation of the fields, Eq. 58, in a more elegant way which is often useful. Let βc be the velocity of a frame F' as seen from a frame F. We can always resolve the fields in both F and F' into vectors parallel to and perpendicular to, respectively, the direction of $\boldsymbol{\beta}$. Thus, using an obvious notation:

$$\begin{aligned} \mathbf{E} &= \mathbf{E}_{\parallel} + \mathbf{E}_{\perp} & \mathbf{E}' &= \mathbf{E}'_{\parallel} + \mathbf{E}'_{\perp} \\ \mathbf{B} &= \mathbf{B}_{\parallel} + \mathbf{B}_{\perp} & \mathbf{B}' &= \mathbf{B}'_{\parallel} + \mathbf{B}'_{\perp} \end{aligned} \tag{59}$$

Then the transformation can be written like this:

$$\boxed{\begin{aligned} \mathbf{E}'_{\parallel} &= \mathbf{E}_{\parallel} & \mathbf{E}'_{\perp} &= \gamma(\mathbf{E}_{\perp} + \boldsymbol{\beta} \times \mathbf{B}_{\perp}) \\ \mathbf{B}'_{\parallel} &= \mathbf{B}_{\parallel} & \mathbf{B}'_{\perp} &= \gamma(\mathbf{B}_{\perp} - \boldsymbol{\beta} \times \mathbf{E}_{\perp}) \end{aligned}} \tag{60}$$

In the special case that led us to Eq. 59, $\boldsymbol{\beta}$ was $\beta\hat{\mathbf{x}}$, $\mathbf{E}_{\perp}$ was $4\pi\sigma\hat{\mathbf{y}}$, and $\mathbf{B}_{\perp}$ was $4\pi\sigma v_0\hat{\mathbf{z}}/c$. The symmetry of the transformation is even more striking in the more general form, Eq. 60.

In SI units, with $\mathbf{E}$ in volts/meter and $\mathbf{B}$ in teslas, the Lorentz transformation of the fields reads as follows:

$$\boxed{\begin{aligned} \mathbf{E}'_{\parallel} &= \mathbf{E}_{\parallel} & \mathbf{E}'_{\perp} &= \gamma\left(\mathbf{E}_{\perp} + \frac{1}{\sqrt{\epsilon_0\mu_0}}\boldsymbol{\beta} \times \mathbf{B}_{\perp}\right) \\ \mathbf{B}'_{\parallel} &= \mathbf{B}_{\parallel} & \mathbf{B}'_{\perp} &= \gamma(\mathbf{B}_{\perp} - \sqrt{\epsilon_0\mu_0}\,\boldsymbol{\beta} \times \mathbf{E}_{\perp}) \end{aligned}} \tag{60'}$$

In this system, unfortunately, the use of different units for **E** and **B** tends to obscure the essential electromagnetic symmetry of the vacuum. The electric and magnetic fields are after all components of one tensor. The Lorentz transformation is something like a rotation, turning **E** partly into **B′**, and **B** partly into **E′**. It seems quite natural and appropriate that the only parameter in Eq. 60 is the dimensionless ratio β. To draw an analogy which is not altogether unfair, imagine that it has been decreed that east-west displacement components must be expressed in centimeters while north-south components are to be in inches. The transformation effecting a rotation of coordinate axes would be, to say the least, aesthetically unappealing. Nor is symmetry restored to Eq. 60′ when **B** is replaced, as is often done, by a vector **H**, which we shall meet in Chapter 11, and which in the vacuum is simply $\mathbf{B}/\mu_0$.

There is a remarkably simple relation between the electric and magnetic field vectors in a special but important class of cases. Suppose a frame exists—let's call it the unprimed frame—in which **B** is zero in some region. Then in *any* other frame F' which moves with velocity βc relative to that special frame, we have according to Eq. 60

$$\mathbf{E}'_{\parallel} = \mathbf{E}_{\parallel} \qquad \mathbf{E}'_{\perp} = \gamma \mathbf{E}_{\perp} \qquad \mathbf{B}'_{\parallel} = 0 \qquad \mathbf{B}'_{\perp} = -\gamma \boldsymbol{\beta} \times \mathbf{E}_{\perp} \tag{61}$$

But $\boldsymbol{\beta} \times \mathbf{E}_{\parallel} = 0$ in any case, for $\mathbf{E}_{\parallel}$ is parallel to $\boldsymbol{\beta}$ by definition. Hence the relation between $\mathbf{E}'$ and $\mathbf{B}'$ reduces simply to

$$\mathbf{B}' = -\boldsymbol{\beta} \times \mathbf{E}' \tag{62}$$

This holds in every frame if $\mathbf{B} = 0$ in one frame. Remember that βc is the velocity of the frame in question with respect to the special frame in which $\mathbf{B} = 0$.

In the same way, we can deduce from Eq. 60 that, if there exists a frame in which $\mathbf{E} = 0$, then in any other frame

$$\mathbf{E}' = \boldsymbol{\beta} \times \mathbf{B}' \tag{63}$$

As before, βc is the velocity of the frame F' with respect to the special frame F in which, in this case, $\mathbf{E} = 0$.

Because Eqs. 62 and 63 involve only quantities measured in the same frame of reference, they are easy to apply, whenever the restriction is met, to fields that vary in space. A good example is the field of a point charge q moving with constant velocity, the problem studied in Chapter 5. Take the unprimed frame to be the frame in which the charge is at rest. In this frame, of course, there is no magnetic field. Equation 61 tells us that in the lab frame, where we find the charge moving with speed v, there must be a magnetic field perpendicular to the electric field and to the direction of motion. We have already worked out the exact form of the electric field in this frame: We know the field is radial from the instantaneous position of the charge, with a magnitude given by Eq. 12 of Chapter 5. The magnetic field lines must be circles around the direction of motion, as indicated crudely in Fig. 6.26. When the velocity of the charge is high, so that $\gamma \gg 1$, the radial "spokes" which are the electric field lines are folded together into a thin disk. The circular magnetic field lines are likewise concentrated in this disk. The magnitude of $\mathbf{B}$ is then nearly equal to the magnitude of $\mathbf{E}$. That is, the magnitude of the magnetic field in gauss is almost exactly the same as the magnitude of the electric field, at the same point and instant of time, in statvolts/cm.

FIGURE 6.26
The electric and magnetic fields, at one instant of time, of a charge in uniform motion.

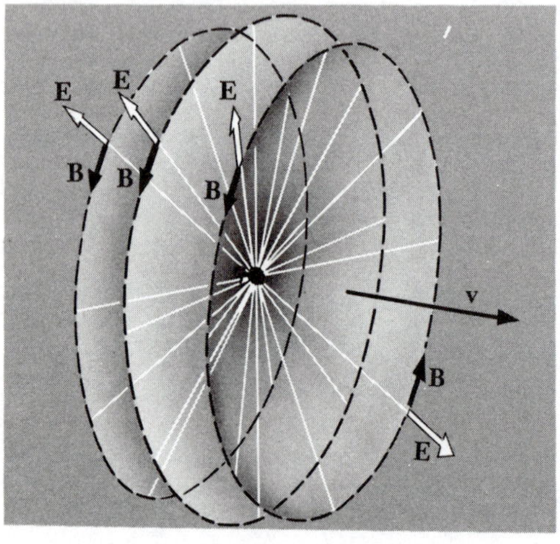

We have come a long way from Coulomb's law in the last two chapters. Yet with each step we have only been following out consistently the requirements of relativity and of the invariance of electric charge. We can begin to see that the existence of the magnetic field and its curiously symmetrical relationship to the electric field is a necessary consequence of these general principles. We remind the reader again that this was not at all the historical order of discovery and elucidation of the laws of electromagnetism. One aspect of the coupling between the electric and magnetic fields which is implicit in Eq. 58 came to light in Michael Faraday's experiments with changing electric currents, which will be described in Chapter 7. That was 75 years before Einstein, in his epochal paper of 1905, first wrote out our Eq. 58.

ROWLAND'S EXPERIMENT

6.8 As we remarked in Sec. 5.9, it was not obvious 100 years ago that a current flowing in a wire and a moving electrically charged object are essentially alike as sources of magnetic field. The unified view of electricity and magnetism which was then emerging from Maxwell's work suggested that any moving charge ought to cause a magnetic field, but experimental proof was hard to come by.

That the motion of an electrostatically charged sheet produces a magnetic field was first demonstrated by Henry Rowland, the great American physicist renowned for his perfection of the diffraction grating. Rowland made many ingenious and accurate electrical measurements, but none that taxed his experimental virtuosity as severely as the detection and measurement of the magnetic field of a rotating charged disk. The field to be detected was something like 10^{-5} of the earth's field in magnitude—a formidable experiment, even with today's instruments! In Fig. 6.27, you will see a sketch of Rowland's apparatus and a reproduction of the first page of the paper in which he described his experiment. Ten years before Hertz' discovery of electromagnetic waves, Rowland's result gave independent, if less dramatic, support to Maxwell's theory of the electromagnetic field.

ELECTRIC CONDUCTION IN A MAGNETIC FIELD: THE HALL EFFECT

6.9 When a current flows in a conductor in the presence of a magnetic field, the force $(q/c)\mathbf{v} \times \mathbf{B}$ acts directly on the moving charge carriers. Yet we observe a force on the conductor as a whole. Let's see how this comes about. Figure 6.28*a* shows a section of a metal bar in which a steady current is flowing. Driven by a field **E**, electrons are drifting to the left with average speed $\bar{v}$, which has the same meaning as the $\bar{u}$ in our discussion of conduction in Chapter 4. The conduction electrons are indicated, very schematically, by the white dots. The black dots are the positive ions which form the rigid framework of the solid metal bar. Since the electrons are negative, we have a current in the y direction. The current density **J** and the field **E** are related by the conductivity of the metal, σ, as usual: $\mathbf{J} = \sigma E$. There is no magnetic field in Fig. 6.28*a* except that of the current itself, which we shall ignore. Now an external field **B** in the x direction is switched on. The state of motion immediately thereafter is shown in Fig. 6.28*b*. The electrons are being deflected downward. But since they cannot escape at the bottom of the bar, they simply pile up there, until the surplus

FIGURE 6.27
The essential parts of Rowland's apparatus. In the tube at the left two short magnetized needles are suspended horizontally.

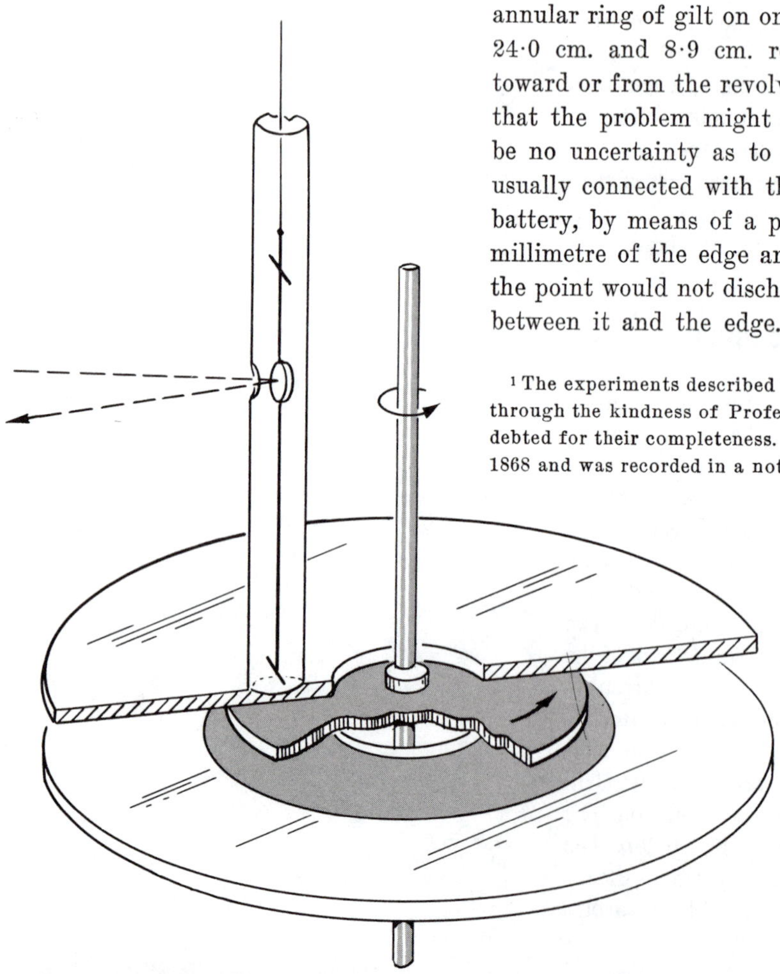

ON THE MAGNETIC EFFECT OF ELECTRIC CONVECTION [1]

[*American Journal of Science* [3], *XV*, 30-38, 1878]

The experiments described in this paper were made with a view of determining whether or not an electrified body in motion produces magnetic effects. There seems to be no theoretical ground upon which we can settle the question, seeing that the magnetic action of a conducted electric current may be ascribed to some mutual action between the conductor and the current. Hence an experiment is of value. Professor Maxwell, in his 'Treatise on Electricity,' Art. 770, has computed the magnetic action of a moving electrified surface, but that the action exists has not yet been proved experimentally or theoretically.

The apparatus employed consisted of a vulcanite disc 21·1 centimetres in diameter and ·5 centimetre thick which could be made to revolve around a vertical axis with a velocity of 61· turns per second. On either side of the disc at a distance of ·6 cm. were fixed glass plates having a diameter of 38·9 cm. and a hole in the centre of 7·8 cm. The vulcanite disc was gilded on both sides and the glass plates had an annular ring of gilt on one side, the outside and inside diameters being 24·0 cm. and 8·9 cm. respectively. The gilt sides could be turned toward or from the revolving disc but were usually turned toward it so that the problem might be calculated more readily and there should be no uncertainty as to the electrification. The outside plates were usually connected with the earth; and the inside disc with an electric battery, by means of a point which approached within one-third of a millimetre of the edge and turned toward it. As the edge was broad, the point would not discharge unless there was a difference of potential between it and the edge. Between the electric battery and the disc,

[1] The experiments described were made in the laboratory of the Berlin University through the kindness of Professor Helmholtz, to whose advice they are greatly indebted for their completeness. The idea of the experiment first occurred to me in 1868 and was recorded in a note book of that date.

of negative charge at the bottom of the bar and the corresponding excess of positive charge at the top create an electric field $\mathbf{E}_t$ in which

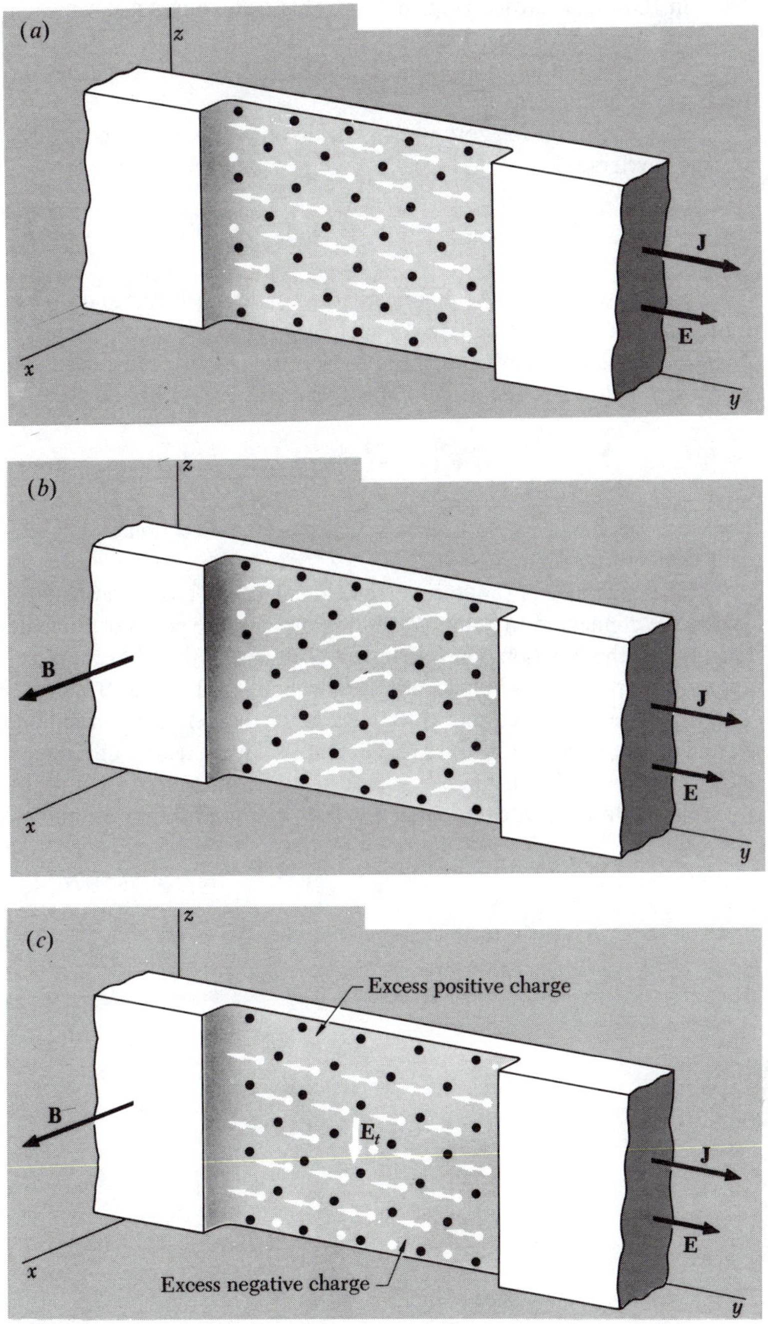

FIGURE 6.28
(*a*) A current flows in a metal bar. Only a short section of the bar is shown. Conduction electrons are indicated (not in true size and number!) by white dots, positive ions of the crystal lattice by black dots. The arrows indicate the average velocity $\bar{v}$ of the electrons. (*b*) A magnetic field is applied to the *x* direction, causing (at first) a downward deflection of the moving electrons. (*c*) The altered charge distribution makes a transverse electric field E_t. In this field the stationary positive ions experience a downward force.

the upward force, of magnitude eE_t, exactly balances the downward force $(e/c)\bar{v}B$. In the steady state (which is attained very quickly!) the average motion is horizontal again, and there exists in the interior of the metal this transverse electric field $\mathbf{E}_t$, as observed in coordinates fixed in the metal lattice (Fig. 6.28*c*). This field causes a downward force on the positive ions. That is how the force, $(-e/c)\bar{\mathbf{v}} \times \mathbf{B}$, on the electrons is passed on to the solid bar. The bar, of course, pushes on whatever is holding *it*.

The condition for zero average transverse force on the moving charge carriers is

$$\mathbf{E}_t + \frac{\bar{\mathbf{v}}}{c} \times \mathbf{B} = 0 \tag{64}$$

Suppose there are m mobile charge carriers per cm^3 and, to be more general, denote the charge of each by q. Then the current density $\mathbf{J}$ is $nq\bar{\mathbf{v}}$. If we now substitute $\mathbf{J}/nq$ for $\bar{\mathbf{v}}$ in Eq. 64, we can relate the transverse field $\mathbf{E}_t$ to the directly measurable quantities $\mathbf{J}$ and $\mathbf{B}$:

$$\mathbf{E}_t = \frac{-\mathbf{J} \times \mathbf{B}}{nqc} \tag{65}$$

For electrons $q = -e$, so $\mathbf{E}_t$ has in that case the direction of $\mathbf{J} \times \mathbf{B}$, as it does in Fig. 6.28*c*.

The existence of the transverse field can easily be demonstrated. Wires are connected to points P_1 and P_2 on opposite edges of the bar (Fig. 6.29), the junction points being carefully located so that they are at the same potential when current is flowing in the bar and $\mathbf{B}$ is zero: The wires are connected to a voltmeter. After the field $\mathbf{B}$ is turned on P_1 and P_2 are no longer at the same potential. The potential difference is E_t times the width of the bar, and in the case illustrated P_1 is positive relative to P_2. A steady current will flow around the external cir-

FIGURE 6.29

The Hall effect. When a magnetic field is applied perpendicular to a conductor carrying current, a potential difference is observed between points on opposite sides of the bar—points which, in the absence of the field, would be at the same potential. This is consistent with the existence of the field E_t inside the bar. By measuring the "Hall voltage" one can determine the number of charge carriers per cubic centimeter, and their sign.

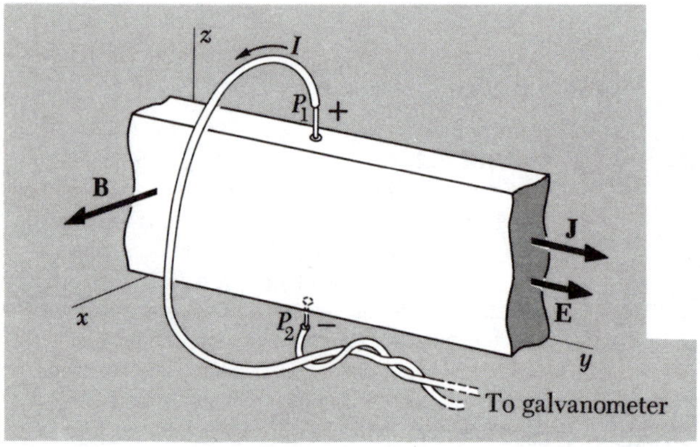

cuit from P_1 to P_2, its magnitude determined by the resistance of the voltmeter. Notice that the potential difference would be reversed if the current **J** consisted of positive carriers moving to the right rather than electrons moving to the left. Here for the first time we have an experiment that promises to tell us the *sign* of the charge carriers in a conductor.

The effect was discovered in 1879 by E. H. Hall who was studying under Rowland at Johns Hopkins. In those days no one understood the mechanism of conduction in metals. The electron itself was unknown. It was hard to make much sense of the results. Generally the sign of the "Hall voltage" was consistent with conduction by negative carriers, but there were exceptions even to that. A complete understanding of the Hall effect in metallic conductors came only with the quantum theory of metals, about 50 years after Hall's discovery.

The Hall effect has proved to be especially useful in the study of semiconductors. There it fulfills its promise to reveal directly both the concentration and the sign of the charge carriers. The *n*-type and *p*-type semiconductors described in Chapter 4 give Hall voltages of opposite sign, as we should expect. As the Hall voltage is proportional to B, an appropriate semiconductor in the arrangement of Fig. 6.29 can serve, once calibrated, as a simple and compact device for measuring an unknown magnetic field. An example is described in Problem 6.35.

PROBLEMS

6.1 Suppose the current I that flows in the circuit in Fig. 5.1*b* is 6×10^{10} esu/sec, or 20 amperes. The distance between the wires is 5 cm. How large is the force, per centimeter of length, that pushes horizontally on one of the wires?

6.2 A current of 8000 amperes flows through an aluminum rod 4 cm in diameter. Assuming the current density is uniform through the cross section, find the strength of the magnetic field at 1 cm, at 2 cm, and at 3 cm from the axis of the rod.

6.3 Consider the magnetic field of a circular current ring, at points on the axis of the ring, given by Eq. 41. Calculate explicitly the line integral of the field along the axis from $-\infty$ to ∞, to check the general formula

$$\int \mathbf{B} \cdot d\mathbf{s} = \frac{4\pi I}{c}$$

Why may we ignore the "return" part of the path which would be necessary to complete a closed loop?

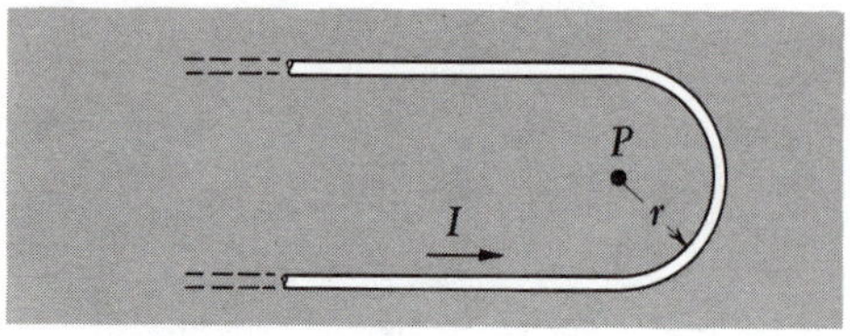

PROBLEM 6.4

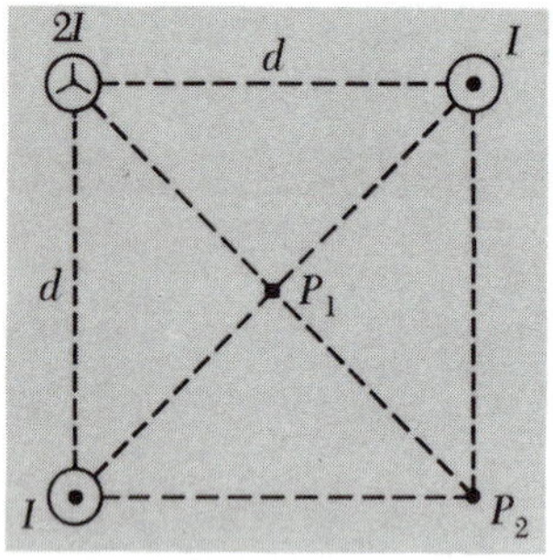

PROBLEM 6.5

6.4 A long wire is bent into the hairpinlike shape shown in the figure. Find an exact expression for the magnetic field at the point P which lies at the center of the half-circle.

6.5 Three long straight parallel wires are located as shown in the diagram. One wire carries current $2I$ into the paper; each of the others carries current I in the opposite direction. What is the strength of the magnetic field at the point P_1 and at the point P_2?

6.6 Suppose that the current I_2 in Fig. 6.4*b* is equal to I, but reversed, so that CD is repelled by GH. Suppose also that AB and EF lie vertically above GH, that the lengths BC and CD are 30 and 15 cm, respectively, and that the conductor $BCDE$, which is 1-mm-diameter copper wire as in (*a*), has a weight of 8 dynes/cm. In equilibrium the deflection of the hanging frame from the vertical is such that $r =$ 1.5 cm. How large is the current in esu/sec and in amperes? Is the equilibrium stable?

6.7 The earth's metallic core extends out to 3000 km, about half the earth's radius. Imagine that the field we observe at the earth's surface, which has a strength of roughly 0.5 gauss at the north magnetic pole, is caused by a current flow around the "equator" of the core. How big would that current be, in amperes?

Ans. 3×10^9 amps.

6.8 A wire carrying current I runs down the y axis to the origin, thence out to infinity along the positive x axis. Show that the magnetic field in the quadrant $x > 0$, $y > 0$ of the xy plane is given by

$$B_z = \frac{I}{c}\left(\frac{1}{x} + \frac{1}{y} + \frac{x}{y\sqrt{x^2 + y^2}} + \frac{y}{x\sqrt{x^2 + y^2}}\right)$$

6.9 Describing the experiment in which he discovered the influence of an electric current on a nearby compass needle, H. C. Oersted wrote: "If the distance of the connecting wire does not exceed three-quarters of an inch from the needle, the declination of the needle makes an angle of about 45°. If the distance is increased the angle diminishes proportionally. The declination likewise varies with the power of the battery." About how large a current, in amperes, must have been flowing in Oersted's "connecting wire"? Assume the horizontal component of the earth's field in Copenhagen in 1820 was the same as it is today, 0.2 gauss.

6.10 A 50-kilovolt direct-current power line consists of two conductors 2 meters apart. When this line is transmitting 10 megawatts, how strong is the magnetic field midway between the conductors?

6.11 A solenoid is made by winding two layers of No. 14 copper wire on a cylindrical form 8 cm in diameter. There are four turns per

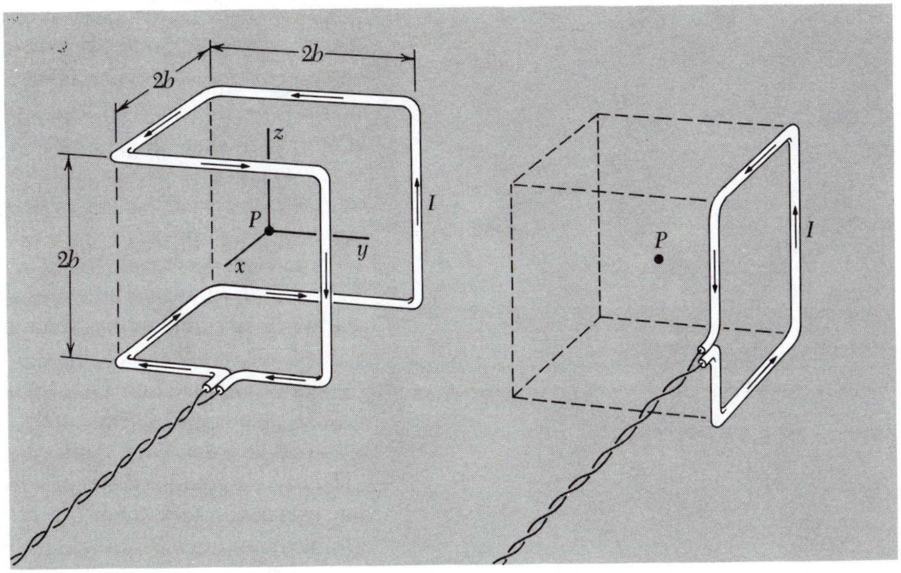

PROBLEM 6.12

centimeter in each layer, and the length of the solenoid is 32 cm. From the wire tables we find that No. 14 copper wire, which has a diameter of 0.163 cm, has a resistance of 0.010 ohm/meter at 75°C. (The coil will run hot!) If the solenoid is connected to a 50-volt generator, what will be the magnetic field strength at the center of the solenoid in gauss and what is the power dissipation in watts?

6.12 Current I flows around the wire frame in the figure.

(a) What is the direction of the magnetic field at P, the center of the cube?

(b) Show by using superposition that the field at P is the same as if the frame were replaced by the single square loop shown on the right.

6.13 One way to produce a very uniform magnetic field is to use a very long solenoid and work only in the middle section of its interior. This is often convenient, wasteful of space and power. Can you suggest ways in which two short coils or current rings might be arranged to achieve good uniformity over a limited region? *Hint:* Consider two coaxial current rings of radius a, separated axially by a distance b. Investigate the uniformity of the field in the vicinity of the point on the axis midway between the two coils. Determine the magnitude of the coil separation b which for given coil radius a will make the field in this region as nearly uniform as possible.

6.14 A coil is wound evenly on a torus of rectangular cross section. There are N turns of wire in all. Only a few are shown in the figure.

PROBLEM 6.14

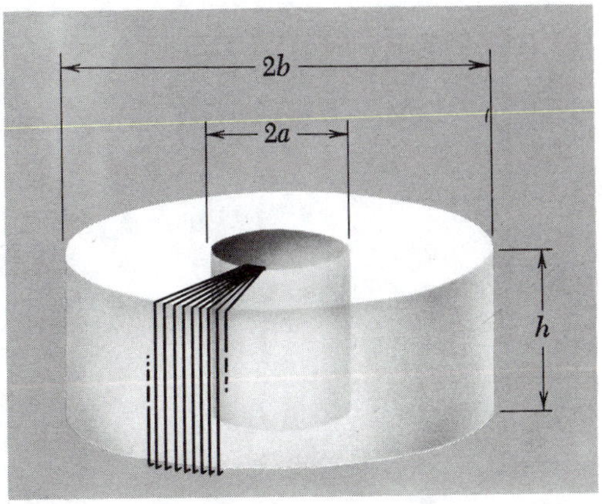

With so many turns, we shall assume that the current on the surface of the torus flows exactly radially on the annular end faces, and exactly longitudinally on the inner and outer cylindrical surfaces. First convince yourself that on this assumption symmetry requires that the magnetic field everywhere should point in a "circumferential" direction, that is, that all field lines are circles about the axis of the torus. Second, prove that the field is zero at all points outside the torus, including the interior of the central hole. Third, find the magnitude of the field inside the torus, as a function of radius.

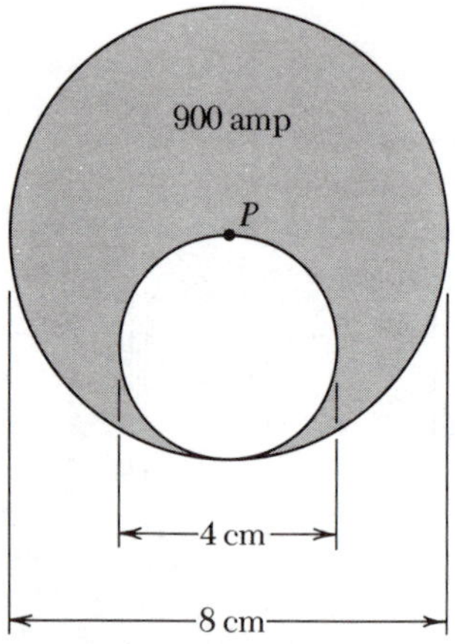
900 amp
P
4 cm
8 cm

PROBLEM 6.16

6.15 For a delicate magnetic experiment, a physicist wants to cancel the earth's field over a volume roughly 30 × 30 × 30 cm in size, so that the residual field in this region will not be greater than 10 milligauss at any point. The strength of the earth's field in this location is 0.55 gauss, making an angle of 30° with the vertical. It may be assumed constant to a milligauss or so over the volume in question. (The earth's field itself would hardly vary that much over a foot or so, but in a laboratory there are often local perturbations.) Suggest an arrangement of coils suitable for the task, and estimate the number of ampere turns required in your compensating system.

6.16 A long copper rod 8 cm in diameter has an off-center cylindrical hole, as shown in the diagram, down its full length. This conductor carries a current of 900 amps flowing in the direction "into the paper." We want to know the direction, and strength in gauss, of the magnetic field at the point P which lies on the axis of the outer cylinder.

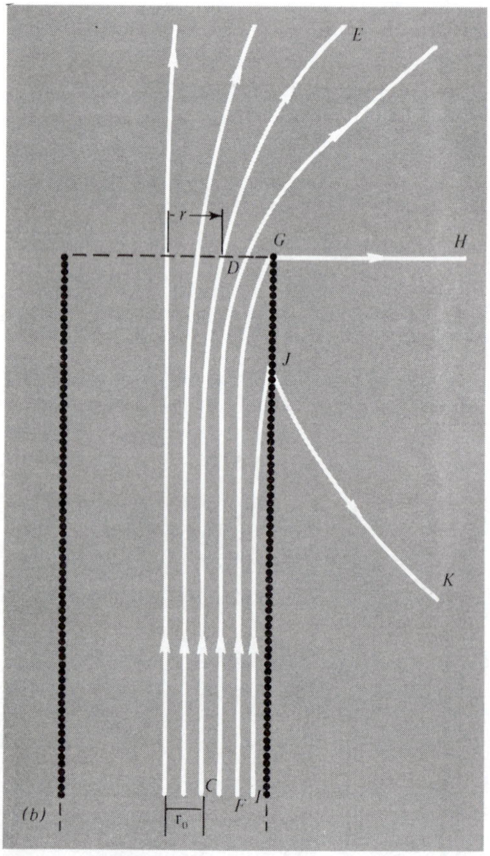
E
r
G
H
D
J
K
C
F
I
(b)
r_0

PROBLEM 6.17

6.17 A number of simple facts about the fields of solenoids can be found by using superposition. The idea is that two solenoids of the same diameter, and length L, if joined end to end, make a solenoid of length $2L$. Two semi-infinite solenoids butted together make an infinite solenoid, and so on. (A semi-infinite solenoid is one that has one end here and the other infinitely far away.) Here are some facts you can prove this way:

(*a*) In the finite-length solenoid in part (*a*) of the figure, the magnetic field on the axis at the point P_2 at one end is approximately half the field at the point P_1 in the center. (Is it slightly more than half, or slightly less than half?)

(*b*) In the semi-infinite solenoid shown in part (*b*) of the figure, the field line FGH which passes through the very end of the winding is a straight line from G out to infinity.

(*c*) The flux of **B** through the end face of the semi-infinite solenoid is just half the flux through the coil at a large distance back in the interior.

(*d*) Any field line which is r_0 cm from the axis far back in the interior of the coil exits from the end of the coil at a radius $r_1 = \sqrt{2}r_0$.

Show that these statements are true. What else can you find out?

6.18 Two long coaxial aluminum cylinders are charged to a potential difference of 50 statvolts. The inner cylinder has an outer diameter of 6 cm, the outer cylinder an inner diameter of 8 cm. With the outer cylinder stationary the inner cylinder is rotated around its axis at a constant speed of 30 revolutions per sec. Describe the magnetic field this produces and determine its intensity in gauss. What if both cylinders are rotated in the same direction at 30 revolutions per sec.

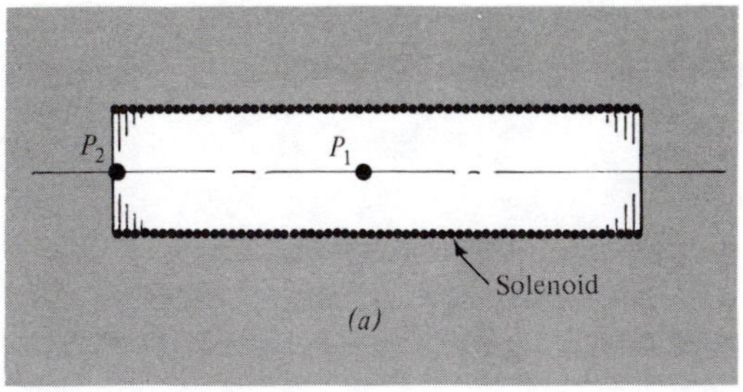

PROBLEM 6.17

6.19 A student said, "You almost convinced me that the force between currents, which I thought was magnetism, is explained by electric fields of moving charges. But if so, why doesn't the metal plate in Fig. 5.1*c* shield one wire from the influence of the other?" Can you explain it?

6.20 Suppose we had a situation in which the component of the magnetic field parallel to the plane of a sheet had the same *magnitude* on both sides, but changed *direction* by 90° in going through the sheet. What is going on here? Would there be a force on the sheet? Should our formula for the force on a current sheet apply to cases like this?

6.21 Since parallel current filaments attract one another, one might think that a current flowing in a solid rod like the conductor in Problem 6.2 would tend to concentrate near the axis of the rod. That is, the conduction electrons, instead of distributing themselves evenly as usual over the interior of the metal, would crowd in toward the axis and most of the current would be there. What do you think prevents this from happening? Ought it happen to any extent at all? Can you suggest an experiment to detect such an effect, if it should exist?

6.22 The main goal of this problem is to find the torque that acts on a current loop in a constant magnetic field. The constant field **B** points in some direction in space. We shall orient our coordinates so that **B** is perpendicular to the x axis, and our current loop lies in the xy plane, as shown in the figure. The shape and size of the loop are arbitrary; we may think of the current as being supplied by twisted leads on which any net force will be zero. Consider some small element of the loop, and work out its contribution to the torque about the x axis. Only the z component of force on it will be involved, and hence only the y component of the field **B**, which we have indicated as $\hat{\mathbf{y}}B_y$ on the diagram. Set up the integral which will give the total torque. Show that this integral will give, except for constant factors, the *area* of the loop. The *magnetic moment* of a current loop is defined as a vector **m** of magnitude Ia/c where I is the current in esu/sec, a is the area of the loop in cm^2, and the direction of the vector is normal to the loop with a right-hand-thread relation to the current, as shown in the figure. (We'll meet the current loop and its magnetic moment

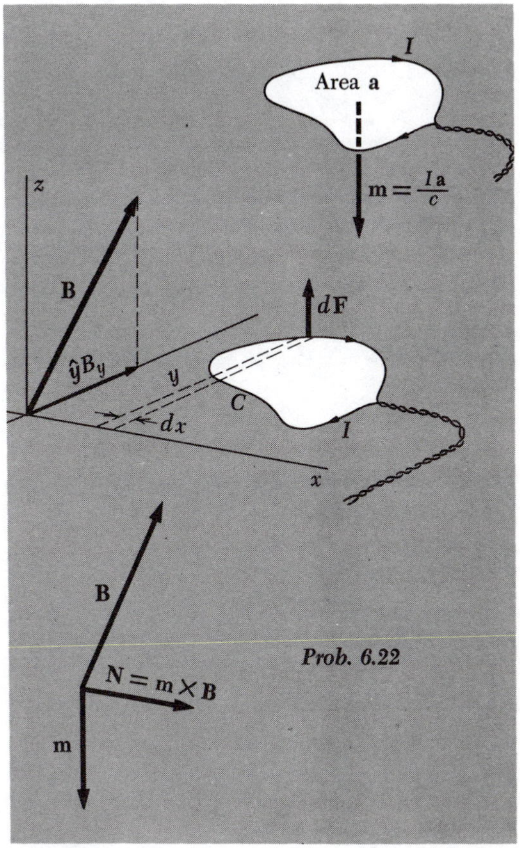

PROBLEM 6.22

again in Chapter 11.) Show now that your result implies that the torque **N** on any current loop is given by the vector equation

$$\mathbf{N} = \mathbf{m} \times \mathbf{B}$$

What about the net *force* on the loop?

6.23 For some purposes it is useful to accelerate negative hydrogen ions in a cyclotron. A negative hydrogen ion, H^-, is a hydrogen atom to which an extra electron has become attached. The attachment is fairly weak; an electric field of only 1.5×10^4 statvolts/cm (a rather small field by atomic standards) will pull an electron loose, leaving a hydrogen atom. If we want to accelerate H^- ions up to a kinetic energy of 1 Gev (10^9 ev), what is the highest magnetic field we dare use to keep them on a circular orbit up to final energy? (To find γ for this problem you only need the rest mass of the H^- ion, which is of course practically the same as that of the proton, approximately 1 Gev.)

6.24 An electron is moving at a speed $0.01c$ on a circular orbit of radius 10^{-8} cm. What is the strength of the resulting magnetic field at the center of the orbit? (The numbers given are typical, in order of magnitude, for an electron in an atom.)

6.25 See if you can devise a vector potential that will correspond to a uniform field in the z direction: $B_x = 0$, $B_y = 0$, $B_z = B_0$.

6.26 A round wire of radius r_0 carries a current I distributed uniformly over the cross section of the wire. Let the axis of the wire be the z axis, with $\hat{\mathbf{z}}$ the direction of the current. Show that a vector potential of the form $\mathbf{A} = constant \times \hat{\mathbf{z}}\,(x^2 + y^2)$ will correctly give the magnetic field **B** of this current at all points inside the wire. What is the value of the constant?

6.27 A particle of charge q and rest mass m is moving with velocity **v** where the magnetic field is **B**. Here **B** is perpendicular to **v**, and there is no electric field. Show that the path of the particle is a curve with radius of curvature R given by $R = pc/qB$, where p is the momentum of the particle, $\beta\gamma mc$. (*Hint:* Note that the force $q\mathbf{v} \times \mathbf{B}/c$ can only change the direction of the particle's momentum, not its magnitude. By what angle $\Delta\theta$ is the direction of **p** changed in a short time Δt?) If **B** is the same everywhere the particle will follow a circular path. Find the time required to complete one revolution.

6.28 A proton with kinetic energy 10^{16} ev ($\gamma = 10^7$) is moving perpendicular to the interstellar magnetic field which in that region of the galaxy has a strength 3×10^{-6} gauss. What is the radius of curvature of its path and how long does it take to complete one revolution? (Use the results for Problem 6.27.)

6.29 A high-energy accelerator produces a beam of protons with kinetic energy 2 Gev (that is, 2×10^9 ev per proton). The current is

1 milliamp. The beam diameter is 2 mm. As measured in the laboratory frame:

(*a*) What is the strength of the electric field caused by the beam 1 cm from the central axis of the beam?

(*b*) What is the strength of the magnetic field at the same distance? Now consider a frame F' which is moving along with the protons. What fields would be measured in F'? For this problem you may assume that the rest energy of a proton is 10^9 ev.

6.30 In the neighborhood of the origin in the coordinate system x, y, z, there is an electric field $\mathbf{E}$ of magnitude 100 statvolts/cm, pointing in a direction that makes angles of 30° with the x axis, 60° with the y axis. The frame F' has its axes parallel to those just described, but is moving, relative to the first frame, with a speed $0.6c$ in the positive y direction. Find the direction and magnitude of the electric field which will be reported by an observer in the frame F'. What magnetic field does this observer report?

6.31 According to observers in the frame F, the following events occurred in the xy plane. A singly charged positive ion which had been moving with the constant velocity $v = 0.6c$ in the $\hat{\mathbf{y}}$ direction passed through the origin at $t = 0$. At the same instant a similar ion which had been moving with the same speed, but in the $-\hat{\mathbf{y}}$ direction, passed the point (2, 0, 0) on the x axis. The distances are in cm.

(*a*) What is the strength and direction of the electric field, at $t = 0$, at the point (3, 0, 0)?

(*b*) What is the strength and direction of the magnetic field at the same place and time?

Ans. $\mathbf{E} = (25/18)e\hat{\mathbf{x}}$; $\mathbf{B} = (2/3)e\hat{\mathbf{z}}$.

6.32 Consider two electrons in a cathode ray tube which are moving on parallel paths, side by side, at the same speed v. The distance between them, a distance measured at right angles to their velocity, is r. What is the force that acts on one of them, owing to the presence of the other, as observed in the laboratory frame? If v were very small compared to c, you could answer e^2/r^2 and let it go at that. But v isn't small, so you have to be careful.

(*a*) The easiest way to get the answer is this: Go to a frame of reference moving with the electrons. In that frame the two electrons are at rest, the distance between them is still r (why?), and the force *is* just e^2/r^2. Now transform the force into the laboratory frame, using the force transformation law, Eq. 14 of Chapter 5. (Be careful about which is the primed system; is the force in the lab frame greater or less than the force in the electron frame?)

(*b*) It should be possible to get the same answer working entirely in the lab frame. In the lab frame, at the instantaneous position of electron 1, there are both electric and magnetic fields arising from electron 2 (see Fig. 6.26). Calculate the net force on electron 1, which

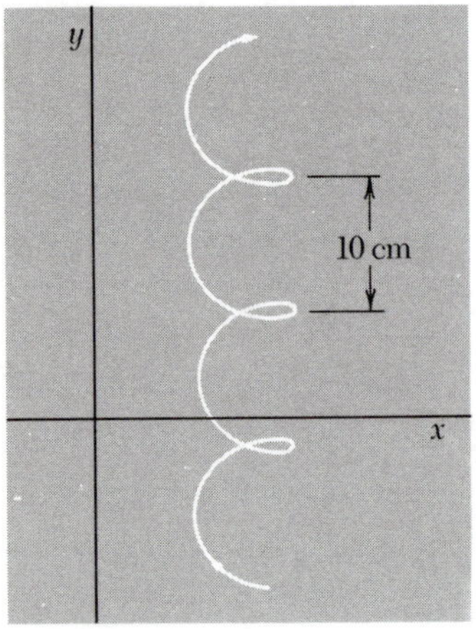

PROBLEM 6.33

is moving through these fields with speed v, and show that you get the same result as in (a). Make a diagram to show the directions of the fields and forces.

(c) In the light of this, what can you say about the force between two side-by-side moving electrons, in the limit $v \to c$?

6.33 The figure shows the path of a positive ion moving in the xy plane. There is a uniform magnetic field of 6000 gauss in the $\hat{z}$ direction. Each period of the ion's cycloidal motion is completed in 1 microsecond. What is the magnitude and the direction of the electric field that must be present? *Hint:* Think about a frame in which the electric field is zero.

Ans. $\mathbf{E} = -2\hat{\mathbf{x}}$ statvolts/cm.

6.34 Calculate approximately the magnetic field to be expected just above the rotating disk in Rowland's experiment. Take the relevant data from the description on the page of his paper that is reproduced in Fig. 6.27. You will need to know also that the potential of the rotating disk, with respect to the grounded plates above and below it, was around 10 kilovolts in most of his runs. This information is of course given later in his paper, as is a description of a crucial part of the apparatus, the "astatic" magnetometer shown in the vertical tube on the left. This is an arrangement in which two magnetic needles, oppositely oriented, are rigidly connected together on one suspension so that the torques caused by the earth's field cancel one another. The field produced by the rotating disk, acting mainly on the nearer needle, can then be detected in the presence of a very much stronger uniform field. That is by no means the only precaution Rowland had to take.

6.35 A Hall probe for measuring magnetic fields is made from arsenic-doped silicon which has 2×10^{15} conduction electrons per cm^3 and a resistivity of 1.6 ohm-cm. The Hall voltage is measured across a ribbon of this n-type silicon which is 0.2 cm wide, 0.005 cm thick, and 0.5 cm long between thicker ends at which it is connected into a 1-volt battery circuit. What voltage will be measured across the 0.2 cm dimension of the ribbon when the probe is inserted into a field of 1 kilogauss?

Ans. 7.8 millivolts.

6.36 Show that the SI version of Eq. 65 must read $\mathbf{E}_t = -\mathbf{J} \times \mathbf{B}/nq$, where E_t is in volts/meter, B in teslas, n in m^{-3}, and q in coulombs.

6.37 Consider two solenoids, one of which is a tenth-scale model of the other. The larger solenoid is 2 meters long, and 1 meter in diameter and is wound with 1-cm-diameter copper wire. When the coil is connected to a 120-volt direct-current generator, the magnetic field at its center is 1000 gauss. The scaled-down model is exactly one-tenth the size in every linear dimension, including the diameter of the wire.

The number of turns is the same, and it is designed to provide the same central field.

(*a*) Show that the voltage required is the same, namely, 120 volts.

(*b*) Compare the coils with respect to the power dissipated and the difficulty of removing this heat by some cooling means.

6.38 This problem concerns the electrically charged interstellar dust grain that was the subject of Problem 2.22. Its mass, which was not involved in that problem, may be taken as 10^{-13} gm. Suppose it is moving quite freely, with speed $v \ll c$, in a plane perpendicular to the interstellar magnetic field which in that region has a strength of 3×10^{-6} gauss. How many years will it take to complete a circular orbit?

7

ELECTROMAGNETIC INDUCTION

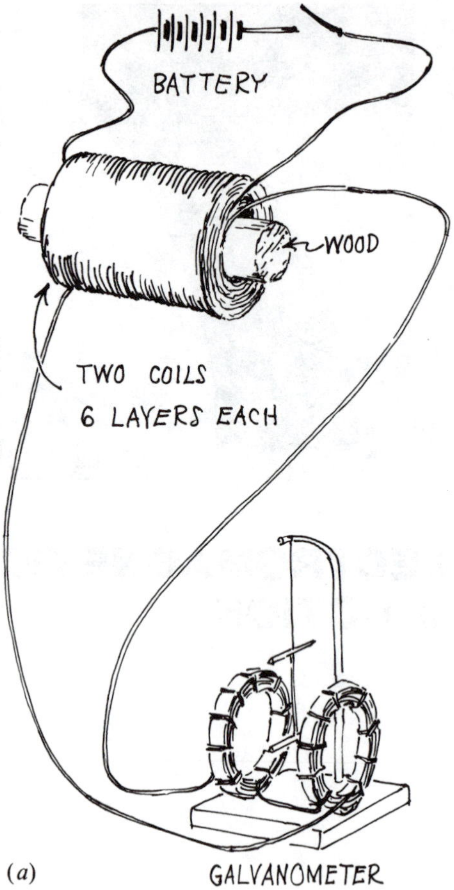

(*a*)

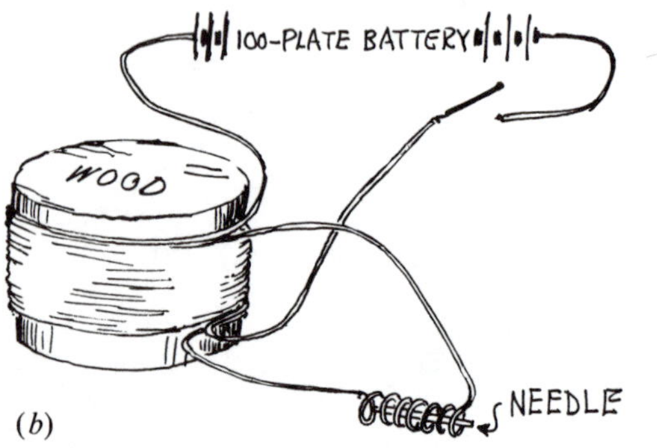

(*b*)

FIGURE 7.1
(*a–e*) Interpretation by the author of some of Faraday's experiments described in his "Experimental Researches in Electricity," London, 1839.

FARADAY'S DISCOVERY 7.1

1. The power which electricity of tension possesses of causing an opposite electrical state in its vicinity has been expressed by the general term Induction; which, as it has been received into scientific language, may also, with propriety, be used in the same general sense to express the power which electrical currents may possess of inducing any particular state upon matter in their immediate neighbourhood, otherwise indifferent. It is with this meaning that I purpose using it in the present paper.

2. Certain effects of the induction of electrical currents have already been recognised and described: as those of magnetization; Ampère's experiments of bringing a copper disc near to a flat spiral; his repetition with electromagnets of Arago's extraordinary experiments, and perhaps a few others. Still it appeared unlikely that these could be all the effects which induction by currents could produce; especially as, upon dispensing with iron, almost the whole of them disappear, whilst yet an infinity of bodies, exhibiting definite phenomena of induction with electricity of tension, still remain to be acted upon by the induction of electricity in motion.

3. Further: Whether Ampère's beautiful theory were adopted, or any other, or whatever reservation were mentally made, still it appeared very extraordinary, that as every electric current was accompanied by a corresponding intensity of magnetic action at right angles to the current, good conductors of electricity, when placed within the sphere of this action, should not have any current induced through them, or some sensible effect produced equivalent in force to such a current.

4. These considerations, with their consequence, the hope of obtaining electricity from ordinary magnetism, have stimulated me at various times to investigate experimentally the inductive effect of electric currents. I lately arrived at positive results; and not only had my hopes fulfilled, but obtained a key which appeared to me to open out a full explanation of Arago's magnetic phenomena, and also to discover a new state, which may probably have great influence in some of the most important effects of electric currents.

5. These results I purpose describing, not as they were obtained, but in such a manner as to give the most concise view of the whole.

So begins Michael Faraday's account of the discovery of electromagnetic induction. This passage was part of a paper Faraday presented in 1831. It is quoted from his "Experimental Researches in

Electricity," published in London in 1839. There follows in the paper a description of a dozen or more experiments, through which Faraday brought to light every essential feature of the production of electric effects by magnetic action.

By "electricity of tension" Faraday meant electrostatic charges, and the induction he refers to in the first sentence involves nothing more than we have studied in Chapter 3: The presence of a charge causes a redistribution of charges on conductors nearby. Faraday's question was, why does not an electric current cause another current in nearby conductors?

The production of magnetic fields by electric currents had been thoroughly investigated after Oersted's discovery. The familiar laboratory source of these "galvanic" currents was the voltaic battery. The most sensitive detector of such currents was a galvanometer. It consisted of a magnetized needle pivoted like a compass needle or suspended by a weak fiber between two coils of wire. Sometimes another needle, outside the coil but connected rigidly to the first needle, was used to compensate the influence of the earth's magnetic field (Fig. 7.1*a*). The sketches in Fig. 7.1*b* through *e* represent a few of Faraday's induction experiments. You must read his own account, one of the classics of experimental science, to appreciate the resourcefulness with which he pressed the search, the alert and open mind with which he viewed the evidence.

In his early experiments Faraday was puzzled to find that a steady current had no detectable effect on a nearby circuit. He constructed various coils of wire, of which Fig. 7.1*a* shows an example, winding two conductors so that they should lie very close together while still separated by cloth or paper insulation. One conductor would form a circuit with the galvanometer. Through the other he would send a strong current from a battery. There was, disappointingly, no deflection of the galvanometer. But in one of these experiments he noticed a very slight disturbance of the galvanometer when the current was switched on and another when it was switched off. Pursuing this lead, he soon established beyond doubt that currents in other conductors are induced, not by a *steady* current, but by a *changing* current. One of Faraday's brilliant experimental tactics at this stage was to replace his galvanometer, which he realized was not a good detector for a brief pulse of current, by a simple small coil in which he put an unmagnetized steel needle (Fig. 7.1*b*). He found that the needle was left magnetized by the pulse of current induced when the primary current was switched on—and it could be magnetized in the opposite sense by the current pulse induced when the primary circuit was broken.

Here is his own description of another experiment:

In the preceding experiments the wires were placed near to each

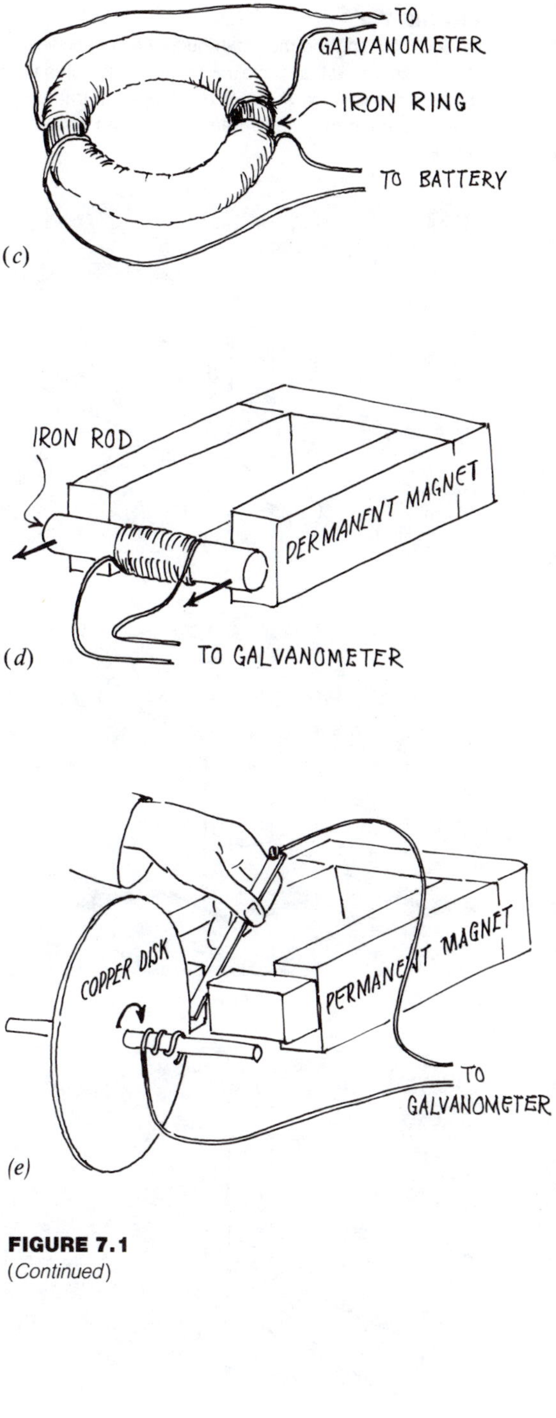

FIGURE 7.1
(*Continued*)

FIGURE 7.2

(*a*) A conducting rod moves through a magnetic field. (*b*) Any charge q that travels with the rod is acted upon by the force $(q/c)\,\mathbf{v} \times \mathbf{B}$. (*c*) The reference frame F' moves with the rod; in this frame there is an electric field $\mathbf{E}'$.

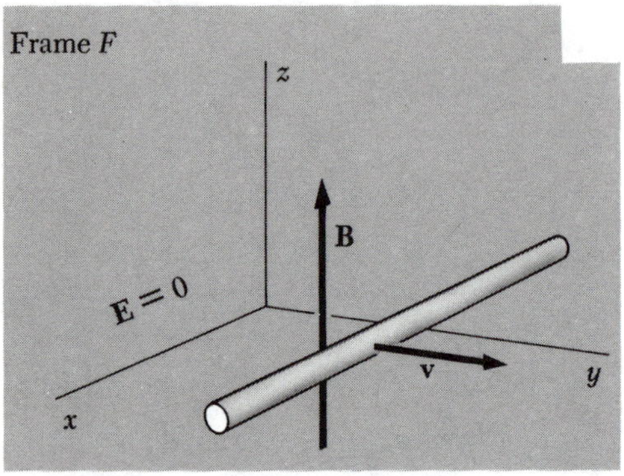

(*a*)

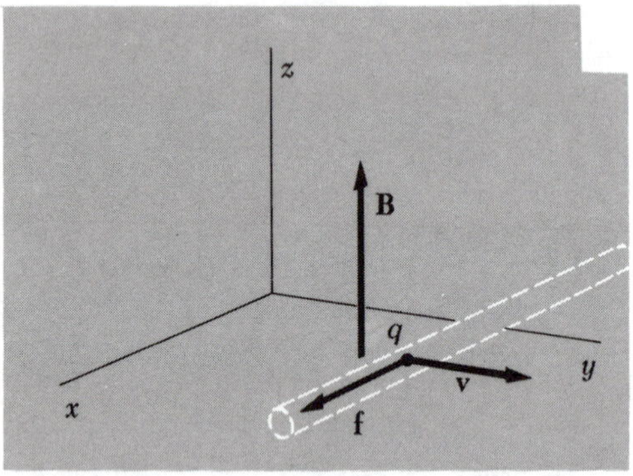

(*b*)

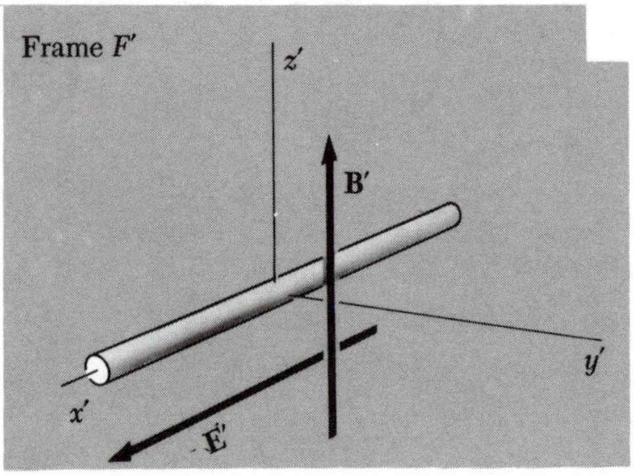

(*c*)

other, and the contact of the inducing one with the battery made when the inductive effect was required; but as the particular action might be supposed to be exerted only at the moments of making and breaking contact, the induction was produced in another way. Several feet of copper wire were stretched in wide zigzag forms, representing the letter W, on one surface of a broad board; a second wire was stretched in precisely similar forms on a second board, so that when brought near the first, the wires should everywhere touch, except that a sheet of thick paper was interposed. One of these wires was connected with the galvanometer, and the other with a voltaic battery. The first wire was then moved towards the second, and as it approached, the needle was deflected. Being then removed, the needle was deflected in the opposite direction. By first making the wires approach and then recede, simultaneously with the vibrations of the needle, the latter soon became very extensive; but when the wires ceased to move from or towards each other, the galvanometer needle soon come to its usual position.

As the wires approximated, the induced current was in the *contrary* direction to the inducing current. As the wires receded, the induced current was in the *same* direction as the inducing current. When the wires remained stationary, there was no induced current.

In this chapter we study the electromagnetic interaction that Faraday explored in those experiments. From our present viewpoint, induction can be seen as a natural consequence of the force on a charge moving in a magnetic field. In a limited sense, we can derive the induction law from what we already know. In following this course we again depart from the historical order of development, but we do so (borrowing Faraday's own words from the end of the passage first quoted) "to give the most concise view of the whole."

A CONDUCTING ROD MOVING THROUGH A UNIFORM MAGNETIC FIELD

7.2 Figure 7.2*a* shows a straight piece of wire, or slender metal rod, supposed to be moving at constant velocity **v** in a direction perpendicular to its length. Pervading the space through which the rod moves there is a uniform magnetic field **B**, constant in time. This could be supplied by a large solenoid enclosing the entire region of the diagram. The reference frame F with coordinates x, y, z is the one in which this solenoid is at rest. In the absence of the rod there is no electric field in that frame, only the uniform magnetic field **B**.

The rod, being a conductor, contains charged particles that will move if a force is applied to them. Any charged particle that is carried along with the rod, such as the particle of charge q in Fig. 7.2*b*, necessarily moves through the magnetic field $\mathbf{B}$ and does therefore experience a force

$$\mathbf{f} = \frac{q}{c}\mathbf{v} \times \mathbf{B} \tag{1}$$

With $\mathbf{B}$ and $\mathbf{v}$ directed as shown in Fig. 7.2, the force is in the positive x direction if q is a positive charge, and in the opposite direction for the negatively charged electrons that are in fact the mobile charge carriers in most conductors. The consequences will be the same, whether negatives or positives, or both, are mobile.

When the rod is moving at constant speed and things have settled down to a steady state, the force $\mathbf{f}$ given by Eq. 1 must be balanced, at every point inside the rod, by an equal and opposite force. This can only arise from an electric field in the rod. The electric field develops in this way: the force $\mathbf{f}$ pushes negative charges toward one end of the rod, leaving the other end positively charged. This goes on until these separated charges themselves cause an electric field $\mathbf{E}$ such that, everywhere in the interior of the rod,

$$q\mathbf{E} = -\mathbf{f} \tag{2}$$

Then the motion of charge relative to the rod ceases. This charge distribution causes an electric field outside the rod, as well as inside. The field outside looks something like that of separated positive and negative charges, with the difference that the charges are not concentrated entirely at the ends of the rod but are distributed along it. The external field is sketched in Fig. 7.3*a*. Figure 7.3*b* is an enlarged view of the positively charged end of the rod, showing the charge distribution on the surface and some field lines both outside and inside the conductor. That is the way things look, at any instant of time, in frame F.

Let us observe this system from a frame F' that moves with the rod. Ignoring the rod for the moment, we see in this frame F', indicated in Fig. 7.2*c*, a magnetic field $\mathbf{B}'$ (not much different from $\mathbf{B}$ if v is small) together with a uniform electric field, as given by Eq. 6.63,

$$\mathbf{E}' = -\frac{\mathbf{v}'}{c} \times \mathbf{B}' = \frac{\mathbf{v}}{c} \times \mathbf{B}' \tag{3}$$

When we add the rod to this system, all we are doing is putting a stationary conducting rod into a uniform electric field. There will be a redistribution of charge on the surface of the rod so as to make the electric field zero inside, as in the case of the metal box of Fig. 3.6, or of any other conductor in an electric field. The presence of the magnetic field $\mathbf{B}'$ has no influence on this static charge distribution. Figure

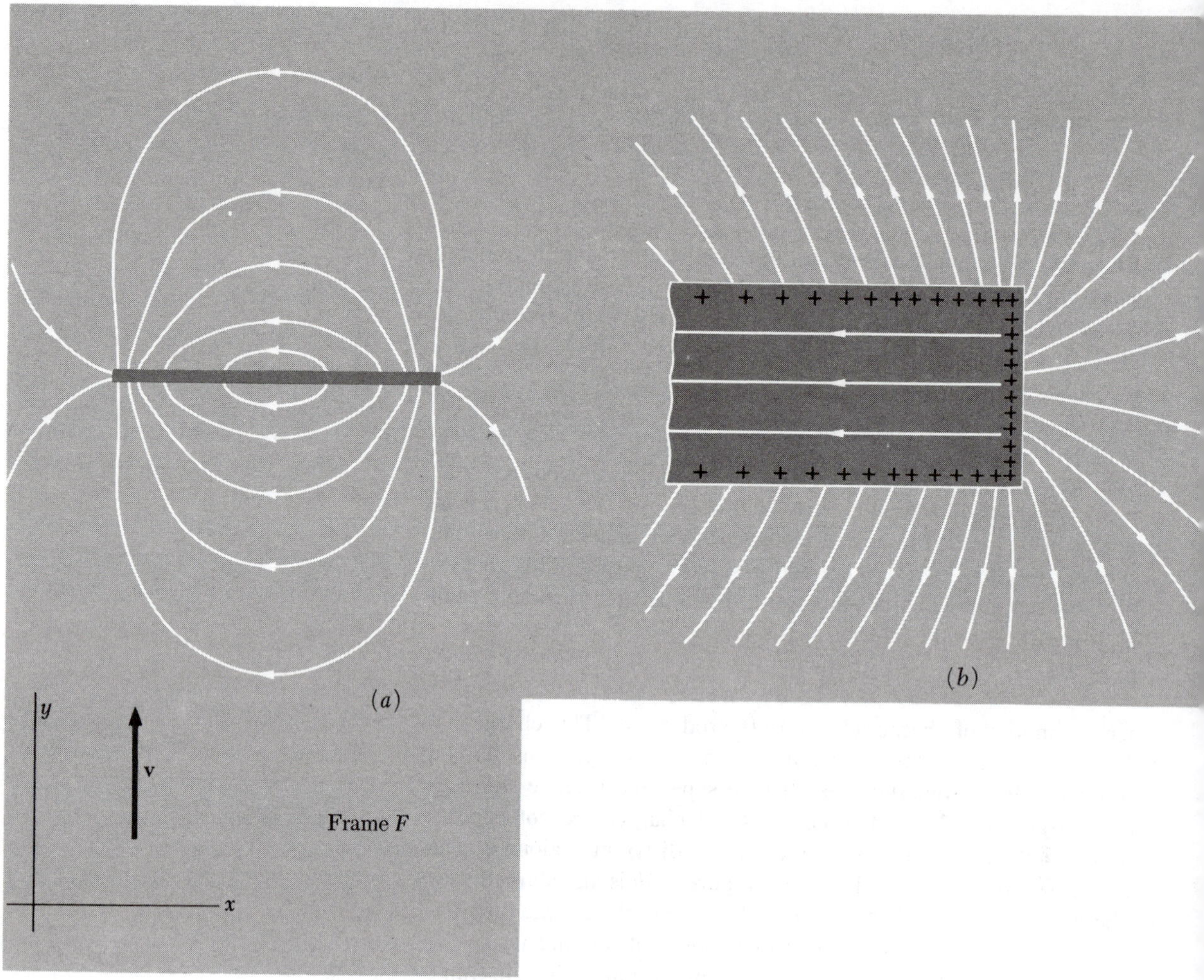

FIGURE 7.3

(*a*) The electric field, as seen at one instant of time, in the frame *F*. There is an electric field in the vicinity of the rod, and also inside the rod. The sources of the field are charges on the surface of the rod, as shown in (*b*), the enlarged view of the right-hand end of the rod.

7.4*a* shows some electric field lines in the frame F', and in the magnified view of the end of the rod in Fig. 7.4*b* we observe that the electric field *inside* the rod is zero.

Except for the Lorentz contraction, which is second order in v/c, the charge distribution seen at one instant in frame F, Fig. 7.3*b*, is the same as that seen in F'. The electric fields differ because the field in Fig. 7.3 is that of the surface charge distribution alone, while the electric field we see in Fig. 7.4 is the field of the surface charge distribution *plus* the uniform electric field that exists in that frame of reference. An observer in F says: "Inside the rod there has developed an electric field $\mathbf{E} = (\mathbf{v}/c) \times \mathbf{B}$, exerting a force $q\mathbf{E} = -q(\mathbf{v}/c) \times \mathbf{B}$ which just balances the force $q(\mathbf{v}/c) \times \mathbf{B}$ that would otherwise cause any charge q to move along the rod." An observer in F' says:

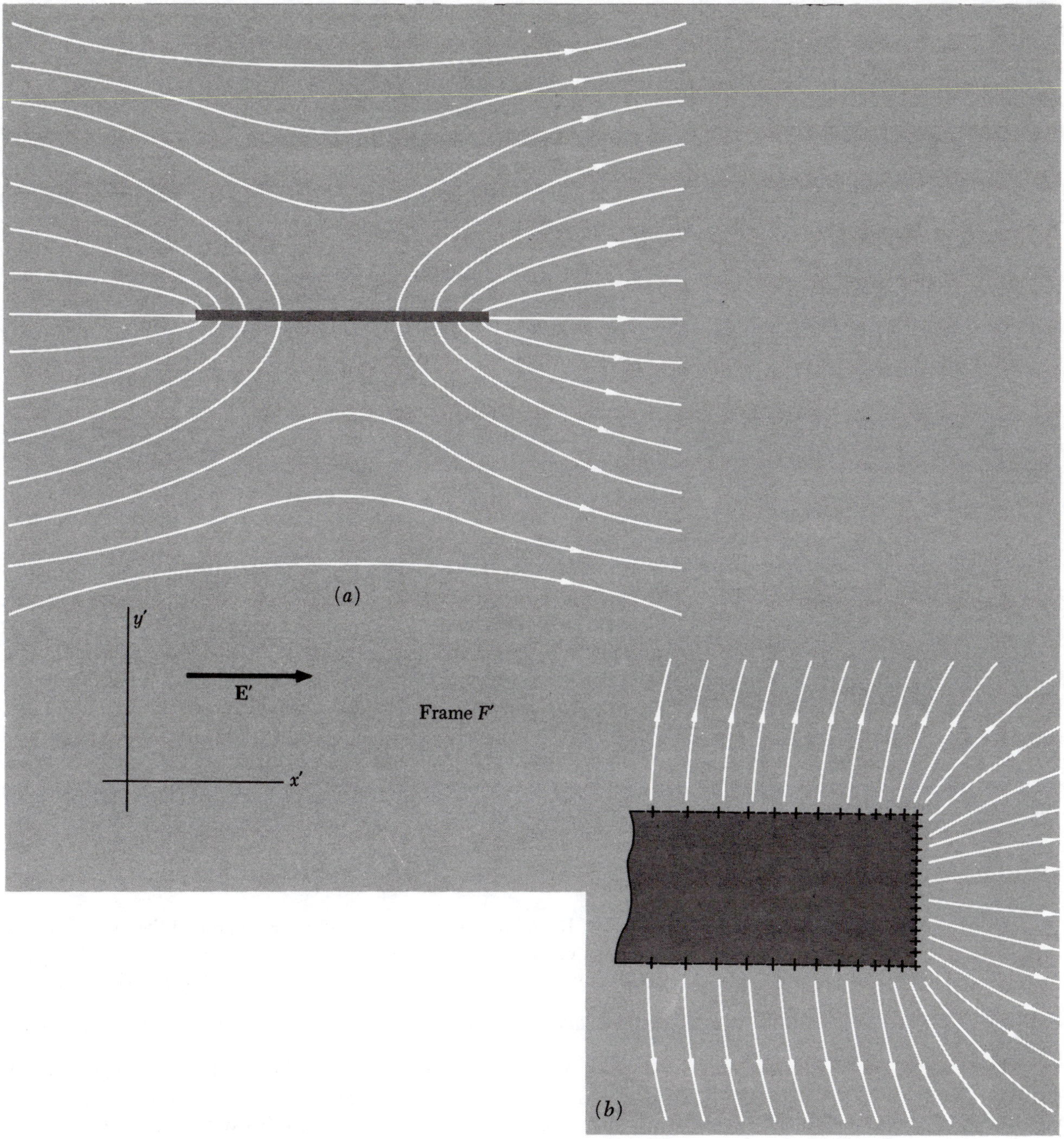

FIGURE 7.4
(a) The electric field in the frame F' in which the rod is at rest. This field is a superposition of a general field $\mathbf{E}'$, uniform throughout space, and the field of the surface charge distribution. The result is zero electric field inside the rod, shown in magnified detail in (b). Compare with Fig. 7.3.

"Inside the rod there is no electric field, and although there is a uniform magnetic field here, no force arises from it because no charges are moving." Each account is correct.

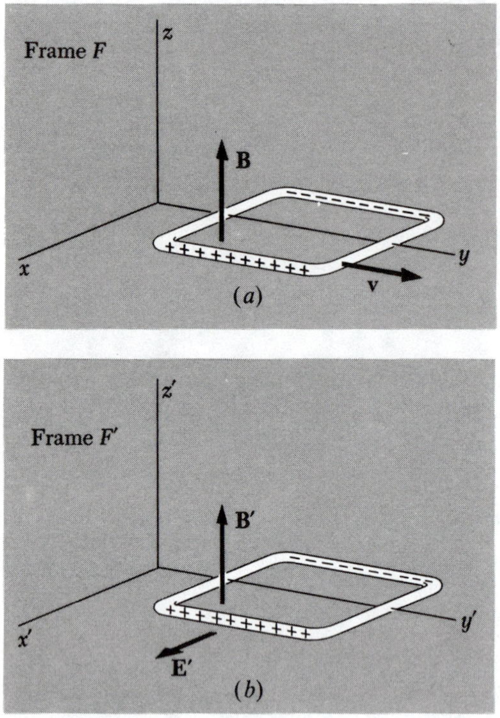

FIGURE 7.5
(*a*) Here the wire loop is moving in a uniform magnetic field **B**. (*b*) Observed in the frame F', in which the loop is at rest, the fields are $\mathbf{B}'$ and $\mathbf{E}'$.

A LOOP MOVING THROUGH A NONUNIFORM MAGNETIC FIELD

7.3 What if we made a rectangular loop of wire, as shown in Fig. 7.5, and moved it at constant speed through the uniform field **B**? To predict what will happen, we need only ask ourselves—adopting the frame F'—what would happen if we put such a loop into a uniform electric field. Obviously two opposite sides of the rectangle would acquire some charge, but that would be all. Suppose, however, that the field **B** in the frame F, though constant in time, is *not uniform* in space. To make this vivid, we show in Fig. 7.6 the field **B** with a short solenoid as its source. This solenoid, together with the battery that supplies its constant current, is fixed near the origin in the frame F. (We said earlier there is no electric field in F; if we really use a solenoid of finite resistance to provide the field, there will be an electric field associated with the battery and this circuit. It is irrelevant to our problem and can be ignored. Or we can pack the whole solenoid, with its battery, inside a metal box, making sure the total charge is zero.)

Now, with the loop moving with speed v in the y direction, in the frame F, let its position at some instant t be such that the magnetic field strength is B_1 at the left side of the loop and B_2 along the right side (Fig. 7.6). Let **f** denote the force which acts on a charge q that rides along with the loop. This force is a function of position on the loop, at this instant of time. Let's evaluate the line integral of **f**, taken around the whole loop: On the two sides of the loop which lie parallel to the direction of motion, **f** is perpendicular to the path element $d\mathbf{s}$, so these give nothing. Taking account of the contributions from the other two sides, each of length w, we have

$$\int \mathbf{f} \cdot d\mathbf{s} = \frac{qv}{c}(B_1 - B_2)w \tag{4}$$

If we imagine a charge q to move all around the loop, in a time short enough so that the position of the loop has not changed appreciably, then Eq. 4 gives the work done by the force **f**. The work done *per unit charge* is $(1/q)\int \mathbf{f} \cdot d\mathbf{s}$. We call this quantity *electromotive force*. We use the symbol $\mathcal{E}$ for it, and often shorten the name to emf. $\mathcal{E}$ has the same dimensions as electric potential. It is measured in statvolts, or ergs per unit charge, in the CGS system. The SI unit is the volt.

$$\mathcal{E} = \frac{1}{q}\int \mathbf{f} \cdot d\mathbf{s} \tag{5}$$

The term *electromotive force* was introduced earlier, in Section 4.10. It was defined as the work per unit charge involved in moving a charge around a circuit containing a voltaic cell. We now broaden the defi-

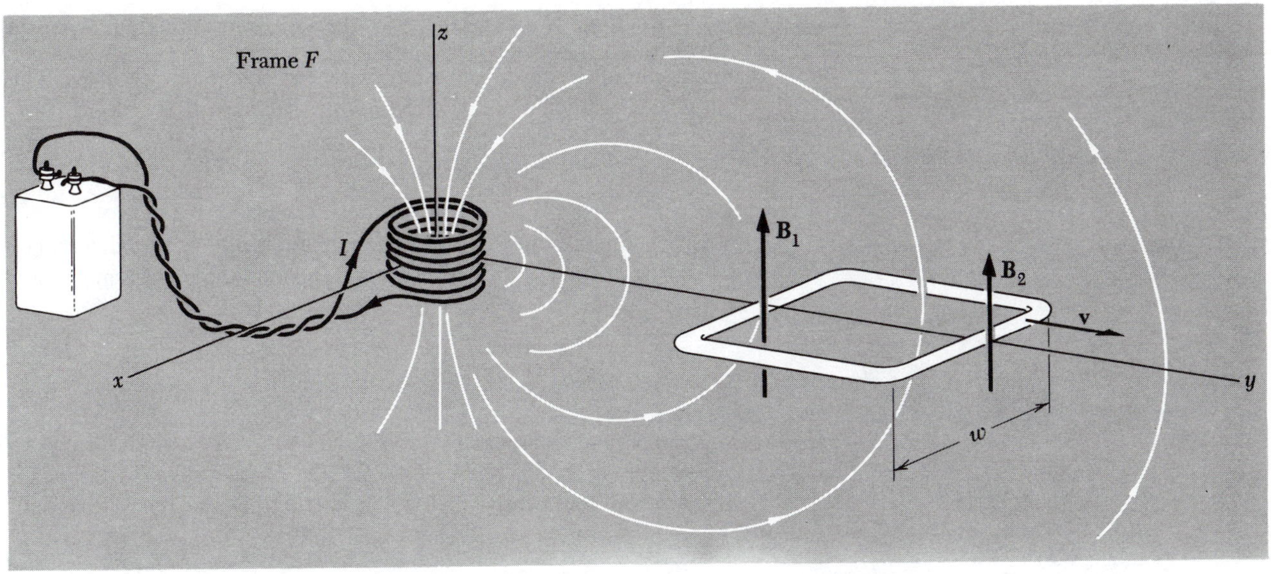

FIGURE 7.6
Here the field **B**, observed in *F*, is not uniform. It varies in both direction and magnitude from place to place.

nition of emf to include any influence that causes charge to circulate around a closed path. If the path happens to be a physical circuit with resistance *R*, then the emf $\mathcal{E}$ will cause a current to flow according to Ohm's law: $I = \mathcal{E}/R$. In the particular case we are considering, **f** is the force that acts on a charge moving in a magnetic field, and $\mathcal{E}$ has the magnitude

$$\mathcal{E} = \frac{vw}{c}(B_1 - B_2) \tag{6}$$

The electromotive force given by Eq. 6 is related in a very simple way to the *rate of change of magnetic flux* through the loop. By the magnetic flux through a loop we mean the surface integral of **B** over a surface which has the loop for its boundary. The flux Φ through the closed curve or loop *C* in Fig. 7.7*a* is given by the surface integral of **B** over S_1:

$$\Phi_{S_1} = \int_{S_1} \mathbf{B} \cdot d\mathbf{a}_1 \tag{7}$$

We could draw infinitely many surfaces bounded by *C*. Figure 7.7*b* shows another one, S_2. Why don't we have to specify which surface to use in computing the flux? It *doesn't make any difference* because $\int \mathbf{B} \cdot d\mathbf{a}$ will have the same value for all surfaces. Let's take a minute to settle this point once and for all. The flux through S_2 will

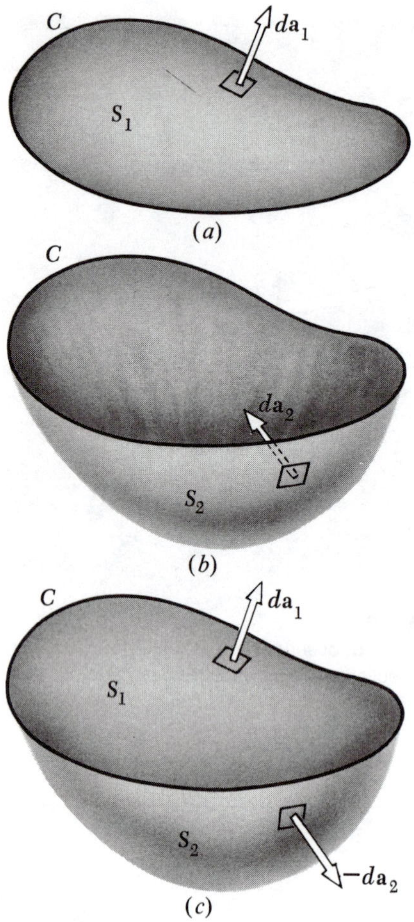

FIGURE 7.7
(*a*) The flux through *C* is
$\Phi = \int_{S1} \mathbf{B} \cdot d\mathbf{a}_1$
(*b*) S_2 is another surface which has *C* as its boundary. This will do just as well for computing Φ.
(*c*) Combining S_1 and S_2 to make a closed surface, for which $\int \mathbf{B} \cdot d\mathbf{a}$ must vanish, proves that $\int_{S1} \mathbf{B} \cdot d\mathbf{a}_1 = \int_{S2} \mathbf{B} \cdot d\mathbf{a}_2$.

be $\int_{S2} \mathbf{B} \cdot d\mathbf{a}_2$. Notice that we let the vector $d\mathbf{a}_2$ stick out from the upper side of S_2, to be consistent with our choice of side of S_1. This will give a positive number if the net flux through C is upward.

$$\Phi_{S_2} = \int_{S_2} \mathbf{B} \cdot d\mathbf{a}_2 \tag{8}$$

We learned in Section 6.2 that the magnetic field has zero divergence: div $\mathbf{B} = 0$. It follows then from Gauss' theorem that, if S is any *closed* surface ("balloon") and V is the volume inside it:

$$\int_S \mathbf{B} \cdot d\mathbf{a} = \int_V \text{div}\, \mathbf{B}\, dv = 0 \tag{9}$$

Apply this to the closed surface, rather like a kettledrum, formed by joining our S_1 to S_2, as in Fig. 7.7*c*. On S_2 the outward normal is *opposite* the vector $d\mathbf{a}_2$ we used in calculating the flux through C. Thus

$$0 = \int_S \mathbf{B} \cdot d\mathbf{a} = \int \mathbf{B} \cdot d\mathbf{a}_1 + \int \mathbf{B} \cdot (-d\mathbf{a}_2)$$

or

$$\int_{S_1} \mathbf{B} \cdot d\mathbf{a}_1 = \int_{S_2} \mathbf{B} \cdot d\mathbf{a}_2 \tag{10}$$

This shows that it doesn't matter which surface we use to compute the flux through C.

This is all pretty obvious if you realize that div $\mathbf{B} = 0$ implies a kind of spatial conservation of flux. As much flux enters any volume as leaves it. (We are considering the situation in the whole space at one instant of time.) It is often helpful to visualize "tubes" of flux. A flux tube (Fig. 7.8) is a surface at every point on which the magnetic field line lies in the plane of the surface. It is a surface through which no flux passes, and we can think of it as containing a certain amount of flux, as a telephone cable contains wires. Through any closed curve drawn tightly around a flux tube, the same flux passes. This could be said about the electric field $\mathbf{E}$ only for regions where there is no electric charge, since div $\mathbf{E} = 4\pi\rho$. The magnetic field always has zero divergence everywhere.

Returning now to the moving rectangular loop, let us find the *rate of change* of flux through the loop. In time dt the loop moves a distance $v\, dt$. This changes in two ways the total flux through the loop, which is $\int \mathbf{B} \cdot d\mathbf{a}$ over a surface spanning the loop. As you can see in Fig. 7.9, flux is gained at the right, in amount $B_2 wv\, dt$, while an

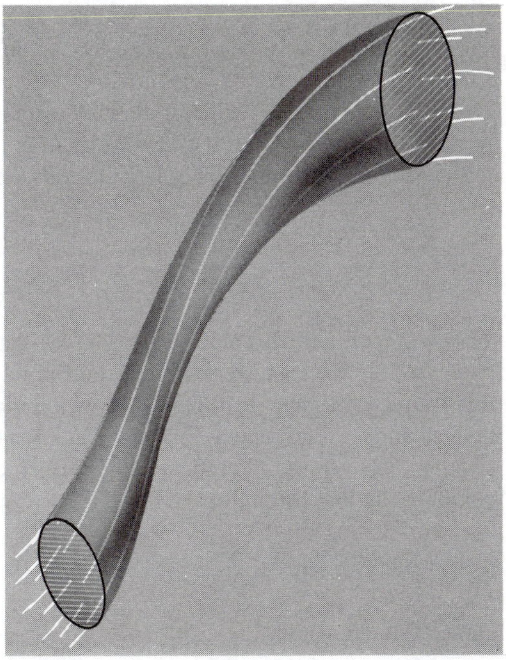

FIGURE 7.8
A flux tube. Magnetic field lines lie in the surface of the tube. The tube encloses a certain amount of flux Φ. No matter where you chop it, you will find that $\int \mathbf{B} \cdot d\mathbf{a}$ over the section has this same value Φ. A flux tube doesn't have to be round. You can start somewhere with any cross section, and the course of the field lines will determine how the section changes size and shape as you go along the tube.

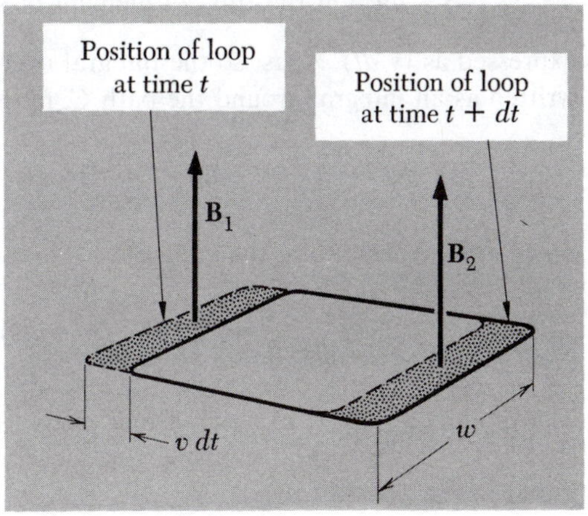

FIGURE 7.9
In the interval dt the loop gains an increment of flux $B_2wv\,dt$ and loses an increment $B_1wv\,dt$.

amount of flux $B_1wv\,dt$ is lost at the left. Hence $d\Phi$, the change in flux through the loop in time dt, is

$$d\Phi = -(B_1 - B_2)wv\,dt \tag{11}$$

Comparing Eq. 11 with Eq. 6, we see that, in this case at least, the electromotive force can be expressed as

$$\mathcal{E} = -\frac{1}{c}\frac{d\Phi}{dt} \tag{12}$$

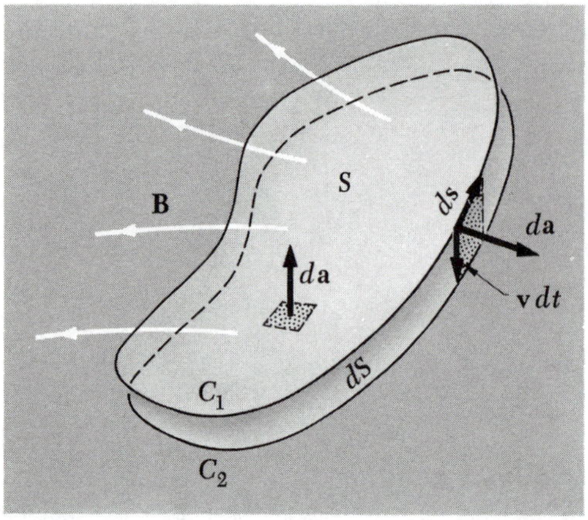

FIGURE 7.10
The loop moves from position C_1 to position C_2 in time *dt*.

We can show that this holds quite generally, for a loop of any shape moving in any manner. The loop C in Fig. 7.10 occupies the position C_1 at time t, and it is moving so that it occupies the position C_2 at time $t + dt$. A particular element of the loop $d\mathbf{s}$ has been transported with velocity $\mathbf{v}$ to its new position. S indicates a surface that spans the loop at time t. The flux through the loop at this instant of time is

$$\Phi(t) = \int_S \mathbf{B} \cdot d\mathbf{a} \tag{13}$$

The magnetic field $\mathbf{B}$ comes from sources that are stationary in our frame of reference and remains constant in time, at any point fixed in this frame. At time $t + dt$ a surface which spans the loop is the original surface S, left fixed in space, augmented by the "rim" dS. (Remember, we are allowed to use *any* surface spanning the loop to compute the flux through it.) Thus

$$\Phi(t + dt) = \int_{S+dS} \mathbf{B} \cdot d\mathbf{a} = \Phi(t) + \int_{dS} \mathbf{B} \cdot d\mathbf{a} \tag{14}$$

Hence the change in flux, in time dt, is just the flux through the rim dS, $\int_{dS} \mathbf{B} \cdot d\mathbf{a}$. On the rim, an element of surface area $d\mathbf{a}$ can be expressed as $(\mathbf{v}\, dt) \times d\mathbf{s}$, so the integral over the surface dS can be written as an integral around the path C, in this way:

$$d\Phi = \int_{dS} \mathbf{B} \cdot d\mathbf{a} = \int_C \mathbf{B} \cdot [(\mathbf{v}\, dt) \times d\mathbf{s}] \tag{15}$$

Since dt is a constant for the integration, we can factor it out and have

$$\frac{d\Phi}{dt} = \int_C \mathbf{B} \cdot (\mathbf{v} \times d\mathbf{s}) \tag{16}$$

The product $\mathbf{a} \cdot (\mathbf{b} \times \mathbf{c})$ of any three vectors satisfies the relation $\mathbf{a} \cdot (\mathbf{b} \times \mathbf{c}) = -(\mathbf{b} \times \mathbf{a}) \cdot \mathbf{c}$. Using this identity to rearrange the integrand in Eq. 16, we have

$$\frac{d\Phi}{dt} = -\int_C (\mathbf{v} \times \mathbf{B}) \cdot d\mathbf{s} \tag{17}$$

Now the force on a charge q which is carried along by the loop is just $q(\mathbf{v} \times \mathbf{B})/c$, so the electromotive force, which is the line integral around the loop of the force per unit charge, is just

$$\mathcal{E} = \frac{1}{c} \int_C (\mathbf{v} \times \mathbf{B}) \cdot d\mathbf{s} \tag{18}$$

Comparing Eq. 17 with Eq. 18 we get the simple relation already given in Eq. 12, but valid now for arbitrary shape and motion of the loop. (We did not even have to assume that **v** is the same for all parts of the loop!) In summary, the line integral around a moving loop of $\mathbf{f}/q$, the force per unit charge, is just $-1/c$ times the rate of change of flux through the loop.

The sense of the line integral and the direction in which flux is called positive are to be related by a right-hand-thread rule. For instance, in Fig. 7.6, the flux is *upward* through the loop and is *decreasing*. Taking the minus sign in Eq. 12 into account, our rule would predict an electromotive force which would tend to drive a positive charge around the loop in a counterclockwise direction, as seen looking down on the loop (Fig. 7.11).

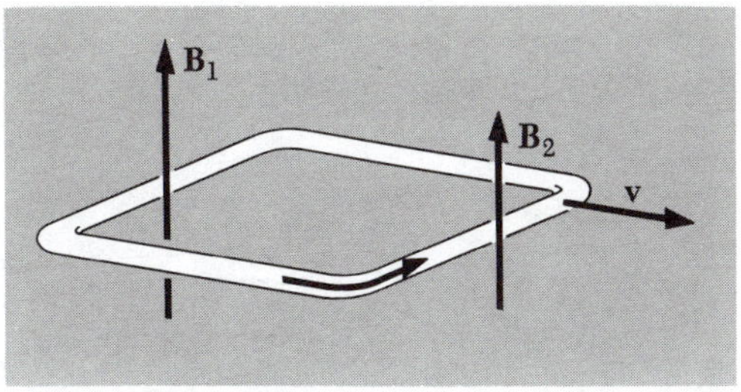
B1
B2
v

FIGURE 7.11
The flux through the loop is upward and is decreasing in magnitude as time goes on. The arrow shows the direction of the electromotive force, that is, the direction in which positive charge tends to be driven.

There is a better way to look at this question of sign and direction. Notice that if a current should flow in the direction of the induced electromotive force, in the situation shown in Fig. 7.11, this current itself would create some flux through the loop in a direction to *counteract* the assumed flux change. That is an essential physical fact, and not the consequence of an arbitrary convention about signs and directions. It is a manifestation of the tendency of systems to resist change. In this context it is traditionally called *Lenz's law.*

Another example of Lenz's law is illustrated in Fig. 7.12. The conducting ring is falling in the magnetic field of the coil. The flux through the ring is *downward* and is *increasing* in magnitude. To counteract this change, some new flux upward is needed. It would take a current flowing around the ring in the direction of the arrows to produce such flux. Lenz's law assures us that the induced emf will be in the right direction to cause such a current.

FIGURE 7.12
As the ring falls, the downward flux through the ring is increasing. Lenz's law tells us that the induced emf will be in the direction indicated by the arrows, for that is the direction in which current must flow to produce upward flux through the ring. The system reacts so as to oppose the change that is occurring.

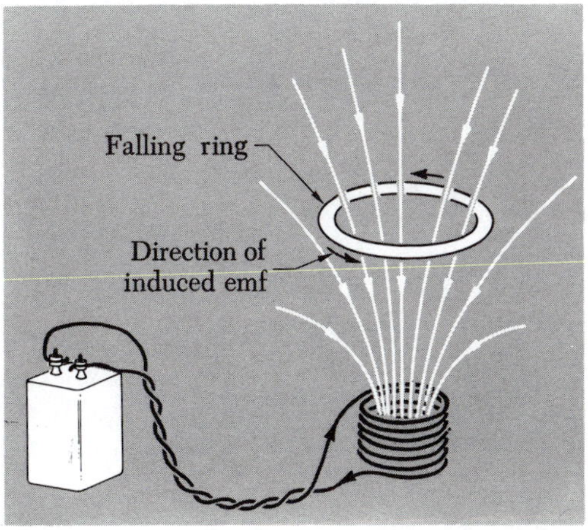
Falling ring
Direction of induced emf

If the electromotive force causes current to flow in the loop which is shown in Figs. 7.6 and 7.11, as it will if the loop has a finite resistance, some energy will be dissipated in the wire. What supplies this energy? To answer that, consider the force that acts on the current in the loop if it flows in the sense indicated by the arrow in Fig. 7.11. The conductor on the right, in the field B_2, will experience a force toward the right, while the opposite side of the loop, in the field B_1, will be pushed toward the left. But B_1 is greater than B_2, so the net force on the loop is toward the left, *opposing the motion.* To keep the loop moving at constant speed some external agency has to do work, and the energy thus invested eventually shows up as heat in the wire. Imagine what would happen if Lenz's law were violated, or if the force on the loop were to act in a direction to assist the motion of the loop!

A very common element in electrical machinery and electrical instruments is a loop or coil that rotates in a magnetic field. Let's apply what we have just learned to the system shown in Fig. 7.13, a single loop rotating at constant speed in a magnetic field that is approximately uniform. The mechanical essentials, shaft, bearings, drive, etc., are not drawn. The field **B** is provided by the two fixed coils.

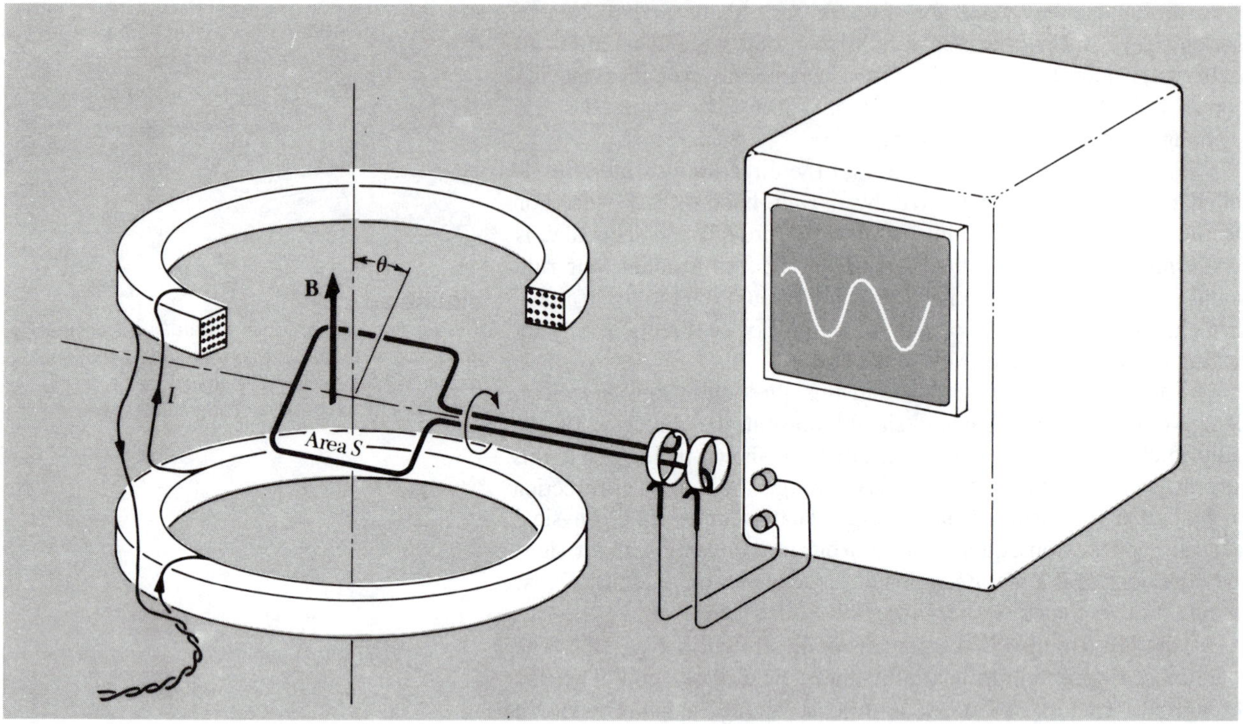

FIGURE 7.13
The two coils produce a magnetic field **B** which is approximately uniform in the vicinity of the loop. In the loop, rotating with angular velocity ω, a sinusoidally varying electromotive force is induced.

Suppose the loop rotates with angular velocity ω, in radians/sec. If its position at any instant is specified by the angle θ, then $\theta = \omega t + \alpha$, where the constant α is simply the position of the loop at $t = 0$. The component of **B** perpendicular to the plane of the loop is $B \sin \theta$. Therefore the flux through the loop at time t is

$$\Phi(t) = SB \sin(\omega t + \alpha) \tag{19}$$

where S is the area of the loop. For the induced electromotive force we then have

$$\mathcal{E} = -\frac{1}{c}\frac{d\Phi}{dt} = -\frac{SB\omega}{c}\cos(\omega t + \alpha) \tag{20}$$

If the loop instead of being closed is connected through slip rings to external wires, as shown in Fig. 7.13, we can detect at these terminals a sinusoidally alternating potential difference.

A numerical example will show how the units work out. Suppose the area of the loop in Fig. 7.13 is 80 cm^2, the field strength B is 50 gauss, and the loop is rotating at 30 revolutions per sec. Then $\omega = 2\pi$

$\times$ 30, or 188 radians/sec. The amplitude, that is, the maximum magnitude of the oscillating electromotive force induced in the loop, is

$$\mathcal{E}_0 = \frac{SB\omega}{c} = \frac{(80\ \text{cm}^2)(50\ \text{gauss})(188\ \text{sec}^{-1})}{3 \times 10^{10}\ \text{cm/sec}} \tag{21}$$
$$= 2.51 \times 10^{-5}\ \text{gauss-cm or statvolt}$$

One gauss-cm is equivalent to 1 statvolt. Remember that electric field E and magnetic field B have the same dimensions in our CGS system, being related by a dimensionless factor v/c.

A STATIONARY LOOP WITH THE FIELD SOURCE MOVING

7.4 We can, if we like, look at the events depicted in Fig. 7.6 from a frame of reference that is moving with the loop. That can't change the physics, only the words we use to describe it. Let F', with coordinates x', y', z', be the frame attached to the loop, which we now regard as stationary (Fig. 7.14). The coil and battery, stationary in frame F, are moving in the $-y'$ direction with velocity $\mathbf{v}' = -\mathbf{v}$. Let B_1' and B_2' be the magnetic field measured at the two ends of the loop by

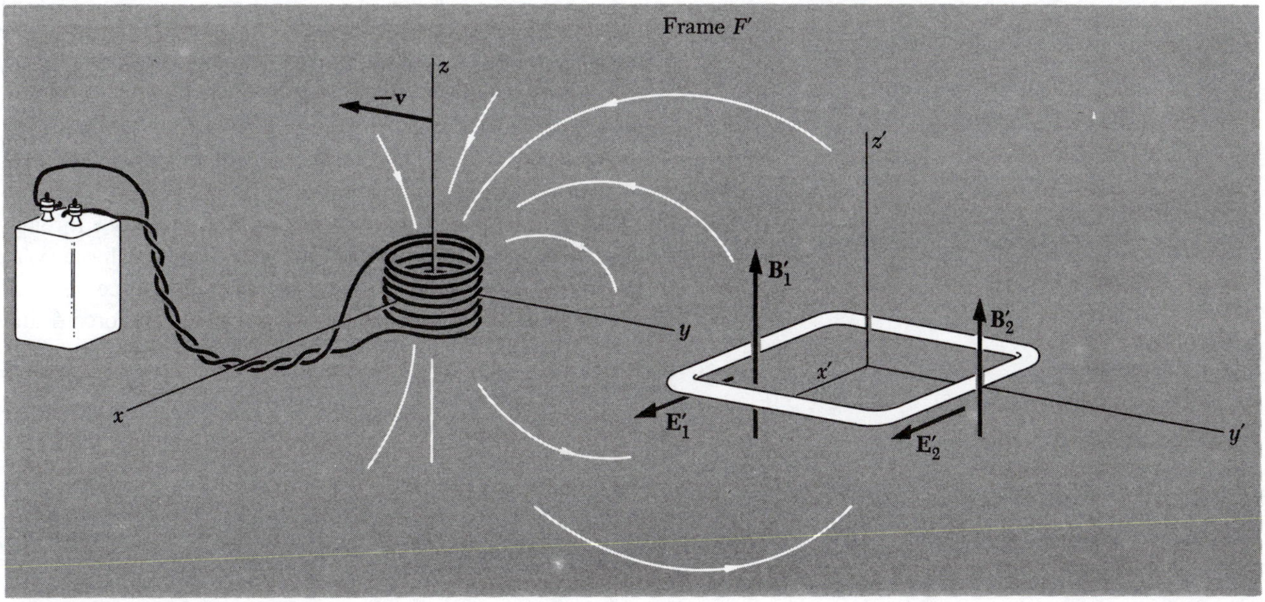

FIGURE 7.14
As observed in the frame F', the loop is at rest, the field source is moving. The fields $\mathbf{B}'$ and $\mathbf{E}'$ are both present and are functions of both position and time.

observers in F' at some instant t'. At these positions there will be an electric field in F'. Equation 6.63 tells us that

$$\mathbf{E}'_1 = -\frac{\mathbf{v}' \times \mathbf{B}'_1}{c} = \frac{\mathbf{v} \times \mathbf{B}'_1}{c} \tag{22}$$
$$\mathbf{E}'_2 = -\frac{\mathbf{v}' \times \mathbf{B}'_2}{c} = \frac{\mathbf{v} \times \mathbf{B}'_2}{c}$$

For observers in F' this is a genuine electric field. It is not an electrostatic field. The line integral of $\mathbf{E}'$ around any closed path in F' is not generally zero. In fact, the line integral of $\mathbf{E}'$ around the rectangular loop is

$$\int \mathbf{E}' \cdot d\mathbf{s}' = \frac{wv}{c}(B'_1 - B'_2) \tag{23}$$

We can call the line integral in Eq. 23 the electromotive force $\mathcal{E}'$ on this path. If a charged particle moves once around the path, $\mathcal{E}'$ is the work done on it, per unit charge. $\mathcal{E}'$ is related to the rate of change of flux through the loop. To see this, note that, while the loop itself is stationary, the *magnetic field pattern* is now moving with the velocity $-\mathbf{v}$ of the source. Hence for the flux lost or gained at either end of the loop, in a time interval dt', we get a result similar to Eq. 11, and we conclude that

$$\mathcal{E}' = -\frac{1}{c}\frac{d\Phi'}{dt'} \tag{24}$$

We can summarize as follows the descriptions in the two frames of reference, F, in which the source of $\mathbf{B}$ is at rest, and F', in which the loop is at rest:

An observer in F says, "We have here a magnetic field which, though it is not uniform spatially, is constant in time. There is no electric field. That wire loop over there is moving with velocity $\mathbf{v}$ through the magnetic field, so the charges in it are acted on by a force $(\mathbf{v}/c) \times \mathbf{B}$ dynes per unit charge. The line integral of this force per unit charge, taken around the whole loop, is the electromotive force $\mathcal{E}$ and it is equal to $-(1/c)(d\Phi/dt)$. The flux Φ is $\int \mathbf{B} \cdot d\mathbf{a}$ over a surface S which, at some instant of time t by my clock, spans the loop."

An observer in F' says, "This loop is stationary, and only an electric field could cause the charges in it to move. But there is in fact an electric field $\mathbf{E}'$. It seems to be caused by that magnetlike object which happens at this moment to be whizzing by with a velocity $-\mathbf{v}$, producing at the same time a rather strong magnetic field $\mathbf{B}'$. The electric field is such that $\int \mathbf{E}' \cdot d\mathbf{s}'$ around this stationary loop is not zero but instead is equal to $-1/c$ times the rate of change of flux through the loop, $d\Phi'/dt'$. The flux Φ' is $\int \mathbf{B}' \cdot d\mathbf{a}'$ over a surface spanning the loop, the values of B' to be measured all over this surface at some one instant t', by my clock."

Our conclusions so far are relativistically exact. They hold for any speed $v \leqq c$ provided we observe scrupulously the distinctions between $\mathbf{B}$ and $\mathbf{B}'$, t and t', etc. If $v \ll c$, so that v^2/c^2 can be neglected, $\mathbf{B}'$ will be practically equal to $\mathbf{B}$, and we can safely ignore also the distinction between t and t'.

A UNIVERSAL LAW OF INDUCTION

7.5 Let's carry out three experiments with the apparatus shown in Fig. 7.15. The tables are on wheels so that they can be easily moved. A sensitive galvanometer has been connected to our old rectangular loop, and to increase any induced electromotive force we put several turns of wire in the loop rather than one. Frankly though, our sensitivity might still be marginal, with the feeble source of magnetic field pictured. Perhaps you can devise a more practical version of the experiment in the laboratory.

Experiment I With constant current in the coil and table 1 stationary, table 2 moves toward the right with speed v. The *galvanometer deflects*. We are not surprised; we have already analyzed this situation in Section 7.3.

FIGURE 7.15
We imagine that either table can move or, with both tables fixed, the current I in the coil can be gradually changed.

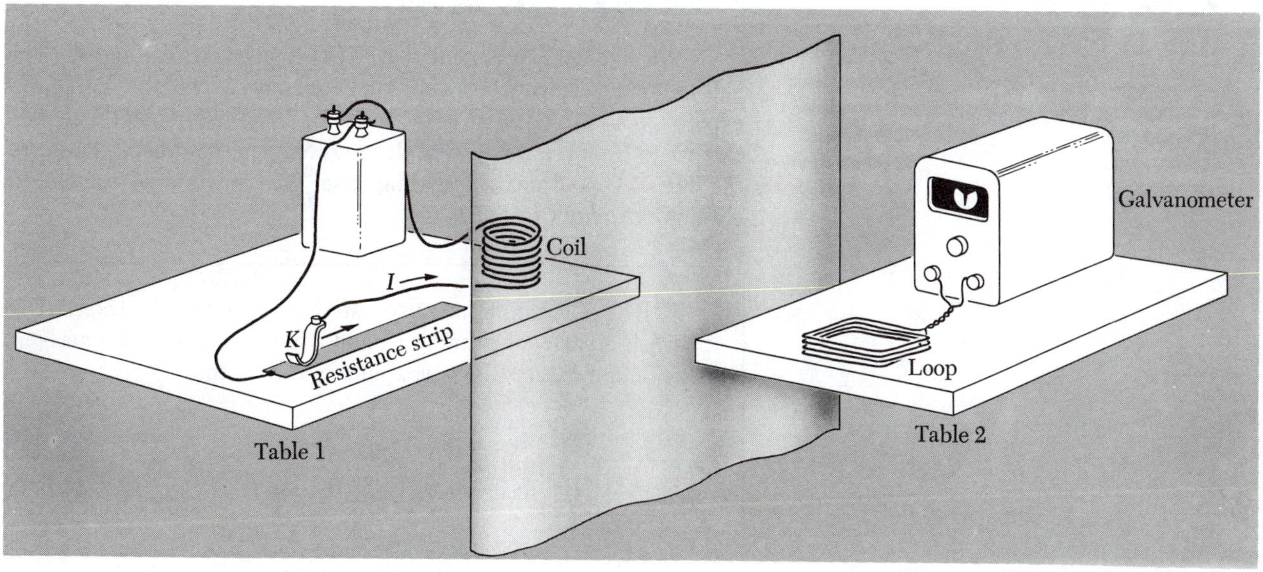

Experiment II With constant current in the coil and table 2 stationary, table 1 moves to the left with speed v. The *galvanometer deflects*. This doesn't surprise us either. We have just discussed the equivalence of Experiments I and II, an equivalence which is an example of Lorentz invariance or, for the low speeds of our tables, Galilean invariance. We know that in both experiments the deflection of the galvanometer can be related to the rate of change of flux of **B** through the loop.

Experiment III Both tables remain at rest, but we vary the current I in the coil by sliding the contact K along the resistance strip. We do this in such a way that the *rate of decrease* of the field **B** at the loop is the same as it was in Experiments I and II. *Does the galvanometer deflect?*

For an observer stationed at the loop on table 2 and measuring the magnetic field in that neighborhood as a function of time and position, there is no way to distinguish among Experiments I, II, and III. Imagine a black cloth curtain between the two tables. Although there might be minor differences between the field configurations for II and III, an observer who did not know what was behind the curtain could not decide, on the basis of local **B** measurements alone, which case it was. Therefore if the galvanometer did *not* respond with the same deflection in Experiment III, it would mean that the relation between the magnetic and electric fields in a region depends on the nature of a remote source. Two magnetic fields essentially similar in their local properties would have associated in one case, but not in the other, an electric field with $\int \mathbf{E} \cdot d\mathbf{s} \neq 0$.

We find by experiment that III *is* equivalent to I and II. The galvanometer deflects, by the same amount as before. Faraday's experiments were the first to demonstrate this fundamental fact. The electromotive force we observe depends only on the rate of change of the flux of **B**, and not on anything else. We can state as a universal relation *Faraday's law of induction:*

> If C is some closed curve, stationary in coordinates x, y, z, if S is a surface spanning C, and if $\mathbf{B}\,(x, y, z, t)$ is the magnetic field measured in x, y, z, at any time t, then
>
> $$\mathcal{E} = \int_C \mathbf{E} \cdot d\mathbf{s} = -\frac{1}{c}\frac{d}{dt}\int_S \mathbf{B} \cdot d\mathbf{a} = -\frac{1}{c}\frac{d\Phi}{dt} \qquad (25)$$

Using the vector derivative curl, we can express this law in differential form. If the relation

$$\int_C \mathbf{E} \cdot d\mathbf{s} = -\frac{1}{c}\frac{d}{dt}\int_S \mathbf{B} \cdot d\mathbf{a} \tag{26}$$

is true for *any* curve C and spanning surface S, as our law asserts, it follows that at any point

$$\text{curl } \mathbf{E} = -\frac{1}{c}\frac{d\mathbf{B}}{dt} \tag{27}$$

To show that Eq. 27 follows from Eq. 26, we proceed as usual to let C shrink down around a point, which we take to be a nonsingular point for the function $\mathbf{B}$. Then in the limit the variation of $\mathbf{B}$ over the small patch of surface $\mathbf{a}$ that spans C will be negligible and the surface integral will approach simply $\mathbf{B} \cdot \mathbf{a}$. Now by definition (Eq. 2.61) the limit approached by $\int_C \mathbf{E} \cdot d\mathbf{s}$ as the patch shrinks is $\mathbf{a} \cdot \text{curl } \mathbf{E}$. Thus we have, in the limit,

$$\mathbf{a} \cdot \text{curl } \mathbf{E} = -\frac{1}{c}\frac{d}{dt}(\mathbf{B} \cdot \mathbf{a}) = \mathbf{a} \cdot \left(-\frac{1}{c}\frac{d\mathbf{B}}{dt}\right) \tag{28}$$

Since this holds for *any* infinitesimal **a**, it must be that†

$$\text{curl } \mathbf{E} = -\frac{1}{c}\frac{d\mathbf{B}}{dt} \tag{29}$$

Recognizing that $\mathbf{B}$ may depend on position as well as time we shall write $\partial\mathbf{B}/\partial t$ in place of $d\mathbf{B}/dt$. We have then these two entirely equivalent statements of the law of induction:

$$\begin{aligned} \int_C \mathbf{E} \cdot d\mathbf{s} &= -\frac{1}{c}\frac{d}{dt}\int_S \mathbf{B} \cdot d\mathbf{a} \\ \text{curl } \mathbf{E} &= -\frac{1}{c}\frac{\partial\mathbf{B}}{\partial t} \end{aligned} \tag{30}$$

In Eq. 30 the electric field $\mathbf{E}$ is to be expressed in our CGS units of statvolts/cm, with $\mathbf{B}$ in gauss, $d\mathbf{s}$ in cm, $d\mathbf{a}$ in cm^2, and c in cm/sec. The electromotive force $\mathcal{E} = \int_C \mathbf{E} \cdot d\mathbf{s}$ will then be given in statvolts.

†If that isn't obvious, note that choosing **a** in the x direction will establish that $(\text{curl } \mathbf{E})_x = -\frac{1}{c}\frac{dB_x}{dt}$, and so on.

In SI units the relation expressed by Eq. 30 looks like this:

$$\int_C \mathbf{E} \cdot d\mathbf{s} = -\frac{d}{dt}\int_S \mathbf{B} \cdot d\mathbf{a}$$
$$\text{curl } \mathbf{E} = -\frac{\partial \mathbf{B}}{\partial t} \qquad (30')$$

Here **E** is in volt m^{-1}, **B** is in teslas, $d\mathbf{s}$ and $d\mathbf{a}$ are in meters and m^2, respectively, with t in sec. The electromotive force $\mathcal{E} = \int_C \mathbf{E} \cdot d\mathbf{s}$ will be given in volts.

The magnetic flux Φ, which is $\int_S \mathbf{B} \cdot d\mathbf{a}$, would be expressed in gauss-cm^2 in our CGS units, and in tesla-m^2, a unit exactly 10^8 times larger, in SI units. (This latter flux unit was assigned a name of its own, the *weber*.)

When in doubt about the units you may find one of the following equivalent statements helpful:

Electromotive force in statvolts equals
$1/c$ times rate of change of flux in gauss-cm^2/sec (31)

Electromotive force in volts equals
rate of change of flux in tesla-m^2/sec (31′)

Electromotive force in volts equals
10^{-8} times rate of change of flux in gauss-cm^2/sec (31″)

The third statement is consistent with the first two because 1 m^2 = 10^4 cm^2 and 1 tesla = 10^4 gauss, exactly. If this seems confusing, don't try to remember it. Just remember that you can look it up on this page.

The differential expression, curl $\mathbf{E} = -(1/c)\partial\mathbf{B}/\partial t$, brings out rather plainly the point we tried to make earlier about the local nature of the field relations. The variation in time of **B** in a neighborhood completely determines curl **E** there—nothing else matters. That does not completely determine **E** itself, of course. Without affecting this relation any electrostatic field, with curl $\mathbf{E} = 0$, could be superposed.

As a concrete example, suppose coils like those in Fig. 7.13 are supplied with 60 cycles per sec alternating current, instead of direct current. The current and the magnetic field vary as sin $(2\pi \cdot 60 \cdot t)$, or sin $377t$. Suppose the amplitude of the current is such that the mag-

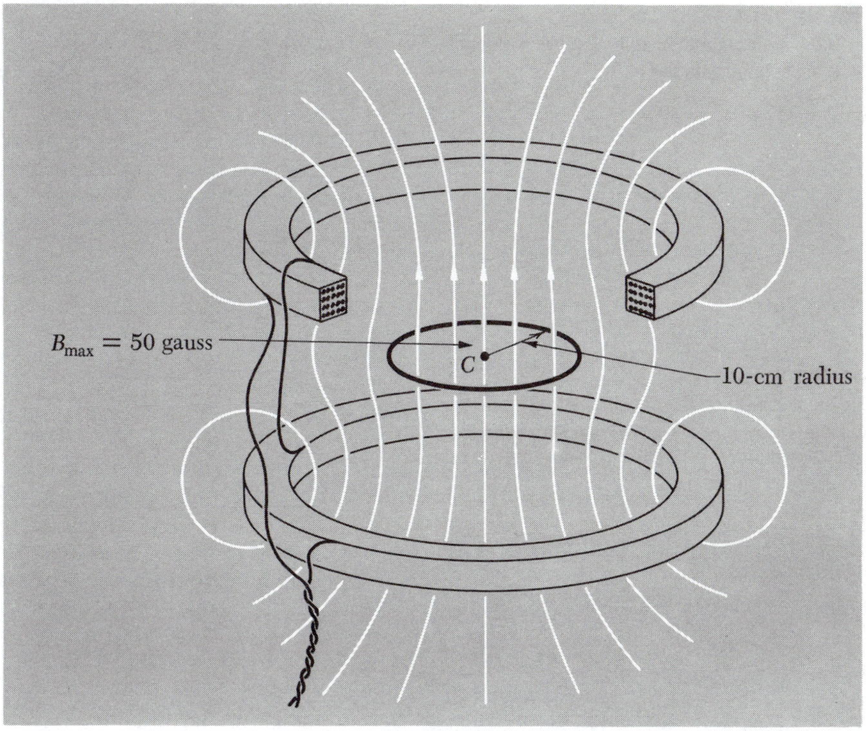

B_{max} = 50 gauss
C
10-cm radius

FIGURE 7.16
Alternating current in the coils produces a magnetic field which, at the center, oscillates between 50 gauss upward and 50 gauss downward. At any instant the field is approximately uniform within the circle C.

netic field **B** in the central region reaches a maximum value of 50 gauss. We want to investigate the induced electric field, and the electromotive force, on the circular path 10 cm in radius shown in Fig. 7.16. We may assume that the field B is practically uniform in the interior of this circle, at any instant of time.

$$B = 50 \sin 377t \tag{32}$$

B is in gauss and t in sec. The flux through the loop C is

$$\begin{aligned} \Phi &= \pi r^2 B = \pi \times 10^2 \times 50 \sin 377t \\ &= 15{,}700 \sin 377t \qquad (\text{gauss-cm}^2) \end{aligned} \tag{33}$$

Using Eq. 31″ to calculate the electromotive force in volts,

$$\begin{aligned} \mathcal{E} &= -(10^{-8}) \frac{d\Phi}{dt} = -(10^{-8})(377)(15{,}700) \cos 377t \\ &= -0.059 \cos 377t \qquad (\text{volts}) \end{aligned} \tag{34}$$

The maximum attained by $\mathcal{E}$ is 59 millivolts. The minus sign will ensure that Lenz' law is respected, if we have defined our directions consistently. The variation of both Φ and $\mathcal{E}$ with time is shown in Fig. 7.17.

What about the electric field itself? Usually we cannot deduce

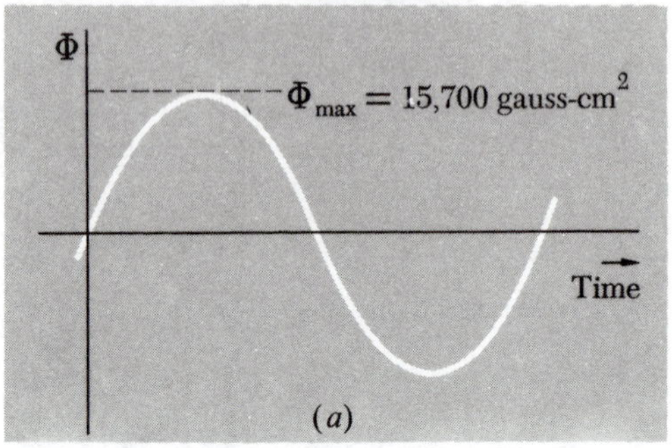

Φ
Φ_{max} = 15,700 gauss-cm^2
Time
(a)

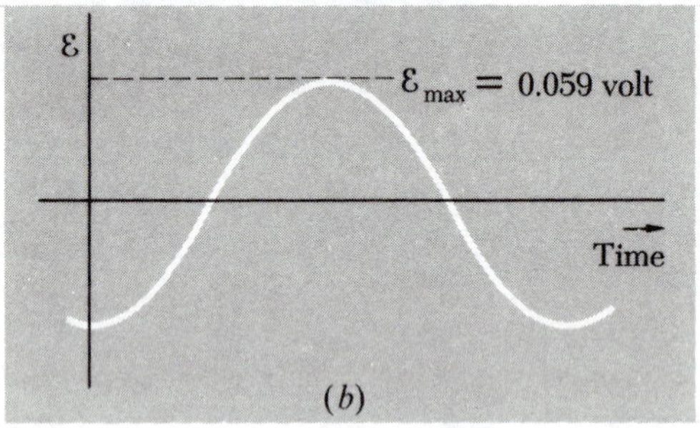

ℰ
$\mathcal{E}_{max}$ = 0.059 volt
Time
(b)

FIGURE 7.17
(*a*) The flux through the circle *C*. (*b*) The electromotive force associated with the path *C*.

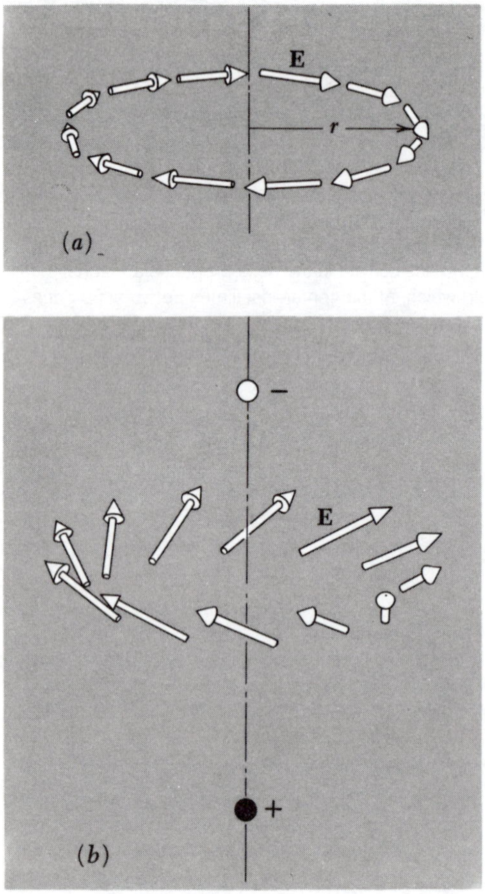

E
r
(a)
E
−
+
(b)

FIGURE 7.18
The electric field on the circular path *C*. (*a*) In the absence of sources other than the symmetrical, oscillating current. (*b*) Including the electrostatic field of two charges on the axis.

E from a knowledge of curl **E** alone. However, our path *C* is here a circle around the center of a symmetrical system. *If there are no other* electric fields around, we may assume that, on the circle *C*, **E** lies in that plane and has a constant magnitude. Then it is a trivial matter to predict its magnitude, since $\int_C \mathbf{E} \cdot d\mathbf{s} = 2\pi rE = \mathcal{E}$, which we have already calculated. In this case, the electric field on the circle might look like Fig. 7.18*a* at a particular instant. But if there are other field sources, it could look quite different. If there happened to be a positive and a negative charge located on the axis as shown in Fig. 7.18*b*, the electric field in the vicinity of the circle would be the superposition of the electrostatic field of the two charges and the induced electric field.

MUTUAL INDUCTANCE

7.6 Two circuits, or loops, C_1 and C_2 are fixed in position relative to one another (Fig. 7.19). By some means, such as a battery and a variable resistance, a controllable current I_1 is caused to flow in circuit C_1. Let $\mathbf{B}_1(x, y, z)$ be the magnetic field that would exist if the current

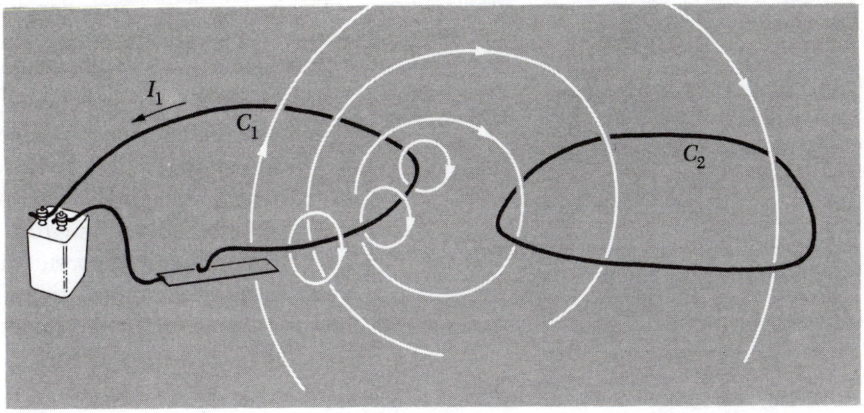

FIGURE 7.19
Current I_1 in loop C_1 causes a certain flux ϕ_{21} through loop C_2.

in C_1 remained constant at the value I_1, and let Φ_{21} denote the flux of $\mathbf{B}_1$ through the circuit C_2. Thus

$$\Phi_{21} = \int_{S_2} \mathbf{B}_1 \cdot d\mathbf{a}_2 \tag{35}$$

where S_2 is a surface spanning the loop C_2. With the shape and relative position of the two circuits fixed, Φ_{21} will be proportional to I_1:

$$\frac{\Phi_{21}}{I_1} = \text{const} \tag{36}$$

Suppose now that I_1 changes with time, but *slowly enough* so that the field $\mathbf{B}_1$ at any point in the vicinity of C_2 and the current I_1 in C_1 at the same instant of time are related as they would be for steady currents. (To see why such a restriction is necessary, imagine that C_1 and C_2 are 10 meters apart and we cause the current in C_1 to double in value in 10 nanoseconds!) The flux Φ_{21} will change in proportion as I_1 changes. There will be an electromotive force induced in circuit C_2, of magnitude

$$\mathcal{E}_{21} = -\frac{\text{const}}{c}\frac{dI_1}{dt} \tag{37}$$

The constant here is the same as the one in Eq. 36. Let's absorb the c in the denominator into a single constant, denoted by M_{21}, and write Eq. 37 in this way:

$$\mathcal{E}_{21} = -M_{21}\frac{dI_1}{dt} \tag{38}$$

We call the constant M_{21} the coefficient of *mutual inductance.* Its value is determined by the geometry of our arrangement of loops.

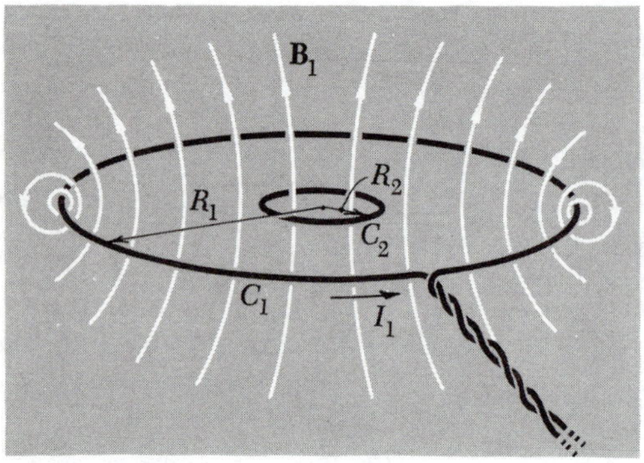

FIGURE 7.20
Current I_1 in ring C_1 causes a field $\mathbf{B}_1$ which is approximately uniform over the region of the small ring C_2.

The units will of course depend on our choice of units for $\mathcal{E}$, I, and t. With $\mathcal{E}$ in statvolts, or esu/cm, and I in esu/sec, the unit for M_{21} is cm^{-1}-sec^2. You are more likely to be working with volts and amperes when you are applying this relation. In SI units, $\mathcal{E}$ in volts and I in amperes, the unit for M_{21} is volt amp^{-1} sec, or ohm-sec. This unit is called the *henry*.† That is, the mutual inductance M_{21} is one henry if a current I_1 changing at the rate of 1 ampere/sec induces an electromotive force of 1 volt in circuit C_2.

As an example, consider the circuits in Fig. 7.20, two coplanar, concentric rings, a small ring C_2 and a much larger ring C_1. What is M_{21} in this case? At the center of C_1, with I_1 flowing, the field B_1 is given by

$$B_1 = \frac{2\pi I_1}{cR_1} \tag{39}$$

with I_1 in esu/sec, B_1 in gauss. Here we are simply applying Eq. 6.42. We assume $R_2 \ll R_1$, so that we can neglect the variation of B_1 over the interior of the small ring. Then the flux through the small ring is

$$\Phi_{21} = (\pi R_2^2)\frac{2\pi I_1}{cR_1} = \frac{2\pi^2 I_1 R_2^2}{cR_1} \tag{40}$$

Thus the "constant" in Eq. 36, in this particular case, has the value $2\pi^2 R_2^2/cR_1$, and the electromotive force induced in C_2 will be

$$\mathcal{E}_{21} = -\frac{1}{c}\frac{2\pi^2 R_2^2}{cR_1}\frac{dI_1}{dt} \tag{41}$$

with $\mathcal{E}_{21}$ in statvolts and I_1 in esu/second. To express the mutual inductance in henrys, we note that a statvolt is 300 volts and I(esu/sec) = I(amps) $\times$ 3 $\times$ 10^9, so that

$$\mathcal{E}(\text{volts}) = -\frac{2\pi^2 R_2^2}{R_1} \times 10^{-9}\frac{dI}{dt} \quad (\text{amps/sec}) \tag{42}$$

Thus the value of M_{21} in henrys, with R_2 and R_1 in cm, is

$$M_{21} = \frac{2\pi^2 \times 10^{-9} R_2^2}{R_1} \tag{43}$$

Incidentally, the minus sign we had been carrying along doesn't tell us much at this stage. If you want to be sure which way the electromotive force will tend to drive current in C_2, Lenz' law is your most reliable guide.

†The unit is named after Joseph Henry (1797–1878), the foremost American physicist of his time. Electromagnetic induction was discovered independently by Henry, practically at the same time as Faraday's experiments. Henry was the first to recognize the phenomenon of self-induction. He developed the electromagnet and the prototype of the electric motor, invented the electric relay, and all but invented telegraphy.

If the circuit C_1 consisted of N_1 turns of wire instead of a single ring, the field B_1 at the center would be N_1 times as strong, for a given current I_1. Also, if the small loop C_2 consisted of N_2 turns, all of the same radius R_2, the electromotive force in each turn would add to that in the next, making the total electromotive force in that circuit N_2 times that of a single turn. Thus for *multiple turns* in each coil the mutual inductance will be given by

$$M_{21} = \frac{2\pi^2 \times 10^{-9} N_1 N_2 R_2^2}{R_1} \tag{44}$$

This assumes that the turns in each coil are neatly bundled together, the cross section of the bundle being small compared with the coil radius. However, the mutual inductance M_{21} has a well-defined meaning for two circuits of any shape or distribution. It is the ratio of the electromotive force in volts in circuit 2, caused by changing current in circuit 1, to the rate of change of current I_1 in amperes/sec. That is,

$$M_{21} = \frac{\mathcal{E}_{21}}{\left(\dfrac{dI_1}{dt}\right)} \tag{45}$$

M_{21} will be in henrys if $\mathcal{E}_{21}$ is in volts and dI_1/dt is in amp/sec.

A RECIPROCITY THEOREM

7.7 In considering the circuits C_1 and C_2 we might have inquired about the electromotive force induced in circuit C_1 by a changing current in circuit C_2. That would involve another coefficient of mutual inductance, M_{12}:

$$M_{12} = \frac{\mathcal{E}_{12}}{\left(\dfrac{dI_2}{dt}\right)} \tag{46}$$

It is a remarkable fact that, for *any* two circuits,

$$M_{12} = M_{21} \tag{47}$$

This is not a matter of geometrical symmetry. Even the simple example in Fig. 7.20 is not symmetrical with respect to the two circuits. Note that R_1 and R_2 enter in different ways into the expression for M_{21}; Eq. 47 asserts that, for these two dissimilar circuits, if

$$M_{21} = \frac{2\pi^2 \times 10^{-9} N_1 N_2 R_2^2}{R_1} \quad \text{then} \quad M_{12} = \frac{2\pi^2 \times 10^{-9} N_1 N_2 R_2^2}{R_1}$$

also—and *not* what we would get by switching 1's and 2's everywhere!

To prove the theorem, Eq. 47, we have to show that the flux Φ_{12} through some circuit C_1 as a result of a current I in a circuit C_2 is equal to the flux Φ_{21} that threads circuit 2 when an *equal* current I flows in circuit C_1. To show this, we use the vector potential.

According to Stokes' theorem:

$$\int_C \mathbf{A} \cdot d\mathbf{s} = \int_S (\text{curl } \mathbf{A}) \cdot d\mathbf{a} \tag{48}$$

In particular, if **A** is the vector potential of a magnetic field **B**, that is, if $\mathbf{B} = \text{curl } \mathbf{A}$, then we have

$$\boxed{\int_C \mathbf{A} \cdot d\mathbf{s} = \int_S \mathbf{B} \cdot d\mathbf{a} = \Phi_S} \tag{49}$$

That is, the line integral of the vector potential around a loop is equal to the flux of **B** *through the loop.*

Now the vector potential is related to its current source as follows, according to Eq. 6.35:

$$\mathbf{A}_{21} = \frac{I}{c} \int_{C_1} \frac{d\mathbf{s}_1}{r_{21}} \tag{50}$$

FIGURE 7.21
Calculation of the flux ϕ_{21} which passes through C_2 as a result of current I flowing in C_1.

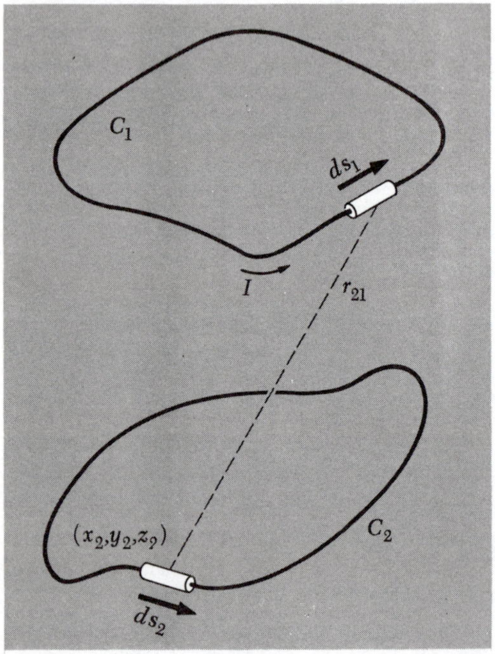

$\mathbf{A}_{21}$ is the vector potential, at some point (x_2, y_2, z_2), of the magnetic field caused by current I(esu/sec) flowing in circuit C_1; $d\mathbf{s}_1$ is an element of the loop C_1; and r_{21} is the magnitude of the distance from that element to the point (x_2, y_2, z_2).

Figure 7.21 shows the two loops C_1 and C_2, with current I flowing in C_1. Let (x_2, y_2, z_2) be a point on the loop C_2. Then the flux through C_2 due to current I in C_1 is

$$\Phi_{21} = \int_{C_2} \mathbf{A}_{21} \cdot d\mathbf{s}_2 = \int_{C_2} d\mathbf{s}_2 \cdot \mathbf{A}_{21} = \frac{I}{c} \int_{C_2} d\mathbf{s}_2 \cdot \int_{C_1} \frac{d\mathbf{s}_1}{r_{21}} \tag{51}$$

Similarly, the flux through C_1 due to current I flowing in C_2 would be given by

$$\Phi_{12} = \frac{I}{c} \int_{C_1} d\mathbf{s}_1 \cdot \int_{C_2} \frac{d\mathbf{s}_2}{r_{12}} \tag{52}$$

Now $r_{12} = r_{21}$, for these are just distance magnitudes, not vectors. The meaning of each of the integrals above is: Take the scalar product of a pair of line elements, one on each loop, divide by the distance between them and sum over all pairs. The only difference between Eqs. 51 and 52 is the *order* in which this operation is carried out, and that cannot affect the final sum. Hence $\Phi_{12} = \Phi_{21}$, from which it follows directly that $M_{12} = M_{21}$. Thanks to this theorem, we need make no distinction between M_{12} and M_{21}. We may speak, henceforth, of *the* mutual inductance M of any two circuits.

Theorems of this sort are often called "reciprocity" theorems. There are some other reciprocity theorems on electric circuits not unrelated to this one. This may remind you of the relation $C_{jk} = C_{kj}$ mentioned in Section 3.6 and treated in Problem 3.27. A reciprocity relation usually expresses some general symmetry law which is *not* apparent in the superficial structure of the system.

SELF-INDUCTANCE

7.8 When the current I_1 is changing, there is a change in the flux through circuit C_1 itself, consequently an electromotive force is induced. Call this $\mathcal{E}_{11}$. The induction law holds, whatever the source of the flux:

$$\mathcal{E}_{11} = -\frac{1}{c}\frac{d\Phi_{11}}{dt} \tag{53}$$

where Φ_{11} is the flux through circuit 1 of the field B_1 due to the current I_1 in circuit 1. The minus sign expresses the fact that the electromotive force is always directed so as to *oppose* the *change* in current—Lenz's law, again. Since Φ_{11} will be proportional to I_1 we can write

$$\mathcal{E}_{11} = -L_1\frac{dI_1}{dt} \tag{54}$$

The constant L_1 is called the *self-inductance* of the circuit.

As an example of a circuit for which L_1 can be calculated, consider the rectangular toroidal coil of Problem 6.14, shown here again in Fig. 7.22. You found (if you worked that problem) that a current I, in esu/sec, flowing in the coil of N turns produces a field the strength of which, at a radial distance r from the axis of the coil, is given by $B = 2NI/cr$. The total flux through one turn of the coil is the integral of this field over the cross section of the coil:

$$\Phi(\text{one turn}) = h\int_a^b \frac{2NI}{cr}\,dr = \frac{2NIh}{c}\ln\left(\frac{b}{a}\right) \tag{55}$$

The flux threading the circuit of N turns is N times as great:

$$\Phi = \frac{2N^2Ih}{c}\ln\left(\frac{b}{a}\right) \tag{56}$$

Hence the induced electromotive force $\mathcal{E}$ is

$$\mathcal{E} = -\frac{1}{c}\frac{d\Phi}{dt} = -\frac{2N^2h}{c^2}\ln\left(\frac{b}{a}\right)\frac{dI}{dt} \tag{57}$$

FIGURE 7.22
A toroidal coil of rectangular cross section. Only a few turns are shown.

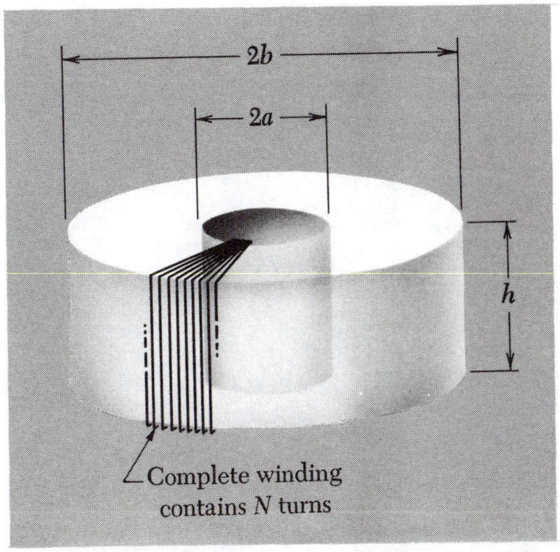
2b
2a
h
Complete winding
contains N turns

Thus the self-inductance of this coil is given by

$$L = \frac{2N^2h}{c^2} \ln\left(\frac{b}{a}\right) \tag{58}$$

Equation 58 is the correct expression for the inductance if I is measured in esu/sec and $\mathscr{E}$ in statvolts. For I in amps and $\mathscr{E}$ in volts the appropriate unit for L is the henry, as in the case of mutual inductance. Converting to these units we get:

$$L(\text{henrys}) = 2 \times 10^{-9} N^2 h \ln\left(\frac{b}{a}\right) \tag{59}$$

You may think that one of the rings we considered earlier would have made a simpler example to illustrate the calculation of self-inductance. However, if we try to calculate the inductance of a simple circular loop of wire, we encounter a puzzling difficulty. It seems a good idea to simplify the problem by assuming that the wire has zero diameter. But we soon discover that, if finite current flows in a filament of zero diameter, the flux threading a loop made of such a filament is infinite! The reason is that the field B, in the neighborhood of a filamentary current, varies as $1/r$, where r is the distance from the filament, and the integral of $B \times$ *area* diverges as $\int (dr/r)$ when we extend it down to $r = 0$. To avoid this we may let the radius of the wire be finite, not zero, which is more realistic anyway. This may make the calculation a bit more complicated, in a given case, but that won't worry us. The real difficulty is that different parts of the wire now appear as *different circuits*, linked by different amounts of flux. We are no longer sure what we mean by *the* flux through *the* circuit. In fact, because the electromotive force is different in the different filamentary loops into which the circuit can be divided, some *redistribution* of current density must occur when rapidly changing currents flow in the ring. Hence the inductance of the circuit may depend somewhat on the rapidity of change of I, and thus not be strictly a constant as Eq. 54 would imply.

We avoided this embarrassment in the toroidal coil example by ignoring the field in the immediate vicinity of the individual turns of the winding. Most of the flux does *not* pass through the wires themselves, and whenever that is the case the effect we have just been worrying about will be unimportant.

A CIRCUIT CONTAINING SELF-INDUCTANCE

7.9 Suppose we connect a battery, providing electromotive force $\mathscr{E}_0$, to a coil, or *inductor,* with self-inductance L, as in Fig. 7.23*a*. The coil itself, the connecting wires, and even the battery will have some resistance. We don't care how this is distributed around the circuit. It can all be lumped together in one resistance R, indicated on the circuit diagram of Fig. 7.23*b* by a resistor symbol with this value. Also, the rest of the circuit, especially the connecting wires, contribute a bit to the self-inductance of the whole circuit; we assume that this is included in L. In other words, Fig. 7.23*b* represents an idealization of the physical circuit: The inductor L, symbolized by , has no resistance; the resistor R has no inductance. It is this idealized circuit that we shall now analyze.

If the current I in the circuit is changing at the rate dI/dt, an electromotive force $L\,dI/dt$ will be induced, in a direction to oppose the change. Also, there is the constant electromotive force $\mathcal{E}_0$ of the battery. If we define the positive current direction as the one in which the battery tends to drive current around the circuit, then the net electromotive force at any instant is $\mathcal{E}_0 - L\,dI/dt$. This drives the current I through the resistor R. That is,

$$\mathcal{E}_0 - L\frac{dI}{dt} = RI \tag{60}$$

We can also describe the situation in this way: The potential difference between points A and B, which we'll call the *voltage across the inductor,* is $L\,dI/dt$, with the upper end of the inductor positive if I in the direction shown is *increasing.* The potential difference between B and C, the voltage across the resistor, is RI, with the upper end of the resistor positive. Hence the sum of the voltage across the inductor and the voltage across the resistor is $L\,dI/dt + RI$. This is the same as the potential difference between the battery terminals, which is $\mathcal{E}_0$ (our idealized battery has no internal resistance). Thus we have

$$\mathcal{E}_0 = L\frac{dI}{dt} + RI \tag{61}$$

which is merely a restatement of Eq. 60.

Before we look at the mathematical solution of Eq. 60, let's predict what ought to happen in this circuit if the switch is closed at $t = 0$. Before the switch is closed, $I = 0$, necessarily. A long time after the switch has been closed, some steady state will have been attained, with current practically constant at some value I_0. Then and thereafter, $dI/dt \approx 0$, and Eq. 60 reduces to

$$\mathcal{E}_0 = RI_0 \tag{62}$$

The transition from zero current to the steady-state current I_0 cannot occur abruptly at $t = 0$, for then dI/dt would be infinite. In fact, just after $t = 0$, the current I will be so small that the second term RI in Eq. 60 can be ignored, giving

$$\frac{dI}{dt} = \frac{\mathcal{E}_0}{L} \tag{63}$$

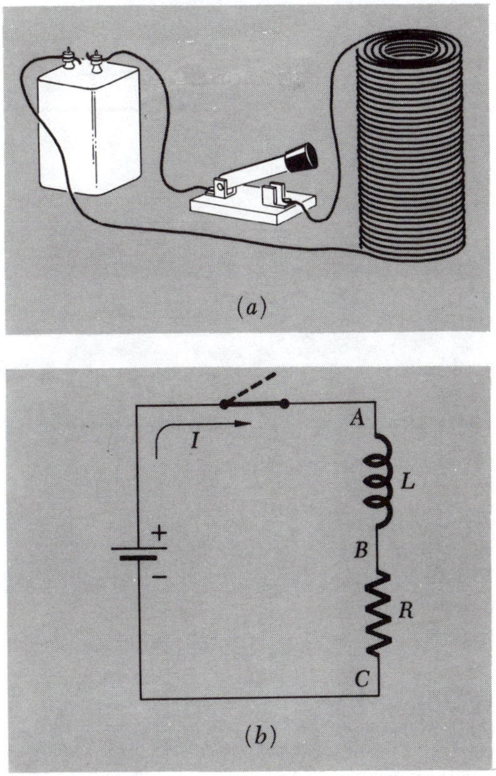

FIGURE 7.23
A simple circuit with inductance (*a*) and resistance (*b*).

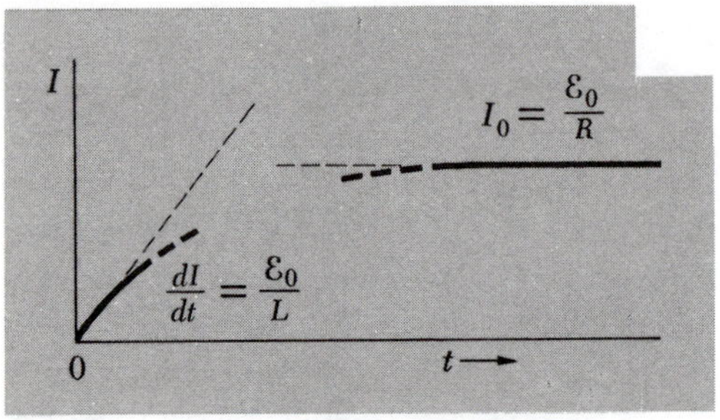

(a)

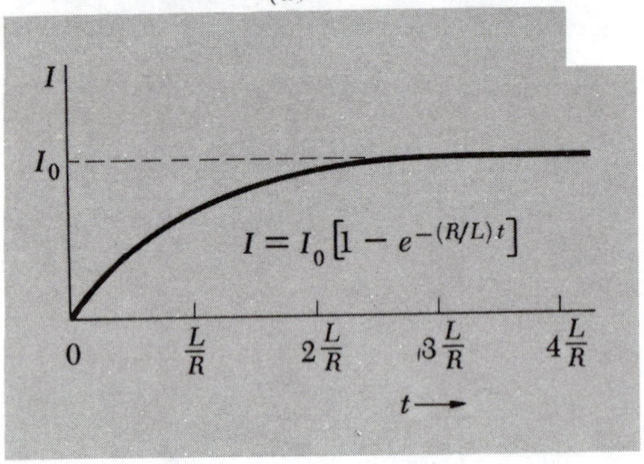

(b)

FIGURE 7.24
(a) How the current must behave initially, and after a very long time has elapsed. (b) The complete variation of current with time in the circuit of Fig. 7.23.

The inductance L limits the rate of rise of the current.

What we now know is summarized in Fig. 7.24*a*. It only remains to find how the whole change takes place. Equation 60 is a differential equation very much like Eq. 29 in Chapter 4. Without further ado we can write down a solution to Eq. 60 which satisfies our initial condition, $I = 0$ at $t = 0$:

$$I = \frac{\mathcal{E}_0}{R}(1 - e^{-(R/L)t}) \tag{64}$$

The graph in Fig. 7.24*b* shows the current approaching its asymptotic value I_0 exponentially. The "time constant" of this circuit is the quantity L/R. If L is measured in henrys and R in ohms, this comes out in sec, since henrys $\sim$ volt amp^{-1} sec, and ohms $\sim$ volt amp^{-1}.

What happens if we open the switch after the current I_0 has been established, thus forcing the current to drop abruptly to zero? That would make the term $L\, dI/dt$ negatively infinite! The catastrophe can be more than mathematical. People have been killed opening switches in highly inductive circuits. What happens generally is that a very high induced voltage causes a spark or arc across the open switch contacts, so that the current continues after all. Let us instead remove the battery from the circuit by closing a conducting path *across* the LR combination, as in Fig. 7.25*a*, at the same time disconnecting the battery. We now have a circuit described by the equation

$$0 = L\frac{dI}{dt} + RI \tag{65}$$

with the initial condition $I = I_0$ at $t = t_1$, where t_1 is the instant at which the short circuit was closed. The solution is the simple exponential decay function

$$I = I_0 e^{-(R/L)(t-t_1)} \tag{66}$$

with the same characteristic time L/R as before.

FIGURE 7.25
(a) LR circuit. (b) Exponential decay of current in the LR circuit.

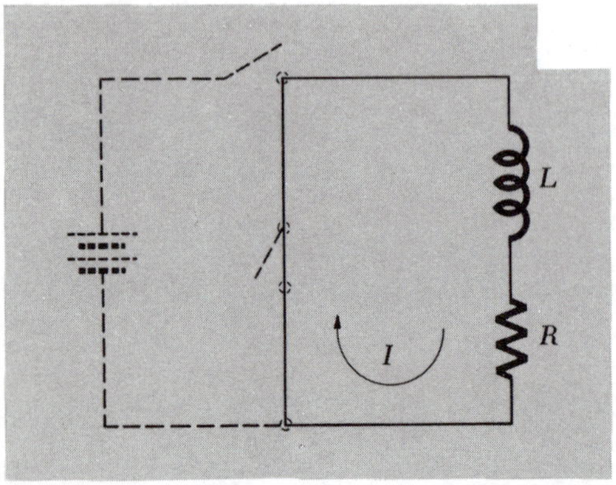

(a)

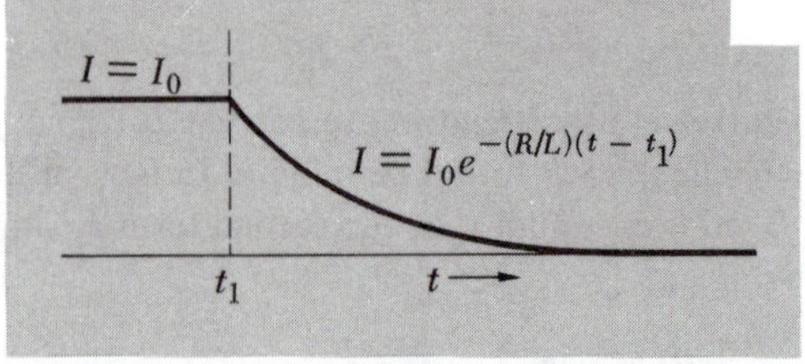

(b)

ENERGY STORED IN THE MAGNETIC FIELD

7.10 During the decay of the current described by Eq. 66 and Fig. 7.25*b*, energy is dissipated in the resistor R. Since the energy dU dissipated in any short interval dt is $RI^2\,dt$, the total energy dissipated after the closing of the switch at time t_1 must be

$$U = \int_{t_1}^{\infty} RI^2\,dt = \int_{t_1}^{\infty} RI_0^2 e^{-(2R/L)(t-t_1)}\,dt \tag{67}$$

With the substitution $x = 2R(t - t_1)/L$ this is easily evaluated:

$$U = RI_0^2\left(\frac{L}{2R}\right)\int_0^{\infty} e^{-x}\,dx = \tfrac{1}{2}\,LI_0^2 \tag{68}$$

The source of this energy was the inductor with its magnetic field. Indeed, exactly that amount of work had been done by the battery to build up the current in the first place—over and above the energy dissipated in the resistor between $t = 0$ and $t = t_1$, which was also provided by the battery. To see that this is a general relation, note that, if we have an increasing current in an inductor, work must be done to drive the current I against the induced electromotive force $L\,dI/dt$. So in time dt the work done is

$$dW = LI\frac{dI}{dt}\,dt = LI\,dI = \tfrac{1}{2}\,L\,d(I^2) \tag{69}$$

Therefore, we may assign a total energy

$$U = \tfrac{1}{2}\,LI^2 \tag{70}$$

to an inductor carrying current I. With the eventual decay of this current, that amount of energy will appear somewhere else.

It is natural to regard this as energy stored in the magnetic field of the inductor, just as we have described the energy of a charged capacitor as stored in its electric field. The energy of a capacitor charged to potential difference V is $\tfrac{1}{2}CV^2$ and is accounted for by assigning to an element of volume dv, where the electric field strength is E, an amount of energy $(1/8\pi)E^2\,dv$. It is pleasant, but hardly surprising, to find that an exactly similar relation holds for the energy stored in an inductor. That is, we can ascribe to the magnetic field an energy density $(1/8\pi)B^2$, and summing the energy of the whole field will give the energy $\tfrac{1}{2}LI^2$.

To show how this works out in one case, we can go back to the toroidal coil whose inductance L we calculated in Section 7.8. We found (Eq. 58)

$$L = \frac{2N^2h}{c^2}\ln\left(\frac{b}{a}\right) \tag{71}$$

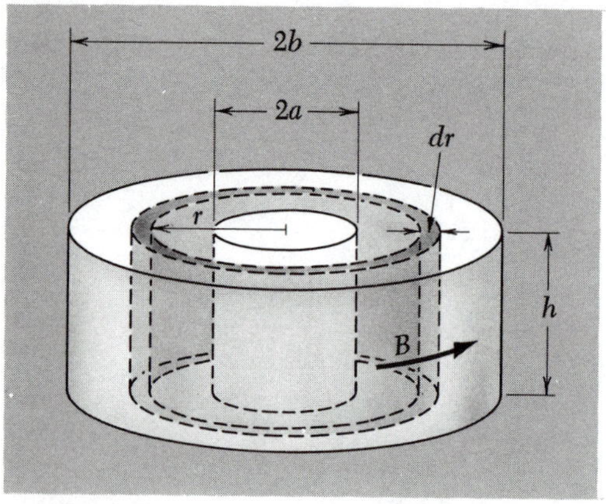

FIGURE 7.26
Calculation of energy stored in the magnetic field of the toroidal coil of Fig. 7.22.

The magnetic field strength B, with current I flowing, was given

$$B = \frac{2NI}{cr} \tag{72}$$

To calculate the volume integral of $B^2/8\pi$ we can use a volume element consisting of the cylindrical shell sketched in Fig. 7.26, with volume $2\pi rh\, dr$. As this shell expands from $r = a$ to $r = b$, it sweeps through all the space that contains magnetic field. (The field B is zero everywhere outside the torus, remember.)

$$\frac{1}{8\pi}\int B^2\, dv = \frac{1}{8\pi}\int_a^b \left(\frac{2NI}{cr}\right)^2 2\pi rh\, dr = \frac{N^2hI^2}{c^2}\ln\left(\frac{b}{a}\right) \tag{73}$$

Comparing this result with Eq. 71, we see that, indeed,

$$\frac{1}{8\pi}\int B^2\, dv = \tfrac{1}{2} LI^2 \tag{74}$$

The more general statement, the counterpart of our statement for the electric field in Eq. 38 of Chapter 1, is that the energy U to be associated with any magnetic field $B(x, y, z)$ is given by:

$$\boxed{U = \frac{1}{8\pi}\int_{\substack{\text{Entire}\\ \text{field}}} B^2\, dv} \tag{75}$$

With B in gauss and v in cm^3, U in Eq. 75 will be given in ergs. In Eq. 70, we may use henrys and amperes, for L and I, and then U will be given in joules. The SI equivalent of Eq. 75 for U in joules, B in teslas, and v in m^3 is

$$U = \frac{1}{2\mu_0}\int_{\substack{\text{Entire}\\ \text{field}}} B^2\, dv \tag{75$'$}$$

PROBLEMS

7.1 What is the maximum electromotive force induced in a coil of 4000 turns, average radius 12 cm, rotating at 30 revolutions per sec in the earth's magnetic field where the field intensity is 0.5 gauss?

Ans. 0.0057 statvolt, or 1.71 volts.

7.2 A long straight wire is parallel to the y axis and passes through the point $z = h$ on the z axis. A current I flows in this wire, returning by a remote conductor whose field we may neglect. Lying in the xy plane is a square loop with two of its sides, of length b, parallel to the long wire. This loop slides with constant speed v in the $\hat{\mathbf{x}}$ direction. Find the magnitude of the electromotive force induced in the loop at the moment when the center of the loop crosses the y axis.

7.3 In the central region of a solenoid which is connected to a radiofrequency power source, the magnetic field oscillates at 2.5×10^6 cycles per sec with an amplitude of 4 gauss. What is the amplitude of the oscillating electric field, in statvolts/cm, at a point 3 cm from the axis? (This point lies within the region where the magnetic field is nearly uniform.)

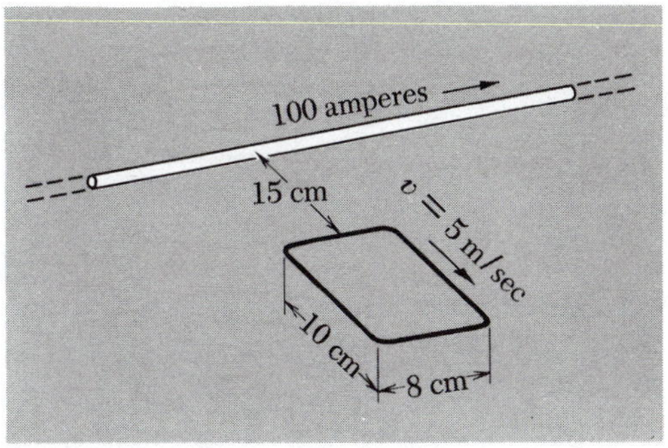

PROBLEM 7.4

7.4 Calculate the electromotive force in the moving loop in the figure at the instant when it is in the position there shown. Assume the resistance of the loop is so great that the effect of the current in the loop itself is negligible. Estimate very roughly how large a resistance would be safe, in this respect. Indicate the direction in which current would flow in the loop, at the instant shown.

7.5 Suppose the loop in Fig. 7.6 has a resistance R. Show that whoever is pulling the loop along at constant speed does an amount of work during the interval dt which agrees precisely with the energy dissipated in the resistance during this interval, providing the self-inductance of the loop can be neglected. What is the source of the energy in Fig. 7.14 where the loop is stationary?

7.6 Does the prediction of a simple sinusoidal variation of electromotive force for the rotating loop in Fig. 7.13 depend on the loop being rectangular, on the magnetic field being uniform, or on both? Explain. Can you suggest an arrangement of rotating loop and stationary coils which will give a definitely nonsinusoidal emf? Sketch the voltage-time curve you would expect to see on the oscilloscope, with that arrangement.

7.7 Calculate the self-inductance of a cylindrical solenoid 10 cm in diameter and 2 meters long. It has a single-layer winding containing a total of 1200 turns. Assume as a first approximation that the magnetic field inside the solenoid is uniform right out to the ends. Estimate roughly the magnitude of the error you will thereby incur. Is the true L larger or smaller than your approximate result?

7.8 How could we wind a resistance coil so that its self-inductance would be *small?*

7.9 Derive an approximate formula for the mutual inductance of two circular rings of the same radius a, arranged like wheels on the same axle with their centers b cm apart. Use an approximation good for $b \gg a$.

7.10 The coils which first produced a slight but detectable kick in Faraday's galvanometer he describes as made of 203 feet of copper

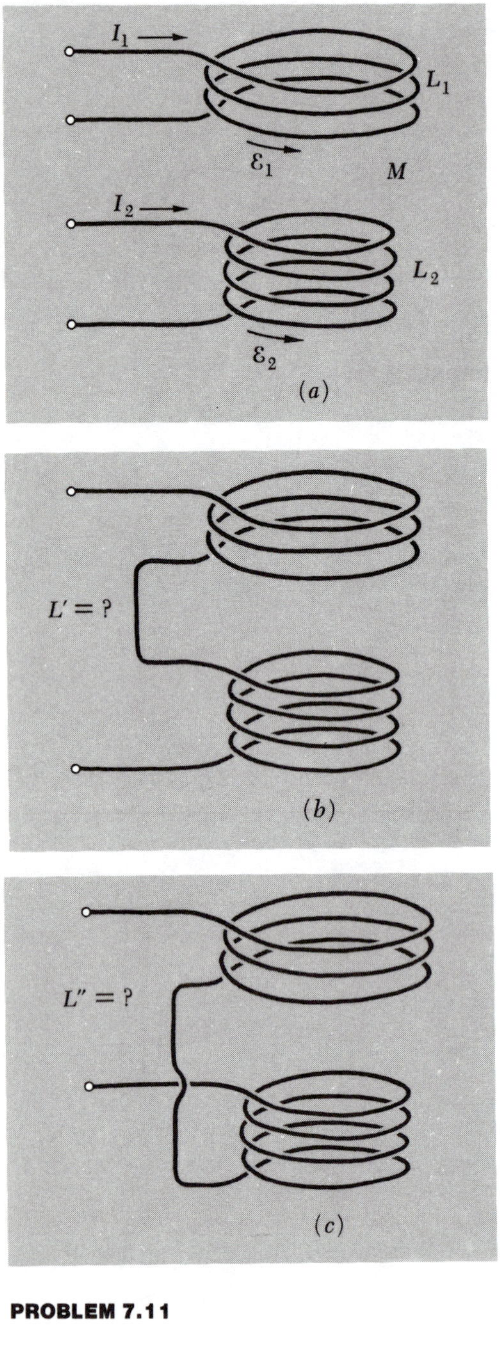

I1
L1
ℰ1
M
I2
L2
ℰ2
(a)
L′ = ?
(b)
L″ = ?
(c)

PROBLEM 7.11

wire each, wound around a large block of wood. The turns of the second spiral (that is, single-layer coil) were interposed between those of the first, but separated from them by twine. The diameter of the copper wire itself was 1/20 inch. He does not give the dimensions of the wooden block or the number of turns in the coils. In the experiment, one of these coils was connected to a "battery of 100 plates." See if you can make a rough estimate of the duration in seconds and magnitude in amperes of the pulse of current that passed through his galvanometer.

7.11 Part (*a*) of the figure shows two coils with self-inductances L_1 and L_2. In the relative position shown their mutual inductance is M. The positive current direction and the positive electromotive force direction in each coil are defined by the arrows in the figure. The equations relating currents and electromotive forces are

$$\mathcal{E}_1 = -L_1 \frac{dI_1}{dt} \underset{?}{\pm} M \frac{dI_2}{dt} \qquad \text{and} \qquad \mathcal{E}_2 = -L_2 \frac{dI_2}{dt} \underset{?}{\pm} M \frac{dI_1}{dt}$$

Given that M is always to be taken as a positive constant, how must the signs be chosen in these equations? What if we had chosen, as we might have, the other direction for positive current, and for positive electromotive force, in the lower coil? Now connect the two coils together as in part (*b*) of the figure to form a single circuit. What is the inductance L' of this circuit, expressed in terms of L_1, L_2, and M? What is the inductance L'' of the circuit formed by connecting the coils as shown in (*c*)? Which circuit, (*b*) or (*c*), has the greater self-inductance? Considering that the self-inductance of any circuit must be a positive quantity (why couldn't it be negative?), see if you can draw a general conclusion, valid for any conceivable pair of coils, concerning the relative magnitude of L_1, L_2, and M.

7.12 An ocean current flows at a speed of 2 knots (approximately 1 meter/sec) in a region where the vertical component of the earth's magnetic field is 0.35 gauss. The conductivity of seawater in that region is 0.04 (ohm-cm)$^{-1}$. On the assumption that there is no other horizontal component of **E** than the motional term $(\mathbf{v}/c) \times \mathbf{B}$, find the density of horizontal electric current in amps/m^2. If you were to carry a bottle of seawater through the earth's field at this speed, would such a current be flowing in it?

7.13 A coil with resistance of 0.01 ohm and self-inductance 0.50 millihenry is connected across a large 12-volt battery of negligible internal resistance. How long after the switch is closed will the current reach 90 percent of its final value? At that time, how much energy, in joules, is stored in the magnetic field? How much energy has been withdrawn from the battery up to that time?

7.14 A metal crossbar of mass m slides without friction on two long parallel conducting rails a distance b apart. A resistor R is connected across the rails at one end; compared with R, the resistance of bar and rails is negligible. There is a uniform field **B** perpendicular to the plane of the figure. At time $t = 0$ the crossbar is given a velocity v_0 toward the right. What happens then?

(*a*) Does the rod ever stop moving? If so when?

(*b*) How far does it go?

(*c*) How about conservation of energy?

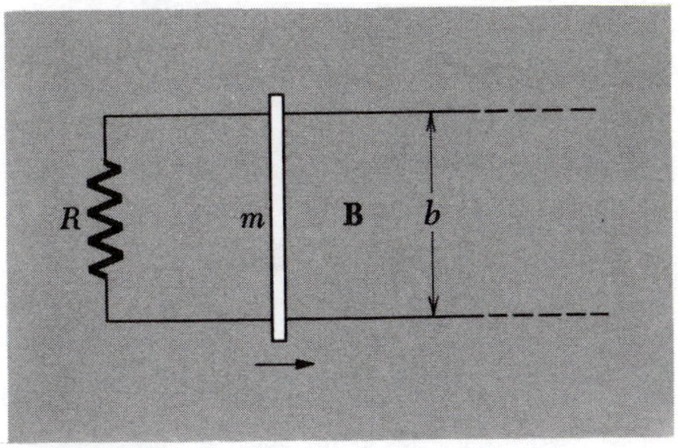

PROBLEM 7.14

7.15 A taut wire passes through the gap of a small magnet, where the field strength is 5000 gauss. The length of wire within the gap is 1.8 cm. Calculate the amplitude of the induced alternating voltage when the wire is vibrating at its fundamental frequency of 2000 Hz with an amplitude of 0.03 cm.

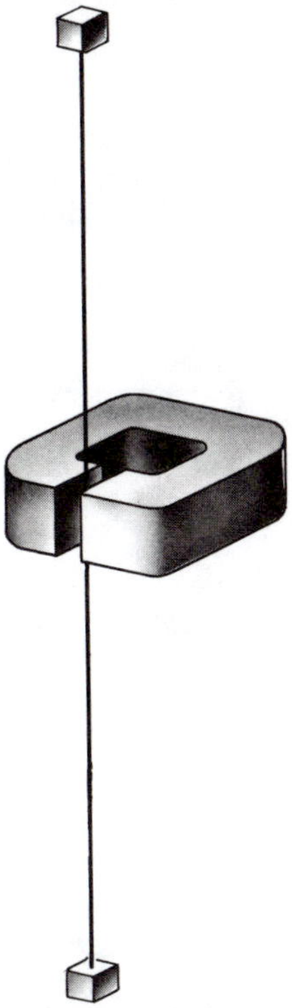

PROBLEM 7.15

7.16 The shaded region represents the pole of an electromagnet where there is a strong magnetic field perpendicular to the plane of the paper. The rectangular frame is made of a 5-mm-diameter aluminum rod, bent and with its ends welded together. Suppose that by applying a steady force of 1 newton, starting at the position shown, the frame can be pulled out of the magnet in 1 sec. Then: If the force is doubled, to 2 newtons, the frame will be pulled out in __________ sec. Brass has about twice the resistivity of aluminum. If the frame had been made of a 5-mm brass rod, the force needed to pull it out in 1 sec would be __________ newtons. If the frame had been made of a 1-cm-diameter aluminum rod, the force required to pull it out in 1 sec would be __________ newtons. You may neglect in all cases the inertia of the frame.

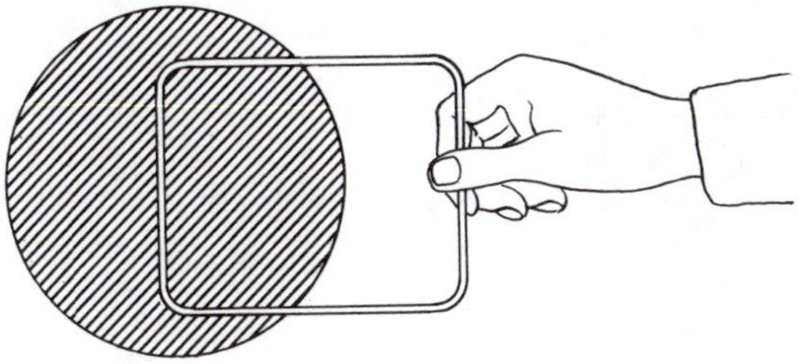

PROBLEM 7.16

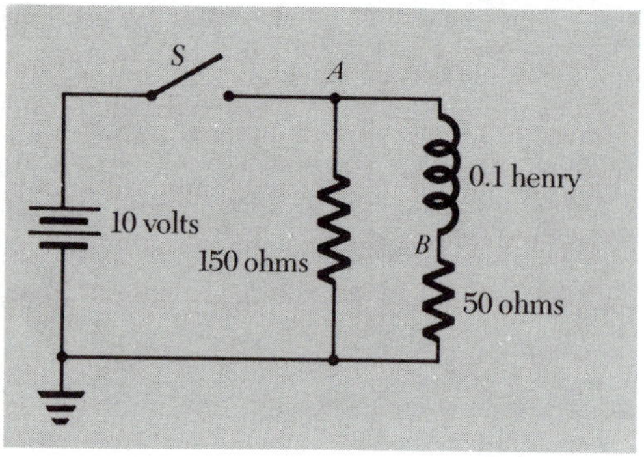

PROBLEM 7.17

7.17 In the circuit shown in the diagram the 10-volt battery has negligible internal resistance. The switch S is closed for several seconds, then opened. Make a graph with the abscissa time in milliseconds, showing the potential of point A with respect to ground, just before and then for 10 milliseconds after the opening of switch S. Show also the variation of the potential at point B in the same period of time.

7.18 A circular coil of wire, with N turns of radius a, is located in the field of an electromagnet. The magnetic field is perpendicular to the coil, and its strength has the constant value B_0 over that area. The coil is connected by a pair of twisted leads to an external resistance. The total resistance of this closed circuit, including that of the coil itself, is R. Suppose the electromagnet is turned off, its field dropping more or less rapidly to zero. The induced electromotive force causes current to flow around the circuit. Derive a formula for the total charge $Q = \int I\, dt$ which passes through the resistor, and explain why it does not depend on the rapidity with which the field drops to zero.

7.19 Discuss the implications of the theorem $\Phi_{12} = \Phi_{21}$ in the case of the large and small concentric rings in Fig. 7.20. With fixed current I_1 in the outer ring, obviously Φ_{21}, the flux through the inner ring, decreases if R_1 is increased, simply because the field at the center gets weaker. But with fixed current in the inner ring, why should Φ_{12}, the flux through the outer ring, *decrease* as R_1 increases, holding R_2 constant? It must do so to satisfy our theorem.

7.20 Can you devise a way to use the theorem $\Phi_{21} = \Phi_{12}$ to find the magnetic field strength due to a ring current at points in the plane of the ring at a distance from the ring much greater than the ring radius? (*Hint:* Consider the effect of a small change ΔR_1 in the radius of the outer ring in Fig. 7.20; it must have the same effect on Φ_{12} as on Φ_{21}.)

PROBLEM 7.21

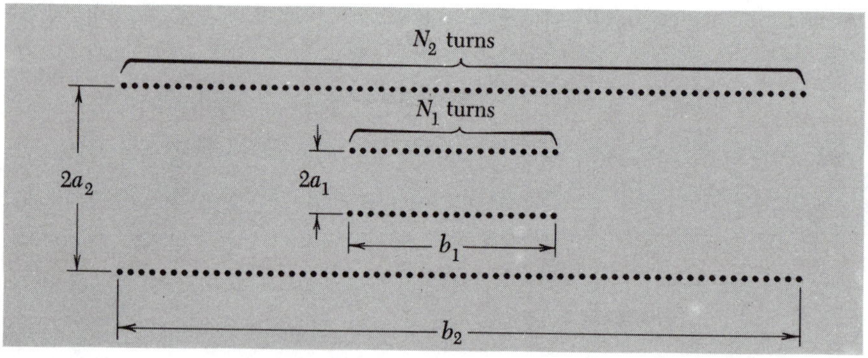

7.21 The figure shows a solenoid of radius a_1 and length b_1 located inside a longer solenoid of radius a_2 and length b_2. The total number of turns is N_1 on the inner coil, N_2 on the outer. Work out a formula for the mutual inductance M.

7.22 A thin ring of radius a carries a static charge q. This ring is in a magnetic field of strength B_0, parallel to the ring's axis, and is supported so that it is free to rotate about that axis. If the field is switched off, how much angular momentum will be added to the ring? Supposing the mass of the ring to be m, show that the ring, if initially at rest, will acquire an angular velocity $\omega = qB_0/2mc$. Notice that, as in Problem 7.18, the result depends only on the initial and final values of the field strength, and not the rapidity of change.

7.23 A magnetic field exists in most of the interstellar space in our galaxy. There is evidence that its strength in most regions is between 10^{-6} and 10^{-5} gauss. Adopting 3×10^{-6} gauss as a typical value, find, in order of magnitude, the total energy stored in the magnetic field of the galaxy. For this purpose you may assume the galaxy is a disk roughly 10^{23} cm in diameter and 10^{21} cm thick. To see whether the magnetic energy amounts to much, on that scale, you might consider the fact that all the stars in the galaxy are radiating about 10^{44} ergs/sec. How many years of starlight is the magnetic energy worth?

7.24 A superconducting solenoid designed for whole-body imaging by nuclear magnetic resonance is 0.9 meters in diameter and 2.2 meters long. The field at its center is 0.4 tesla. Estimate roughly the energy stored in the field of this coil, in joules.

7.25 It has been estimated that the magnetic field strength at the surface of a neutron star, or *pulsar*, may be as high as 10^{12} gauss. What is the energy density in such a field? Express it, using the mass-energy equivalence, in grams per cm^3.

7.26 Faraday describes in the following words an unsuccessful attempt to detect a current induced when part of a circuit consists of water moving through the earth's magnetic field:

> I made experiments therefore (by favour) at Waterloo Bridge, extending a copper wire nine hundred and sixty feet in length upon the parapet of the bridge, and dropping from its extremities other wires with extensive plates of metal attached to them to complete contact with the water. Thus the wire and the water made one conducting circuit; and as the water ebbed or flowed with the tide, I hoped to obtain currents analogous to those of the brass ball. I constantly obtained deflections at the galvanometer, but they were irregular, and were, in succession, referred to other causes than that sought for. The different condition of

> the water as to purity on the two sides of the river; the difference in temperature; slight differences in the plates, in the solder used, in the more or less perfect contact made by twisting or otherwise; all produced effects in turn: and though I experimented on the water passing through the middle arches only; used platina plates instead of copper; and took every other precaution, I could not after three days obtain any satisfactory results. ("Experimental Researches in Electricity," vol. I, London, 1839, p. 55.)

Assume the vertical component of the field was 0.5 gauss, make a reasonable guess about the velocity of tidal currents in the Thames and estimate the magnitude of the induced voltage Faraday was trying to detect.

7.27 We can think of a voltmeter as a device which registers the line integral $\int \mathbf{E} \cdot d\mathbf{s}$ along a path C *from* the clip at the end of its (+) lead *to* the clip at the end of its (−) lead. Part of C lies inside the voltmeter itself. Path C may also be part of a loop which is completed by some external path from the (−) clip to the (+) clip. With that in mind, consider the arrangement in the figure. The solenoid is so long that its external magnetic field is negligible. Its

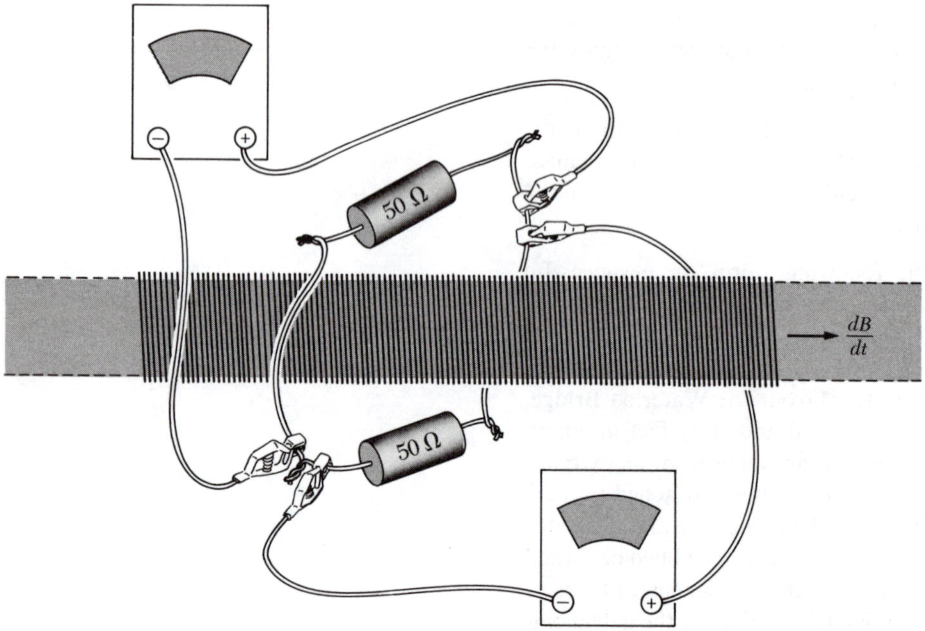

PROBLEM 7.27

cross section is 20 cm^2 in area, and the field inside is toward the right and increasing at the rate of 100 gauss/sec. Two identical voltmeters are connected as shown to points on the loop which encloses the solenoid and contains the two 50-ohm resistors. The voltmeters are capable of reading microvolts and have high internal resistance. What will each voltmeter read? Make sure your answer is consistent, from every point of view, with Eq. 25.

7.28 Magnetic fields inside good conductors cannot change quickly. We found that current in a simple inductive circuit decays exponentially with characteristic time L/R (Eq. 66). In a large conducting body such as the metallic core of the earth the "circuit" is not easy to identify. Nevertheless, we can find the order of magnitude of the decay time, and what it depends on, by making some reasonable approximations. Consider the solid doughnut of square cross section made of material with conductivity σ, in $\sec^{-1}$. A current I flows around it. Of course I is spread out in some manner over the cross section, but we shall assume the resistance is that of a wire of area a^2 and length πa, that is, $R \approx \pi/a\sigma$. For the field B we adopt the field at the center of a ring with current I and radius $a/2$. For the stored energy U a reasonable estimate would be $B^2/8\pi$ times the volume of the doughnut. Since $dU/dt = -I^2R$, the decay time of the energy U will be $\tau \approx U/I^2R$. Show that, except for some numerical factor depending on our various approximations, $\tau \approx \sigma a^2/c^2$. The radius of the earth's core is 3000 km, and its conductivity is believed to be 10^{16} $\sec^{-1}$, roughly one-tenth that of iron at room temperature. Evaluate τ in centuries.

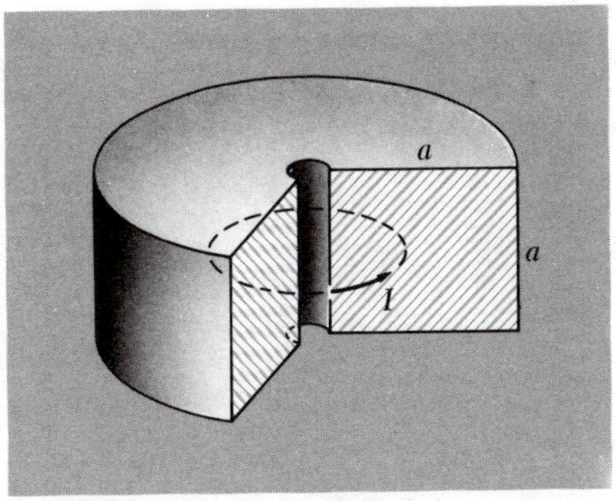

PROBLEM 7.28

7.29 The constant c which turns up in Maxwell's equations can be determined by electrical experiments involving low-frequency fields only. Consider the arrangement shown in the figure. The force between capacitor plates is balanced against the force between parallel wires carrying current in the same direction. A voltage alternating sinusoidally at a frequency f cycles per sec is applied to the parallel-plate capacitor C_1 and also to the capacitor C_2. The charge flowing into and out of C_2 constitutes the current in the rings. Suppose that C_2 and the various distances involved have been adjusted so that the time-average downward force on the upper plate of C_1 exactly balances the time-average downward force on the upper ring. (Of course, the *weights* of the two sides should be adjusted to balance with the voltage turned off.) Show that under these conditions the constant c can be computed from measured quantities as follows:

$$c = (2\pi)^{3/2}\, a \left(\frac{b}{h}\right)^{1/2} \left(\frac{C_2}{C_1}\right) f \qquad \text{(cm/sec)}$$

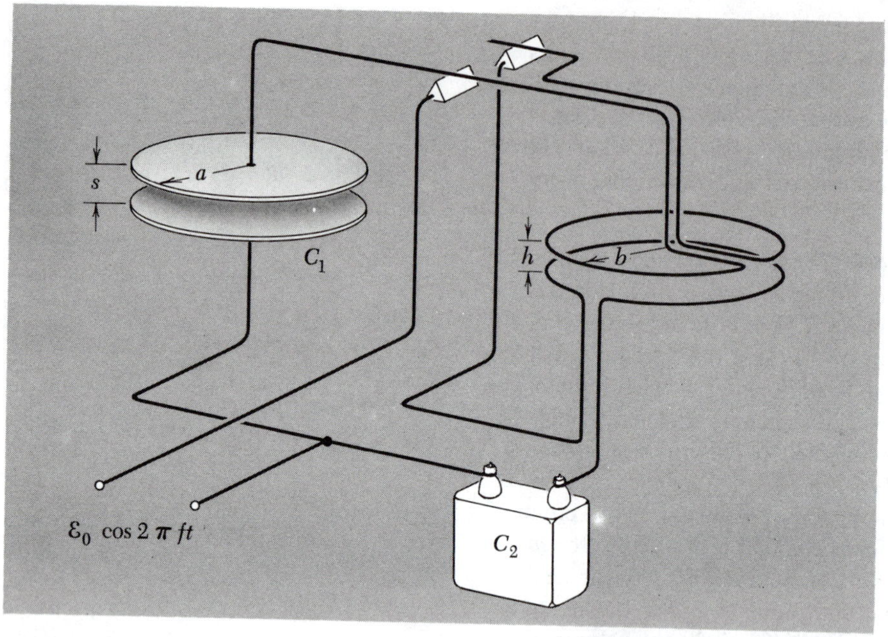

PROBLEM 7.29

Note that only measurements of *distance* and *time* (or frequency) are required, apart from a measurement of the ratio of the two capacitances C_1 and C_2. Electrical units, as such, are not involved in the result. (The experiment is actually feasible at a frequency as low as 60 cycles/sec if C_2 is made, say, 10^6 times C_1 and the current rings are made with several turns to multiply the effect of a small current.)

7.30 Consider the loop of wire shown in the figure. Suppose we want to calculate the flux of **B** through this loop. *Two* surfaces bounded by the loop are shown, in parts (*a*) and (*b*), respectively.

PROBLEM 7.30

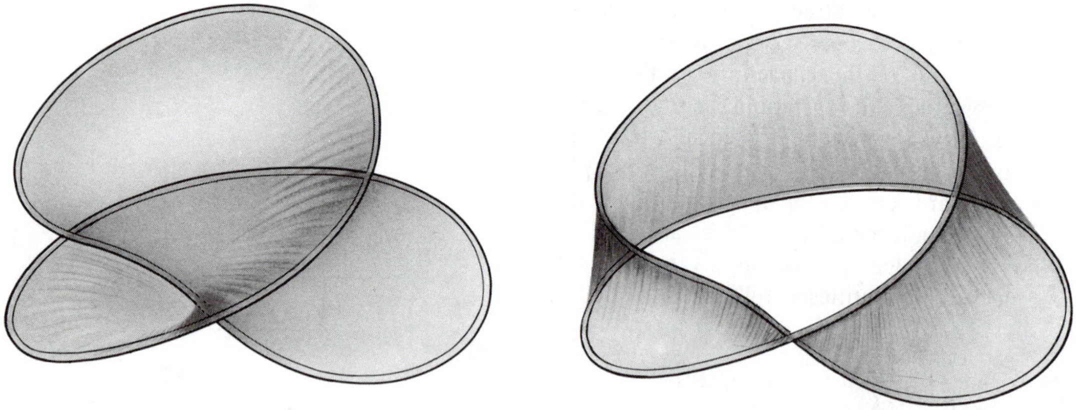

(*a*) (*b*)

What is the essential difference between them? Which, if either, is the correct surface to use in performing the surface integral $\int \mathbf{B} \cdot d\mathbf{a}$ to find the flux? Describe the corresponding surface for a three-turn coil. Show that this is all consistent with our previous assertion that for a compact coil of N turns the electromotive force is just N times what it would be for a single loop of the same size and shape.

7.31 In this question the term *dynamo* will be used for a generator which works in the following way. By some external agency—the shaft of a steam turbine, for instance—a conductor is driven through a magnetic field, inducing an electromotive force in a circuit of which that conductor is part. The source of the magnetic field is the current which is caused to flow in that circuit by that electromotive force. An electrical engineer would call it a self-excited dc generator. One of the simplest dynamos conceivable is sketched below. It has only two essential parts. One part is a solid metal disk and axle which can be driven in rotation. The other is a two-turn "coil" which is stationary but is connected by sliding contacts, or "brushes," to the axle and to the rim of the revolving disk. One of the two devices pictured is, at least potentially, a dynamo. The other is not. Which is the dynamo? Note that the answer to this question cannot depend on any convention about handedness or current directions. An intelligent extraterrestrial being inspecting the sketches could give the answer, provided only that it knows about arrows! What do you think determines the direction of the current in such a dynamo? What will determine the magnitude of the current?

7.32 A dynamo like the one in the preceding problem has a certain critical speed ω_0. If the disk revolves with an angular velocity less than ω_0, nothing happens. Only when that speed is attained is the induced

PROBLEM 7.31

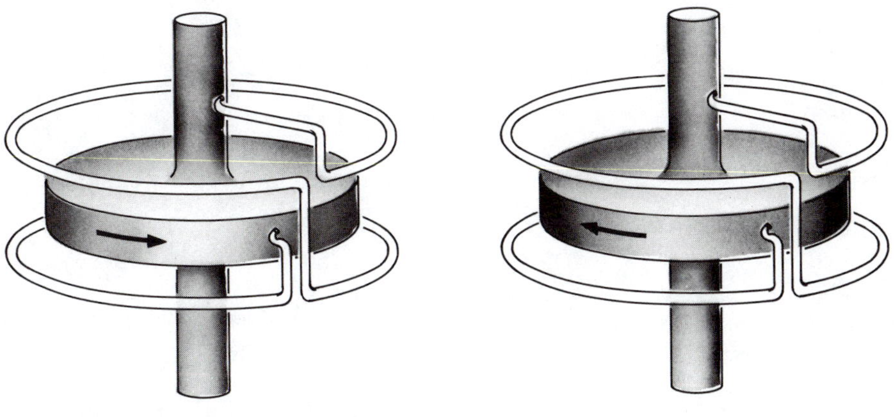

$\mathcal{E}$ large enough to make the current large enough to make the magnetic field large enough to induce an $\mathcal{E}$ of that magnitude. The critical speed can only depend on the size and shape of the conductors and the conductivity σ. Remember that σ has the dimensions $\sec^{-1}$. Let d be some characteristic dimension expressing the size of the dynamo, such as the radius of the disk in our example. Show by a dimensional argument that ω_0 must be given by a relation of this form: $\omega_0 = Kc^2/d^2\sigma$, where K is some dimensionless numerical factor that depends only on the arrangement and *relative* size of the various parts of the dynamo. For a dynamo of modest size made wholly of copper, the critical speed ω_0 would be practically unattainable. It is ferromagnetism that makes possible the ordinary dc generator by providing a magnetic field much stronger than the current in the coils, unaided, could produce. For an earth-sized dynamo, however, with d measured in hundreds of kilometers rather than meters, the critical speed is very much smaller. The earth's magnetic field is almost certainly produced by a nonferromagnetic dynamo involving motions in the fluid metallic core. That fluid happens to be molten iron, but it is not even slightly ferromagnetic because it is too hot. (That will be explained in Chapter 11.) We don't know how the conducting fluid moves, or what configuration of electric currents and magnetic fields its motion generates in the core. The magnetic field we observe at the earth's surface is the external field of the dynamo in the core. The direction of the earth's field a million years ago is preserved in the magnetization of rocks that solidified at that time. That magnetic record shows that the field has reversed its direction nearly 200 times in the last 100 million years. Although a reversal cannot have been instantaneous (see Problem 7.28), it was a relatively sudden event on the geological time scale. The immense value of *paleomagnetism* as an indelible record of our planet's history is well explained in Chapter 18 of *Earth,* by Frank Press and Raymond Siever, second edition, 1978 (W. H. Freeman).

8

ALTERNATING-CURRENT CIRCUITS

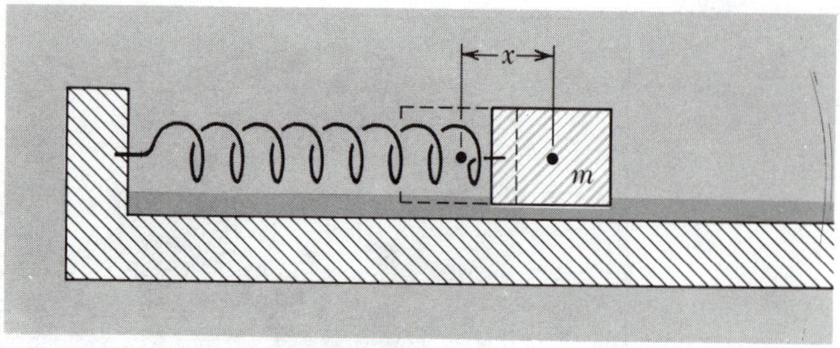

FIGURE 8.1
A mechanical damped harmonic oscillator.

A RESONANT CIRCUIT

8.1 A mass attached to a spring is a familiar example of an oscillator. If the amplitude of oscillation is not too large, the motion will be a sinusoidal function of the time. In that case, we call it a *harmonic oscillator.* The characteristic feature of any mechanical harmonic oscillator is a restoring force proportional to the displacement of a mass m from its position of equilibrium, $F = -kx$ (Fig. 8.1). In the absence of other external forces the mass, if initially displaced, will oscillate with unchanging amplitude at the angular frequency, $\omega = \sqrt{k/m}$. But usually some kind of friction will bring it eventually to rest. The simplest case is that of a retarding force proportional to the velocity of the mass, dx/dt. Motion in a viscous fluid provides an example. A system in which the restoring force is proportional to some displacement x and the retarding force is proportional to the time derivative dx/dt is called a *damped harmonic oscillator.*

An electric circuit containing capacitance and inductance has the essentials of a harmonic oscillator. Ohmic resistance makes it a damped harmonic oscillator. Indeed, thanks to the extraordinary linearity of actual electric circuit elements, the electrical damped harmonic oscillator is more nearly ideal than most mechanical oscillators. The system we shall study first is the "series *RLC*" circuit diagrammed in Fig. 8.2.

FIGURE 8.2
A "series *RLC*" circuit.

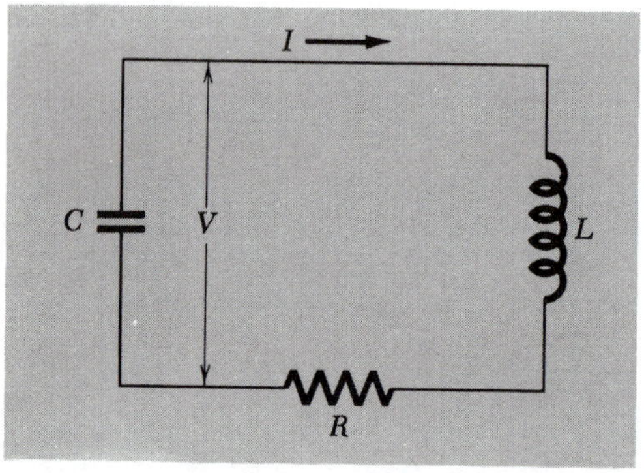

Let Q be the charge, at time t, on the capacitor in this circuit. The potential difference, or voltage across the capacitor, is V, which obviously is the same as the voltage across the series combination of inductor L and resistor R. We take V to be positive when the upper capacitor plate is positively charged, and we define the positive current direction by the arrow in Fig. 8.2. With the signs chosen that way, the relations connecting charge Q, current I, and voltage across the capacitor V are

$$I = -\frac{dQ}{dt} \qquad Q = CV \qquad V = L\frac{dI}{dt} + RI \qquad (1)$$

We want to eliminate two of the three variables Q, I, and V. From the first two equations we obtain $I = -C\,dV/dt$, and the third equation becomes $V = -LC(d^2V/dt^2) - RC(dV/dt)$, or

$$\frac{d^2V}{dt^2} + \left(\frac{R}{L}\right)\frac{dV}{dt} + \left(\frac{1}{LC}\right)V = 0 \tag{2}$$

This is a second-order differential equation with constant coefficients. We shall try a solution of the form

$$V = Ae^{-\alpha t}\cos\omega t \tag{3}$$

where A, α, and ω are constants. The first and second derivatives of this function are

$$\frac{dV}{dt} = Ae^{-\alpha t}[-\alpha\cos\omega t - \omega\sin\omega t] \tag{4}$$

$$\frac{d^2V}{dt^2} = Ae^{-\alpha t}[(\alpha^2 - \omega^2)\cos\omega t + 2\alpha\omega\sin\omega t] \tag{5}$$

Substituting back into Eq. 2, we cancel out the common factor $Ae^{-\alpha t}$ and are left with

$$(\alpha^2 - \omega^2)\cos\omega t + 2\alpha\omega\sin\omega t - \frac{R}{L}(\alpha\cos\omega t + \omega\sin\omega t) + \frac{1}{LC}\cos\omega t = 0 \tag{6}$$

This will be satisfied for all t if, and only if, the coefficients of $\sin\omega t$ and $\cos\omega t$ are both zero. That is, we must require

$$2\alpha\omega - \frac{R\omega}{L} = 0 \tag{7}$$

and

$$\alpha^2 - \omega^2 - \alpha\frac{R}{L} + \frac{1}{LC} = 0 \tag{8}$$

The first of these equations gives a condition on α:

$$\alpha = \frac{R}{2L} \tag{9}$$

while the second equation requires that

$$\omega^2 = \frac{1}{LC} - \alpha\frac{R}{L} + \alpha^2 = \frac{1}{LC} - \frac{R^2}{4L^2} \tag{10}$$

Since our constant ω is a real number, ω^2 cannot be negative. Therefore we succeed in obtaining a solution of the form assumed in Eq. 3 only if $R^2/4L^2 \leqq 1/LC$. In fact it is the case of "light damping," that is, low resistance, that we want to examine, so we shall assume that the values of R, L, and C in the circuit are such that the inequality $R < 2\sqrt{L/C}$ holds.

The function $Ae^{-\alpha t}\cos \omega t$ is not the only possible solution. $Be^{-\alpha t}\sin \omega t$ works just as well, with the same requirements, Eq. 9 and Eq. 10, on α and ω, respectively. The general solution is the sum of these:

$$V(t) = e^{-\alpha t}(A\cos \omega t + B\sin \omega t) \tag{11}$$

The arbitrary constants A and B could be adjusted to fit initial conditions. That is not very interesting. Whether the solution in any given case involves the sine or the cosine function, or some superposition, is a trivial matter of how the clock is set. The essential phenomenon is a damped sinusoidal oscillation.

The variation of voltage with time is shown in Fig. 8.3*a*. Of course, this cannot really hold for all *past* time. At some time in the past the circuit must have been provided with energy somehow, and then left running. For instance, the capacitor might have been charged, with the circuit open, and then connected to the coil.

In Fig. 8.3*b* the time scale has been expanded and the dotted curve showing the variation of the current I has been added. For V let us take the damped cosine, Eq. 3. Then the current as a function of time is given by

$$I = -C\frac{dV}{dt} = AC\omega\left(\sin \omega t + \frac{\alpha}{\omega}\cos \omega t\right)e^{-\alpha t} \tag{12}$$

The ratio α/ω is a measure of the damping. If α/ω is very small, many oscillations occur while the amplitude is decaying only a little. For Fig. 8.3 we chose a case in which $\alpha/\omega \approx 0.04$. Then the cosine term in Eq. 12 doesn't amount to much. All it does, in effect, is shift the phase by a small angle, $\tan^{-1}(\alpha/\omega)$. So the current oscillation is almost exactly one-quarter cycle out of phase with the voltage oscillation.

The oscillation involves a transfer of energy back and forth from the capacitor to the inductor, or from electric field to magnetic field. At the times marked 1 in Fig. 8.3*b* all the energy is in the electric field. A quarter-cycle later, at 2, the capacitor is discharged and nearly all this energy is found in the magnetic field of the coil. Meanwhile, the circuit resistance R is taking its toll, and as the oscillation goes on, the energy remaining in the fields gradually diminishes.

The relative damping in an oscillator is often expressed by giving a number called Q. This number Q (not to be confused with the charge on the capacitor!) is said to stand for *quality* or *quality factor*. In fact, no one calls it that; we just call it Q. The less the damping, the larger the number Q. For an oscillator with frequency ω, Q is the dimensionless ratio formed as follows:

$$Q = \omega\frac{\text{energy stored}}{\text{average power dissipated}} \tag{13}$$

FIGURE 8.3
(*a*) The damped sinusoidal oscillation of voltage in the *RLC* circuit. (*b*) A portion of (*a*) with the time scale expanded and the graph of the current *I* included. (*c*) The periodic transfer of energy from electric field to magnetic field and back again. Each picture represents the condition at times marked by the corresponding number in (*b*).

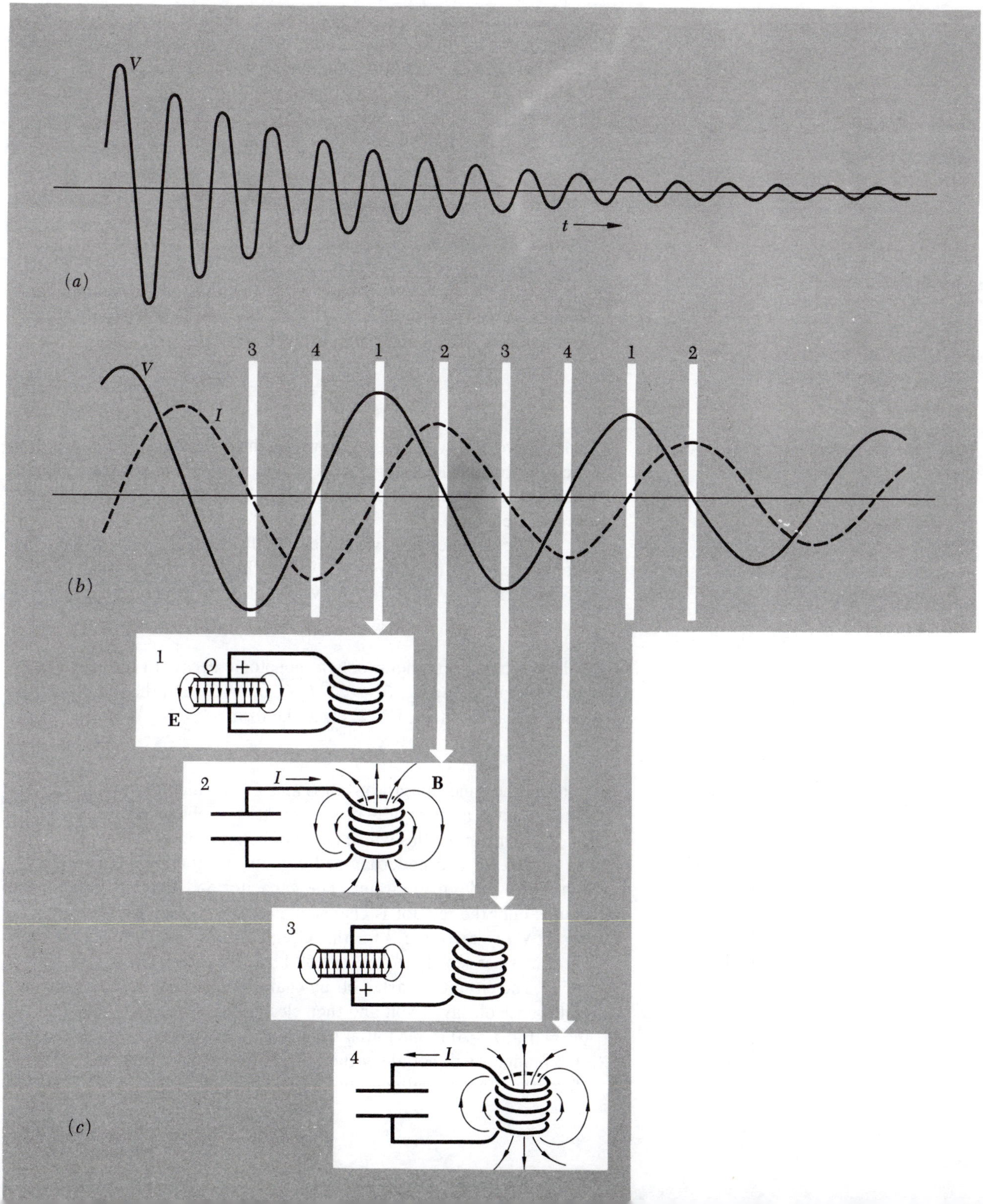
V
t
(a)
3 4 1 2 3 4 1 2
V
I
(b)
1
Q
+
−
E
2
I
B
3
−
+
4
I
(c)

Or you may prefer to remember that Q is the number of radians of the argument ωt (that is, 2π times the number of cycles) required for the energy in the oscillator to diminish by the factor $1/e$.

In our circuit the stored energy is proportional to V^2 or I^2 and, therefore, to $e^{-2\alpha t}$. The energy decays by $1/e$ in a time $= 1/2\alpha$, which covers $\omega/2\alpha$ radians. Hence for our RLC circuit

$$Q = \frac{\omega}{2\alpha} = \frac{\omega L}{R} \tag{14}$$

As a rough estimate, what is the Q of the oscillation represented in Fig. 8.3?

Clearly, the general case we have just studied includes some simple special cases. If $R = 0$, we have the completely undamped oscillator, whose frequency ω_0 is given by

$$\omega_0 = \frac{1}{\sqrt{LC}} \tag{15}$$

Mostly we deal with systems in which the damping is small enough to be ignored in calculating the frequency. As we can see from Eq. 10, and as Problem 8.9 will demonstrate, light damping has only a second-order effect on ω.

For completeness we review briefly what goes on in the overdamped circuit, in which $R > 2\sqrt{L/C}$. Equation 2 then has a solution of the form $V = Ae^{-\beta t}$ for two values of β, the general solution being

$$V(t) = Ae^{-\beta_1 t} + Be^{-\beta_2 t} \tag{16}$$

There are no oscillations, only a monotonic decay. In the special case of "critical" damping, $R = 2\sqrt{L/C}$, $\beta_1 = \beta_2$, and the solution of the differential equation, Eq. 2, takes the form

$$V(t) = (A + Bt)e^{-\beta t} \tag{17}$$

This is the condition, for given L and C, in which the total energy in the circuit is most rapidly dissipated. (See Problem 8.8.)

You can see this whole range of behavior in Fig. 8.4, where $V(t)$ is plotted for two underdamped circuits, a critically damped circuit, and an overdamped circuit. The capacitor and inductor remain the same; only the resistor is changed. The natural angular frequency $\omega_0 = 1/\sqrt{LC}$ is 10^6 sec^{-1} for this circuit, corresponding to a frequency in cycles per sec of $10^6/2\pi$, or 159 kilocycles per sec.

The circuit is started off by charging the capacitor to a potential difference of, say, 1 volt and then closing the switch at $t = 0$. That is, $V = 1$ at $t = 0$ is one initial condition. Also, $I = 0$ at $t = 0$, because the inductor will not allow the current to rise discontinuously. Therefore, the other initial condition on V is $dv/dt = 0$, at $t = 0$. Notice

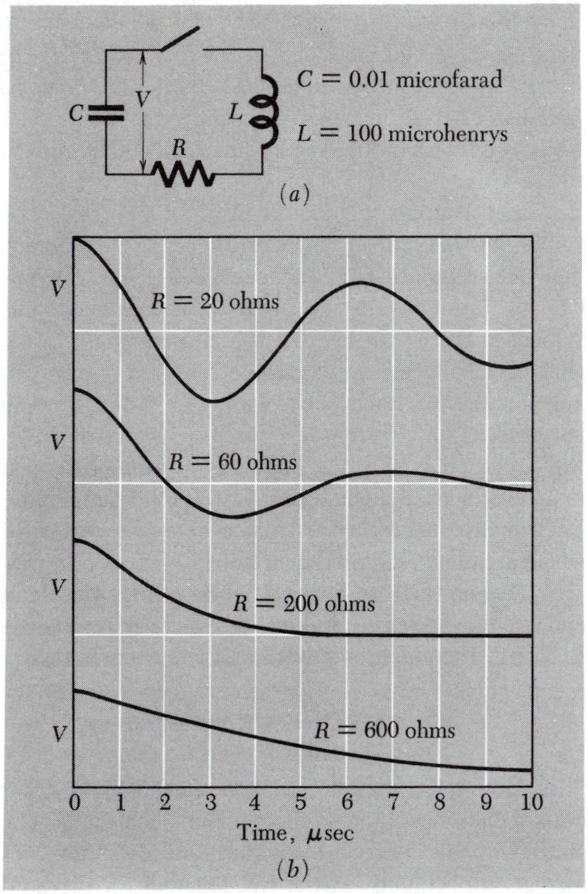

FIGURE 8.4
(*a*) With the capacitor charged, the switch is closed at $t = 0$. (*b*) Four cases are shown, one of which, $R = 200$ ohms, is the case of critical damping.

that all four decay curves start the same way. In the heavily damped case ($R = 600$ ohms) most of the decay curve looks like the simple exponential decay of an *RC* circuit. Only the very beginning where the curve is rounded over so that it starts with zero slope, betrays the presence of the inductance *L*.

ALTERNATING CURRENT

8.2 The resonant circuit we have just discussed contained no source of energy and was, therefore, doomed to a *transient* activity, an oscillation that must sooner or later die out. In an alternating-current circuit we are concerned with a *steady state,* a current and voltage oscillating sinusoidally without change in amplitude. Some oscillating electromotive force drives the system.

The frequency f of an alternating current is ordinarily expressed

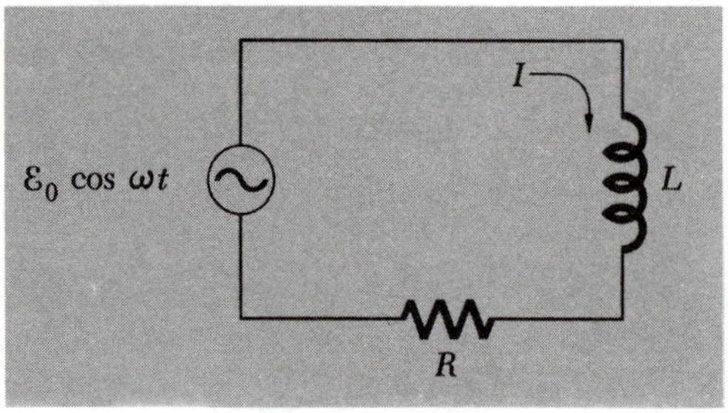

FIGURE 8.5
A circuit with inductance driven by an alternating electromotive force.

in cycles per sec [or Hertz (Hz), after the discoverer† of electromagnetic waves]. The angular frequency $\omega = 2\pi f$ is the quantity that usually appears in our equations. It will always be assumed to be in radians/sec. That unit has no special name; we'll write it simply sec^{-1}. Thus our familiar (in North America) 60-Hz current has $\omega = 377$ sec^{-1}.

Let us apply an electromotive force $\mathcal{E} = \mathcal{E}_0 \cos \omega t$ to a circuit containing inductance and resistance. We might generate $\mathcal{E}$ by a machine schematically like the one in Fig. 7.13, having provided some engine or motor to turn the shaft at the constant angular speed ω. The symbol at the left in Fig. 8.5 is a conventional way to show the presence of an alternating electromotive force in a circuit. It suggests a generator connected in series with the rest of the circuit. But you need not think of an electromotive force as located at a particular place in the circuit. It is only the line integral around the whole circuit that matters. Figure 8.5 could just as well represent a circuit in which the electromotive force arises from a changing magnetic field over the whole area enclosed by the circuit.

We set the sum of potential drops over the elements of this circuit equal to the electromotive force $\mathcal{E}$, exactly as we did in developing Eq. 7.61. The equation governing the current is then

$$L\frac{dI}{dt} + RI = \mathcal{E}_0 \cos \omega t \tag{18}$$

Now there may be some transient behavior, depending on the initial conditions, that is, on how and when the generator is switched on. But we are interested only in the steady state, when the current is oscillating obediently at the frequency of the driving force, with the amplitude and phase necessary to keep Eq. 18 satisfied. To show that this is possible, consider a current described by

$$I = I_0 \cos(\omega t + \varphi) \tag{19}$$

To determine the constants I_0 and φ, we put this into Eq. 18:

$$-LI_0\omega \sin(\omega t + \varphi) + RI_0 \cos(\omega t + \varphi) = \mathcal{E}_0 \cos \omega t \tag{20}$$

The functions $\cos \omega t$ and $\sin \omega t$ can be separated out:

$$\begin{aligned} -LI_0\omega\, &(\sin \omega t \cos \varphi + \cos \omega t \sin \varphi) \\ &+ RI_0 (\cos \omega t \cos \varphi - \sin \omega t \sin \varphi) = \mathcal{E}_0 \cos \omega t \end{aligned} \tag{21}$$

†In 1887 Heinrich Hertz demonstrated electromagnetic waves generated by oscillating currents in a macroscopic electric circuit. The frequencies were around 10^9 cycles per sec, corresponding to wavelengths around 30 cm. Although Maxwell's theory, developed 15 years earlier, had left little doubt that light must be an electromagnetic phenomenon, in the history of electromagnetism Hertz's experiments were an immensely significant turning point.

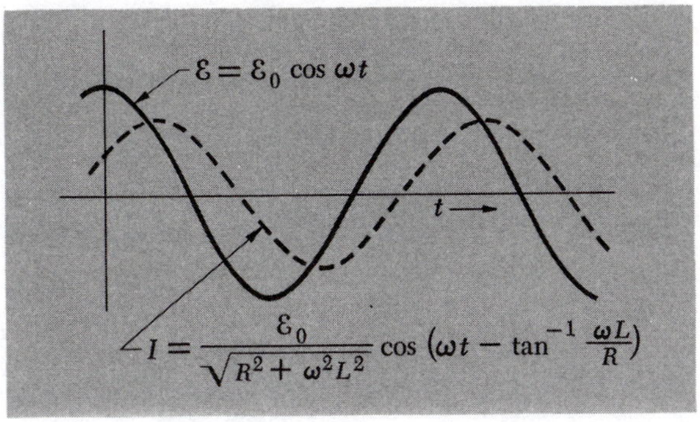

FIGURE 8.6
The current I, in the circuit of Fig. 8.5, plotted along with the electromotive force $\mathcal{E}$ on the same time scale. Note the phase difference.

Setting the coefficients of cos ωt and sin ωt separately equal to zero,

$$-LI_0\,\omega \cos \varphi - RI_0 \sin \varphi = 0 \tag{22}$$

which gives

$$\tan \varphi = -\frac{\omega L}{R} \tag{23}$$

$$-LI_0\,\omega \sin \varphi + RI_0 \cos \varphi - \mathcal{E}_0 = 0 \tag{24}$$

which gives

$$\begin{aligned} I_0 &= \frac{\mathcal{E}_0}{R \cos \varphi - \omega L \sin \varphi} \\ &= \frac{\mathcal{E}_0}{R\,(\cos \varphi + \sin \varphi \tan \varphi)} = \frac{\mathcal{E}_0 \cos \varphi}{R} \end{aligned} \tag{25}$$

or since

$$\cos \varphi = \frac{R}{\sqrt{R^2 + \omega^2 L^2}} \qquad \text{(from Eq. 23)} \tag{26}$$

$$I_0 = \frac{\mathcal{E}_0}{\sqrt{R^2 + \omega^2 L^2}} \tag{27}$$

In Fig. 8.6 the oscillations of $\mathcal{E}$ and I are plotted on the same graph. Since φ is a negative angle, the current reaches its maximum a bit *later* than the electromotive force. One says, "The current lags the voltage in an inductive circuit." The quantity ωL, which has the dimensions of resistance and can be expressed in ohms is called the *inductive reactance.*

If we replace the inductor L by a capacitor C, as in Fig. 8.7, we have a circuit governed by the equation

$$-\frac{Q}{C} + RI = \mathcal{E}_0 \cos \omega t \tag{28}$$

FIGURE 8.7
An alternating electromotive force in a circuit containing resistance and capacitance.

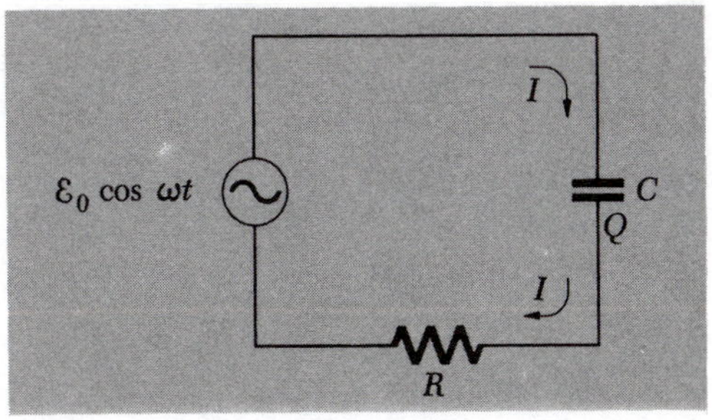

We consider the steady-state solution

$$I = I_0 \cos(\omega t + \varphi) \tag{29}$$

Since $I = -dQ/dt$, we have

$$Q = -\int I\, dt = -\frac{I_0}{\omega} \sin(\omega t + \varphi) \tag{30}$$

Note that, in going from I to Q by integration, there is no question of adding a constant of integration, for we know that Q must oscillate symmetrically about zero in the steady state.

Substituting back into Eq. 28 leads to

$$\frac{I_0}{\omega C} \sin(\omega t + \varphi) + RI_0 \cos(\omega t + \varphi) = \mathcal{E}_0 \cos \omega t \tag{31}$$

Just as before, we obtain conditions on φ and I_0 by requiring that the coefficients of $\cos \omega t$ and $\sin \omega t$ separately vanish. In this case, the results are

$$\tan \varphi = \frac{1}{R\omega C} \tag{32}$$

and

$$I_0 = \frac{\mathcal{E}_0}{\sqrt{R^2 + (1/\omega C)^2}} \tag{33}$$

Notice that the phase angle is now positive. As the saying goes, the current "leads the voltage" in a capacitive circuit. What this means is apparent in the graph of Fig. 8.8.

Mathematically speaking, the function

$$I = \frac{\mathcal{E}_0}{\sqrt{R^2 + \omega^2 L^2}} \cos\left(\omega t - \tan^{-1} \frac{\omega L}{R}\right) \tag{34}$$

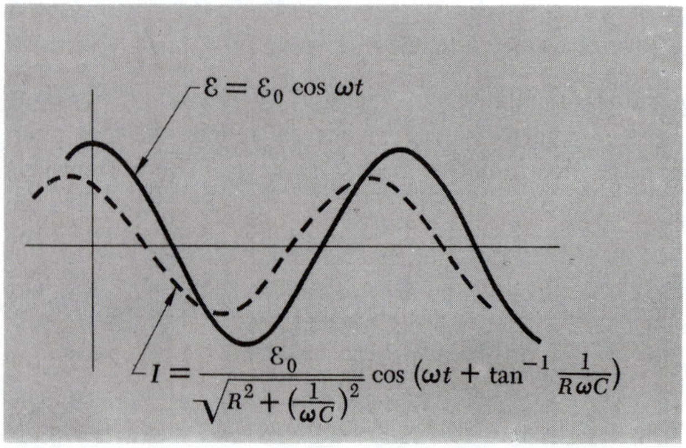

FIGURE 8.8
The current in the RC circuit. Compare the phase shift here with the phase shift in the inductive circuit in Fig. 8.6. The maximum in I occurs here a little earlier than the maximum in $\mathcal{E}$.

is a *particular integral* of the differential equation, Eq. 18. To this could be added a *complementary function,* that is, any solution of the homogeneous differential equation

$$L\frac{dI}{dt} + RI = 0 \tag{35}$$

Now this is just Eq. 65 of Chapter 7, whose solution we found, in Section 7.9, to be an exponentially decaying function,

$$I \sim e^{-(R/L)t} \tag{36}$$

The physical significance is this: A transient, determined by some initial conditions, is represented by a decaying component of $I(t)$, of the form of Eq. 36. After a time $t \gg L/R$, this will have vanished leaving only the steady sinusoidal oscillation at the driving frequency, represented by the particular integral, Eq. 34.

The similarity of our results for the RL circuit and the RC circuit suggests a way to look at the inductor and capacitor in series. Suppose an alternating current $I = I_0 \cos(\omega t + \varphi)$ is somehow caused to flow through such a combination (shown in Fig. 8.9). The voltage across the inductor, V_L, will be

$$V_L = L\frac{dI}{dt} = -I_0\omega L \sin(\omega t + \varphi) \tag{37}$$

The voltage V_C across the capacitor, with sign consistent with the sign of V_L, is

$$V_C = -\frac{Q}{C} = \frac{1}{C}\int I\,dt = \frac{I_0}{\omega C}\sin(\omega t + \varphi) \tag{38}$$

The voltage across the combination is then

$$V = V_L + V_C = -\left(\omega L - \frac{1}{\omega C}\right) I_0 \sin(\omega t + \varphi) \tag{39}$$

For a given ω, the combination is evidently equivalent to a single element, either an inductor or a capacitor, depending on whether the quantity $\omega L - 1/\omega C$ is positive or negative. Suppose, for example, that $\omega L > 1/\omega C$. Then the combination is equivalent to an inductor L' such that

$$\omega L' = \omega L - \frac{1}{\omega C} \tag{40}$$

Equivalence means *only* that the relation between current and voltage, for steady oscillation at the particular frequency ω, is the same. This allows us to replace L and C by L' in any circuit driven at this frequency.

This can be applied to the simple RLC circuit in Fig. 8.10. We

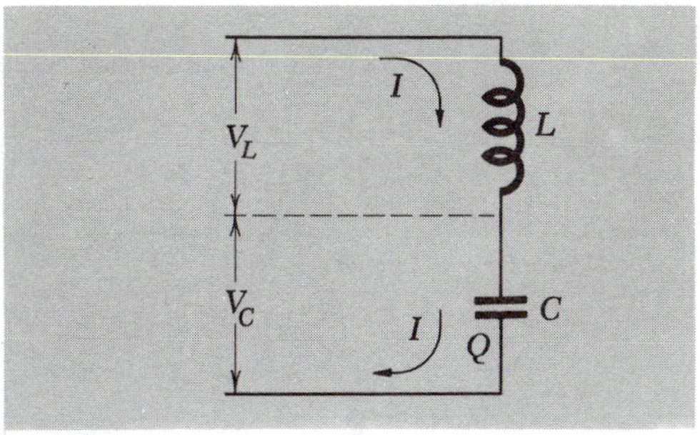

FIGURE 8.9
The inductor and capacitor in series are equivalent to a single reactive element which is either an inductor or a capacitor depending on whether $\omega^2 LC$ is greater or less than 1.

FIGURE 8.10
The RLC circuit driven by a sinusoidal electromotive force.

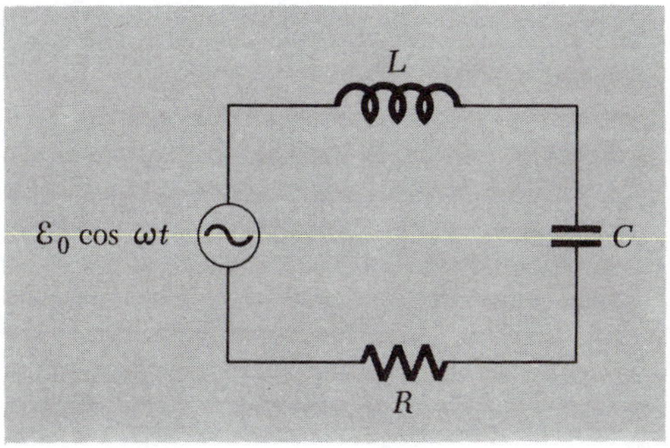

need only recall Eqs. 23 and 27, the solution for the RL circuit driven by the electromotive force $\mathcal{E}_0 \cos \omega t$, and replace ωL by $\omega L - 1/\omega C$:

$$I = \frac{\mathcal{E}_0}{\sqrt{R^2 + (\omega L - 1/\omega C)^2}} \cos(\omega t + \varphi) \tag{41}$$

$$\tan \varphi = \frac{1}{R\omega C} - \frac{\omega L}{R} \tag{42}$$

For fixed amplitude $\mathcal{E}_0$ of the electromotive force, and given circuit elements L, C, and R, we get the greatest current when the driving frequency ω is such that

$$\omega L - \frac{1}{\omega C} = 0 \tag{43}$$

which is the same as saying that $\omega = 1/\sqrt{LC} = \omega_0$, the resonant frequency of the undamped LC circuit. In that case Eq. 41 reduces to

$$I = \frac{\mathcal{E}_0 \cos \omega t}{R} \tag{44}$$

That is exactly the current that would flow if the circuit contained the resistor alone.

As an example, consider the circuit of Fig. 8.4*a*, connected now to a source or generator of alternating emf, $\mathcal{E} = \mathcal{E}_0 \cos \omega t$. The driving frequency ω may be different from the resonant frequency $\omega_0 = 1/\sqrt{LC}$, which, for the given capacitance (0.01 microfarads) and the inductance (100 microhenrys), is 10^6 radians/sec (or $10^6/2\pi$ cycles per sec). Figure 8.11 shows the amplitude of the oscillating current, as a function of the driving frequency ω, for three different values of the circuit resistance R. It is assumed that the amplitude $\mathcal{E}_0$ of the emf is 100 volts in each case. Notice the resonance peak at $\omega = \omega_0$, which is most prominent and sharp for the lowest resistance value. This is the same value of R for which, running as a damped oscillator without any driving emf, the circuit behaved as shown in the top graph of Fig. 8.4*b*.

The Q of the circuit, defined by Eq. 14 as $\omega_0 L/R$,† is $(10^6 \times 10^{-4})/20$, or 5, in this case. Generally speaking, the higher the Q of a circuit, the narrower and higher the peak of its response as a function of driving frequency ω. To be more precise, consider frequencies in the neighborhood of ω_0, writing $\omega = \omega_0 + \Delta\omega$. Then to first order in $\Delta\omega/\omega_0$, the expression $\omega L - 1/\omega C$ which occurs in the denominator in Eq. 41 can be approximated this way:

†The ω in Eq. 14 was the frequency of the freely decaying damped oscillator, practically the same as ω_0 for moderate or light damping. We use ω_0 here in the definition of Q. In the present discussion ω is *any* frequency we may choose to apply to this circuit.

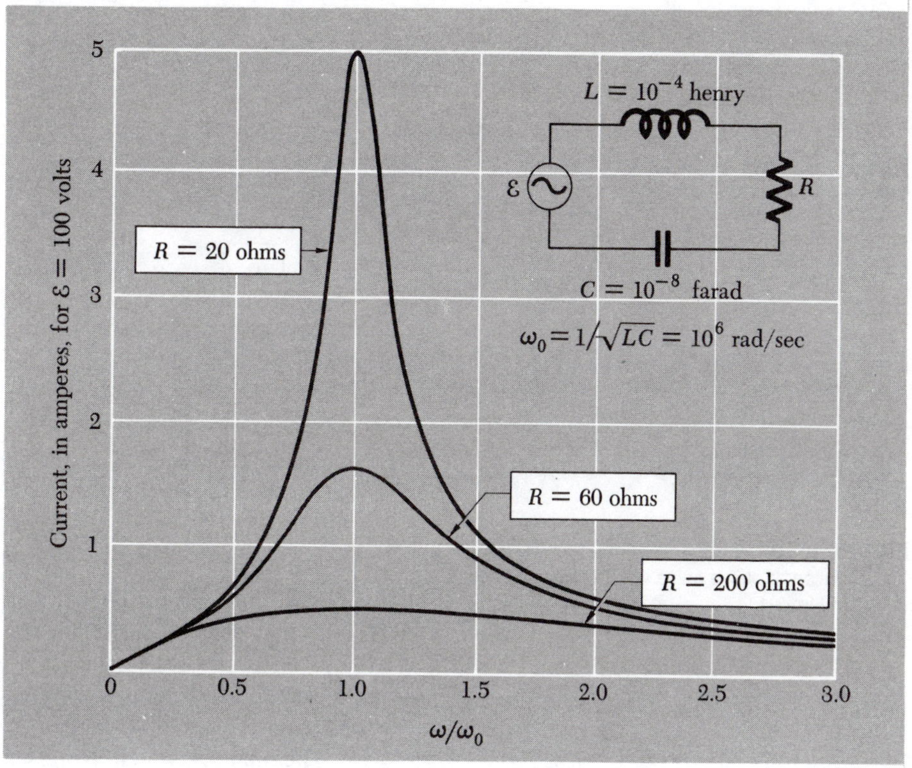

FIGURE 8.11
An emf of 100 volts amplitude is applied to a series *RLC* circuit. The circuit elements are the same as in the example of the damped circuit in Fig. 8.4. Current amplitude is calculated by Eq. 41 and plotted, as a function of ω/ω_0, for three different resistance values.

$$\omega L - \frac{1}{\omega C} = \omega_0 L\left(1 + \frac{\Delta\omega}{\omega_0}\right) - \frac{1}{\omega_0 C(1 + \Delta\omega/\omega_0)} \tag{45}$$

and since ω_0 is $1/\sqrt{LC}$, this becomes

$$\omega_0 L\left(1 + \frac{\Delta\omega}{\omega_0} - \frac{1}{1 + \Delta\omega/\omega_0}\right) \approx \omega_0 L\left(2\frac{\Delta\omega}{\omega_0}\right) \tag{46}$$

Exactly at resonance, the quantity inside the square root sign in Eq. 41 is just R^2. As ω is shifted away from resonance the quantity under the square root will have doubled when $|\omega L - 1/\omega C| = R$, or when, approximately,

$$\frac{2|\Delta\omega|}{\omega_0} = \frac{R}{\omega_0 L} = \frac{1}{Q} \tag{47}$$

This means that the current amplitude will have fallen to $1/\sqrt{2}$ times the peak when $|\Delta\omega/\omega_0| = 1/2Q$. These are the "half-power" points, because the energy or power is proportional to the amplitude squared, as we shall explain in Section 8.5. One often expresses the width of a resonance peak by giving the full width between half-power points. Evidently that is just $1/Q$ times the resonant frequency itself. Circuits

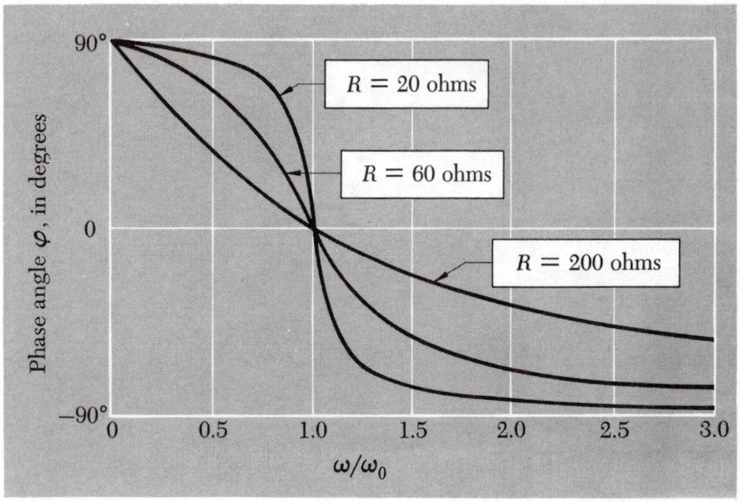

FIGURE 8.12
The variation of phase angle with frequency, in the circuit of Fig. 8.11.

with very much higher Q than this one are quite common. A radio receiver may select a particular station and discriminate against others by means of a resonant circuit with a Q of several hundred. It is quite easy to make a microwave resonant circuit with a Q of 10^4, or even 10^5.

The angle φ, which expresses the relative phase of the current and emf oscillations, varies with frequency in the manner shown in Fig. 8.12. At a very low frequency the capacitor is the dominant hindrance to current flow, and φ is positive. At resonance, $\varphi = 0$. The higher the Q, the more abruptly φ shifts from positive to negative angles as the frequency is raised through ω_0.

FIGURE 8.13
An alternating-current network.

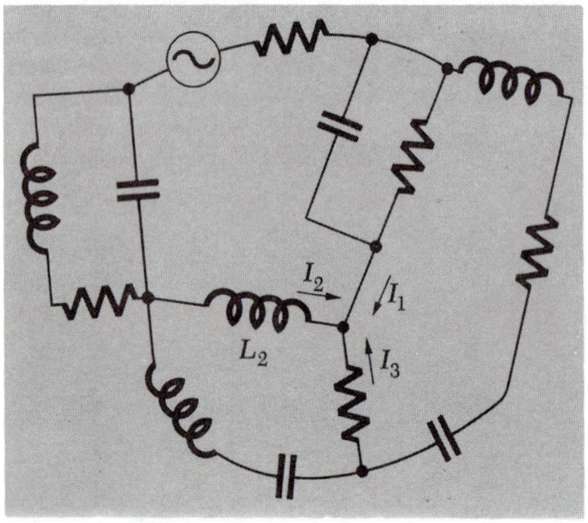

ALTERNATING-CURRENT NETWORKS

8.3 An alternating-current network is any collection of resistors, capacitors, and inductors in which currents flow that are oscillating steadily at the constant frequency ω. One or more electromotive forces, at this frequency, drive the oscillation. Figure 8.13 is a diagram of one such network. The source of alternating electromotive force is represented by the symbol -(~)-. In a branch of the network, for instance the branch that includes the inductor L_2, the current as a function of time is

$$I_2 = I_{02} \cos(\omega t + \varphi_2) \tag{48}$$

Since the frequency is a constant for the whole network, two numbers, such as the amplitude I_{02} and the phase constant φ_2 above, are enough to determine for all time the current in a particular branch. Similarly, the voltage across a branch oscillates with a certain amplitude and phase

$$V_2 = V_{02} \cos (\omega t + \theta_2) \tag{49}$$

If we have determined the currents and voltages in all branches of a network, we have analyzed it completely. To find them by constructing and solving all the appropriate differential equations is possible, of course; and if we were concerned with the transient behavior of the network, we might have to do something like that. For the steady state at some given frequency ω, we can use a far simpler and more elegant method. It is based on two ideas:

1. An alternating current or voltage can be represented by a complex number.

2. Any one branch or element of the circuit can be characterized, at a given frequency, by the relation between the voltage and current in that branch.

The first idea exploits that remarkable mathematical identity

$$e^{i\theta} = \cos \theta + i \sin \theta \tag{50}$$

with $i^2 = -1$. To carry it out we adopt the following *rule* for the representation:

> An alternating current $I_0 \cos (\omega t + \varphi)$ is to be *represented by* the complex number $I_0 e^{i\varphi}$, that is, the number whose real part is $I_0 \cos \varphi$ and whose imaginary part is $I_0 \sin \varphi$.
>
> Going the other way, if the complex number $x + iy$ *represents* a current I, then the current as a function of time is given by the real part of the product $(x + iy)e^{i\omega t}$.

Figure 8.14 is a reminder of this two-way correspondence. Since a complex number $z = x + iy$ can be graphically represented on the two-dimensional plane, it is easy to visualize the phase constant as the angle $\tan^{-1} y/x$, and the amplitude I_0 as the modulus $\sqrt{x^2 + y^2}$.

What makes all this useful is the following fact: *The representation of the sum of two currents is the sum of their representations.* Consider the sum of two currents I_1 and I_2 that meet at a junction of wires in Fig. 8.13. At any instant of time t the sum of the currents is

$$\begin{aligned} I_1 + I_2 &= I_{01} \cos (\omega t + \varphi_1) + I_{02} \cos (\omega t + \varphi_2) \\ &= (I_{01} \cos \varphi_1 + I_{02} \cos \varphi_2) \cos \omega t \\ &\quad - (I_{01} \sin \varphi_1 + I_{02} \sin \varphi_2) \sin \omega t \end{aligned} \tag{51}$$

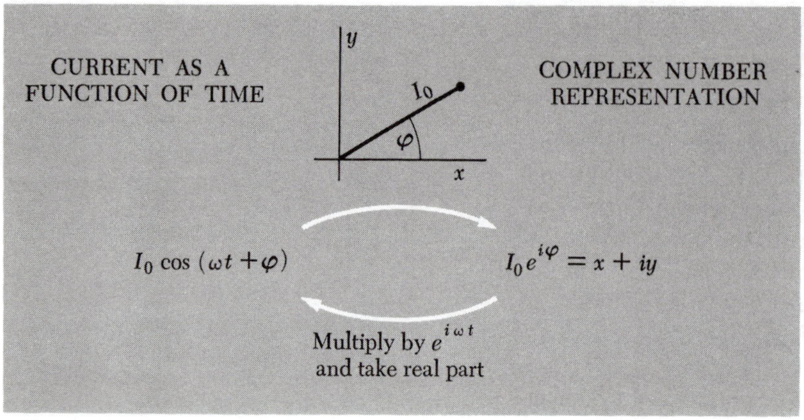

FIGURE 8.14
Rules for representing an alternating current by a complex number.

On the other hand, the sum of the complex numbers that, according to our rule, represent I_1 and I_2 is

$$I_{01}e^{i\varphi_1} + I_{02}e^{i\varphi_2} = (I_{01}\cos\varphi_1 + I_{02}\cos\varphi_2) + i(I_{01}\sin\varphi_1 + I_{02}\sin\varphi_2) \tag{52}$$

If you multiply the right-hand side of Eq. 52 by $\cos \omega t + i \sin \omega t$ and take the real part of the result, you will get just what appears on the right in Eq. 51.

This means that, instead of adding or subtracting the periodic functions of time themselves, we can add or subtract the complex numbers that represent them. Or putting it another way, the algebra of alternating currents turns out to be the same as the algebra of complex numbers in respect to addition. The correspondence does *not* extend to multiplication. The complex number $I_{01}I_{02}e^{i(\varphi_1+\varphi_2)}$ does *not* represent the product of the two current functions in Eq. 51.

However, it is only addition of currents and voltages that we need to carry out in analyzing the network. For example, at the junction where I_1 meets I_2 in Fig. 8.13, there is the physical requirement that *at every instant* the net flow of current into the junction shall be zero. Hence the condition

$$I_1 + I_2 + I_3 = 0 \tag{53}$$

must hold, where I_1, I_2, and I_3 are the *actual periodic functions of time.* Thanks to our correspondence, this can be expressed in the simple algebraic statement that the sum of three complex numbers is zero. Voltages can be handled in the same way. Instantaneously, the sum of voltage drops around any loop in the network must equal the electromotive force in the loop at that instant. This condition relating periodic voltage functions can likewise be replaced by a statement about the sum of some complex numbers, the representations of the various oscillating functions, $V_1(t)$, $V_2(t)$, etc.

ADMITTANCE AND IMPEDANCE

8.4 The relation between current flow in a circuit element and the voltage across the element can be expressed as a relation between the complex numbers that represent the voltage and the current. Look at the inductor-resistor combination in Fig. 8.5. The voltage oscillation is represented by $\mathcal{E}_0$ and the current by $I_0 e^{i\varphi}$, where $I_0 = \mathcal{E}_0/\sqrt{R^2 + \omega^2 L^2}$ and $\tan\varphi = -\omega L/R$. The phase difference φ and the ratio of current amplitude to voltage amplitude are properties of the circuit at this frequency. We define a complex number Y as follows:

$$Y = \frac{e^{i\varphi}}{\sqrt{R^2 + \omega^2 L^2}} \qquad \text{with} \qquad \varphi = \tan^{-1}\left(-\frac{\omega L}{R}\right) \tag{54}$$

Then the relation

$$I = YV \tag{55}$$

holds, where V is the complex number that represents the voltage across the series combination of R and L, and I is the complex number that represents the current. Y is called the *admittance*. The same relation can be expressed with the reciprocal of Y, denoted by Z and called the *impedance:*

$$V = \left(\frac{1}{Y}\right) I = ZI \tag{56}$$

Here we do make use of the product of two complex numbers, but only one of the numbers is the representation of an alternating current or voltage. The other is the impedance or admittance. Our algebra thus contains two categories of complex numbers, those that represent impedances, for example, and those that represent currents. The product of two "impedance numbers," like the product of two "current numbers," doesn't represent anything.

The impedance is measured in ohms. Indeed, if the circuit element had consisted of the resistance R alone, the impedance would be real and equal simply to R, so that Eq. 56 would resemble Ohm's law for a direct-current circuit; $V = RI$.

The admittance of a resistanceless inductor is the imaginary quantity $Y = -i/\omega L$. This can be seen by letting R go to zero in Eq. 54. The factor $-i$ shows that the current oscillation lags the voltage oscillation by $\pi/2$ in phase. On the complex number diagram, if the voltage is represented by V (Fig. 8.15*b*), the current might be represented by I, located as shown there. For the capacitor, $Y = i\omega C$, as

FIGURE 8.15
V and I are complex numbers that represent the voltage across a circuit element and the current through it. The relative phase of current and voltage oscillation is manifest here in the angle between the "vectors." (*a*) In the resistor, current and voltage are in phase. (*b*) In the inductor, current lags the voltage. (*c*) In the capacitor, current leads the voltage.

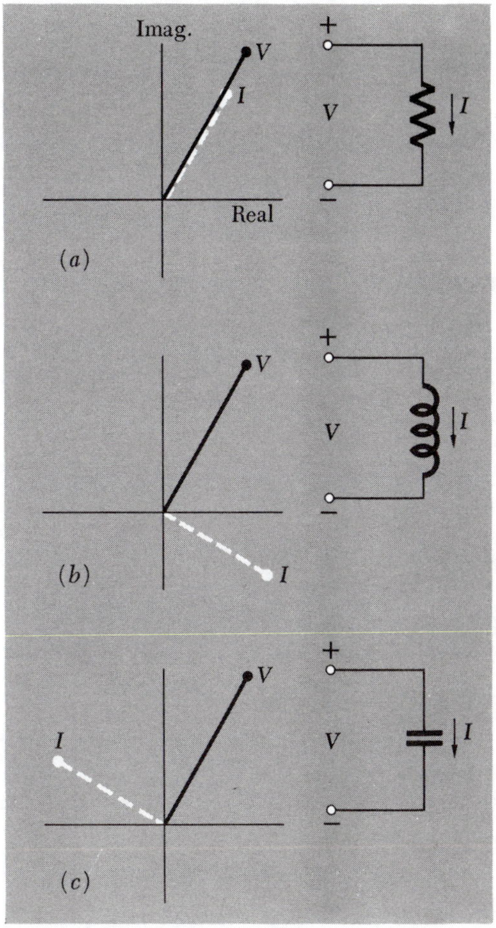

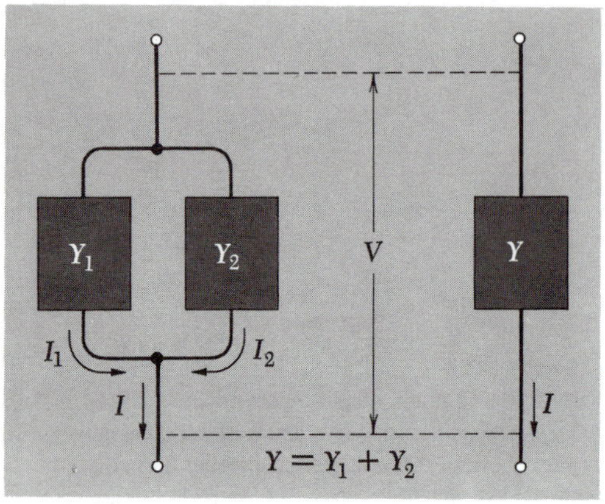

FIGURE 8.16
Combining admittances in parallel.

can be seen from the expression for the current in Fig. 8.8. In this case V and I are related as indicated in Fig. 8.15*c*. The inset in each of the figures shows how the relative sign of V and I is to be specified. Unless that is done consistently, *leading* and *lagging* are meaningless. Note that we always define the positive current direction so that a positive voltage applied to a resistor causes positive current (Fig. 8.15*a*).

The properties of the three basic circuit elements are summarized below.

Symbol	Admittance, Y	Impedance, $Z = \frac{1}{Y}$
R	$\frac{1}{R}$	R
L	$\frac{-i}{\omega L}$	$i\omega L$
C	$i\omega C$	$\frac{-i}{\omega C}$
	$I = YV$	$V = ZI$

We can build up any circuit from these elements. When elements or combinations of elements are connected in parallel, it is convenient to use the admittance, for in that case admittances add. In Fig. 8.16 two black boxes with admittances Y_1 and Y_2 are connected in parallel. We have then

$$I = I_1 + I_2 = Y_1 V + Y_2 V = (Y_1 + Y_2)V \tag{57}$$

which implies that the equivalent single black box has an admittance $Y = Y_1 + Y_2$. From Fig. 8.17 it will be obvious that the *impedances* add for elements connected in *series*. It sounds as if we are talking about a direct-current network! In fact, we have now reduced the ac network problem to the dc network problem, with only this difference: The numbers we deal with are complex numbers.

As an example, let's look at the "parallel *RLC*" circuit in Fig. 8.18. The combined admittance of the three parallel branches is

$$Y = \frac{1}{R} + i\omega C - \frac{i}{\omega L} \tag{58}$$

The voltage is simply $\mathcal{E}_0$, so the complex current is

$$I = YV = \mathcal{E}_0 \left[\frac{1}{R} + i\left(\omega C - \frac{1}{\omega L} \right) \right] \tag{59}$$

The amplitude of the current oscillation is the modulus of the complex

FIGURE 8.17
Combining impedances in series.

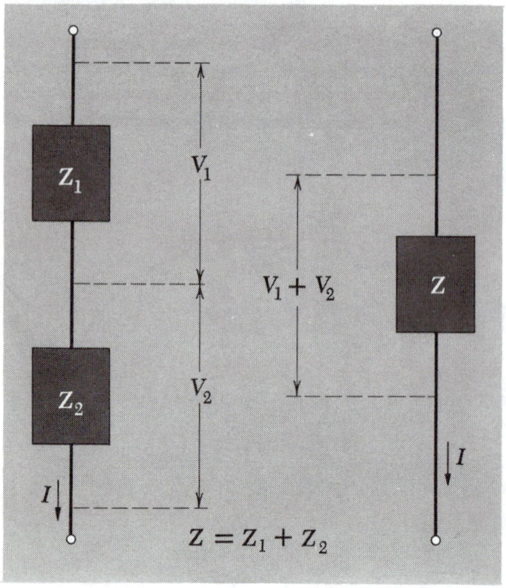

number I, which is $\mathcal{E}_0[(1/R)^2 + (\omega C - 1/\omega L)^2]^{1/2}$, and the phase angle is $\tan^{-1}(R\omega C - R/\omega L)$.

We can only deal in this way with *linear* circuit elements, elements in which the current is proportional to the voltage. In other words, our circuit must be described by a linear differential equation. You can't even define an impedance for a nonlinear element. Nonlinear circuit elements are very important and interesting devices. If you have studied some in the laboratory, you can see why they will not yield to this kind of analysis.

This is all predicated, too, on continuous oscillation at constant frequency. The transient behavior of the circuit is a different problem. However, for linear circuits the tools we have just developed have some utility, even for transients. The reason is that by superposing steady oscillations of many frequencies we can represent a nonsteady behavior, and the response to each of the individual frequencies can be calculated as if that frequency were present alone.

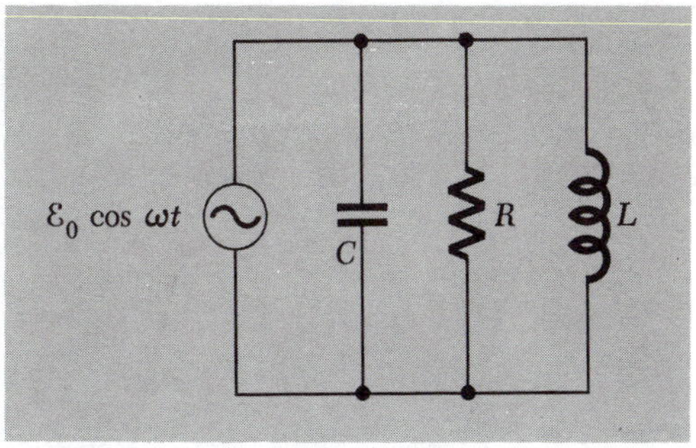

FIGURE 8.18
A parallel resonant circuit. Add the complex admittances of the three elements, as in Eq. 58.

POWER AND ENERGY IN ALTERNATING-CURRENT CIRCUITS

8.5 If the voltage across a resistor R is $V_0 \cos \omega t$, the current is $I = (V_0/R)\cos \omega t$. The instantaneous power, that is, the instantaneous rate at which energy is being dissipated in the resistor, is

$$P = RI^2 = \frac{V_0^2}{R}\cos^2 \omega t \tag{60}$$

Since the average of $\cos^2 \omega t$ over many cycles is ½, the average power dissipated in the circuit is

$$\overline{P} = \frac{1}{2}\frac{V_0^2}{R} \tag{61}$$

It is customary to express voltage and current in ac circuits by giving not the amplitude but $1/\sqrt{2}$ times the amplitude. This is often called the *root-mean-square* (rms) value. That takes care of the factor ½ in Eq. 61, so that

$$\overline{P} = \frac{V_{\text{rms}}^2}{R} \tag{62}$$

For example, the common domestic line voltage in North America, 120 volts, corresponds to an *amplitude* $120\sqrt{2}$ volts. The potential difference between the terminals of the electric outlet in your room (if the voltage is up to normal) is

$$V(t) = 170 \cos 377t \tag{63}$$

with V in volts and t in seconds. An ac ammeter is calibrated to read 1 amp when the current amplitude is 1.414 amps.

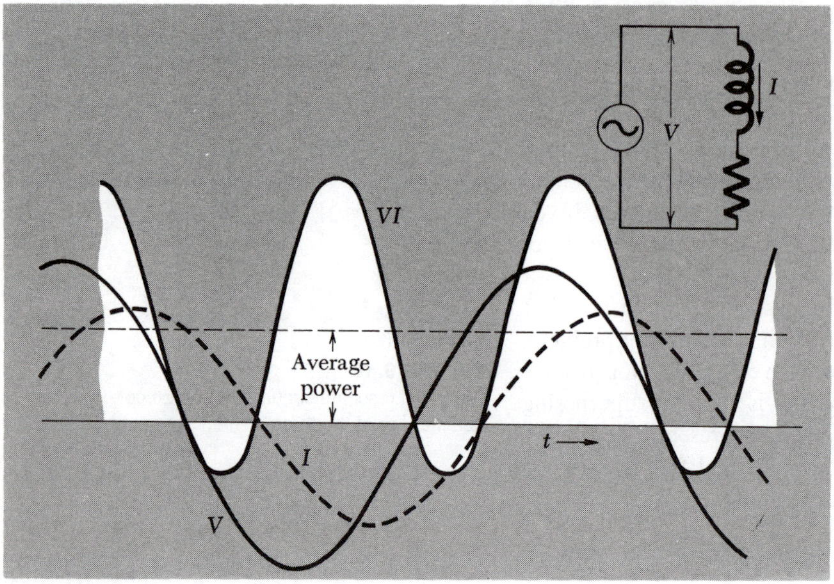

FIGURE 8.19
The instantaneous power VI is the rate at which energy is being transferred from the source of electromotive force on the left to the circuit elements on the right. The time average of this is indicated by the horizontal dashed line.

In general, the instantaneous rate at which energy is delivered to a circuit element is VI, the product of the instantaneous voltage and current, with due regard to sign. Consider this aspect of the current flow in the simple LR circuit in Fig. 8.5. In Fig. 8.19 we have redrawn the current and voltage graphs and added a curve proportional to the product VI. Positive VI means energy is being transferred into the LR combination from the source of electromotive force, or generator. Notice that VI is negative in certain parts of the cycle. In those periods some energy is being returned to the generator. This is explained by the oscillation in the energy stored in the magnetic field of the inductor. This stored energy, $\frac{1}{2}LI^2$, goes through a maximum twice in each full cycle.

The *average* power $\overline{P}$ corresponds to the horizontal dashed line. To calculate its value, let's take a look at the product VI, with $V = \mathcal{E}_0 \cos \omega t$ and $I = I_0 \cos (\omega t + \varphi)$:

$$\begin{aligned} VI &= \mathcal{E}_0 I_0 \cos \omega t \cos (\omega t + \varphi) \\ &= \mathcal{E}_0 I_0 (\cos^2 \omega t \cos \varphi - \cos \omega t \sin \omega t \sin \varphi) \end{aligned} \tag{64}$$

The term proportional to $\cos \omega t \sin \omega t$ has a time average zero, as is obvious if you write it as $\frac{1}{2} \sin 2\omega t$, while the average of $\cos^2 \omega t$ is $\frac{1}{2}$. Thus for the time average we have

$$\overline{P} = \overline{VI} = \tfrac{1}{2}\mathcal{E}_0 I_0 \cos \varphi \tag{65}$$

If both current and voltage are expressed as rms values, in volts and amps, respectively,

$$\overline{P} = V_{\text{rms}} I_{\text{rms}} \cos \varphi \tag{66}$$

In this circuit all the energy dissipated goes into the resistance R. Naturally, any real inductor has some resistance. For the purpose of analyzing the circuit, we included that with the resistance R. Of course the heat evolves at the actual site of the resistance.

To practice with the methods we developed in Section 8.4, we'll analyze the circuit in Fig. 8.20*a*. A 10,000-ohm, 1-watt resistor has been connected up with two capacitors of capacitance 0.2 and 0.5 microfarads. We propose to plug this into the 120-volt, 60-Hz outlet. *Question:* Will the 1-watt resistor get too hot? In the course of finding out whether the average power dissipated in R exceeds the 1-watt rating, we'll calculate some of the currents and voltages we might expect to measure in this circuit. One way to work through the circuit is outlined below.

$$\text{Admittance of } C_2 = i\omega C_2 = (377)(2 \times 10^{-7})i = 0.754 \times 10^{-4} i \text{ ohm}^{-1}$$

$$\text{Admittance of the resistor} = \frac{1}{R} = 10^{-4} \text{ ohm}^{-1}$$

Admittance of 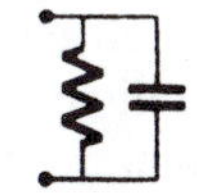 $= 10^{-4}(1 + 0.754i) \text{ ohm}^{-1}$

Impedance of 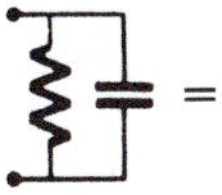 $= \dfrac{1}{10^{-4}(1 + 0.754i)} = \dfrac{10^{4}(1 - 0.754i)}{1^2 + 0.754^2} = (6360 - 4800i) \text{ ohms}$

$$\text{Impedance of } C_1 = -\frac{i}{\omega C} = -\frac{i}{(377)(5 \times 10^{-7})} = -5300i \text{ ohms}$$

$$\text{Impedance of entire circuit} = (6360 - 10{,}100i) \text{ ohms}$$

$$I_1 = \frac{120}{6360 - 10{,}100i} = \frac{120\,(6360 + 10{,}100i)}{(6360)^2 + (10{,}100)^2} = (5.37 + 8.53i) \times 10^{-3} \text{ amp}$$

Since we have used 120 volts, which is the rms voltage, we obtain the rms current. That is, the modulus of the complex number I_1, which is $[(5.37)^2 + (8.53)^2]^{1/2} \times 10^{-3}$ amp or 10.0 milliamps, is the rms current. An ac milliammeter inserted in series with the line would read

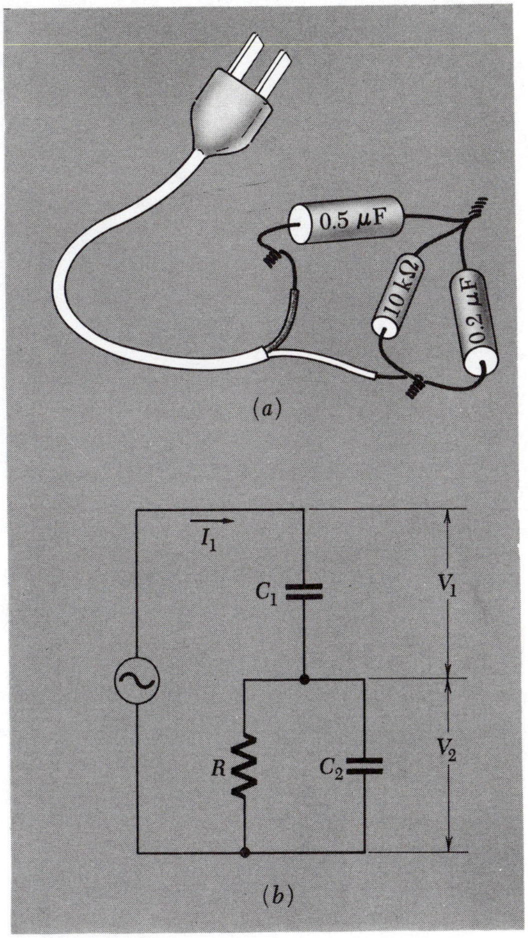

FIGURE 8.20
An actual network (*a*) ready to be connected to a source of electromotive force, and (*b*) the circuit diagram.

10 milliamps. This current has a phase angle $\varphi = \tan^{-1}(0.853/0.537)$ or 1.01 radians with respect to the line voltage. The average power delivered to the entire circuit is then

$$\overline{P} = (120 \text{ volts})(0.010 \text{ amp}) \cos 1.01 = 0.64 \text{ watt}$$

In this circuit the resistor is the only dissipative element, so this must be the average power dissipated in it. Just as a check, we can find the voltage V_2 across the resistor:

$$V_1 = I_1\left(\frac{-i}{\omega C}\right) = (5.37 + 8.53i)(-5300i)10^{-3}$$
$$= (45.2 - 28.4i) \text{ volts}$$
$$V_2 = 120 - V_1 = (74.8 + 28.4i) \text{ volts}$$

The current I_2 in R will be in phase with V_2, of course, so the average power in R will be

$$\overline{P} = \frac{V_2^2}{R} = \frac{(74.8)^2 + (28.4)^2}{10^4} = 0.64 \text{ watt} \tag{67}$$

which checks.

Thus the rating of the resistor isn't exceeded, for what that assurance is worth. Actually, whether the resistor will get too hot depends not only on the average power dissipated in it but also on how easily it can get rid of the heat. The power rating of a resistor is only a rough guide.

PROBLEMS

8.1 How large an inductance, in henrys, should be connected in series with a 120-volt, 60-watt light bulb if it is to operate normally when the combination is connected across a 240-volt, 60-Hz line? (First determine the inductive reactance required. You may neglect the resistance of the inductor and the inductance of the light bulb.)

8.2 A 2000-ohm resistor and a 1-microfarad capacitor are connected in series across a 120-volt (rms), 60-Hz line.

(*a*) What is the impedance?

(*b*) What is the rms value of the current?

(*c*) What is the power dissipated in the circuit?

(*d*) What will be the reading of an ac voltmeter connected across the resistor? Across the capacitor?

(*e*) The horizontal plates of a cathode ray tube are connected across the resistor and the vertical plates across the capacitor. Sketch the pattern that you expect to see on the screen.

8.3 A 1000-ohm resistor, a 500-picofarad capacitor, and a 2-millihenry inductor are connected all in parallel. What is the impedance of this combination at a frequency of 10 kilocycles per sec? At a frequency of 10 megacycles per sec? What is the frequency at which the absolute value of the impedance is greatest?

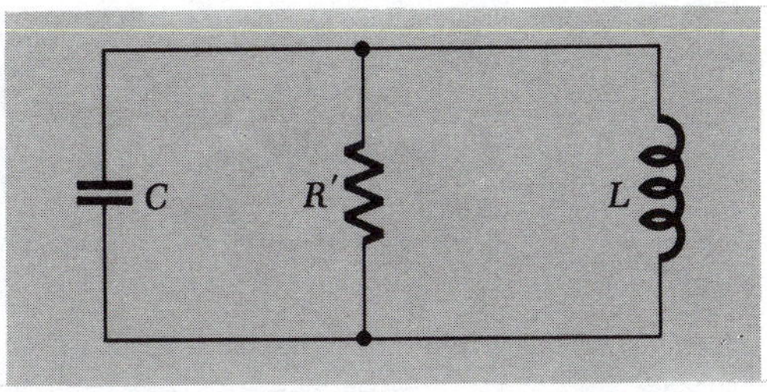

PROBLEM 8.4

8.4 In the resonant circuit of the figure the dissipative element is a resistor R' connected in parallel, rather than in series, with the LC combination. Work out the equation, analogous to Eq. 2, which applies to this circuit. Find also the conditions on the solution analogous to those that hold in the series RLC circuit. If a series RLC and a parallel $R'LC$ circuit have the same L, C, and Q, how must R' be related to R?

8.5 The coil in the circuit shown in the diagram is known to have an inductance of 0.01 henry. When the switch is closed, the oscilloscope sweep is triggered.

(*a*) Determine as well as you can the value of the capacitance C.

(*b*) Estimate the value of the resistance R of the coil.

(*c*) What is the magnitude of the voltage across the oscilloscope input a long time, say 1 second, after the switch has been closed?

8.6 For the circuit in Fig. 8.4*a*, determine the values of β_1 and β_2 for the overdamped case, with $R = 600$ ohms. Determine also the ratio of B to A, the constants in Eq. 16.

PROBLEM 8.5

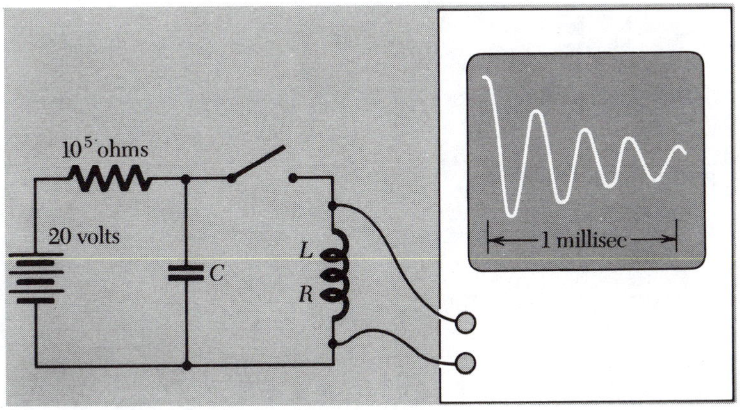

8.7 A resonant cavity of the form illustrated in the figure is an essential part of many microwave oscillators. It can be regarded as a simple LC circuit. The inductance is that of a toroid with one turn; this inductor is connected directly to parallel-plate capacitors. Find an expression for the resonant frequency of this circuit and show by a sketch the configuration of the magnetic and electric fields.

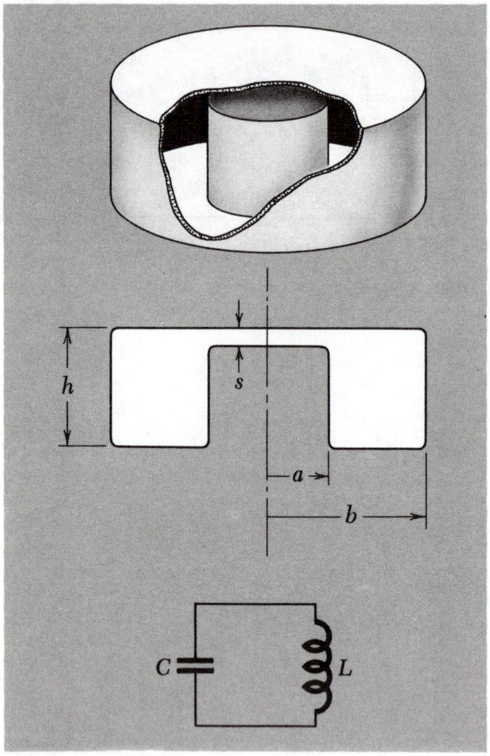

PROBLEM 8.7

8.8 For the damped RLC circuit of Fig. 8.3, work out an expression for the total energy stored in the circuit, the energy in the capacitor plus the energy in the inductor, at any time t. Show that the critical damping condition, $R = 2\sqrt{L/C}$, is the one in which the total energy is most quickly dissipated.

8.9 Using Eqs. 10 and 13, express the effect of damping on the frequency of a series RLC circuit. Let $\omega_0 = 1/\sqrt{LC}$ be the frequency of the undamped circuit. Suppose enough resistance is added to bring Q from ∞ down to 1000. By what percentage is the frequency ω thereby shifted from ω_0?

8.10 Is it possible to find a frequency at which the impedance at the terminals of this circuit will be purely real?

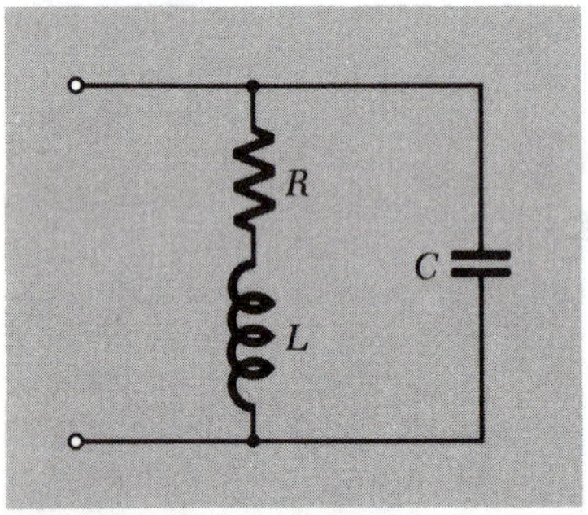

PROBLEM 8.10

8.11 An alternating voltage $V_0 \cos \omega t$ is applied to the terminals at A. The terminals at B are connected to an audio amplifier of very high input impedance. (That is, current flow into the amplifier is negligible.) Calculate the ratio $|V_1|^2/V_0^2$. Here $|V_1|$ is the absolute value of the complex voltage amplitude at terminals B. Choose values for R and C to make $|V_1|^2/V_0^2 = 0.1$ for a 5000-Hz signal. This circuit is the most primitive of "low-pass" filters, providing attenuation that increases with increasing frequency. Show that, for sufficiently high frequencies, the signal power is reduced by a factor ¼ for every doubling of the frequency. Can you devise a filter with a more drastic cutoff—such as a factor 1/16 per octave?

PROBLEM 8.11

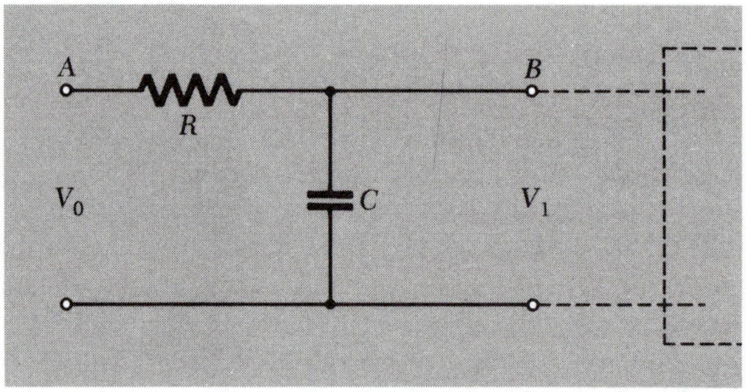

8.12 Let $V_{AB} = V_B - V_A$, in this circuit. Show that $|V_{AB}|^2 = V_0^2$ for any frequency ω. Find the frequency for which V_{AB} is 90° out of phase with V_0.

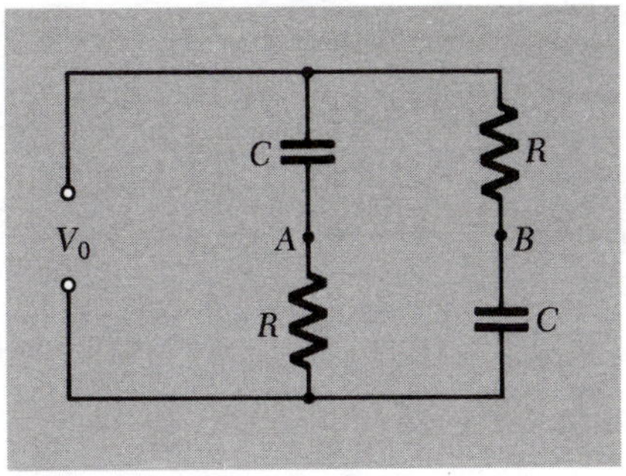

PROBLEM 8.12

8.13 Show that, if the condition $R_1R_2 = L/C$ is satisfied by the components of the circuit below, the difference in voltage between points A and B will be zero at any frequency. Discuss the suitability of this circuit as an ac bridge for measurement of an unknown inductance.

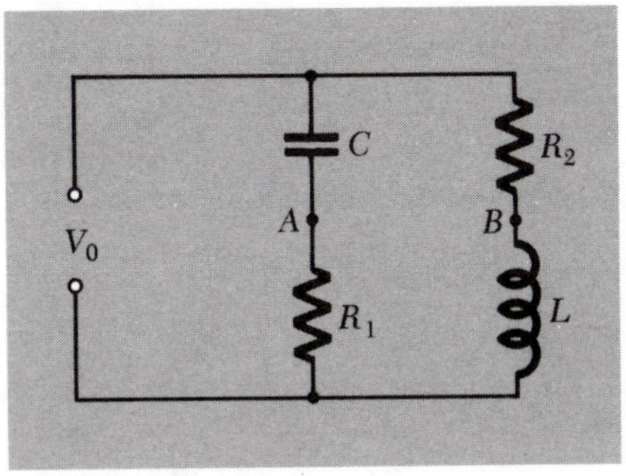

PROBLEM 8.13

8.14 In the laboratory you find an inductor of unknown inductance L and unknown internal resistance R. Using a dc ohmmeter, an ac voltmeter of high impedance, a 1-microfarad capacitor, and a 1000-Hz signal generator, determine L and R as follows: According to the ohmmeter, R is 35 ohms. You connect the capacitor in series with the inductor and the signal generator. The voltage across both is 10.1 volts. The voltage across the capacitor alone is 15.5 volts. You note also, as a check, that the voltage across the inductor alone is 25.4 volts. How large is L? Is the check consistent?

8.15 Show that the impedance Z at the terminals of each of the two circuits below is

$$\frac{5000 + 16 \times 10^{-3}\omega^2 - 16i\omega}{1 + 16 \times 10^{-6}\,\omega^2}$$

Since they present, at any frequency, the identical impedance, the two black boxes are completely equivalent and indistinguishable from the outside. See if you can discover the general rules for constructing the box on the right, given the values of the resistances and capacitance in the box on the left.

PROBLEM 8.15

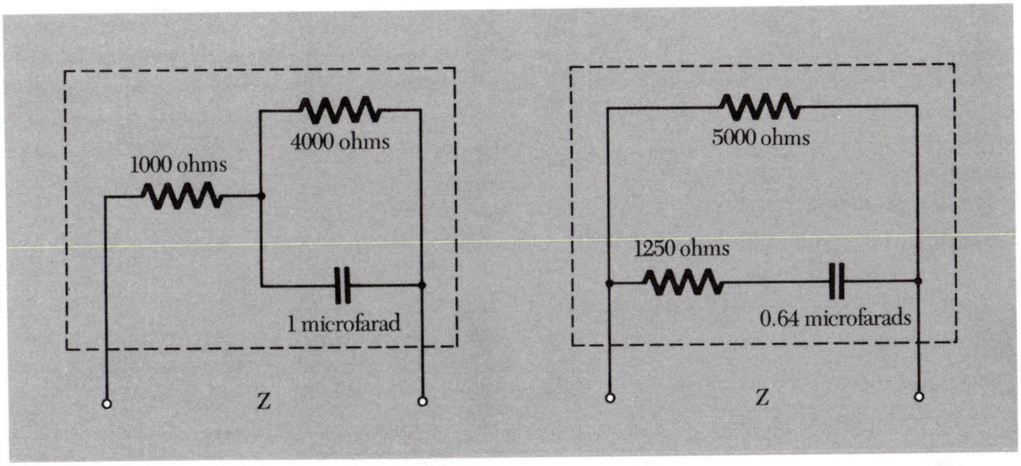

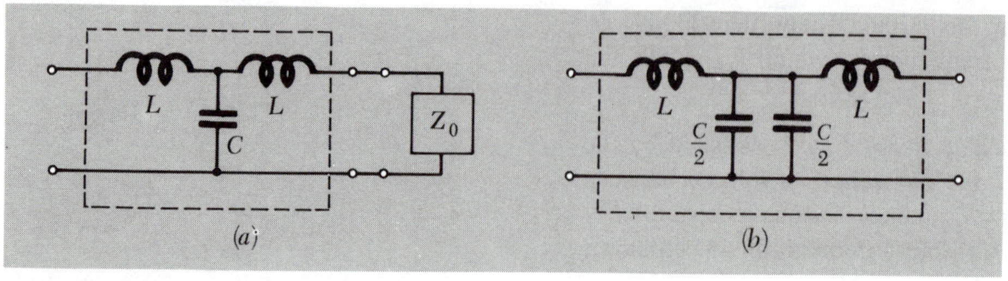

PROBLEM 8.16

8.16 The box (a) with four terminals contains a capacitor C and two inductors of equal inductance L connected as shown. An impedance Z_0 is to be connected to the terminals on the right. For given frequency ω find the value which Z_0 must have if the resulting impedance between the terminals on the left (the "input" impedance) is to be equal to Z_0. You will find that the required value of Z_0 is a pure resistance R_0 providing that $\omega^2 < 2/LC$. A chain of such boxes could be connected together to form a ladder network resembling the ladder of resistors in Problem 4.32. If the chain is terminated with a resistor of the correct value R_0, its input impedance at frequency ω will be R_0, no matter how many boxes make up the chain.

What is Z_0 in the special case $\omega = \sqrt{2/LC}$? It helps in understanding that case to note that the contents of the box (a) can be equally well represented by box (b).

9

MAXWELL'S EQUATIONS AND ELECTROMAGNETIC WAVES

"SOMETHING IS MISSING"

9.1 Let us review the relations between charges and fields. As we learned in Chapter 2, a statement equivalent to Coulomb's law is the differential relation

$$\text{div}\,\mathbf{E} = 4\pi\rho \tag{1}$$

connecting the electric charge density ρ and the electric field $\mathbf{E}$. This holds for moving charges as well as stationary charges. That is, ρ can be a function of time as well as position. As we emphasized in Chapter 5, the fact that Eq. 1 holds for moving charges is consistent with *charge invariance:* No matter how an isolated charged particle may be moving, its charge, as measured by the integral of $\mathbf{E}$ over a surface surrounding it, appears the same in every frame of reference.

Electric charge in motion is electric current. Because charge is never created or destroyed, the charge density ρ and the current density $\mathbf{J}$ always satisfy the condition

$$\text{div}\,\mathbf{J} = -\frac{\partial\rho}{\partial t} \tag{2}$$

We first wrote down this "equation of continuity" as Eq. 9 in Chapter 4.

If the current density $\mathbf{J}$ is constant in time, we call it a *stationary current distribution.* The magnetic field of a stationary current distribution satisfies the equation

$$\text{curl}\,\mathbf{B} = \frac{4\pi}{c}\mathbf{J} \tag{3}$$

We worked with this relation in Chapter 6.

Now we are interested in charge distributions and fields that are changing in time. Suppose we have a charge distribution $\rho(x, y, z, t)$ with $\partial\rho/\partial t \neq 0$. For instance, we might have a capacitor which is discharging through a resistor. According to Eq. 2, $\partial\rho/\partial t \neq 0$ implies

$$\text{div}\,\mathbf{J} \neq 0$$

But according to Eq. 3, since the divergence of the curl of *any* vector function is identically zero (see Problem 2.16),

$$\text{div}\,\mathbf{J} = \frac{c}{4\pi}\,\text{div}\,(\text{curl}\,\mathbf{B}) = 0 \tag{4}$$

The contradiction shows that Eq. 3 *cannot be correct* for a system in which the charge density is varying in time. Of course, no one claimed it was; a stationary current distribution, for which Eq. (3) *does* hold, is one in which not even the current density $\mathbf{J}$, let alone the charge density ρ, is time-dependent.

The problem can be posed in somewhat different terms by considering the line integral of magnetic field around the wire which carries charge away from the capacitor plate in Fig. 9.1. According to Stokes' theorem,

$$\int_C \mathbf{B} \cdot d\mathbf{l} = \int_S \text{curl}\, \mathbf{B} \cdot d\mathbf{a} \qquad (5)$$

The surface S passes right through the conductor in which a current I is flowing. Inside this conductor, curl **B** has a finite value, namely, $4\pi\mathbf{J}/c$, and the integral on the right comes out equal to $4\pi I/c$. That is to say, if the curve C is close to the wire and well away from the capacitor gap, the magnetic field there is not different from the field around any wire carrying the same current. Now the surface S' in Fig. 9.2 is also a surface spanning C, and has an equally good claim to be used in the statement of Stokes' theorem, Eq. 5. Through this surface, however, there flows *no current at all!* Nevertheless, curl **B** cannot be zero over all of S' without violating Stokes' theorem. Therefore, on S', curl **B** must depend on something other than the current density **J**.

We can only conclude that Eq. 3 has to be replaced by some other relation, in the more general situation of changing charge distributions. Let's write instead

$$\text{curl}\, \mathbf{B} = \frac{4\pi}{c}\mathbf{J} + (?) \qquad (6)$$

and see if we can discover what (?) must be.

Another line of thought suggests the answer. Remember that

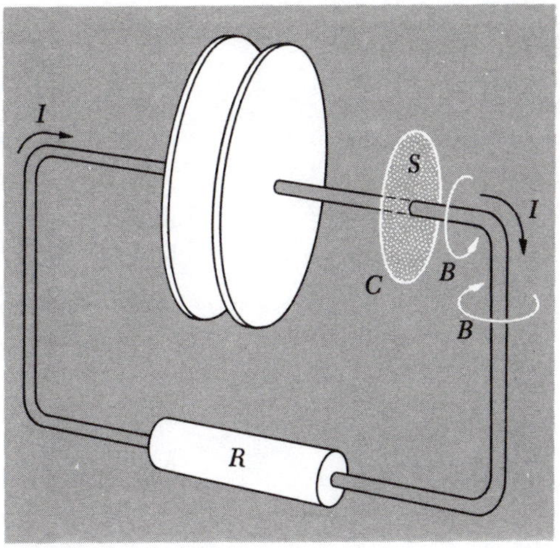

FIGURE 9.1
Having been charged with the right-hand plate positive, the capacitor is being discharged through the resistor. There is a magnetic field **B** around the wire. The integral of curl **B**, over the surface S which passes through the wire, has the value $4\pi I/c$.

FIGURE 9.2
The white arrows show the current flow in the conductors. The surface S', which like S has the curve C for its edge, has no current passing through it.

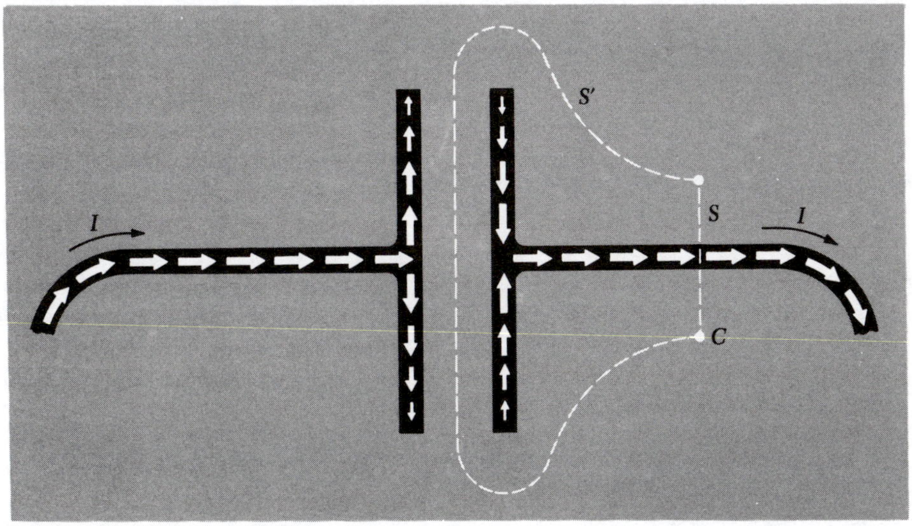

the transformation laws of the electromagnetic field, Eq. 58 of Chapter 6, are quite symmetrical in **E** and **B**. Now in Faraday's induction phenomenon a *changing magnetic field* is accompanied by an *electric field*, in a manner described by Eq. 30 of Chapter 7:

$$\text{curl } \mathbf{E} = -\frac{1}{c}\frac{\partial \mathbf{B}}{\partial t} \tag{7}$$

This is a local relation connecting the electric and magnetic fields in empty space—charges are not directly involved. If symmetry with respect to **E** and **B** is to prevail, we must expect that a *changing electric field* can give rise to a *magnetic field*. There ought to be an induction phenomenon described by an equation like Eq. 7, but with the roles of **E** and **B** switched. It will turn out that we need to change the sign too, but that is all:

$$\text{curl } \mathbf{B} = \frac{1}{c}\frac{\partial \mathbf{E}}{\partial t} \tag{8}$$

This provides the missing term that is called for in Eq. 6. To try it out, write

$$\text{curl } \mathbf{B} = \frac{4\pi}{c}\mathbf{J} + \frac{1}{c}\frac{\partial \mathbf{E}}{\partial t} \tag{9}$$

and take the divergence of both sides:

$$\text{div (curl } \mathbf{B}) = \text{div}\left(\frac{4\pi}{c}\mathbf{J}\right) + \text{div}\left(\frac{1}{c}\frac{\partial \mathbf{E}}{\partial t}\right) \tag{10}$$

The left side is necessarily zero, as already remarked. In the second term on the right we can interchange the order of differentiation with respect to space coordinates and time. Thus

$$\text{div}\left(\frac{1}{c}\frac{\partial \mathbf{E}}{\partial t}\right) = \frac{1}{c}\frac{\partial}{\partial t}(\text{div } \mathbf{E}) = \frac{4\pi}{c}\frac{\partial \rho}{\partial t} \tag{11}$$

by Eq. 1. The right-hand side of Eq. 10 now becomes

$$\frac{4\pi}{c}\text{div } \mathbf{J} + \frac{4\pi}{c}\frac{\partial \rho}{\partial t} \tag{12}$$

which is zero by virtue of the continuity condition, Eq. 2.

The new term resolves the difficulty raised in Fig. 9.2. As charge flows out of the capacitor, the electric field, which at any instant has the configuration in Fig. 9.3, *diminishes* in intensity. In this case, $\partial \mathbf{E}/\partial t$ points opposite to **E**. The vector function $\frac{1}{c}\frac{\partial \mathbf{E}}{\partial t}$ is represented by the black arrows in Fig. 9.4. With curl $\mathbf{B} = \frac{4\pi}{c}\mathbf{J} + \frac{1}{c}\frac{\partial \mathbf{E}}{\partial t}$, the

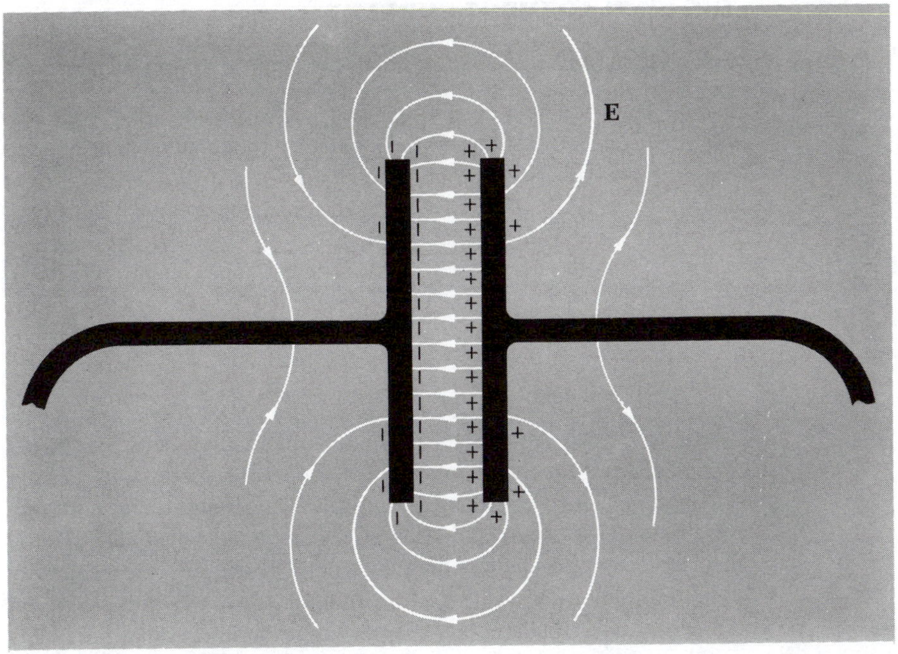

FIGURE 9.3
The electric field at a particular instant. The magnitude of **E** is decreasing everywhere as time goes on.

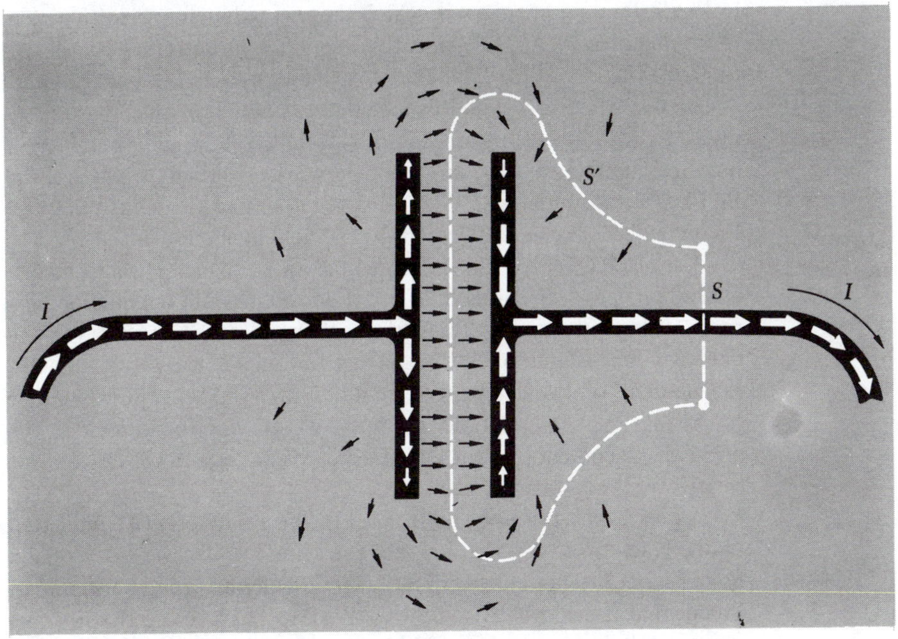

FIGURE 9.4
The conduction current (white arrows) and the displacement current (black arrows).

integral of curl **B** over S' now has the same value as it does over S. On S' the second term contributes everything; on S the first term, the term with **J**, is practically all that counts.

THE DISPLACEMENT CURRENT

9.2 Observe that the vector field $\frac{1}{c}\frac{\partial \mathbf{E}}{\partial t}$ appears to form a *continuation* of the conduction current distribution. Maxwell called it the *displacement current,* and the name has stuck although it no longer seems very appropriate. To be precise, we can define a *displacement current density* $\mathbf{J}_d$, to be distinguished from the conduction current density $\mathbf{J}$, by writing Eq. 84 this way:

$$\text{curl } \mathbf{B} = \frac{4\pi}{c}(\mathbf{J} + \mathbf{J}_d) \tag{13}$$

and defining $\mathbf{J}_d \equiv \frac{1}{4\pi}\frac{\partial \mathbf{E}}{\partial t}$.

We needed the new term to make the relation between current and magnetic field consistent with the continuity equation, in the case of conduction currents changing in time. If it belongs there, it implies the existence of a new induction effect in which a changing electric field is accompanied by a magnetic field. If the effect is real, why didn't Faraday discover it? For one thing, he wasn't looking for it, but there is a more fundamental reason why experiments like Faraday's could not have revealed any new effects attributable to the last term in Eq. 9. In any apparatus in which there are changing electric fields, there are present at the same time conduction currents, charges in motion. The magnetic field $\mathbf{B}$, everywhere around the apparatus, is just about what you would expect those conduction currents to produce. In fact, it is almost exactly the field you would calculate if, ignoring the fact that the circuits may not be continuous, you use the Biot-Savart formula, Eq. 38 of Chapter 6, to find the contribution of each conduction current element to the field at some point in space.

Consider, for example, the point P in the space between our discharging condenser plates, Fig. 9.5. Each element of conduction current, in the wires and on the surface of the plates, contributes to the field at P, according to the Biot-Savart formula. Must we include also the elements of displacement current density $\mathbf{J}_d$? The answer is rather surprising. We *may* include $\mathbf{J}_d$; but if we are careful to include the *entire* displacement current distribution, its net effect will be *zero* for relatively slowly varying fields.

To see why this is so, notice that the vector function $\mathbf{J}_d$, indicated by the black arrows in Fig. 9.4, has the same form as the electric field $\mathbf{E}$ in Fig. 9.3. This electric field is practically an electrostatic field, except that it is slowly dying away. We expect therefore that its curl is practically zero, which would imply that curl $\mathbf{J}_d$ must be practically zero. More precisely, we have curl $\mathbf{E} = -\frac{1}{c}\frac{\partial \mathbf{B}}{\partial t}$ and with the displace-

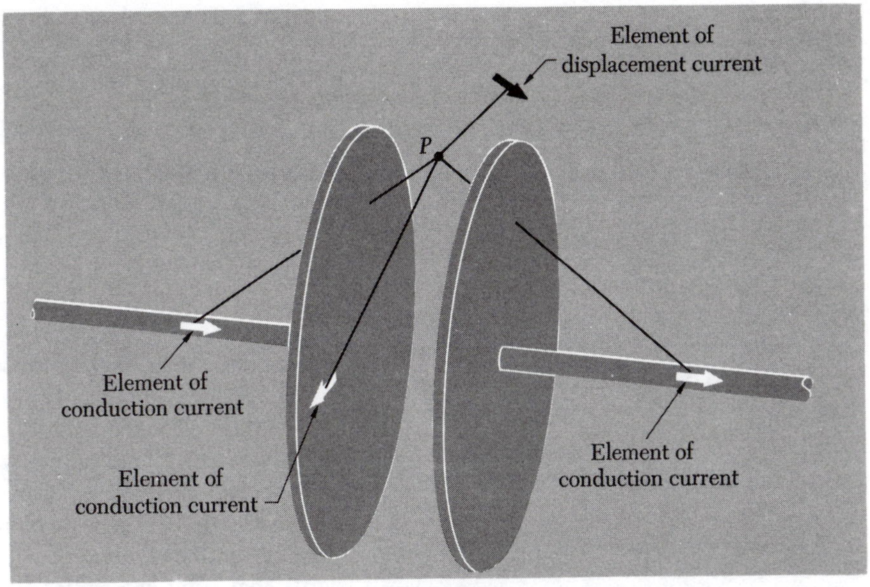

FIGURE 9.5
In the case of slowly varying fields, the total contribution to the magnetic field at any point, from all displacement currents, is zero. The magnetic field at P can be calculated by the Biot-Savart formula applied to conduction current elements only.

ment current $\mathbf{J}_d = \frac{1}{4\pi}\frac{\partial \mathbf{E}}{\partial t}$, we get, by interchanging the order of differentiation,

$$\text{curl } \mathbf{J}_d = \frac{1}{4\pi}\text{curl}\left(\frac{\partial \mathbf{E}}{\partial t}\right) = \frac{1}{4\pi}\frac{\partial}{\partial t}(\text{curl } \mathbf{E}) = -\frac{1}{4\pi c}\frac{\partial^2 \mathbf{B}}{\partial t^2} \qquad (14)$$

This will be negligible for sufficiently slow changes in field. We may call a slowly changing field *quasi-static*. Now if $\mathbf{J}_d$ is a vector field without any curl, it can be made up, in the same way that the electrostatic field can be made of the fields of point charges, by superposing radial currents flowing outward from point sources or in toward point "sinks" (Fig. 9.6). But the magnetic field of any *radial,* symmetrical current distribution, however calculated, must be zero by symmetry, for there is no unique direction anywhere, except the radial direction itself.

In the quasi-static field, then, the conduction currents alone are the only *sources* needed to account for the magnetic field. In other words, if Faraday had arranged something like Fig. 9.5, and had been able to measure the magnetic field at P, by using a compass needle say, he would not have been surprised. He would not have needed to invent a displacement current to explain it.

To see this new induction effect, we need rapidly changing fields. In fact, we need changes to occur in the time it takes light to cross the

FIGURE 9.6
Showing what is meant by a radial current distribution. The current density **J** for the point source in (*a*), or for the point "sink" in (*b*) is like the electric field of a point charge. Any current distribution with curl **J** = 0 could be made by superposing such sources and sinks, and must therefore have zero magnetic field.

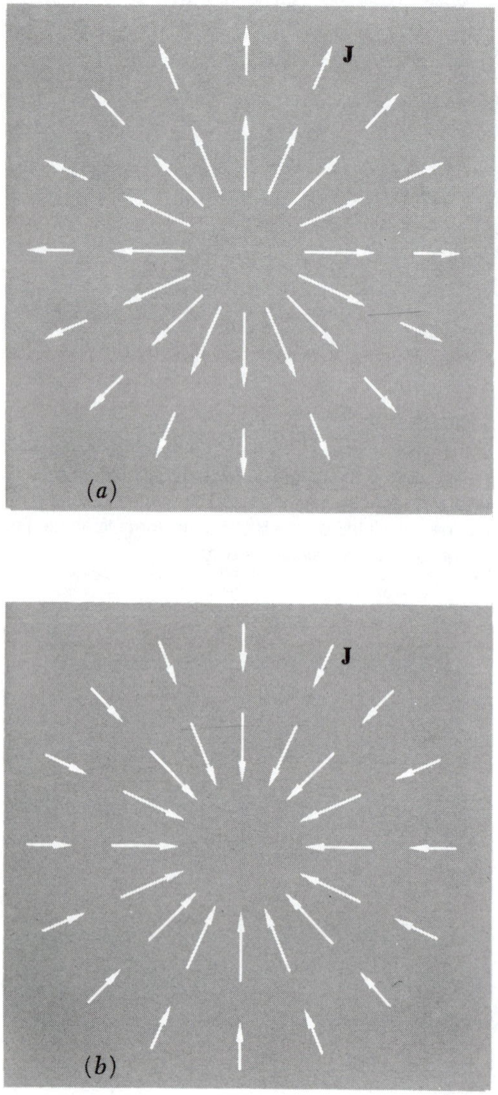

apparatus. That is why the direct demonstration had to wait for Hertz, whose experiment came many years after the law itself had been worked out by Maxwell.

MAXWELL'S EQUATIONS

9.3 James Clerk Maxwell (1831–1879), after immersing himself in the accounts of Faraday's electrical researches, set out to formulate mathematically a theory of electricity and magnetism. Maxwell could not exploit relativity—that came 50 years later. The electrical constitution of matter was a mystery, the relation between light and electromagnetism unsuspected. Many of the arguments that we have used to make our next step seem obvious were unthinkable then. Nevertheless, as Maxwell's theory developed, the term we have been discussing, $\partial \mathbf{E}/\partial t$, appeared quite naturally in his formulation. He called it the displacement current. Maxwell was concerned with electric fields in solid matter as well as in vacuum, and when he talks about a displacement current he is often including some charge-in-motion, too. We'll clarify that point in Chapter 10 when we study electric fields in matter. Indeed, Maxwell thought of space itself as a medium, the "aether," so that even in the absence of solid matter the displacement current was occurring *in* something. But never mind—his mathematical equations were perfectly clear and unambiguous, and his introduction of the displacement current was a *theoretical* discovery of the first rank.

Maxwell's description of the electromagnetic field was essentially complete. We have arrived by different routes at various pieces of it, which we shall now assemble in the form traditionally called *Maxwell's equations:*

$$\begin{aligned} \text{curl } \mathbf{E} &= -\frac{1}{c}\frac{\partial \mathbf{B}}{\partial t} \\ \text{curl } \mathbf{B} &= \frac{1}{c}\frac{\partial \mathbf{E}}{\partial t} + \frac{4\pi}{c}\mathbf{J} \\ \text{div } \mathbf{E} &= 4\pi\rho \\ \text{div } \mathbf{B} &= 0 \end{aligned} \tag{15}$$

These are written for the fields in vacuum, in the presence of electric charge of density ρ and electric current, that is, charge in motion, of density **J**.

The first equation is Faraday's *law of induction*. The second expresses the dependence of the magnetic field on the *displacement current* density, or rate of change of electric field, and on the *conduc-*

tion current density, or rate of motion of charge. The third equation is equivalent to Coulomb's law. The fourth equation states that there are no sources of magnetic field *except* currents. We shall have more to say about this aspect of Nature in Chapter 11.

Notice that the lack of symmetry in these equations, with respect to **B** and **E**, is entirely due to the presence of electric charge and electric conduction current. In empty space, the terms with ρ and **J** are zero, and Maxwell's equations become

$$\begin{aligned} \text{curl } \mathbf{E} &= -\frac{1}{c}\frac{\partial \mathbf{B}}{\partial t} & \text{div } \mathbf{E} &= 0 \\ \text{curl } \mathbf{B} &= \frac{1}{c}\frac{\partial \mathbf{E}}{\partial t} & \text{div } \mathbf{B} &= 0 \end{aligned} \tag{16}$$

Here the displacement current term is all-important. Its presence, along with its counterpart in the first equation, implies the possibility of *electromagnetic waves*. Recognizing that, Maxwell went on to develop with brilliant success an electromagnetic theory of light.

In SI units Maxwell's equations look like this:

$$\begin{aligned} \text{curl } \mathbf{E} &= -\frac{\partial \mathbf{B}}{\partial t} \\ \text{curl } \mathbf{B} &= \mu_0\epsilon_0\frac{\partial \mathbf{E}}{\partial t} + \mu_0\mathbf{J} \\ \text{div } \mathbf{E} &= \epsilon_0\rho \\ \text{div } \mathbf{B} &= 0 \end{aligned} \tag{15'}$$

A reminder about the units: E is in volts/meter, B in teslas, ρ in coulombs/m^3 and J in amps/m^2. In these equations c does not appear. But if one drops the charge and current terms from Eq. 15′ and compares what is left with our "empty space" Maxwell's Equations (16), it becomes obvious that c must be hidden in the constant $\mu_0\epsilon_0$. In fact $\mu_0\epsilon_0 = 1/c^2$. The way in which this fixes the exact value of ϵ_0 is explained in Appendix E.

AN ELECTROMAGNETIC WAVE

9.4 We are going to construct a rather simple electromagnetic field that will satisfy Maxwell's equations for empty space, Eqs. 16. Suppose there is an electric field **E**, everywhere parallel to the z axis, whose intensity depends only on the space coordinate y and the time t. Let the dependence have this particular form:

$$\mathbf{E} = \hat{\mathbf{z}} E_0 \sin(y - vt) \tag{17}$$

in which E_0 and v are simply constants. This field fills all space—at least all the space we are presently concerned with. We'll need a magnetic field, too. We shall assume it has an x component only, with a dependence on y and t similar to that of E_z:

$$\mathbf{B} = \hat{\mathbf{x}} B_0 \sin(y - vt) \tag{18}$$

where B_0 is another constant.

Figure 9.7 may help you to visualize these fields. It is difficult to represent graphically two such fields filling all space. Remember that nothing varies with x or z; whatever is happening at a point on the y axis is happening everywhere on the perpendicular plane through that point. As time goes on the entire field pattern slides steadily to the right, thanks to the particular form of the argument of the sine function in Eqs. 17 and 18. For that argument $y - vt$ has the same value at $y + \Delta y$ and $t + \Delta t$ as it had at y and t, providing $\Delta y = v\,\Delta t$. In other words, we have here a plane wave traveling with the constant speed v in the $\hat{\mathbf{y}}$ direction.

We'll show now that this electromagnetic field satisfies Maxwell's equations if certain conditions are met. It is easy to see that div $\mathbf{E}$ and div $\mathbf{B}$ are both zero for this field. The other derivatives involved are

$$\begin{aligned}
\text{curl}\,\mathbf{E} &= \hat{\mathbf{x}}\frac{\partial E_z}{\partial y} = \hat{\mathbf{x}} E_0 \cos(y - vt)\\
\frac{\partial \mathbf{E}}{\partial t} &= -v\hat{\mathbf{z}} E_0 \cos(y - vt)\\
\text{curl}\,\mathbf{B} &= -\hat{\mathbf{z}}\frac{\partial B_x}{\partial y} = -\hat{\mathbf{z}} B_0 \cos(y - vt)\\
\frac{\partial \mathbf{B}}{\partial t} &= -v\hat{\mathbf{x}}\, B_0 \cos(y - vt)
\end{aligned} \tag{19}$$

Substituting into the two "induction" equations of Eq. 16 and cancelling the common factor, $\cos(y - vt)$, we find the conditions that must be satisfied,

$$E_0 = \frac{vB_0}{c} \quad \text{and} \quad B_0 = \frac{vE_0}{c} \tag{20}$$

which together require that

$$v = \pm c \quad \text{and} \quad B_0 = E_0 \tag{21}$$

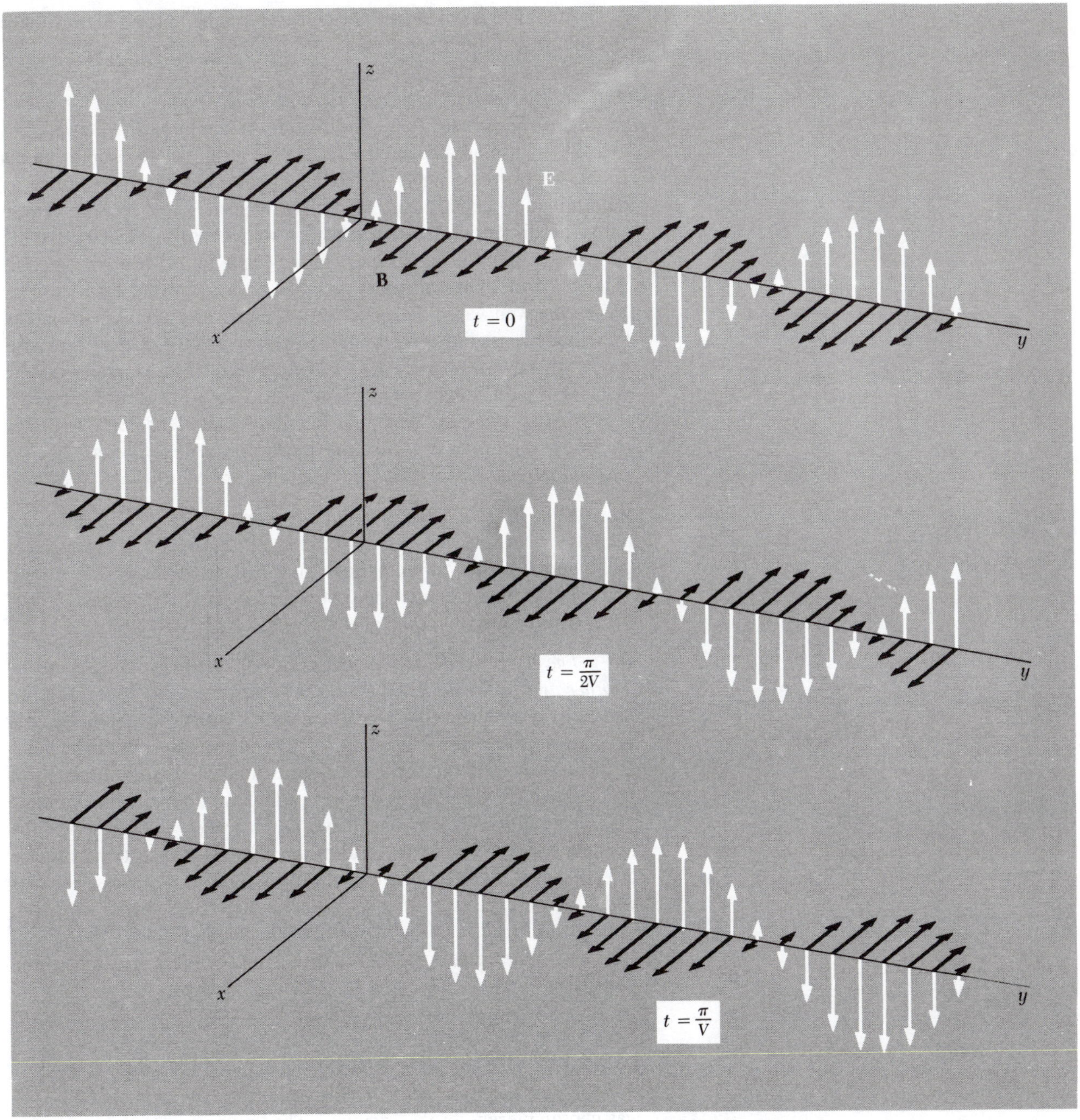

FIGURE 9.7
The wave described by Eqs. 17 and 18 is shown at three different times. It is traveling to the right, in the positive *y* direction.

We have now learned that our electromagnetic wave must have the following properties:

1. *The field pattern travels with speed c.* In the case $v = -c$ it travels in the opposite, or $-\hat{\mathbf{y}}$, direction. When in 1862 Maxwell first arrived (by a more obscure route) at this result, the constant c in his equations expressed only a relation among electrical quantities as determined by experiments with capacitors, coils, and resistors. To be sure, the dimensions of this constant were those of velocity, but its connection with the actual speed of light had not yet been recognized. The speed of light had most recently been measured by Fizeau in 1857. Maxwell wrote, "The velocity of transverse undulations in our hypothetical medium, calculated from the electro-magnetic experiments of MM. Kohlrausch and Weber, agrees so exactly with the velocity of light calculated from the optical experiments of M. Fizeau, that we can scarcely avoid the inference that *light consists in the transverse undulations of the same medium which is the cause of electric and magnetic phenomena.*" The italics are Maxwell's.

2. *At every point in the wave at any instant of time, the electric and magnetic field strengths are equal.* In our CGS units B is expressed in gauss and E in statvolts/cm; but B and E have the same dimensions, and these units are equivalent. If the electric field strength is 10 statvolts/cm, the associated magnetic field strength is 10 gauss.

3. *The electric field and the magnetic field are perpendicular to one another and to the direction of travel, or propagation.* To be sure, we had already assumed that when we constructed our example, but it is not hard to show that it is a necessary condition, given that the fields do not depend on the coordinates perpendicular to the direction of propagation. Notice that, if $v = -c$, which would make the direction of propagation $-\hat{y}$, we must have $B_0 = -E_0$. This preserves the handedness of the essential triad of directions, the direction of **E**, the direction of **B**, and the direction of propagation. We can describe that without reference to a particular coordinate frame in this way: The wave always travels in the direction of the vector $\mathbf{E} \times \mathbf{B}$.

Any plane electromagnetic wave in empty space has these three properties.

OTHER WAVEFORMS; SUPERPOSITION OF WAVES

9.5 In the example we have just studied the function $\sin(y - vt)$ was chosen merely for its simplicity. The "waviness" of the sinusoidal function has *nothing to do* with the essential property of wave motion, which is the propagation unchanged of a form or pattern—*any* pattern. It was not the nature of the function but the way y and t were combined in its argument that caused the pattern to propagate. If we replace the sine function by *any* other function, $f(y - vt)$, we'll get a pattern that travels with speed v in the $\hat{\mathbf{y}}$ direction. Moreover, Eqs. 20 will apply as before, and our wave will have the three general properties just listed.

Here is another example, the plane electromagnetic wave pictured in Fig. 9.8, which is described mathematically as follows:

$$\mathbf{E} = \frac{5\hat{\mathbf{y}}}{1 + (x + ct)^2} \qquad \mathbf{B} = \frac{-5\hat{\mathbf{z}}}{1 + (x + ct)^2} \tag{22}$$

This electromagnetic field satisfies Maxwell's equations, Eqs. 16. It is a *plane* wave because nothing depends on y or z. It is traveling in the direction $-\hat{\mathbf{x}}$, as we recognize at once from the + sign in the argument $x + ct$. That is indeed the direction of $\mathbf{E} \times \mathbf{B}$. In this wave nothing is oscillating or alternating; it is simply an electromagnetic pulse with long tails. At time $t = 0$ the maximum field strengths, E

FIGURE 9.8
The wave described by Eq. 22 is traveling in the negative x direction. It is shown 3 nanoseconds before its peak passed the origin.

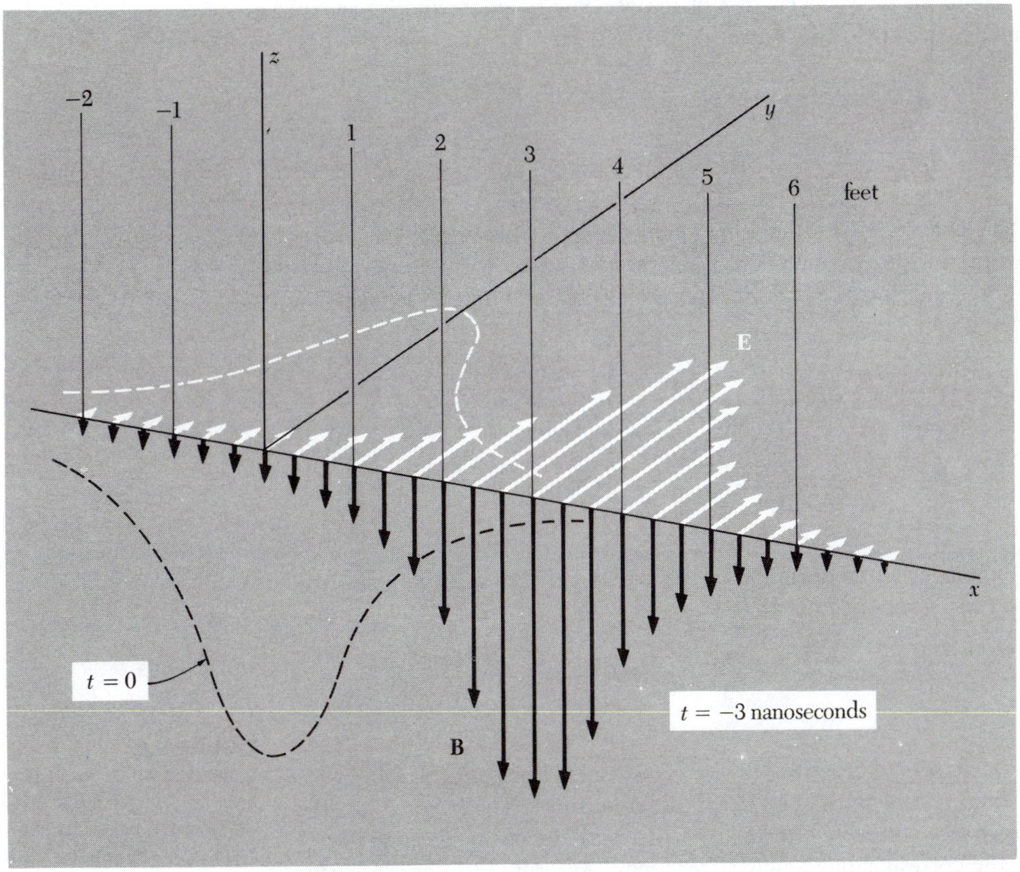

= 5 statvolts/cm and B = 5 gauss, will be experienced by an observer at the origin, or at any other point on the yz plane. In Fig. 9.8 we have shown the field as it was at $t = -3$ nanoseconds, with the distances marked off in feet. (The speed of light is very nearly 1 foot/nanosecond.)

Maxwell's equations for **E** and **B** in empty space are linear. The superposition of two solutions is also a solution. Any number of electromagnetic waves can propagate through the same region without affecting one another. The field **E** at a space-time point is the vector sum of the electric fields of the individual waves, and the same goes for **B**.

An important example is the superposition of two similar plane waves traveling in opposite directions. Consider a wave traveling in the $\hat{\mathbf{y}}$ direction, described by

$$\mathbf{E}_1 = \hat{\mathbf{z}}E_0 \sin\frac{2\pi}{\lambda}(y - ct) \qquad \mathbf{B}_1 = \hat{\mathbf{x}}E_0 \sin\frac{2\pi}{\lambda}(y - ct) \tag{23}$$

This wave differs in only minor ways from our first example. We have introduced the wavelength λ of the periodic function, and we set the magnetic field amplitude explicitly equal to the electric field amplitude.

Now consider another wave:

$$E_2 = \hat{\mathbf{z}}E_0 \sin\frac{2\pi}{\lambda}(y + ct) \qquad B_2 = -\hat{\mathbf{x}}E_0 \sin\frac{2\pi}{\lambda}(y + ct) \tag{24}$$

This is a wave with the same amplitude and wavelength, but propagating in the $-\hat{\mathbf{y}}$ direction. With the two waves both present, Maxwell's equations are still satisfied, the electric and magnetic fields now being

$$\mathbf{E} = \mathbf{E}_1 + \mathbf{E}_2 = \hat{\mathbf{z}}E_0\left[\sin\left(\frac{2\pi y}{\lambda} - \frac{2\pi ct}{\lambda}\right) + \sin\left(\frac{2\pi y}{\lambda} + \frac{2\pi ct}{\lambda}\right)\right]$$
$$\mathbf{B} = \mathbf{B}_1 + \mathbf{B}_2 = \hat{\mathbf{x}}E_0\left[\sin\left(\frac{2\pi y}{\lambda} - \frac{2\pi ct}{\lambda}\right) - \sin\left(\frac{2\pi y}{\lambda} + \frac{2\pi ct}{\lambda}\right)\right] \tag{25}$$

Remembering the formula for the sine of the sum of two angles, you can easily reduce Eqs. 25 to

$$\mathbf{E} = 2\hat{\mathbf{z}}E_0 \sin\frac{2\pi y}{\lambda}\cos\frac{2\pi ct}{\lambda} \qquad \mathbf{B} = -2\hat{\mathbf{x}}E_0 \cos\frac{2\pi y}{\lambda}\sin\frac{2\pi ct}{\lambda} \tag{26}$$

The field described by Eqs. 26 is called a *standing wave.* Figure 9.9 suggests what it looks like at different times. The factor c/λ is the *frequency* with which the field oscillates at any position x, and $2\pi c/\lambda$ is the corresponding angular frequency. According to Eqs. 26 whenever $2ct/\lambda$ equals an integer, which happens every half-period, the

FIGURE 9.9
A standing wave, resulting from the superposition of a wave traveling in the positive y direction (Eq. 23) and a similar wave traveling in the negative y direction (Eq. 24). Beginning with the top figure, the fields are shown at four different times, separated successively by one-eighth of a full period.

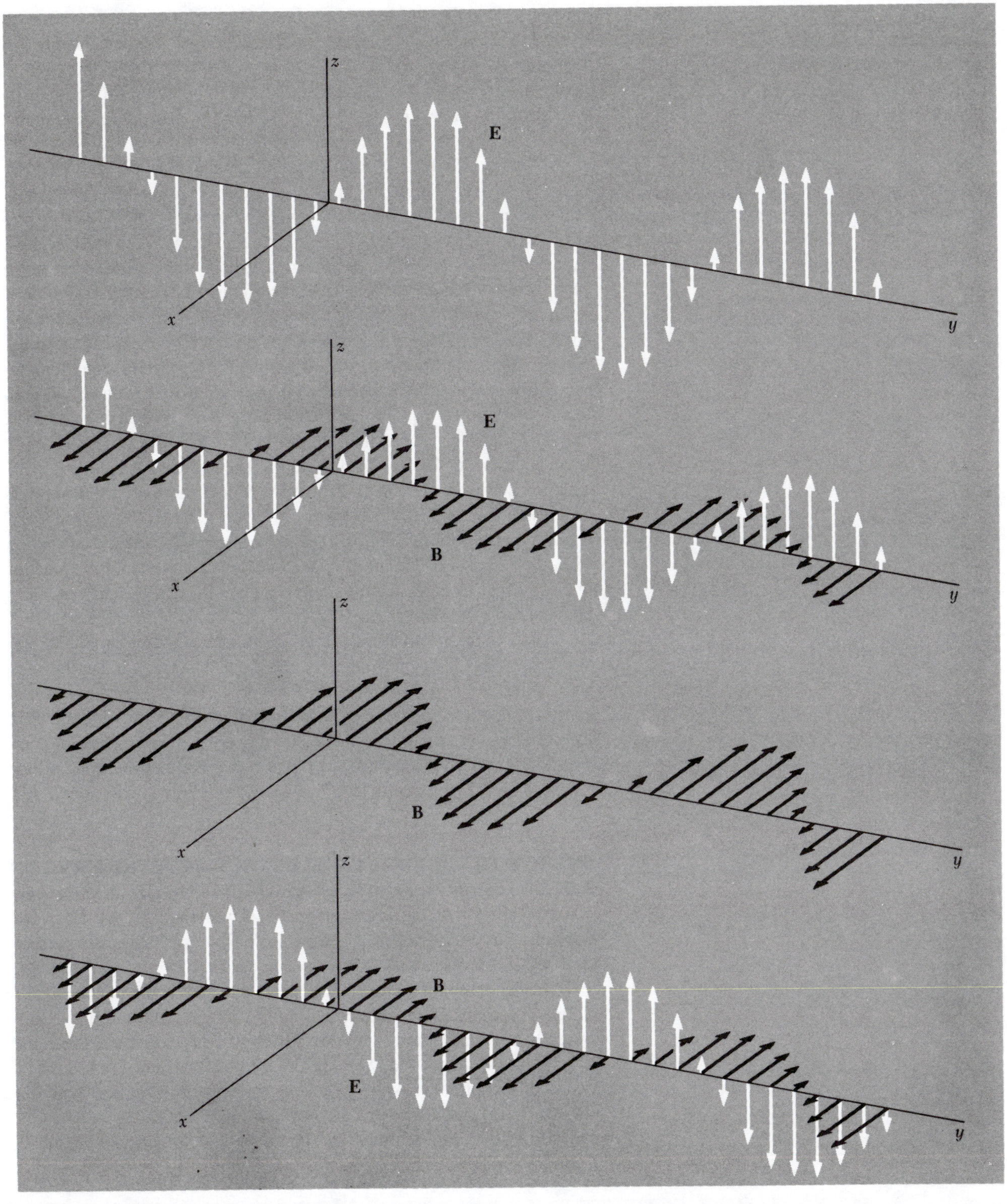
z
E
x
y
z
E
B
x
y
z
B
x
y
z
B
E
x
y

magnetic field **B** vanishes *everywhere*. On the other hand, whenever $2ct/\lambda$ equals an integer plus one-half, $\cos 2\pi ct/\lambda = 0$ and the electric field vanishes everywhere. The maxima of **B** and the maxima of **E** occur at different places as well as at different times. In contrast to the traveling wave, the standing wave has its electric and magnetic fields "out of step" in both space and time.

Notice that $\mathbf{E} = 0$ *at all times* on the plane $y = 0$ and on every other plane for which y equals an integral number of half-wavelengths. Imagine that we could cover the xz plane at $y = 0$ with a sheet of perfectly conducting metal. At the surface of a perfect conductor the electric field component parallel to the surface must be zero—otherwise an infinite current would flow. That imposes a drastic *boundary condition* on any electromagnetic field in the surrounding space. But our standing wave, which is described by Eqs. 26, *already satisfies* that condition, as well as satisfying Maxwell's equations in the entire space $y > 0$. Therefore it provides a ready-made solution to the problem of a plane electromagnetic wave reflected, at normal incidence, from a flat conducting mirror. (See Fig. 9.10.) The incident wave is described by Eq. 24, for $y > 0$, the reflected wave by Eq. 23. There is no field at all behind the mirror, or if there is, it has nothing to do with the field in front. Immediately in front of the mirror there is a magnetic field parallel to the surface, given by Eqs. 26: $\mathbf{B} = -2\hat{\mathbf{x}}E_0 \sin(2\pi ct/\lambda)$. The jump in **B** from this value in front of the conducting sheet to zero behind shows that an alternating current must be flowing in the sheet.

You could install a conducting sheet at any other plane where **E** as given by Eqs. 26 is permanently zero, and thus trap an electromagnetic standing wave between two mirrors. That arrangement has many applications, including lasers. In fact, with an understanding of the properties of the simple plane electromagnetic wave, you can analyze a surprisingly wide variety of electromagnetic devices, including interferometers, rectangular hollow wave guides, and strip lines.

ENERGY TRANSPORT BY ELECTROMAGNETIC WAVES

9.6 The energy the earth receives from the sun has traveled through space in the form of electromagnetic waves that satisfy Eqs. 16. Where *is* this energy when it is traveling? How is it deposited in matter when it arrives?

In the case of a static electric field, such as the field between the plates of a charged capacitor, we found that the total energy of the system could be calculated by attributing to every volume element dv an amount of energy $(E^2/8\pi)\,dv$ and adding it all up. Look back at Eq. 38 of Chapter 1. Likewise the energy invested in the creation of a magnetic field could be calculated by assuming that every volume element dv in the field contains $(B^2/8\pi)\,dv$ units of energy. See Eq. 75

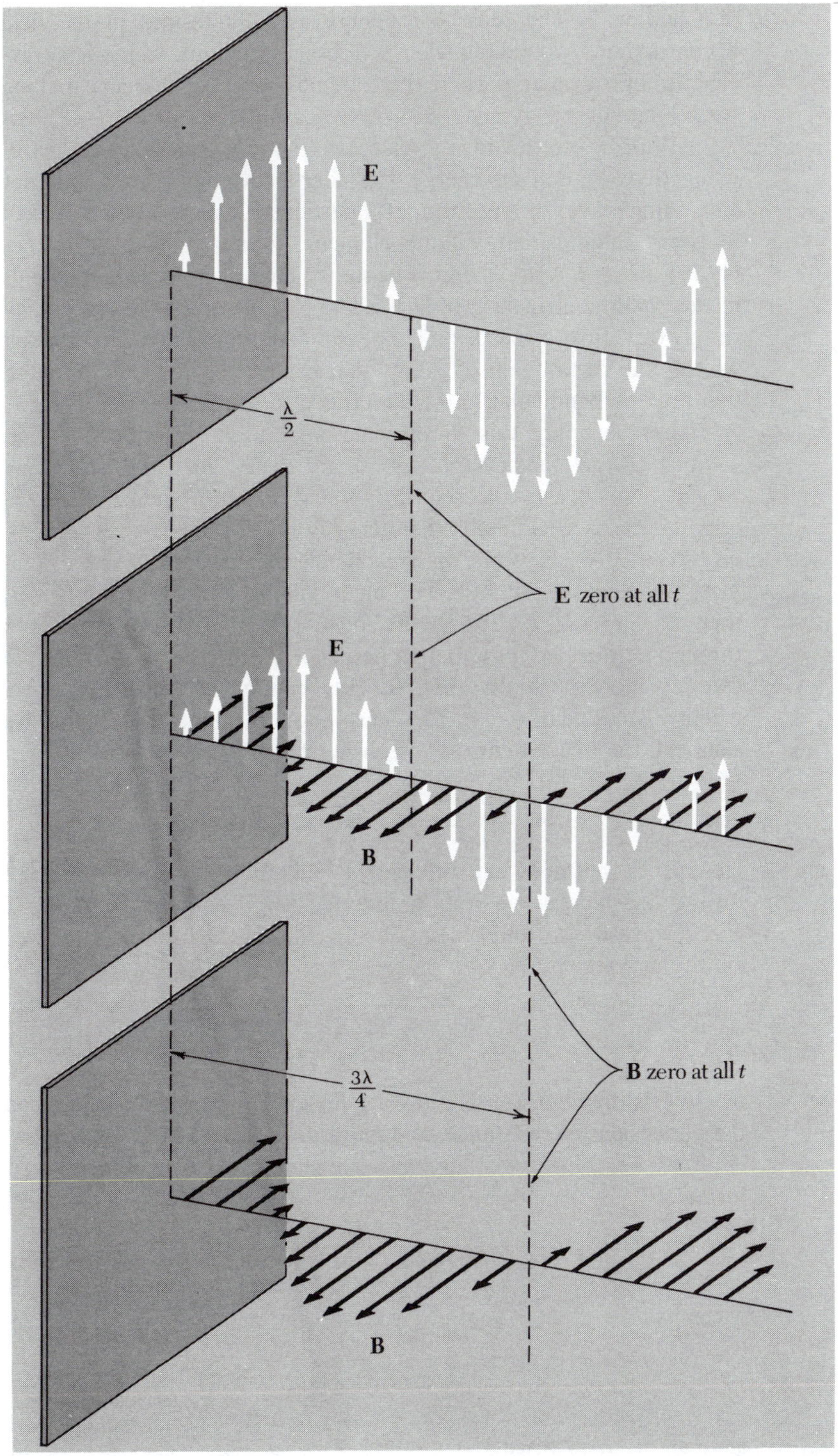

FIGURE 9.10
A standing wave produced by reflection at a perfectly conducting plate.

of Chapter 7. The idea that energy actually resides in the field becomes more compelling when we observe sunlight, which has traveled through a vacuum where there are no charges or currents, making something hot.

We can use this idea to calculate the rate at which an electromagnetic wave delivers energy. Consider a traveling plane wave (not a standing wave) of any form, at a particular instant of time. Assign to every infinitesimal volume element dv an amount of energy $(1/8\pi)(E^2 + B^2)\,dv$, **E** and **B** being the electric and magnetic fields in that volume element at that instant. Now assume that this energy simply travels with speed c in the direction of propagation. In this way we can find the amount of energy that passes, per unit time, through unit area perpendicular to the direction of propagation.

Let us apply this to the sinusoidal wave described by Eqs. 17 and 18. At the instant $t = 0$, $E^2 = E_0^2 \sin^2 y$. Also, $B^2 = E_0^2 \sin^2 y$, since, as we subsequently found, B_0 must equal E_0. The energy density in this field is therefore $(1/8\pi)(E_0^2 \sin^2 y + E_0^2 \sin^2 y)$, or $E_0^2 \sin^2 y/4\pi$. Now the mean value of $\sin^2 y$ averaged over a complete wavelength is just ½. The mean energy density in the field is then $E_0^2/8\pi$, and $E_0^2 c/8\pi$ is the mean rate at which energy flows through a "window" of unit area perpendicular to the y direction. We can say more generally that, for any continuous, repetitive wave, whether sinusoidal or not, the energy flow per unit area, which we shall call the *power density S,* is given by

$$S = \frac{\overline{E^2}c}{4\pi} \tag{27}$$

Here $\overline{E^2}$ is the mean square electric field strength, which was $E_0^2/2$ for the sinusoidal wave of amplitude E_0. S will be in ergs/cm²-sec if E is in statvolts/cm and c in cm/sec. In SI units the formula for power density in watts/m² is

$$S = \frac{\overline{E^2}}{\sqrt{\mu_0/\epsilon_0}} \tag{28}$$

where E is the field strength in volts/meter. The constant $\sqrt{\mu_0/\epsilon_0}$ has the dimensions of resistance, and its value is 376.73 ohms. Rounding it off to 377 ohms, we have a convenient and easily remembered formula:

$$S(\text{watts/m}^2) = \frac{\overline{E^2}(\text{volts/meter})^2}{377 \text{ ohms}} \tag{29}$$

$$\text{watts} = \frac{\text{volt}^2}{\text{ohm}}$$

as in an ordinary resistor.

When the electromagnetic wave encounters an electrical conductor, the electric field causes currents to flow. This generally results in energy being dissipated within the conductor at the expense of the energy in the wave. The total reflection of the incident wave in Fig. 9.10 was a special case in which the conductivity of the reflecting surface was infinite. If the resistivity of the reflector is not zero, the amplitude of the reflected wave will be less than that of the incident wave. Aluminum, for example, reflects visible light, at normal incidence, with about 92 percent efficiency. That is, 92 percent of the incident energy is reflected, the amplitude of the reflected wave being $\sqrt{0.92}$ or 0.96 times that of the incident wave. The lost 8 percent of the incident energy ends up as heat in the aluminum, where the current driven by the electric field of the wave encountered ohmic resistance. What counts, of course, is the resistivity of aluminum at the frequency of the light wave, in this case about 5×10^{14} Hz. That may be somewhat different from the dc or low-frequency resistivity of the metal. Still, the reflectivity of most metals for visible light is essentially due to the same highly mobile conduction electrons that make metals good conductors of steady current. It is no accident that good conductors are generally shiny. But why clean copper looks reddish while aluminum looks "silvery" can't be explained without a detailed theory of each metal's electronic structure.

Energy can also be absorbed when an electromagnetic wave meets nonconducting matter. Little of the light that strikes a black rubber tire is reflected, although the rubber is an excellent insulator for low-frequency electric fields. Here the dissipation of the electromagnetic energy involves the action of the high-frequency electric field on the electrons in the molecules of the material. In the broadest sense that applies to the absorption of light in everything around us, including the retina of the eye.

Some insulators transmit electromagnetic waves with very little absorption. The transparency of glass for visible light, with which we are so familiar, is really a remarkable property. In the purest glass fibers used for optical transmission of audio and video signals, a wave travels as much as a kilometer, or more than 10^9 wavelengths, before most of the energy is lost. However transparent a material medium may be, the propagation of an electromagnetic wave within the medium differs in essential ways from propagation through the vacuum. The matter interacts with the electromagnetic field. To take that interaction into account Eqs. 16 must be modified in a way that will be explained in the next chapter.

HOW A WAVE LOOKS IN A DIFFERENT FRAME

9.7 A plane electromagnetic wave is traveling through the vacuum. Its direction of travel, with respect to a certain inertial frame F, is given by a unit vector $\hat{\mathbf{n}}$. Let $\mathbf{E}$ and $\mathbf{B}$ be the electric and magnetic fields measured at some place and time in F, by an observer in F. What field will be measured by an observer in a different frame who happens to be passing that point at that time? Suppose that frame F' is moving with speed v in the $\hat{\mathbf{x}}$ direction relative to F, with its axes parallel to those of F, as in Fig. 6.25. Let us choose $\hat{\mathbf{n}}$ in the $\hat{\mathbf{x}}$ direction also. We can now turn to Eqs. 58 in Chapter 6 for the transformation of the field components. Let us write them out again:

$$\begin{aligned} E'_x &= E_x \qquad & E'_y &= \gamma(E_y - \beta B_z) \qquad & E'_z &= \gamma(E_z + \beta B_y) \\ B'_x &= B_x \qquad & B'_y &= \gamma(B_y + \beta E_z) \qquad & B'_z &= \gamma(B_z - \beta E_y) \end{aligned} \tag{30}$$

The key to our problem is the way two particular scalar quantities transform, namely, $\mathbf{E} \cdot \mathbf{B}$ and $E^2 - B^2$. Let us use Eqs. 30 to calculate $\mathbf{E}' \cdot \mathbf{B}'$ and see how it is related to $\mathbf{E} \cdot \mathbf{B}$.

$$\begin{aligned} \mathbf{E}' \cdot \mathbf{B}' &= E'_x B'_x + E'_y B'_y + E'_z B'_z \\ &= E_x B_x + \gamma^2(E_y B_y + \beta E_y E_z - \beta B_y B_z - \beta^2 E_z B_z) \\ &\quad + \gamma^2(E_z B_z - \beta E_y E_z + \beta B_y B_z - \beta^2 E_y B_y) \\ &= E_x B_x + \gamma^2(1 - \beta^2)(E_y B_y + E_z B_z) = \mathbf{E} \cdot \mathbf{B} \end{aligned} \tag{31}$$

The scalar product $\mathbf{E} \cdot \mathbf{B}$ is *not changed* in the Lorentz transformation of the fields; it is an invariant. A similar calculation, which will be left to the reader as problem 9.13, shows that $E_x^2 + E_y^2 + E_z^2 - (B_x^2 + B_y^2 + B_z^2)$ is also an invariant, that is,

$$E'^2 - B'^2 = E^2 - B^2 \tag{32}$$

The invariance of these two quantities is an important general property of any electromagnetic field, not just the field of an electromagnetic wave with which we are concerned at the moment. For the wave field its implications are especially simple and direct. We know that the plane wave has $\mathbf{B}$ perpendicular to $\mathbf{E}$ and $B = E$. Each of our two invariants, $\mathbf{E} \cdot \mathbf{B}$ and $E^2 - B^2$, is therefore zero. And if an invariant is zero in any frame, it must be zero in all frames. We see that *any* Lorentz transformation of the wave will leave E and B perpendicular and equal in magnitude. *A light wave looks like a light wave in any inertial frame of reference.* That should not surprise us. It could be said that we have merely come full circle, back to the postulates of relativity, Einstein's starting point. Indeed, according to Einstein's own autobiographical account, he had begun 10 years earlier (at age 16!) to wonder what one would observe if one could "catch up" with a light wave. With Eqs. 30, which were given in just that form in Einstein's 1905 paper, the question can be answered. Let $E_y = E_0$, $E_x = E_z = 0$, $B_z = E_0$, $B_x = B_y = 0$. That is a wave traveling the

$\hat{\mathbf{x}}$ direction, as we can tell from the fact that $\mathbf{E} \times \mathbf{B}$ points in that direction. Using Eqs. 30 and the identity $\gamma^2(1 - \beta^2) = 1$, we find that

$$E'_y = E_0 \sqrt{\frac{1-\beta}{1+\beta}} \qquad B'_z = E_0 \sqrt{\frac{1-\beta}{1+\beta}} \tag{33}$$

As observed in F' the amplitude of the wave is reduced. The wave velocity, of course, is c in F', as it is in F. The electromagnetic wave has no rest frame. In the limit $\beta = 1$, the amplitudes E'_y and B'_z observed in F' are reduced to zero. The wave has vanished!

PROBLEMS

9.1 If the electric field in free space is $\mathbf{E} = E_0(\hat{\mathbf{x}} + \hat{\mathbf{y}}) \sin (2\pi/\lambda)(z + ct)$, with $E_0 = 2$ statvolts/cm, the magnetic field, not including any static magnetic field, must be what?

9.2 The power density in sunlight, at the earth, is roughly 1 kilowatt/meter^2. How large is the rms magnetic field strength?

Ans. 0.02 gauss or 2×10^{-6} tesla.

9.3 A free proton was at rest at the origin before the wave described by Eq. 22 came past. Where would you expect to find the proton at time $t = 1$ microsecond? The pulse amplitude is in statvolts/cm. Proton mass $= 1.6 \times 10^{-24}$ gm. *Hint:* Since the duration of the pulse is only a few nanoseconds, you can neglect the displacement of the proton during the passage of the pulse. Also, if the velocity of the proton is not too large, you may ignore the effect of the magnetic field on its motion. The first thing to calculate is the momentum acquired by the proton during the pulse.

9.4 Suppose that in the preceding problem the effect of the magnetic field was not entirely negligible. How would it change the direction of the proton's final velocity?

9.5 Here is a particular electromagnetic field in free space:

$$E_x = 0 \qquad E_y = E_0 \sin(kx + \omega t) \qquad E_z = 0$$

$$B_x = 0 \qquad B_y = 0 \qquad B_z = -E_0 \sin(kx + \omega t)$$

(*a*) Show that this field can satisfy Maxwell's equations if ω and k are related in a certain way.

(*b*) Suppose $\omega = 10^{10}$ sec^{-1} and $E_0 = 0.05$ statvolt/cm. What is the wavelength in cm? What is the energy density in ergs/cm^3, averaged over a large region? From this calculate the power density, the energy flow in ergs/cm^2-sec.

9.6 Start with the source free, or "empty space" Maxwell's equations in SI units, obtained by dropping the terms with ρ and $\mathbf{J}$ from Eq. 15′. Consider a wave described by Eqs. 17 and 18, but now with E_0 in volts/meter and B_0 in teslas. What conditions must E_0, B_0, and v meet to satisfy Maxwell's equations?

9.7 Write out formulas for **E** and **B** that specify a plane electromagnetic sinusoidal wave with the following characteristics. The wave is traveling in the direction $-\hat{\mathbf{x}}$; its frequency is 100 megahertz (MHz, 10^8 cycles per sec); the electric field is perpendicular to the $\hat{\mathbf{z}}$ direction.

9.8 Show that the electromagnetic field described by

$$\mathbf{E} = E_0\hat{\mathbf{z}} \cos kx \cos ky \cos \omega t$$

$$\mathbf{B} = B_0(\hat{\mathbf{x}} \cos kx \sin ky - \hat{\mathbf{y}} \sin kx \cos ky) \sin \omega t$$

will satisfy Eqs. 16 if $E_0 = \sqrt{2}B_0$ and $\omega = \sqrt{2}ck$. This field can exist inside a square metal box, of dimension π/k in the x and y directions and arbitrary height. What does the magnetic field look like?

9.9 Of all the electromagnetic energy in the universe, by far the largest amount is in the form of waves with wavelengths in the millimeter range. This is the cosmic microwave background radiation discovered by Penzias and Wilson in 1965. It apparently fills all space, including the vast space between galaxies, with an energy density of 4×10^{-13} erg/cm^3. Calculate the rms electric field strength in this radiation, in statvolts/cm, and convert it to volts/meter. Roughly how far away from a 1-kilowatt radio transmitter would you find a comparable electromagnetic wave intensity?

Ans. 0.06 volt/meter; 3 km.

PROBLEM 9.10

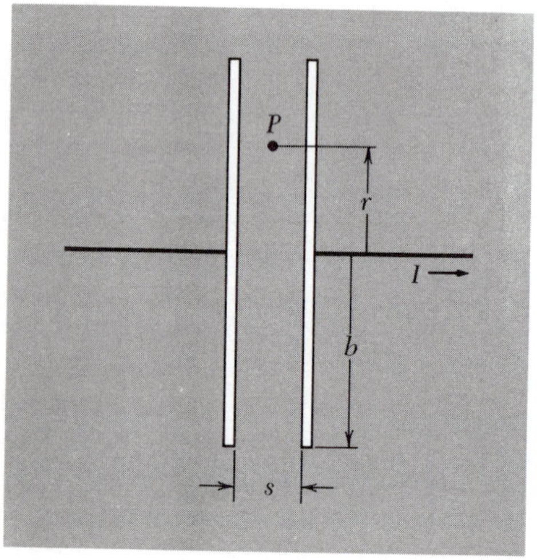

9.10 The magnetic field inside the discharging capacitor shown in Fig. 9.1 can in principle be calculated by summing the contributions from all elements of conduction current, as indicated in Fig. 9.5. That might be a long job. If we can assume symmetry about this axis, it is very much easier to find the field **B** at a point by using the integral law

$$\int_C \mathbf{B} \cdot d\mathbf{s} = \frac{1}{c}\int_S \left(\frac{\partial \mathbf{E}}{\partial t} + 4\pi \mathbf{J}\right) \cdot d\mathbf{a}$$

applied to a circular path through the point. We need only know the total current enclosed by this path. Use this to find the field at P, which is midway between the capacitor plates and a distance r from the axis of symmetry. (Compare this with the calculation of the induced electric field **E**, in the example of Fig. 7.16.)

$$\textit{Ans. } 2\pi r B = \frac{4\pi I}{c}\frac{r^2}{b^2}, \quad B = \frac{2Ir}{cb^2}.$$

9.11 From a satellite in stationary orbit a signal is beamed earthward with a power of 10 kilowatts and a beam width covering a region roughly circular and 1000 km in diameter. What is the electric field strength at the receivers, in millivolts/meter?

9.12 A sinusoidal wave is reflected at the surface of a medium whose properties are such that half the incident energy is absorbed. Consider the field that results from the superposition of the incident and the reflected wave. An observer stationed somewhere in this field finds the local electric field oscillating with a certain amplitude E. What is the ratio of the largest such amplitude noted by an observer to the smallest amplitude noted by any observer? (This is called the *voltage standing wave ratio,* in laboratory jargon, VSWR.)

9.13 Starting from the field transformation given by Eq. 60 of Chapter 6, show that the scalar quantity $E^2 - B^2$ is invariant under the transformation. In other words, show that $E'^2 - B'^2 = E^2 - B^2$. You can do this using only vector algebra, without writing out x, y, z components of anything. (The resolution into parallel and perpendicular vectors is convenient for this, since $\mathbf{E}_\perp \cdot \mathbf{E}_\parallel = 0$, $\mathbf{B}_\parallel \times \mathbf{E}_\parallel = 0$, etc.)

10

ELECTRIC FIELDS IN MATTER

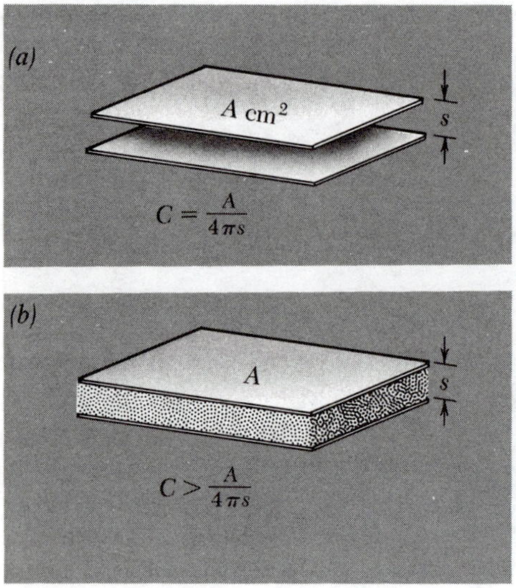

FIGURE 10.1
(*a*) A capacitor formed by parallel conducting plates.
(*b*) The same plates with a slab of insulator in between.

DIELECTRICS

10.1 The capacitor we studied in Chapter 3 consisted of two conductors, insulated from one another, with nothing in between. The system of two conductors was characterized by a certain capacitance C, a constant relating the magnitude of the charge Q on the capacitor (positive charge Q on one plate, equal negative charge on the other) to the difference in electric potential between the two conductors, $\varphi_1 - \varphi_2$. Let's denote the potential difference by φ_{12}:

$$C = \frac{Q}{\varphi_{12}} \tag{1}$$

For the parallel-plate capacitor, two flat plates each of area A cm^2 and separated by a distance s, we found that the capacitance is given by

$$C = \frac{A}{4\pi s} \tag{2}$$

Capacitors like this can be found in some electrical apparatus. They are called *vacuum capacitors* and consist of plates enclosed in a highly evacuated bottle. They are used chiefly where extremely high and rapidly varying potentials are involved. Far more common, however, are capacitors in which the space between the plates is filled with some nonconducting solid or liquid substance. Most of the capacitors you have worked with in the laboratory are of that sort; there are dozens of them in any television receiver. For conductors embedded in a material medium, Eq. 2 doesn't agree with experiment. Suppose we fill the space between the two plates shown in Fig. 10.1*a* with a slab of plastic, as in Fig. 10.1*b*. Experimenting with this new capacitor, we still find a simple proportionality between charge and potential difference, so that we can still *define* a capacitance by Eq. 1. But we find C to be substantially *larger* than Eq. 2 would have predicted. That is, we find more charge on each of the plates, for the same potential difference, plate area, and distance of separation. The plastic slab must be the cause of this.

It is not hard to understand in a general way how this comes about. The plastic slab consists of molecules, the molecules are composed of atoms which in turn are made of electrically charged particles, electrons, and atomic nuclei. The electric field between the capacitor plates acts on those charges, pulling the negative charges up, if the upper plate is positive as in Fig. 10.2, and pushing the positive charges down. Nothing moves very far. (There are no free electrons around, already detached from atoms and ready to travel, as there would be in a metallic conductor.) There will be some slight displacement of the charges nevertheless, for an atom is not an infinitely rigid structure. The effect of this within the plastic slab is that the negative charge distribution, viewed as a whole, and the total positive charge distribution (the atomic nuclei) are very slightly displaced relative to

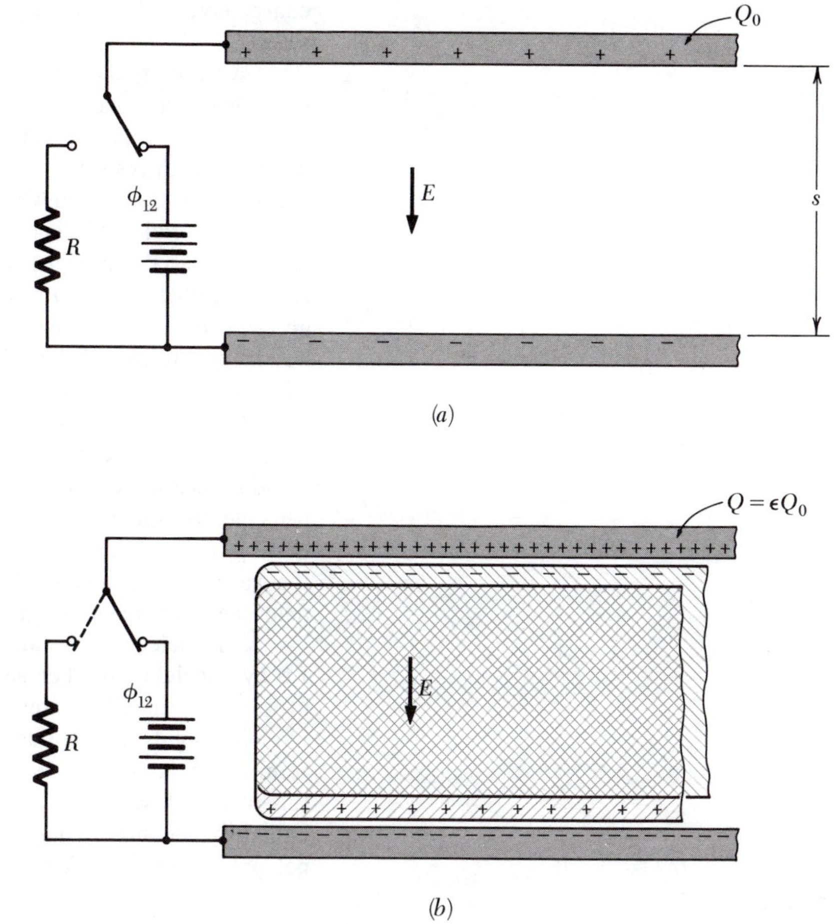

FIGURE 10.2
How a dielectric increases the charge on the plates of a capacitor. (*a*) Space between the plates empty; $Q_0 = C_0\phi_{12}$. (*b*) Space between the plates filled with a nonconducting material, that is, a dielectric. Electric field pulls negative charges up, pushes positive charges down, exposing layer of uncompensated negative charge on upper surface of dielectric, uncompensated positive charge on lower surface. Total charge at the top, *including* charge Q on upper plate, is same as in (*a*). Q itself is now greater than Q_0. $Q = \epsilon Q_0$, Q is the amount of charge that will flow through the resistor R if the capacitor is discharged by throwing the switch.

one another, as indicated in Fig. 10.2*b*. The interior of the block remains electrically neutral, but a thin layer of uncompensated negative charge has emerged at the top, with a corresponding layer of uncompensated positive charge at the bottom.

In the presence of the induced layer of negative charge below the upper plate, the charge Q on the plate itself will increase. In fact, Q must increase until the total charge at the top, the algebraic sum of Q and the induced charge layer, equals Q_0. We shall be able to prove that when we return to this problem in Section 10.8 after settling some questions about the electric field inside matter. The important point now is that the charge Q in Fig. 10.2*b* is *larger* than Q_0 and that this Q *is* the charge of the capacitor in the relation $Q = C\varphi_{12}$. It is the charge that came out of the battery, and it is the amount of charge that would flow through the resistor R were we to discharge the capacitor by throwing the switch in the diagram. If we did that, the induced charge layer, which is not part of Q, would simply disappear into the slab.

According to this explanation the ability of a particular material to increase the capacitance ought to depend on the amount of electric charge in its structure and the ease with which the electrons can be displaced with respect to the atomic nuclei. The factor by which the capacitance is increased when an empty capacitor is filled with a particular material, Q/Q_0 in our example, is called the *dielectric constant* of that material. The symbol ϵ is usually used for it. The material itself is often called a *dielectric* when we are talking about its behavior in an electric field. But any homogeneous nonconducting substance can be so characterized. Table 10.1 lists the measured values of the dielectric constant for a miscellaneous assortment of substances.

Every dielectric constant in the table is larger than 1. We should expect that if our explanation is correct. The presence of a dielectric could *reduce* the capacitance below that of the empty capacitor only if its electrons moved, when the electric field was applied, in a direction opposite to the resulting force. For oscillating electric fields, by the way, some such behavior would not be absurd. But for the steady fields we are here considering it can't work that way.

The dielectric constant of a perfect vacuum is, of course, exactly 1.0 by our definition. For gases under ordinary conditions ϵ is only a little larger than 1.0, simply because a gas is mostly empty space. Ordinary solids and liquids usually have dielectric constants ranging from 2 to 6 or so. But notice that liquid ammonia is an exception to this rule, and water is a spectacular exception. Actually liquid water is slightly conductive, but that, as we shall have to explain later, does not prevent our defining and measuring its dielectric constant. The

TABLE 10.1

Dielectric Constants of Various Substances

Substance	Conditions	Dielectric Constant
Air	Gas, 0°C, 1 atm	1.00059
Methane, CH_4	Gas, 0°C, 1 atm	1.00088
Hydrogen chloride, HCl	Gas, 0°C, 1 atm	1.0046
Water, H_2O	Gas, 110°C, 1 atm	1.0126
	Liquid, 20°C	80.4
Benzene, C_6H_6	Liquid, 20°C	2.28
Methanol, CH_3OH	Liquid, 20°C	33.6
Ammonia, NH_3	Liquid, −34°C	22.6
Mineral oil	Liquid, 20°C	2.24
Sodium chloride, NaCl	Solid, 20°C	6.12
Sulfur, S	Solid, 20°C	4.0
Silicon, Si	Solid, 20°C	11.7
Polyethylene	Solid, 20°C	2.25–2.3
Porcelain	Solid, 20°C	6.0–8.0
Paraffin wax	Solid, 20°C	2.1–2.5
Pyrex glass 7070	Solid, 20°C	4.00

ionic conductivity of the liquid is not the reason for the gigantic dielectric constant of water. You can discern this extraordinary property of water in the dielectric constant of the vapor if you remember that it is really the *difference* between ϵ and 1 that reveals the electrical influence of the material. Compare the values of ϵ given in the table for water vapor and for air.

Once the dielectric constant of a particular material has been determined, perhaps by measuring the capacitance of one capacitor filled with it, we are able to predict the behavior, not merely of two-plate capacitors, but of *any* electrostatic system made up of conductors and pieces of that dielectric of any shape. That is, we can predict all electric fields which will exist in the vacuum outside the dielectrics for given charges or potentials on the conductors in the system.

The theory which enables us to do this was fully worked out by the physicists of the nineteenth century. Lacking a complete picture of the atomic structure of matter, they were more or less obliged to adopt a macroscopic description. From that point of view, the interior of a dielectric is a featureless expanse of perfectly smooth "mathematical jelly" whose single electrical property distinguishing it from a vacuum is a dielectric constant different from unity.

If we develop only a macroscopic description of matter in an electric field, we shall find it hard to answer some rather obvious-sounding questions—or rather, hard to ask these questions in such a way that they can be meaningfully answered. For instance, what is the strength of the electric field *inside* the plastic slab of Fig. 10.1*b* when there are certain charges on the plates? Electric field strength is defined by the force on a test charge. How can we put a test charge inside a perfectly dense solid, without disturbing anything, and measure the force on it? What would that force mean, if we did measure it? You might think of boring a hole and putting the test charge in the hole with some room to move around, so that you can measure the force on it as on a free particle. But then you will be measuring, not the electric field in the dielectric, but the electric field in a cavity in the dielectric, which is quite a different thing.

Fortunately another line of attack is available to us, one that leads up from the microscopic or *atomic* level. We know that matter is made of atoms and molecules; these in turn are composed of elementary charged particles. We know something about the size and structure of these atoms, and we know something about their arrangement in crystals and fluids and gases. Instead of describing our dielectric slab as a volume of structureless but nonvacuous jelly, we shall describe it as a collection of molecules inhabiting a vacuum. If we can find out what the electric charges in *one* molecule do when that molecule is all by itself in an electric field, we should be able to understand the behavior of two such molecules a certain distance apart in a vacuum. It will only be necessary to include the influence, on each molecule, of any electric field arising from the other. This is a vacuum problem. Now all we have to do is extend this to a population of say 10^{20} molecules occupying a cubic centimeter or so of vacuum, and we have our real dielectric. We hope to do this without generating 10^{20} separate problems.

This program if carried through will reward us in two ways. We shall be able at last to say something meaningful about the electric and magnetic fields inside matter, answering questions such as the one raised above. What is more valuable, we shall understand how the macroscopic electric and magnetic phenomena in matter arise from, and therefore reveal, the nature of the underlying atomic structure. We are going to study electric and magnetic effects separately. We begin with dielectrics. Since our first goal is to describe the electric field produced by an atom or molecule, it will help to make some general observations about the electrostatic field external to any small system of charges.

FIGURE 10.3
Calculation of the potential, at a point A, of a molecular charge distribution.

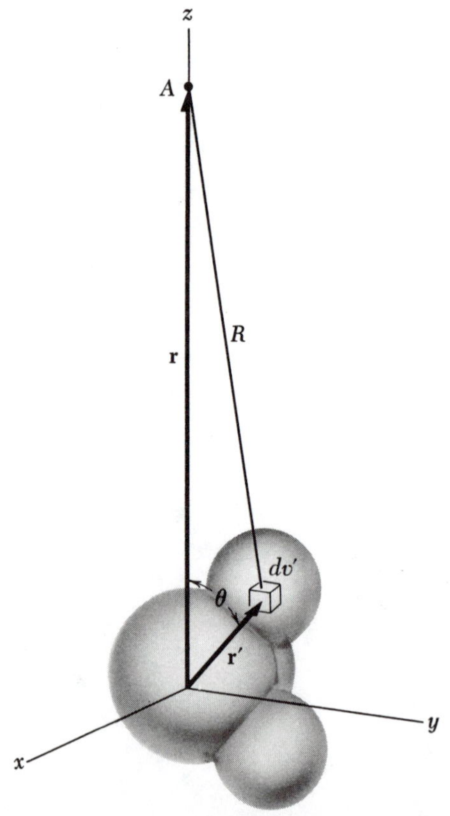

THE MOMENTS OF A CHARGE DISTRIBUTION

10.2 An atom or molecule consists of some electric charges occupying a small volume, perhaps a few cubic angstroms (10^{-24} cm^3) of space. We are interested in the electric field outside that volume, which arises from this rather complicated charge distribution. We shall be particularly concerned with the field far away from the source, by which we mean far away compared with the size of the source itself. What features of the charge structure mainly determine the field at remote points? To answer this, let's look at some arbitrary distribution of charges and see how we might go about computing the field at a point outside it. Figure 10.3 shows a charge distribution of some sort located in the neighborhood of the origin of coordinates. It might be a molecule consisting of several positive nuclei and quite a large number of electrons. In any case we shall suppose it is described by a given charge density function $\rho(x, y, z)$. ρ is negative where the electrons are and positive where the nuclei are. To find the electric field at distant points we can begin by computing the potential of the charge distribution. To illustrate, let's take some point A out on the z axis. (Since we are not assuming any special symmetry in the charge distribution, there is nothing special about the z axis.) Let r be the distance of A from the origin. The electric potential at A, denoted by φ_A, is obtained as usual by adding the contributions from all elements of the charge distribution:

$$\varphi_A = \int \frac{\rho(x', y', z')\, dv'}{R} \tag{3}$$

In the integrand dv' is an element of volume within the charge distribution, $\rho(x', y', z')$ is the charge density there, and R in the denominator is the distance from A to this particular charge element. The integration is carried out in the coordinates x', y', z', of course, and is extended over all the region containing charge. We can express R in terms of r and the distance r' from the origin to the charge element. Using the law of cosines with θ the angle between r' and the axis on which A lies:

$$R = (r^2 + r'^2 - 2rr' \cos\theta)^{1/2} \tag{4}$$

With this substitution for R the integral becomes

$$\varphi_A = \int \rho \, dv'(r^2 + r'^2 - 2rr' \cos\theta)^{-1/2} \tag{4a}$$

Now we want to take advantage of the fact that, for a distant point like A, r' is much smaller than r for all parts of the charge distribution. This suggests that we should expand the square root in Eq. 4 in powers of r'/r. Writing

$$(r^2 + r'^2 - 2rr' \cos\theta)^{-1/2} = \frac{1}{r}\left[1 + \left(\frac{r'^2}{r^2} - \frac{2r'}{r}\cos\theta\right)\right]^{-1/2} \tag{5}$$

and using the expansion $(1 + \delta)^{-1/2} = 1 - \frac{1}{2}\delta + \frac{3}{8}\delta^2 \ldots$, we get, after collecting together terms of the same power in r'/r:

$$(r^2 + r'^2 - 2rr' \cos\theta)^{-1/2} =$$

$$\frac{1}{r}\left[1 + \frac{r'}{r}\cos\theta + \left(\frac{r'}{r}\right)^2 \frac{(3\cos^2\theta - 1)}{2} + \left(\begin{matrix}\text{terms of} \\ \text{higher power}\end{matrix}\right)\right] \tag{6}$$

Now r is a constant in the integration, so we can take it outside and write the prescription for the potential at A as follows:

$$\varphi_A = \frac{1}{r}\underbrace{\int \rho \, dv'}_{K_0} + \frac{1}{r^2}\underbrace{\int r' \cos\theta \, \rho \, dv'}_{K_1}$$

$$+ \frac{1}{r^3}\underbrace{\int r'^2 \frac{(3\cos^2\theta - 1)}{2} \rho \, dv'}_{K_2} + \cdots \tag{7}$$

Each of the integrals above, K_0, K_1, K_2, and so on, has a value that depends only on the structure of the charge distribution, not on the distance to point A. Hence the potential for all points along the z axis can be written as a power series in $1/r$ with constant coefficients:

$$\varphi_A = \frac{K_0}{r} + \frac{K_1}{r^2} + \frac{K_2}{r^3} + \cdots \tag{8}$$

To finish the problem we would have to get the potential φ at all other points, in order to calculate the electric field as $-\text{grad}\,\varphi$. We

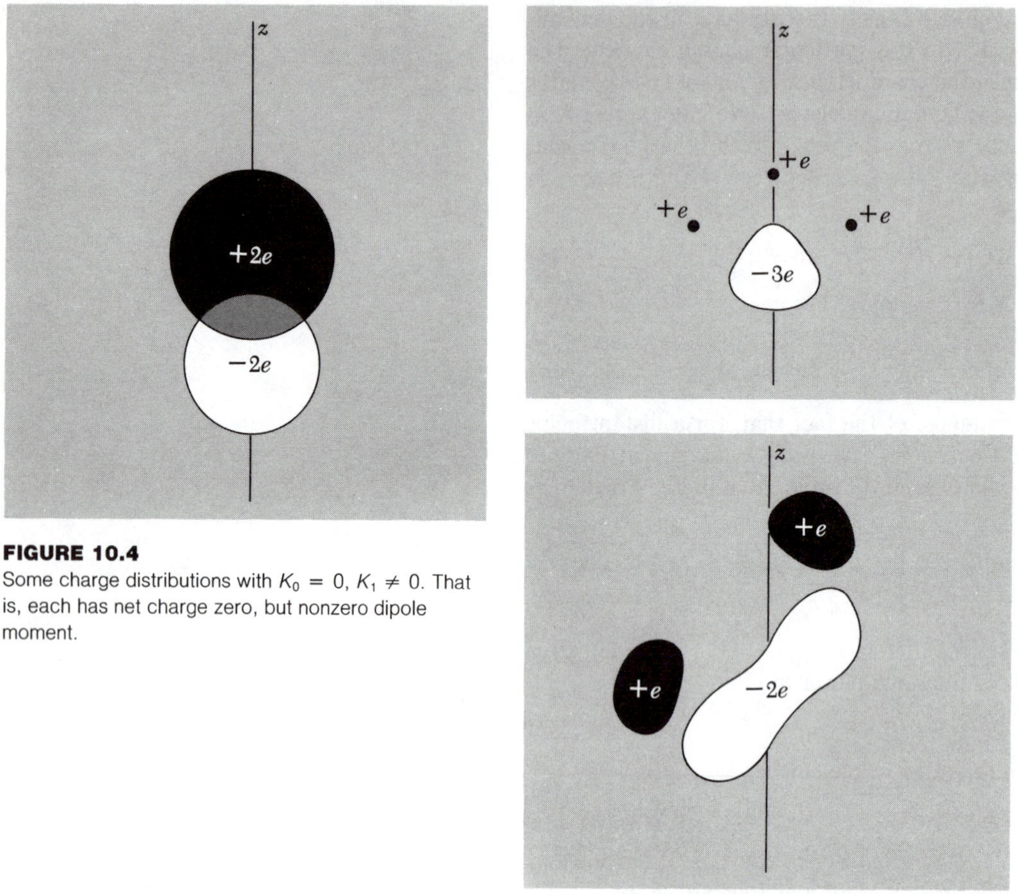

FIGURE 10.4
Some charge distributions with $K_0 = 0$, $K_1 \neq 0$. That is, each has net charge zero, but nonzero dipole moment.

have gone far enough, though, to bring out the essential point: The behavior of the potential at large distances from the source will be dominated by the first term in this series whose coefficient is not zero.

Let us look at these coefficients more closely. The coefficient K_0 is $\int \rho\, dv'$, which is nothing but the total charge in the distribution. If we have equal amounts of positive and negative charge, as in a neutral molecule, K_0 will be zero. For a singly ionized molecule K_0 will have the value e. If K_0 is not zero, then no matter how large K_1, K_2, etc., may be, if we go out to a sufficiently large distance, the term K_0/r will win out. Beyond that, the potential will approach that of a point charge at the origin and so will the field. This is hardly surprising.

Suppose we have a neutral molecule, so that K_0 is equal to zero. Our interest now shifts to the second term, with coefficient $K_1 = \int r' \cos\theta\, \rho\, dv'$. Since $r' \cos\theta$ is simply z', this term measures the relative displacement, in the direction toward A, of the positive and negative charge. It has a nonzero value for the distributions sketched in Fig. 10.4, where the densities of positive and of negative charge have been indicated separately. In fact, all the distributions shown there have approximately the same value of K_1. Furthermore—and this is a crucial point—if any charge distribution is neutral the value of K_1 is independent of the position chosen as origin. That is, if we replace z' by $z' + z_0'$, in effect shifting the origin, the value of the integral is not changed: $\int (z' + z_0')\rho\, dv' = \int z'\rho\, dv' + z_0' \int \rho\, dv'$ and the latter integral is always zero for a neutral distribution.

Evidently if $K_0 = 0$ and $K_1 \neq 0$, the potential along the z axis will vary asymptotically (that is, with ever-closer approximation as we go out to larger distances) as $1/r^2$. We expect the electric field strength, then, to behave asymptotically like $1/r^3$, in contrast to the $1/r^2$ dependence of the field from a point charge. Of course we have discussed only the potential on the z axis. We will return to the question of the exact form of the field after getting a general view of the situation.

If K_0 and K_1 are both zero, and K_2 is not, the potential will behave like $1/r^3$ at large distances, and the field strength will fall off with the inverse fourth power of the distance. Figure 10.5 shows a charge distribution for which K_0 and K_1 are both zero (and would be zero no matter what direction we had chosen for the z axis) while K_2 is not zero.

The quantities $K_0, K_1, K_2, \ldots$ are related to what are called the *moments* of the charge distribution. Using this language, we call K_0, which is simply the net charge, the *monopole moment,* or *monopole strength.* K_1 is one component of the *dipole moment* of the distribution. The dipole moment has the dimensions *charge* $\times$ *displacement;* it is a *vector,* and our K_1 is its z component. The third constant K_2 is related to the *quadrupole moment* of the distribution, the next to the *octupole moment,* and so on. The quadrupole moment is not a vector, but a tensor. The charge distribution shown in Fig. 10.5 has a nonzero quadrupole moment.

The advantage to us of describing a charge distribution by this hierarchy of moments is that it singles out just those features of the charge distribution which determine the field at a great distance. If we were concerned only with the field in the immediate neighborhood of the distribution, it would be a fruitless exercise. For our main task, understanding what goes on in a dielectric, it turns out that *only* the monopole strength (the net charge) and the dipole strength of the molecular building blocks matter. We can ignore all other moments. And if the building blocks are neutral, we have only their dipole moments to consider.

FIGURE 10.5
For this distribution of charge, $K_0 = K_1 = 0$, but $K_2 \neq 0$. It is a distribution with nonzero quadrupole moment.

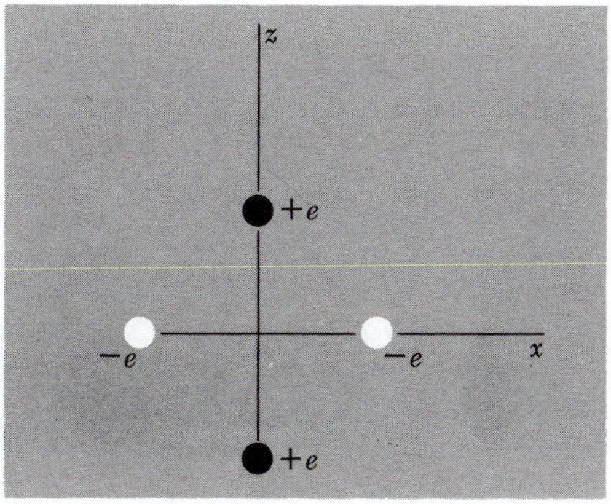

THE POTENTIAL AND FIELD OF A DIPOLE

10.3 The dipole contribution to the potential at the point A, distance r from the origin, was given by $(1/r^2)\int r' \cos\theta\, \rho\, dv'$. We can write $r' \cos\theta$, which is just the projection of $\mathbf{r}'$ on the direction toward A, as $\hat{\mathbf{r}} \cdot \mathbf{r}'$. Thus we can write the potential without reference to any arbitrary axis as

$$\varphi_A = \frac{1}{r^2}\int \hat{\mathbf{r}} \cdot \mathbf{r}'\rho\, dv' = \frac{\hat{\mathbf{r}}}{r^2} \cdot \int \mathbf{r}'\rho\, dv' \tag{9}$$

which will serve to give the potential at any point. The integral on the right in Eq. 9 is the *dipole moment* of the charge distribution. It is a vector, obviously, with the dimensions *charge* × *distance*. We shall denote the dipole moment vector by **p**:

$$\mathbf{p} = \int \mathbf{r}'\rho\, dv' \tag{10}$$

Using the dipole moment **p**, we can rewrite Eq. 9 as

$$\varphi(\mathbf{r}) = \frac{\hat{\mathbf{r}} \cdot \mathbf{p}}{r^2} \tag{11}$$

The electric field is the negative gradient of this potential. To see what the dipole field is like, locate a dipole **p** at the origin, pointing in the z direction (Fig. 10.6). With this arrangement,

$$\varphi = \frac{p \cos\theta}{r^2} \tag{12}$$

The potential and the field are, of course, symmetrical around the z axis. Let's work in the xz plane, where $\cos\theta = z/(x^2 + z^2)^{1/2}$. In that plane, then

$$\varphi = \frac{pz}{(x^2 + z^2)^{3/2}} \tag{13}$$

The components of the electric field are readily derived:

$$\begin{aligned} E_x &= -\frac{\partial\varphi}{\partial x} = \frac{3pxz}{(x^2 + z^2)^{5/2}} = \frac{3p \sin\theta \cos\theta}{r^3} \\ E_z &= -\frac{\partial\varphi}{\partial z} = p\left[\frac{3z^2}{(x^2 + z^2)^{5/2}} - \frac{1}{(x^2 + z^2)^{3/2}}\right] \\ &= \frac{p(3\cos^2\theta - 1)}{r^3} \end{aligned} \tag{14}$$

The dipole field can be described more simply in the polar coordinates r and θ. Let E_r be the component of **E** in the direction of $\hat{\mathbf{r}}$,

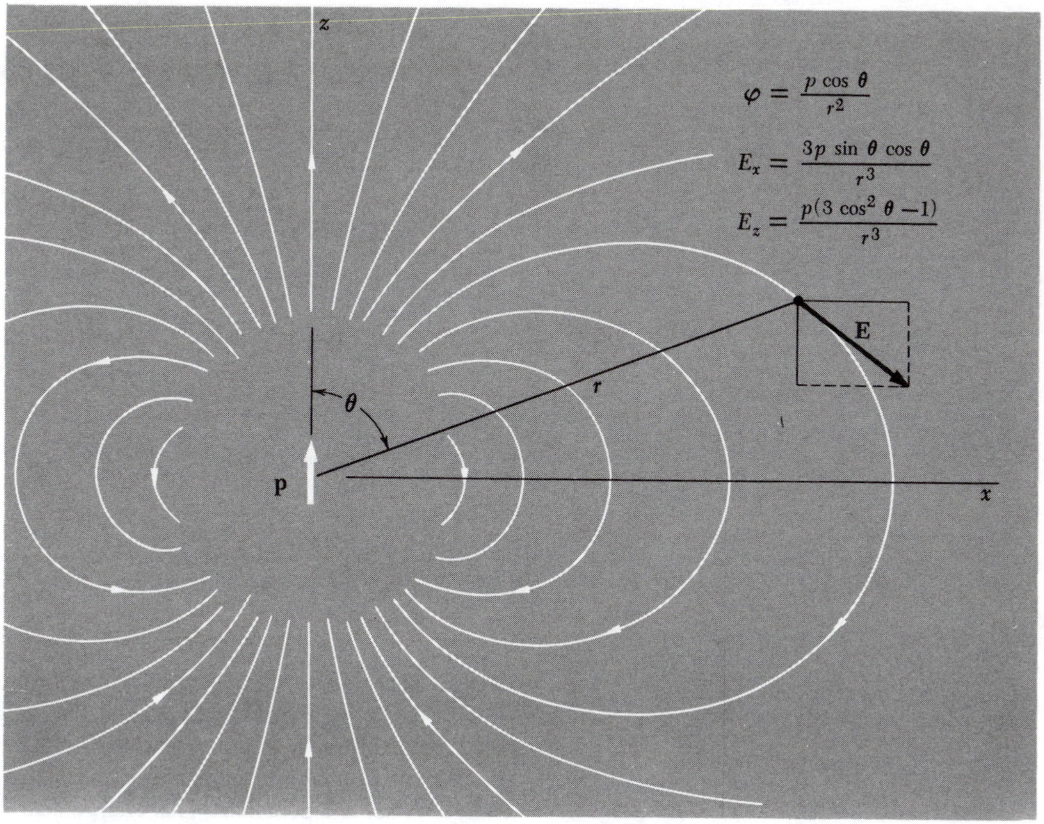
z
$\varphi = \frac{p \cos\theta}{r^2}$
$E_x = \frac{3p \sin\theta \cos\theta}{r^3}$
$E_z = \frac{p(3\cos^2\theta - 1)}{r^3}$
E
r
θ
p
x

FIGURE 10.6
The electric field of a dipole, indicated by some field lines.

and E_θ the component perpendicular to $\hat{\mathbf{r}}$ in the direction of increasing θ. Then

$$E_r = \frac{2p}{r^3}\cos\theta \qquad E_\theta = \frac{p}{r^3}\sin\theta \tag{15}$$

You can check this against Eq. 14, or if you know how to find a gradient in polar coordinates, you can get it directly as the negative gradient of the potential φ given by Eq. 12.

Proceeding out in any direction from the dipole, we find the electric field strength falling off as $1/r^3$, as we had anticipated. Along the z axis the field is parallel to the dipole moment $\mathbf{p}$, with magnitude $2p/r^3$. In the equatorial plane the field points antiparallel to $\mathbf{p}$ and has the value $-\mathbf{p}/r^3$.

This field may remind you of one we have met before. Remember the point charge over the conducting plane, with its image charge. Perhaps the simplest charge distribution with a dipole moment is two point charges, $+q$ and $-q$, separated by a distance s. For a system

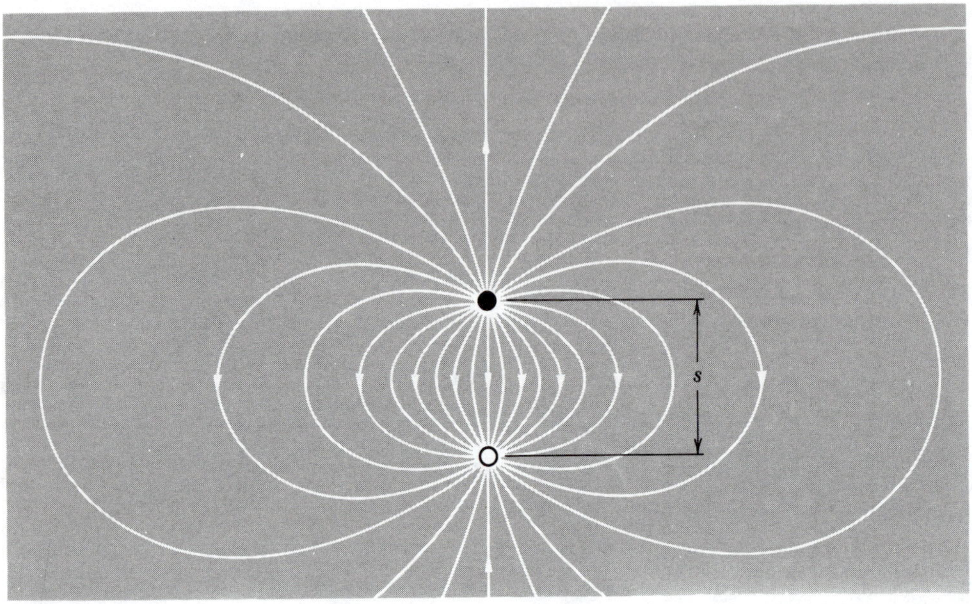

FIGURE 10.7
The electric field of a pair of equal and opposite point charges approximates the field of a dipole for distances large compared to the separation *s*.

of point charges Eq. 10 takes the form of a sum. The dipole moment of our point-charge pair is just qs, and the vector points in the direction from negative charge to positive. In Fig. 10.7 we show the field of this pair of charges, mainly to emphasize that the field near the charges is *not* a dipole field. This charge distribution has many multipole moments, indeed infinitely many, so it is only the far field at distances $r \gg s$ that can be represented as a dipole field.

To generate a complete dipole field right into the origin we would have to let s shrink to zero while increasing q without limit so as to keep $p = qs$ finite. This highly singular abstraction is not very interesting. We know that our molecular charge distribution will have complicated near fields, so we could not easily represent the near region in any case. Fortunately we shall not need to.

THE TORQUE AND THE FORCE ON A DIPOLE IN AN EXTERNAL FIELD

10.4 Suppose two charges, q and $-q$, are mechanically connected so that s, the distance between them, is fixed. You may think of the charges as stuck on the end of a short nonconducting rod of length s. We shall call this object a dipole. Its dipole moment p is simply qs. Let us put the dipole in an external electric field, that is, the field from some other source. The field of the dipole itself does not concern us now. Consider first a uniform electric field, as in Fig. 10.8*a*. The positive end of the dipole is pulled toward the right, the negative end toward the left, by a force of strength Eq. The net force on the object is zero and so is the torque, in this position. A dipole which makes some angle θ with the field direction as in Fig. 10.8*b* obviously experiences a torque. In general, torque $\mathbf{N}$ around an axis through some chosen origin is $\mathbf{r} \times \mathbf{F}$, where $\mathbf{F}$ is the force applied at a distance $\mathbf{r}$ from the origin. Taking the origin in the center of the dipole, so that $r = s/2$, we have

$$\mathbf{N} = \mathbf{r} \times \mathbf{F}_+ + (-\mathbf{r}) \times \mathbf{F}_- \tag{16}$$

$\mathbf{N}$ is a vector perpendicular to the figure, and its magnitude is

$$N = \frac{s}{2} Eq \sin\theta + \frac{s}{2} Eq \sin\theta = sqE \sin\theta = pE \sin\theta \tag{17}$$

This can be written simply

$$\mathbf{N} = \mathbf{p} \times \mathbf{E} \tag{18}$$

When the total force on the dipole is zero, as it is in this case, the torque is independent of the choice of origin, which therefore need not be specified.

The orientation of the dipole in Fig. 10.8*a* has the lowest energy. Work has to be done to rotate it into any other position. Let us calculate the work required to rotate the dipole from a position parallel to the field, through some angle θ_0, as shown in Fig. 10.8*c*. Rotation through an infinitesimal angle $d\theta$ requires an amount of work $N\,d\theta$. Thus the total work done is

$$\int_0^{\theta_0} N\,d\theta = \int_0^{\theta_0} pE \sin\theta\,d\theta = pE(1 - \cos\theta_0) \tag{19}$$

To reverse the dipole, turning it end for end, corresponds to $\theta_0 = \pi$ and requires an amount of work equal to $2pE$.

The net force on the dipole in any *uniform* field is zero, obviously, regardless of its orientation. In a nonuniform field the forces on the two ends of the dipole will generally not be exactly equal and opposite, and there will be a net force on the object. A simple example is a dipole in the field of a point charge Q. If the dipole is oriented radially as in Fig. 10.9*a*, with the positive end nearer the positive charge Q, the net force will be outward, and its magnitude will be

$$F = (q)\frac{Q}{r^2} + (-q)\frac{Q}{(r+s)^2} \tag{20}$$

For $s \ll r$, we need only evaluate this to first order in s/r, which we do as follows:

$$F = \frac{qQ}{r^2}\left[1 - \frac{1}{\left(1 + \dfrac{s}{r}\right)^2}\right] \approx \frac{qQ}{r^2}\left[1 - \frac{1}{1 + \dfrac{2s}{r}}\right] \approx \frac{2sqQ}{r^3} \tag{21}$$

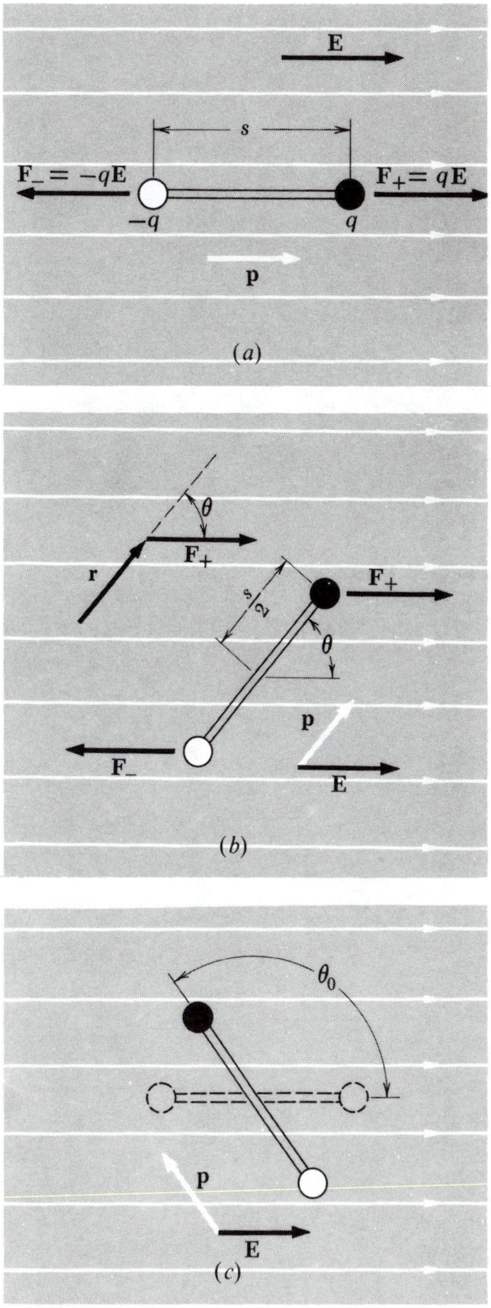

FIGURE 10.8

(*a*) A dipole in a uniform field. (*b*) The torque on the dipole is $\mathbf{N} = \mathbf{p} \times \mathbf{E}$; $\mathbf{N}$ is a vector pointing down into the page. (*c*) The work done in turning the dipole from an orientation parallel to the field to the orientation shown is $pE(1 - \cos\theta_0)$

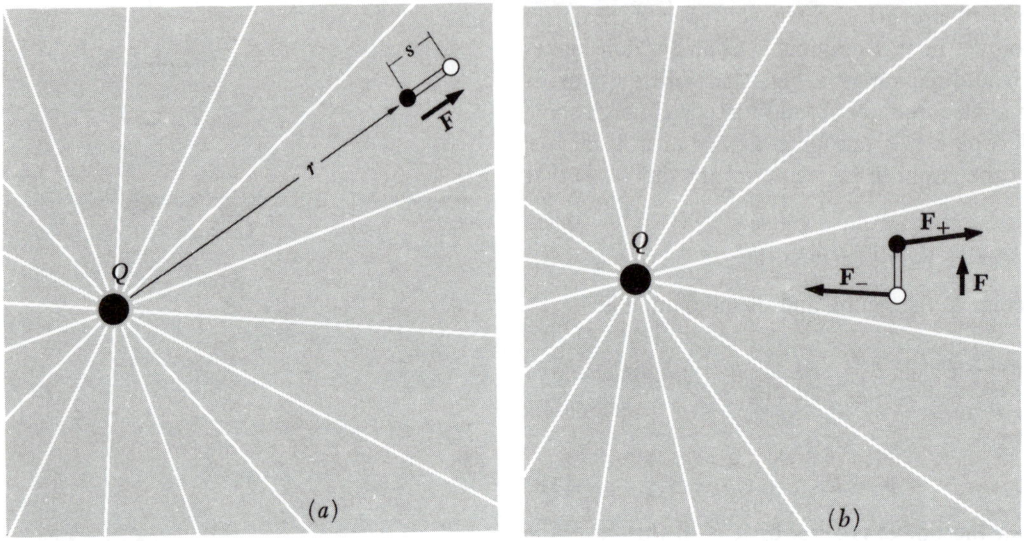

FIGURE 10.9
The force on a dipole in a nonuniform field. (*a*) The net force on the dipole in this position is radially outward. (*b*) The net force on the dipole in this position is upward.

In terms of the dipole moment *p*, this is simply

$$F = \frac{2pQ}{r^3} \tag{22}$$

With the dipole at right angles to the field, as in Fig. 10.9*b*, there is also a force. Now the forces on the two ends, though equal, are not exactly opposite in direction.

It is not hard to work out a general formula for the force on a dipole in a nonuniform electric field. The force depends essentially on the *gradients* of the various components of the field. In general, the x component of the force on a dipole of moment $\mathbf{p}$ is

$$F_x = \mathbf{p} \cdot \text{grad } E_x \tag{23}$$

with corresponding formulas for F_y and F_z.

ATOMIC AND MOLECULAR DIPOLES; INDUCED DIPOLE MOMENTS

10.5 Consider the simplest atom, the hydrogen atom, which consists of a nucleus and one electron. If you imagine the negatively charged electron revolving around the positive nucleus like a planet around the sun—as in the original atomic model of Niels Bohr—you will conclude that the atom has, at any one instant of time, an electric dipole moment. The dipole moment vector $\mathbf{p}$ points parallel to the electron-proton radius vector, and its magnitude is e times the electron-proton distance. The direction of this vector will be continually changing as the electron, in this picture of the atom, circles around its orbit.

To be sure, the *time average* of **p** will be zero for a circular orbit, but we should expect the periodically changing dipole moment components to generate rapidly oscillating electric fields and electromagnetic radiation. The absence of such radiation in the normal hydrogen atom was one of the baffling paradoxes of early quantum physics. Modern quantum mechanics tells us that it is better to think of the hydrogen atom in its lowest energy state (the usual condition of most of the hydrogen atoms in the universe) as a spherically symmetrical structure with the electronic charge distributed, in the time average, over a cloud surrounding the nucleus. Nothing is revolving in a circle or oscillating. If we could take a snapshot with an exposure time shorter than 10^{-16} sec, we might discern an electron localized some distance away from the nucleus. But for processes involving times much longer than that we have, in effect, a smooth distribution of negative charge surrounding the nucleus and extending out in all directions with steadily decreasing density. The total charge in this distribution is just $-e$, the charge of one electron. Roughly half of it lies within a sphere of radius 0.5 angstrom (0.5×10^{-8} cm). The density decreases exponentially outward; a sphere only 2.2 angstroms in radius contains 99 percent of the charge. The electric field in the atom is just what a stationary charge distribution of this form, together with the positive nucleus, would produce.

A similar picture is the best one to adopt for other atoms and molecules. We can treat the nuclei in molecules as point charges; for our present purposes their size is too small to matter. The entire electronic structure of the molecule is to be pictured as a single cloud of negative charge of smoothly varying density. The shape of this cloud and the variation of charge density within it will of course be different for different molecules. But at the fringes of the cloud the density will always fall off exponentially, so that it makes some sense to talk of the size and shape of the molecular charge distribution.

Quantum mechanics makes a crucial distinction between stationary states and time-dependent states of an atom. The state of lowest energy is a time-independent structure, a stationary state. It has to be. It is that state of the atom or molecule that concerns us here. Of course, atoms *can* radiate electromagnetic energy. That happens with the atom in a nonstationary state in which there is an oscillating electric charge.

Figure 10.10 represents the charge distribution in the normal hydrogen atom. It is a cross section through the spherically symmetrical cloud, with the density suggested by shading. Obviously the dipole moment of such a distribution is zero. The same is true of any atom in its state of lowest energy, no matter how many electrons it contains, for in all such states the electron distribution has spherical symmetry. It is also true of any ionized atom, though an ion of course has a monopole moment, that is, a net charge.

FIGURE 10.10
The time-average distribution in the normal hydrogen atom. Shading represents density of electronic (negative) charge.

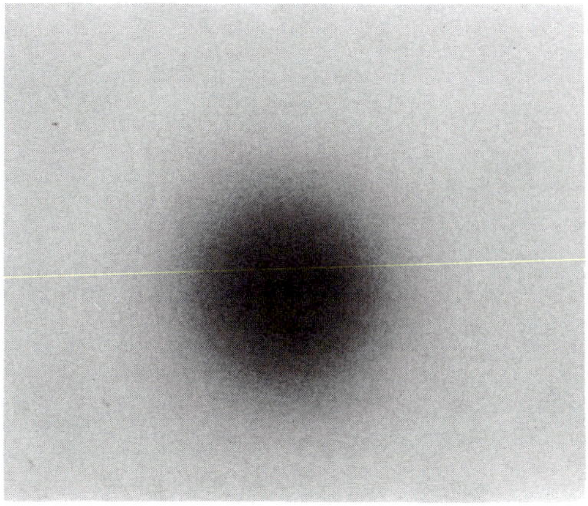

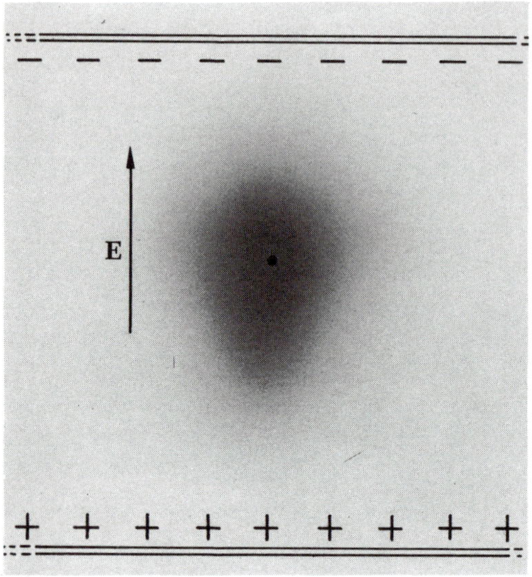

FIGURE 10.11
In an electric field the negative charge is pulled one way, the positive nucleus is pulled the other way. The distortion is grossly exaggerated in this picture. To distort the atom that much would require a field of 10^8 volts/cm.

So far we have nothing very interesting. But now let us put the hydrogen atom in an electric field supplied by some external source, as in Fig. 10.11. The electric field distorts the atom, pulling the negative charge down and pushing the positive nucleus up. The distorted atom will have an electric dipole moment because the "center of gravity" of the negative charge will no longer coincide with the positive nucleus, but will be displaced from the nucleus by some small distance Δz. The electric dipole moment of the atom is now $e\,\Delta z$.

How much distortion will be caused by a field of given strength E? Remember that electric fields already exist in the unperturbed atom, of strength e/a^2 in order of magnitude, where a is a typical atomic dimension. We should expect the relative distortion of the atom's structure, measured by the ratio $\Delta z/a$, to have the same order of magnitude as the ratio of the perturbing field E to the internal fields that hold the atom together. We predict, in other words, that

$$\frac{\Delta z}{a} \approx \frac{E}{e/a^2} \tag{24}$$

Now a is a length of order 10^{-8} cm, and e/a^2 is approximately 10^7 statvolts/cm, a field thousands of times more intense than any large-scale steady field we could make in the laboratory. Evidently the distortion of the atom is going to be very slight indeed, in any practical case. If Eq. 24 is correct, it follows that the dipole moment p of the distorted atom, which is just $e\,\Delta z$, will be

$$p = e\,\Delta z \approx a^3 E \tag{25}$$

Since the atom was spherically symmetrical before the field $\mathbf{E}$ was applied, the dipole moment vector $\mathbf{p}$ will be in the direction of $\mathbf{E}$. The factor that relates $\mathbf{p}$ to $\mathbf{E}$ is called the *atomic polarizability,* and is usually denoted by α.

$$\mathbf{p} = \alpha \mathbf{E} \tag{26}$$

Notice that α has the dimensions of a volume. According to our estimate it is in order of magnitude an atomic volume, something like 10^{-24} cm^3. Its value for a particular atom will depend on the details of the atom's electronic structure. An exact quantum-mechanical calculation of the polarizability of the hydrogen atom predicts $\alpha = \frac{9}{2}a_0^3$, where a_0 is the *Bohr radius,* 0.52×10^{-8} cm, the characteristic distance in the H-atom structure in its normal state. The electric polarizabilities of several species of atoms, experimentally determined, are given in Table 10.2. The examples given are arranged in order of increasing number of electrons. Notice the wide variations in α. If you are acquainted with the periodic table of the elements, you may discern something systematic here. Hydrogen and the alkali metals, lithium, sodium, and potassium, which occupy the first column of the periodic table, have large values of α, and these increase steadily with

TABLE 10.2

Atomic Polarizabilities, in Units of 10^{-24} cm^3

Element	H	He	Li	Be	C	Ne	Na	A	K
α	0.66	0.21	12	9.3	1.5	0.4	27	1.6	34

increasing atomic number, from hydrogen to potassium. The noble gases have much smaller atomic polarizabilities, but these also increase as we proceed, within the family, from helium to neon to krypton. Apparently the alkali atoms, as a class, are easily deformed by an electric field, whereas the electronic structure of a noble gas atom is much stiffer. It is the loosely bound outer, or "valence," electron in the alkali atom structure that is responsible for the easy polarizability.

A molecule, too, develops an induced dipole moment when an electric field is applied to it. The methane molecule depicted in Fig. 10.12 is made from four hydrogen atoms arranged at the corners of a tetrahedron around the central carbon atom. This object has an electrical polarizability, determined experimentally, of

$$2.6 \times 10^{-24} \text{ cm}^3$$

It is interesting to compare this with the sum of the polarizabilities of a carbon atom and four isolated hydrogen atoms. Taking the data from Table 10.2, we find $\alpha_C + 4\alpha_H = 4.1 \times 10^{-24}$ cm^3. Evidently the binding of the atoms into a molecule has somewhat altered the electronic structure. Measurements of atomic and molecular polarizabilities have long been used by chemists as clues to molecular structure.

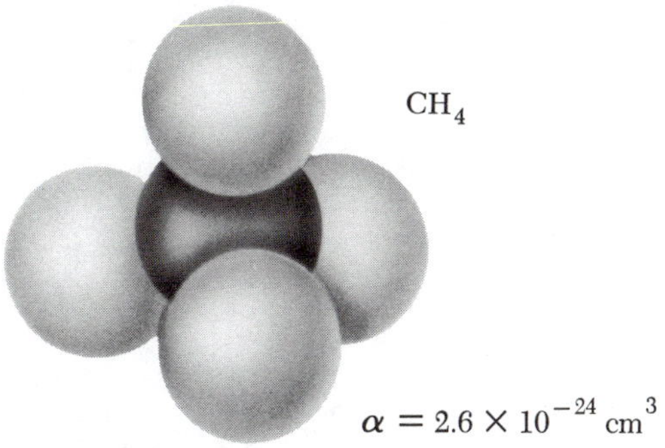

FIGURE 10.12
The methane molecule, made of four hydrogen atoms and a carbon atom.

PERMANENT DIPOLE MOMENTS

10.6 Some molecules are so constructed that they have electric dipole moments even in the absence of an electric field. They are unsymmetrical in their normal state. The molecule shown in Fig. 10.13 is an example. A simpler example is provided by any diatomic molecule made out of dissimilar atoms, such as hydrogen chloride, HCl. There is no point on the axis of this molecule about which the molecule is symmetrical fore and aft; the two ends of the molecule are physically different. It would be a pure accident if the center of gravity of the positive charge and that of the negative charge happened to fall at the same point along the axis. When the HCl molecule is formed from the originally spherical H and Cl atoms, the electron of the H atom shifts partially over to the Cl structure, leaving the hydrogen nucleus partially denuded. So there is some excess of positive charge at the hydrogen end of the molecule and a corresponding excess of

FIGURE 10.13
A molecule with no symmetry whatever, bromochlorofluoromethane. This is methane with three different halogens substituted for three of the hydrogens. The bond lengths and the tetrahedron edges are all a bit different.

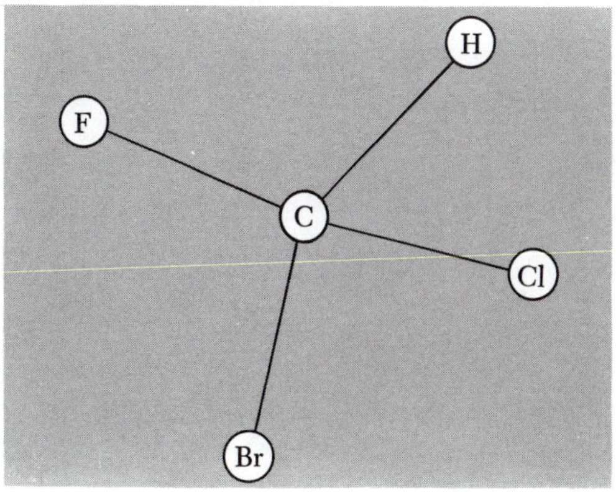

negative charge at the chlorine end. The magnitude of the resulting electric dipole moment, 1.03×10^{-18} esu-cm, is equivalent to shifting one electron about 0.2 angstrom. By contrast the hydrogen atom in a field of 30 kilovolts/cm, with the polarizability listed in Table 10.2, acquires an induced moment less than 10^{-22} esu-cm. Permanent dipole moments, when they exist, are as a rule enormously larger than any moment that can be induced by ordinary laboratory electric fields.† Because of this, the distinction between *polar* molecules, as molecules with "built-in" dipole moments are called, and *nonpolar* molecules is very sharp.

We said at the beginning of Section 10.5 that the hydrogen atom had, at any instant of time, a dipole moment. But then we dismissed it as being zero in the time average, on account of the rapid motion of the electron. Now we seem to be talking about molecular dipole moments as if a molecule were an ordinary stationary object like a baseball bat whose ends could be examined at leisure to see which was larger! Molecules move more slowly than electrons, but their motion is rapid by ordinary standards. Why can we credit them with "permanent" electric dipole moments? If this inconsistency was bothering you, you are to be commended. The full answer can't be given without some quantum mechanics, but the difference essentially involves the time scale of the motion. The time it takes a molecule to interact with its surroundings is generally *shorter* than the time it takes the intrinsic motion of the molecule to average out the dipole moment smoothly. Hence the molecule *really acts* as if it had the moment we have been talking about. A very short time qualifies as permanent in the world of one molecule and its neighbors.

Some common polar molecules are shown in Fig. 10.14, with the direction and magnitude of the permanent dipole moment indicated for each. The water molecule has an electric dipole moment because it is bent in the middle, the O—H axes making an angle of about 105° with one another. This is a structural oddity with the most far-reaching consequences. The dipole moment of the molecule is largely responsible for the properties of water as a solvent, and it plays a decisive role in chemistry that goes on in an aqueous environment. It is hard to imagine what the world would be like if the H_2O molecule, like the CO_2 molecule, had its parts arranged in a straight line; probably we wouldn't be here to observe it. We hasten to add that the shape of the H_2O molecule is not a capricious whim of Nature. Quantum mechanics has revealed clearly why a molecule made of an eight-electron atom joined to two one-electron atoms must prefer to be bent.

†There is a good reason for this. The internal electric fields in atoms and molecules, as we remarked in the last section, are naturally on the order of $e/(10^{-8}\text{ cm})^2$ which is roughly 10^9 volts/cm! We cannot apply such a field to matter in the laboratory for the closely related reason that it would tear the matter to bits.

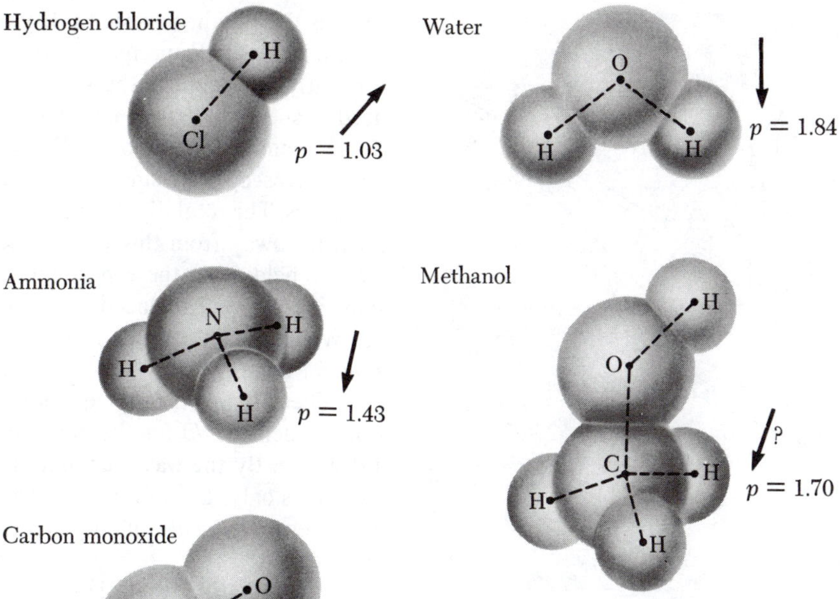

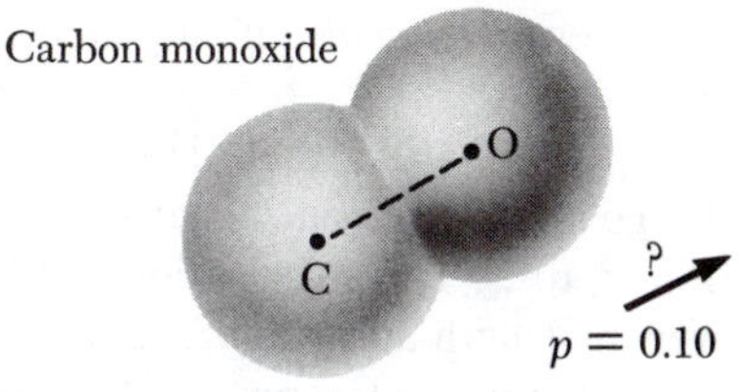

FIGURE 10.14
Some well-known polar molecules. The observed magnitude of the permanent dipole moment p is given in units of 10^{-18} esu-cm.

The behavior of a polar substance as a dielectric is strikingly different from that of material composed of nonpolar molecules. The dielectric constant of water is about 80, that of methyl alcohol 33, while a typical nonpolar liquid might have a dielectric constant around 2. In a nonpolar substance the application of an electric field induces a slight dipole moment in each molecule. In the polar substance dipoles are already present in great strength but, in the absence of a field, are pointing in random directions so that they have no large-scale effect. An applied electric field merely *aligns* them to a certain degree. In either process, however, the macroscopic effects will be determined by the net amount of polarization per unit volume.

THE ELECTRIC FIELD CAUSED BY POLARIZED MATTER

10.7 Suppose we build up a block of matter by assembling a very large number of molecules in a previously empty region of space. Suppose too that each of these molecules is polarized in the same direction. For the present we need not concern ourselves with the nature of the molecules or with the means by which their polarization is maintained. We are interested only in the electric field *they* produce when they are in this condition; later we can introduce any fields from other

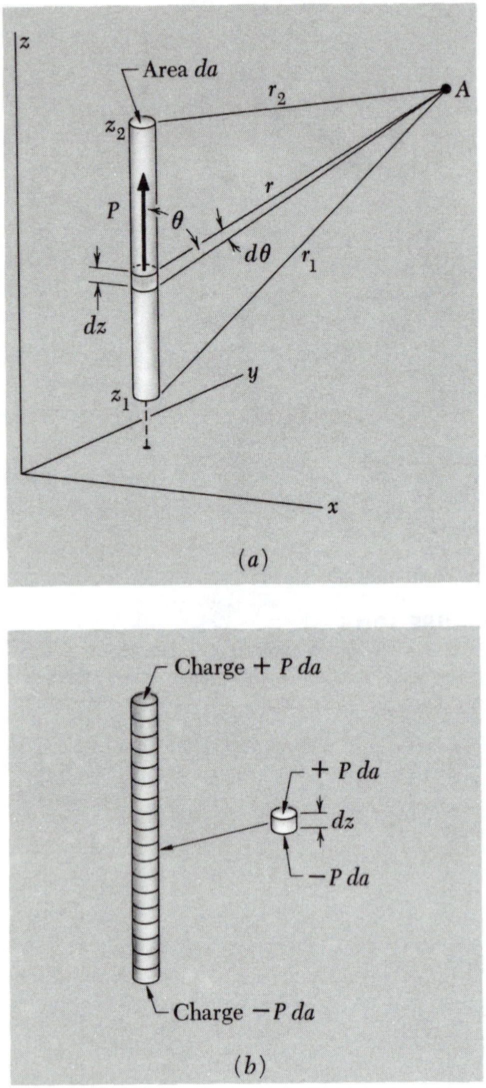

FIGURE 10.15
A column of polarized material (*a*) produces the same field, at any external point *A*, as two charges, one at each end of the column (*b*).

sources that might be around. If you like, you can imagine that these are molecules with permanent dipole moments that have been lined up neatly, all pointing the same way, and frozen in position. All we need to specify is N, the number of dipoles per cubic centimeter, and the moment of each dipole $\mathbf{p}$. We shall assume that N is so large that any macroscopically small volume dv contains quite a large number of dipoles. The total dipole strength in such a volume is $\mathbf{p}N\,dv$. At any point far away from this volume element compared with its size, the electric field from these particular dipoles would be practically the same if they were replaced by a single dipole moment of strength $\mathbf{p}N\,dv$. We shall call $\mathbf{p}N$ the density of polarization, and denote it by $\mathbf{P}$, a vector quantity with the dimensions charge-cm/cm^3, or charge per cm^2. Then $\mathbf{P}\,dv$ is the dipole moment to be associated with any small-volume element dv for the purpose of computing the electric field at a distance. By the way, our matter has been assembled from neutral molecules only; there is no net charge in the system or on any molecule, so we have *only* the dipole moments to consider as sources of a distant field.

In Fig. 10.15 there is shown a slender column, or cylinder, of this polarized material. Its cross section is da, and it extends vertically from z_1 to z_2. The polarization density $\mathbf{P}$ within the column is uniform over the length and points in the positive z direction. We are about to calculate the electric potential, at some external point, of this column of polarization. An element of the cylinder, of height dz, has a dipole moment $\mathbf{P}\,dv = \mathbf{P}\,da\,dz$. Its contribution to the potential at the point A can be written down by referring back to our formula Eq. 12 for the potential of a dipole.

$$d\varphi_A = \frac{P\,da\,dz\cos\theta}{r^2} \tag{27}$$

The potential due to the entire column is

$$\varphi_A = P\,da\int_{z_1}^{z_2}\frac{dz\cos\theta}{r^2} \tag{28}$$

This is simpler than it looks: $dz\cos\theta$ is just $-dr$, so that the integrand is a perfect differential, $d(1/r)$. The result of the integration is then

$$\varphi_A = P\,da\left(\frac{1}{r_2} - \frac{1}{r_1}\right) \tag{29}$$

Equation 29 is precisely the same as the expression for the potential at A that would be produced by two point charges, a positive charge of magnitude $P\,da$ sitting on top of the column at a distance r_2 from A, and a negative charge of the same magnitude at the bottom of the column. The source consisting of a column of uniformly polarized matter is equivalent, at least so far as its field at all *external* points is concerned, to two concentrated charges.

We can prove this in another way without any mathematics. Consider a small section of the column of height dz, containing dipole moment in amount $P\,da\,dz$. Let us make an imitation or substitute for this by taking an unpolarized insulator of the same size and shape and sticking a charge $P\,da$ on top of it, and a charge $-P\,da$ on the bottom. This little block now has the same dipole moment as that bit of our original column, and therefore it will make an identical contribution to the field at any remote point A. (The field inside our substitute, or very close to it, may be different from the field of the original—we don't care about that.) Now make a whole set of such blocks and stack them up to imitate the polarized column. They must give the same field at A as the whole column does, for each block gave the same contribution as its counterpart in the original (Fig. 10.15*b*). Now see what we have! At every joint the positive charge on the top of one block coincides with the negative charge on the bottom of the block above it, making charge zero. The only charges left uncompensated are the negative charge $-P\,da$ on the bottom of the bottom block and the positive charge $P\,da$ on the top of the top block. Seen from a distant point such as A, these look like point charges. We conclude, as before, that two such charges produce at A exactly the same field as does our whole column of polarized material.

With no further calculation we can extend this to a slab, or right cylinder, of any proportions uniformly polarized in a direction perpendicular to its parallel faces (Fig. 10.16*a*). The slab can simply be subdivided into a bundle of columns, and the potential outside will be the sum of the contributions of the columns, each of which can be replaced by a charge at either end. The charges on the top, $P\,da$ on each column end of area da, make up a uniform sheet of surface charge of density $\sigma = P$ esu per unit area. We conclude that the potential everywhere *outside* a uniformly polarized slab or cylinder is precisely what would result from two sheets of surface charge located where the top and bottom surfaces of the slab were located, carrying the constant surface charge density $\sigma = P$ and $\sigma = -P$, respectively (Fig. 10.16*b*).

FIGURE 10.16
A block of polarized material (*a*) is equivalent to two sheets of charge (*b*), as far as the field outside is concerned.

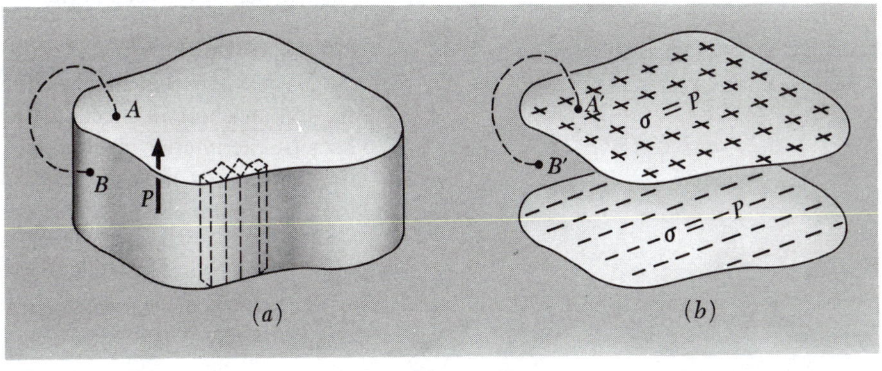

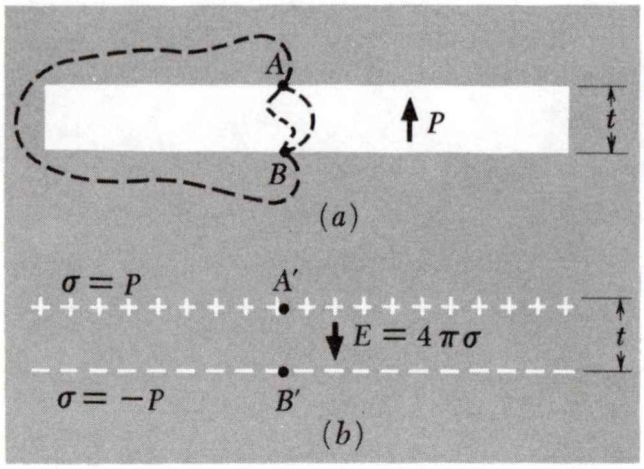

FIGURE 10.17
(*a*) The line integral of **E** from *A* to *B* must be the same over all paths, internal or external, for the internal microscopic or atomic electric fields also are conservative (curl **E** = 0). The equivalent charge sheets (*b*) have the same external field.

We are not quite ready to say anything about the field *inside* the slab. However, we do know the potential at all points on the surface of the slab, top, bottom, or sides. Any two such points, *A* and *B*, can be connected by a path running entirely through the external field, so that the line integral $\int \mathbf{E} \cdot d\mathbf{s}$ is entirely determined by the external field. It must be the same as the integral along the path *A′B′* in Fig. 10.16*b*. A point literally on the surface of the dielectric might be within range of the intense molecular fields, the near field of the molecule that we have left out of account. Let's agree to define the boundary of the dielectric as a surface far enough out from the outermost atomic nucleus—10 or 20 angstroms would be margin enough—so that at any point outside this boundary, the near fields of the individual atoms make a negligible contribution to the whole line integral from *A* to *B*.

With this in mind, let's look at a rather thin, wide plate of polarized material, of thickness *t* shown in cross section in Fig. 10.17*a*. Figure 10.17*b* shows, likewise in cross section, the equivalent sheets of charge. For the system of two charge sheets we know the field, of course, in the space both outside and between the sheets. The field strength inside, well away from the edges, must be just $4\pi\sigma$, pointing down, and the potential difference between points *A′* and *B′* is therefore $4\pi\sigma t$ statvolts. The *same potential difference* must exist between corresponding points *A* and *B* on our polarized slab, because the entire *external* field is the same in the two systems.

Is the field identical inside, too? Certainly *not,* because the slab is full of positive nuclei and electrons, with fields of millions of volts per centimeter pointing in one direction here, another direction there. But one thing *is* the same: The line integral of the field, reckoned over *any* internal path from *A* to *B*, must be just $\varphi_B - \varphi_A$, which as we have seen is the same as $\varphi_{B'} - \varphi_{A'}$, which is equal to $4\pi\sigma t$, or $4\pi Pt$. This must be so because the introduction of atomic charges, no matter what their distribution, cannot destroy the conservative property of the electric field, expressed in the statement that $\int \mathbf{E} \cdot d\mathbf{s}$ is independent of path, or curl $\mathbf{E} = 0$.

We know that in Fig. 10.17*b* the potential difference between the top and bottom sheets is nearly constant, except near the edges, because the interior electric field is practically uniform. Therefore in the central area of our polarized plate the potential difference between top and bottom must likewise be constant. In this region the line integral $\int_A^B \mathbf{E} \cdot d\mathbf{s}$ taken from *any* point *A* on top of the slab to *any* point *B* on the bottom, by *any* path, must always yield the same value

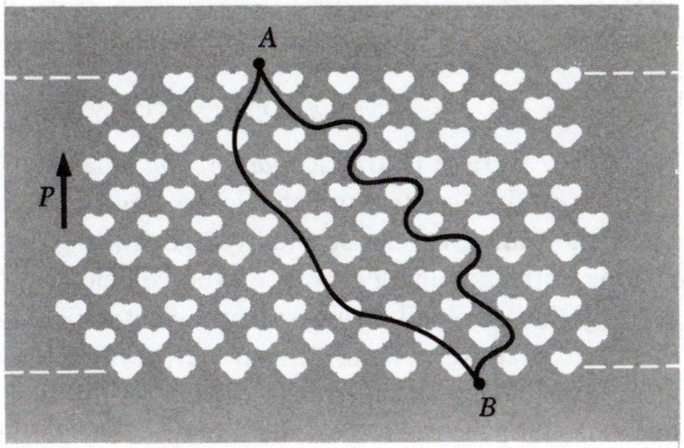

FIGURE 10.18
Over any path from *A* to *B*, the line integral of the actual microscopic field is the same.

$4\pi Pt$. Figure 10.18 is a "magnified view" of the central region of the slab, in which the polarized molecules have been made to look something like H_2O molecules all pointing the same way. We have not attempted to depict the very intense fields that exist between the molecules, and inside them. (Ten angstroms distant from a water molecule its field amounts to several hundred kilovolts per centimeter, as you can discover from Fig. 10.14 and Eq. 15.) You must imagine some rather complicated field configurations in the neighborhood of each molecule. Now the **E** in $\int \mathbf{E} \cdot d\mathbf{s}$ represents the *total electric field* at a given point in space, inside or outside a molecule; it includes these complicated and intense fields just mentioned. We have reached the remarkable conclusion that *any* path through this welter of charges and fields, whether it dodges molecules or penetrates them, must yield the same value for the path integral, namely, the value we find in the system of Fig. 10.17*b* where the field is quite uniform and has the strength $4\pi P$.

This tells us that the *spatial average* of the electric field within our polarized slab must be $-4\pi\mathbf{P}$. By the spatial average of a field **E** over some volume V, which we might denote by $\langle \mathbf{E} \rangle_V$, we mean precisely this:

$$\langle \mathbf{E} \rangle_V = \frac{1}{V} \int_V \mathbf{E}\, dv \tag{30}$$

One way to sample impartially the field in many equal small dv's into which V might be divided would be to measure the field along each line in a bundle of closely spaced parallel lines. We have just seen that the line integral of **E** along any or all such paths is the same as if we were in a constant electric field of strength $-4\pi\mathbf{P}$. That is the justification for the conclusion that within the polarized dielectric slab of Figs. 10.17 and 10.18 the spatial average of the field due to all the charges that belong to the dielectric is

$$\langle \mathbf{E} \rangle = -4\pi \mathbf{P} \tag{31}$$

This average field is a *macroscopic* quantity. The volume over which we take the average should be large enough to include very many molecules, otherwise the average will fluctuate from one such volume to the adjoining one. The average field $\langle \mathbf{E} \rangle$ defined by Eq. 30 is really the only kind of *macroscopic* electric field in the interior of a dielectric that we can talk about. It provides the only satisfactory answer, in the context of a macroscopic description of matter, to the question, What is the electric field inside a dielectric material?

The $\mathbf{E}$ in the integrand on the right, in Eq. 30, we may call the *microscopic* field. If we send someone out to measure the field values we need for the path integral, he will be measuring electric fields in vacuum, in the presence, of course, of electric charge. He will need very tiny instruments, for he may be called on to measure the field at a particular point just inside one end of a certain molecule. Have we any right to talk in this way about taking the line integral of $\mathbf{E}$ along some path that skirts the southwest corner of a particular molecule and then tunnels through its neighbor? Yes. The justification is the massive evidence that the laws of electromagnetism work down to a scale of distances much smaller than atomic size. We can even describe an experiment which would serve to measure the average of the microscopic electric field along a path defined well within the limits of atomic dimensions. All we have to do is shoot an energetic charged particle, an alpha particle for example, through the material. From the net change in its momentum the average electric field that acted on it, over its whole path, could be inferred.

Let us review the properties of the average, or macroscopic field $\langle \mathbf{E} \rangle$, defined by Eq. 30. Its line integral $\int_A^B \langle \mathbf{E} \rangle \cdot d\mathbf{s}$ between any two points A and B which are reasonably far apart is independent of the path. It follows that curl $\langle \mathbf{E} \rangle = 0$ and that $\langle \mathbf{E} \rangle$ is the negative gradient of a potential $\langle \varphi \rangle$. This potential function $\langle \varphi \rangle$ is itself a smoothed-out average, in the sense of Eq. 30, of the microscopic potential φ. (The latter rises to several million volts in the interior of every atomic nucleus!) The surface integral of $\langle \mathbf{E} \rangle$, $\int \langle \mathbf{E} \rangle \cdot d\mathbf{a}$, over any surface that encloses a reasonably large volume, is equal to 4π times the charge within that volume.† That is to say, $\langle \mathbf{E} \rangle$ obeys

†We state this without proof, postponing consideration of the relation of the surface integral of an average field to the average of surface integrals of the microscopic field to the next chapter, where the question arises in Section 11.8 in connection with the magnetic field inside matter. (See Fig. 11.18.)

Gauss's law, a statement we can also make in differential form: $\text{div}\langle \mathbf{E} \rangle = 4\pi\langle \rho \rangle$, with the understanding that $\langle \rho \rangle$ too is a local average over a suitably macroscopic volume. In short, the spatial average quantities $\langle \mathbf{E} \rangle$, $\langle \varphi \rangle$, and $\langle \rho \rangle$ are related to one another in the same way as are electric field, potential, and charge density in vacuum.

From now on, when we speak of the electric field **E** inside any piece of matter much larger than a molecule we will mean an average, or macroscopic, field as defined by Eq. 30, even when the brackets $\langle \quad \rangle$ are omitted.

ANOTHER LOOK AT THE CAPACITOR

10.8 At the beginning of this chapter we explained in a qualitative way how the presence of a dielectric between the plates of a capacitor increases its capacitance. Now we are ready to analyze quantitatively the dielectric-filled capacitor. What we have just learned about the electric field inside matter is the key to the problem. We identified as the macroscopic field **E,** the spatial average of the microscopic field. The line integral of that macroscopic **E** between any two points A and B is path-independent and equal to the potential difference. Looking back at Fig. 10.2a we observe that the field **E** in the empty capacitor must have had the value φ_{12}/s. But the potential difference between the plates, φ_{12}, which was established by the battery, was exactly the same in the dielectric-filled capacitor (Fig. 10.2b). Hence the field **E** in the dielectric, understood now as the macroscopic field, must have had the same value too, for it extends and is uniform over the same distance s. (The layers in the diagram are actually negligible in thickness compared with s.) Then the total charge on and near the top plate must be the same as it was in the empty capacitor, namely, Q_0. To prove that we need only invoke Gauss' law for a suitable imaginary box enclosing the charge layers, as indicated in Fig. 10.19. The charge is made up of two parts, the charge on the plate Q (which will flow off when the capacitor is discharged) and Q', the charge that belongs to the dielectric. Now $Q = \epsilon Q_0$. That was our definition of ϵ. Therefore, if $Q + Q' = Q_0$ as we have just concluded, we must have

$$Q' = Q_0 - Q = Q_0(1 - \epsilon) \qquad (32)$$

We can think of this system as the superposition of a vacuum capacitor and a polarized dielectric slab, Fig. 10.19a and b. In the vacuum capacitor with charge ϵQ_0 the electric field E'' would be ϵ times the field E. In the isolated polarized dielectric slab the field E' is $-4\pi P$, as stated in Eq. 31. The superposition of these two objects creates the actual field E.

$$E = E'' + E' = \epsilon E + 4\pi P \qquad (33)$$

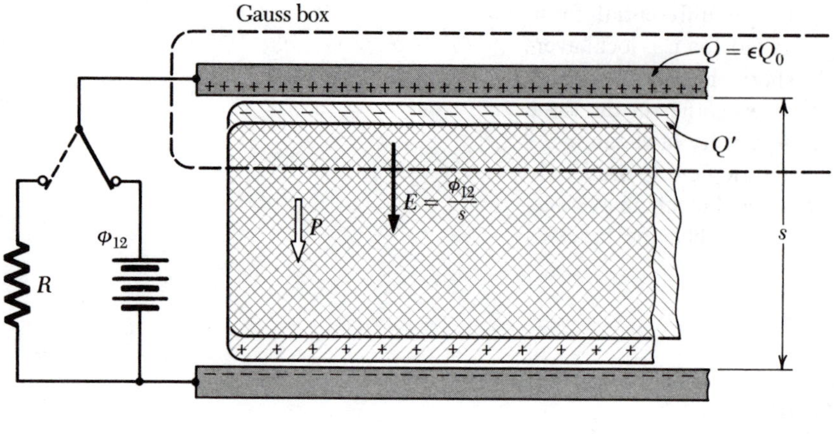

(a)

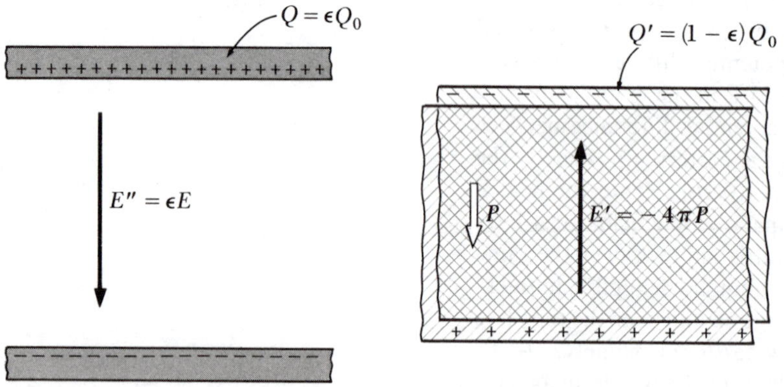

(b) Plates alone. (c) Dielectric alone.

FIGURE 10.19
The dielectric-filled capacitor of Fig. 10.2*b*. The field E which is the average, or macroscopic, field in the dielectric, is ϕ_{12}/s, equal to the field in the empty capacitor of Fig. 10.2*a*. The charge inside the Gauss box must equal Q_0, the charge on the plate of the empty capacitor. The system can be regarded as the superposition of a vacuum capacitor (*b*) and a polarized dielectric (*c*).

Equation 33 can be rearranged like this:

$$\frac{P}{E} = \frac{\epsilon - 1}{4\pi} \tag{34}$$

The ratio P/E is called the electric susceptibility of the dielectric material denoted by χ_e (Greek chi).

In most materials under ordinary circumstances it is the field **E** in the dielectric that *causes* **P**. The relation is quite linear. That is to say, the electric susceptibility χ_e is a constant characteristic of the particular material and not dependent on the strength of the electric field or the size or shape of the electrodes. Cases are known, however, usually involving materials composed of polar molecules, in which polarization can be literally frozen in. A block of ice polarized by an externally applied electric field and then cooled in liquid helium will retain its polarization indefinitely after the external field is removed, thus providing a real example of the hypothetical polarized slab in Fig. 10.18.

Strictly speaking, filling a vacuum capacitor with dielectric material increases its capacitance by the precise factor ϵ characteristic of that material only if we fill all the surrounding space too, or at least all the space where there is any electric field. In the example we discussed it was tacitly assumed that the plates were so large compared with their distance of separation that "edge effects," including the small amount of charge that would be on the outside of the plates near the edge (see Fig. 3.12*b*), could be neglected. A quite general statement can be made about a system of conductors of any shape or arrangement which is entirely immersed in a homogeneous dielectric—for instance, in a large tank of oil. With any charges whatever, Q_1, Q_2, etc., on the various conductors, the macroscopic electric field $\mathbf{E}_{\text{med}}$ at any location in the medium is just $1/\epsilon$ times the field $\mathbf{E}_{\text{vac}}$ that would exist at that location with the same charges on the same conductors in vacuum (Fig. 10.20). This has important consequences in semiconductors. When silicon, for example, is doped with phosphorus to make an *n*-type semiconductor, the high dielectric constant of the silicon crystal (see Table 10.1) greatly reduces the electrical attraction between the outermost electron of the phosphorus atom and the rest of the atom. This makes it easy for the electron to leave the residual P^- ion and join the conduction band, as in Fig. 4.11*a*.

This brings us to a more general problem. What if the space in our system is partly filled with dielectric and partly empty, with electric fields in both parts? We'll begin with a somewhat artificial but instructive example, a polarized solid sphere in otherwise empty space.

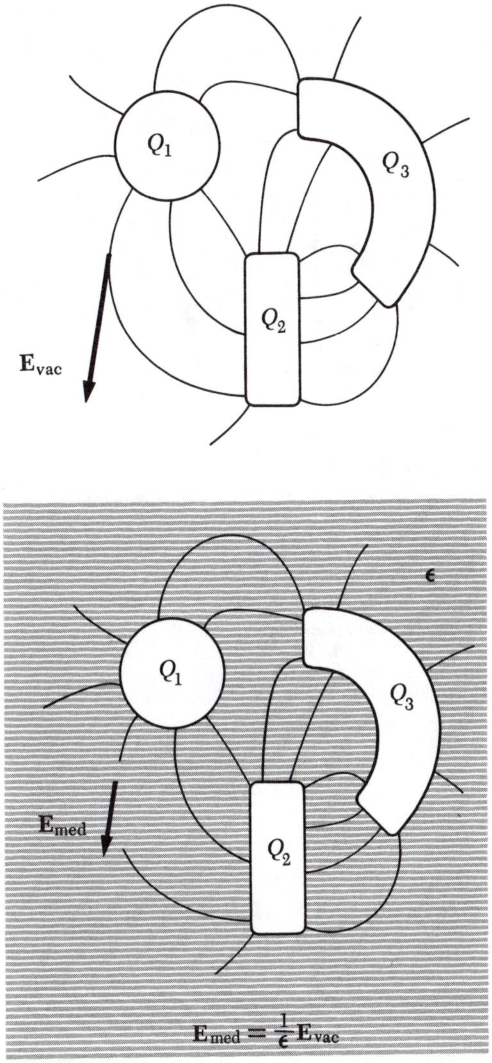

FIGURE 10.20
For the same charges on the conductors, the presence of the dielectric medium reduces all electric field intensities (and hence all potential differences) by the factor $1/\epsilon$. The charges Q_1, Q_2, and Q_3 are the charges that would actually flow off the conductors if we were to discharge the system.

THE FIELD OF A POLARIZED SPHERE

10.9 The solid sphere in Fig. 10.21*a* is supposed to be uniformly polarized, as if it had been carved out of the substance of the slab in Fig. 10.16*a*. What must the electric field be like, both inside and outside the sphere? **P** as usual will denote the density of polarization, constant in magnitude and direction throughout the volume of the sphere. The polarized material could be divided, like the slab in Fig. 10.16*a*, into columns parallel to *P*, and each of these replaced by a charge of magnitude (*P* × *column cross section*) at top and bottom.

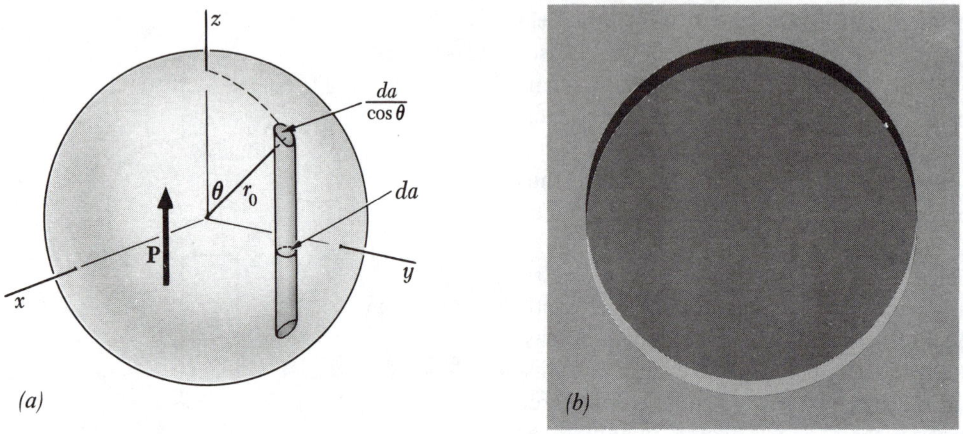

FIGURE 10.21
(*a*) Divide the polarized sphere into polarized rods, and replace each rod by patches of charge in the surface of the sphere. (*b*) A ball of positive volume charge density and a ball of negative volume charge density, slightly displaced, are equivalent to a distribution of charge on the spherical surface.

Thus the field we seek is that of a surface charge distribution spread over a sphere with density $\sigma = P \cos \theta$. The factor $\cos \theta$ enters, as should be obvious from the figure, because a column of cross section da intercepts on the sphere a patch of surface of area $da/\cos \theta$. Figure 10.21*b* is a cross section through this shell of equivalent surface charge in which the density of charge has been indicated by the varying thickness of the black semicircle above (positive charge density) and the light semicircle below (negative charge density).

If it has not already occurred to you, this figure may suggest that we think of the polarization **P** as having arisen from the slight upward displacement of a ball filled uniformly with positive charge of volume density ρ, relative to a ball of negative charge of density $-\rho$. That would leave uncompensated positive charge poking out at the top and negative charge showing at the bottom, varying in amount precisely as $\cos \theta$ over the whole boundary. In the interior, where the positive and negative charge densities still overlap, they would exactly cancel one another. Taking this view, we see a very easy way to calculate the field *outside* the shell of surface charge. Any spherical charge distribution, as we know, has an external field the same as if its entire charge were concentrated at the center. So the superposition of two spheres of total charge Q and $-Q$, respectively, with their centers separated by a small displacement s, will produce an external field the same as that of two point charges Q and $-Q$, s cm apart. That is just a dipole with dipole moment $p_0 = Qs$.

A microscopic description of the polarized substance leads us to the same conclusion. In Fig. 10.22*a* the molecular dipoles actually responsible for the polarization **P** have been crudely represented as consisting individually of a pair of charges q and $-q$, s cm apart, to make a dipole moment $p = qs$. With N of these per cubic centimeter, $P = Np = Nqs$, and the total number of such dipoles in the sphere is $(4\pi/3)r_0^3N$. The positive charges, considered separately (Fig.

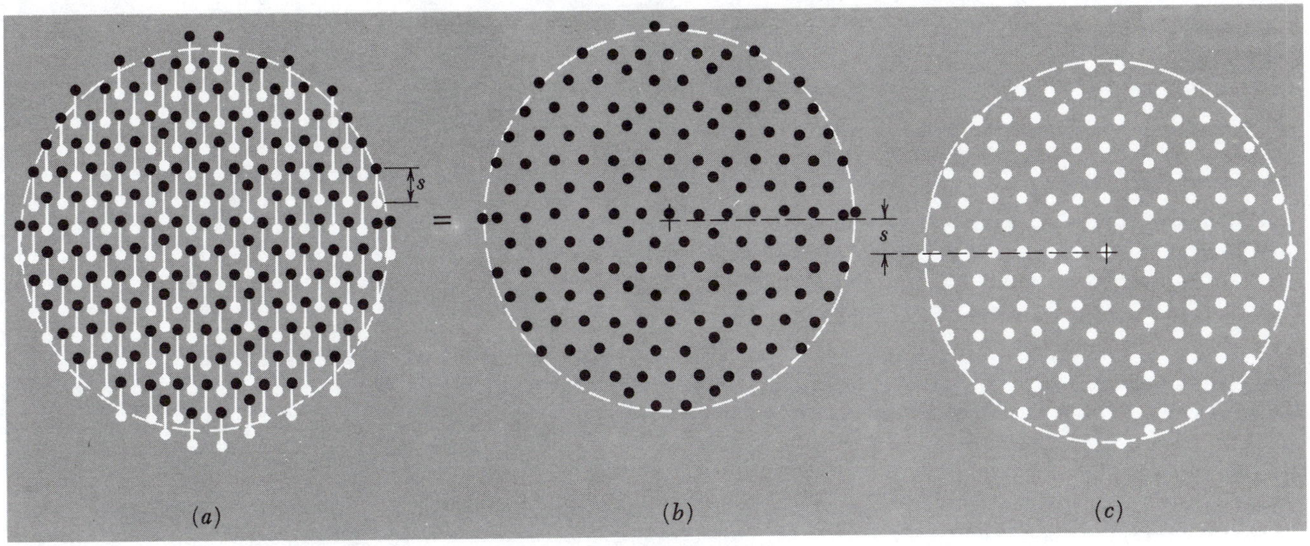

10.22*b*), are distributed throughout a sphere with total charge content $Q = (4\pi/3)r_0^3 Nq$, and the negative charges occupy a similar sphere with its center displaced (Fig. 10.22*c*). Clearly each of these charge distributions can be replaced by a point charge at its center, if we are concerned with the field well outside the distribution. "Well outside" means far enough away from the surface so that the actual graininess of the charge distribution doesn't matter, and of course that is something we always have to ignore when we speak of the macroscopic fields. So for present purposes the picture of overlapping spheres of uniform charge density and the description in terms of actual dipoles in a vacuum are equivalent,† and show that the field outside the distribution is the same as that of a single dipole located at the center. The moment of this dipole p_0 is simply the total polarization in the sphere:

FIGURE 10.22
A sphere of lined-up molecular dipoles (*a*) is equivalent to superposed, slightly displaced, spheres of positive (*b*) and of negative (*c*) charges.

$$p_0 = Qs = \frac{4\pi}{3} r_0^3 Nqs = \frac{4\pi}{3} r_0^3 P \qquad (35)$$

The quantities Q and s have, separately, no significance and may now be dropped from the discussion.

The external field of the polarized sphere is that of a central dipole p_0, not only at a great distance from the sphere; it is the pure dipole field right down to the surface, macroscopically speaking. All we had to do to construct Fig. 10.23, a representation of the external field lines, was to block out a circular area from Fig. 10.6.

†This may have been obvious enough, but we have labored the details in this one case to allay any suspicion that the "smooth-charge-ball" picture, which is so different from what we know the interior of a real substance to be like, might be leading us astray.

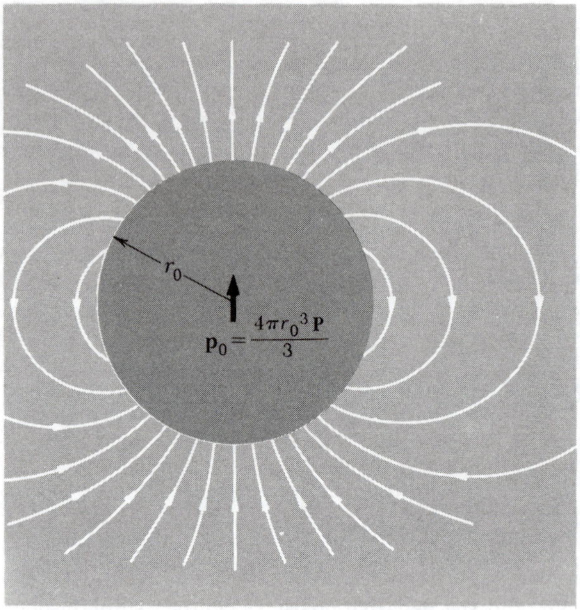

FIGURE 10.23
The field outside a uniformly polarized sphere is exactly the same as that of a dipole located at the center of the sphere.

The internal field is a different matter. Let's look at the electric potential, $\varphi(x, y, z)$. We know the potential at all points on the spherical boundary because we know the external field. It is just the dipole potential, $p_0 \cos\theta / r^2$, which on the spherical boundary of radius r_0 becomes

$$\varphi = p_0 \frac{\cos\theta}{r_0^2} = \frac{4\pi}{3} P r_0 \cos\theta \tag{36}$$

Since $r_0 \cos\theta = z$, we see that the potential of a point on the sphere depends only on its z coordinate:

$$\varphi = \frac{4\pi}{3} Pz \tag{37}$$

The problem of finding the internal field has boiled down to this: Equation 37 gives the potential at every point on the boundary of the region, inside which φ must satisfy Laplace's equation. According to the uniqueness theorem we proved in Chapter 3, that suffices to determine φ throughout the interior. If we can find *a* solution, it must be *the* solution. Now the function Cz, where C is any constant, satisfies Laplace's equation, so Eq. 37 has actually handed us the solution to the potential *in the interior* of the sphere. It is the potential of a uniform electric field in the $-z$ direction:

$$E_z = -\frac{\partial \varphi_{\text{in}}}{\partial z} = -\frac{\partial}{\partial z}\left(\frac{4\pi Pz}{3}\right) = -\frac{4\pi P}{3} \tag{38}$$

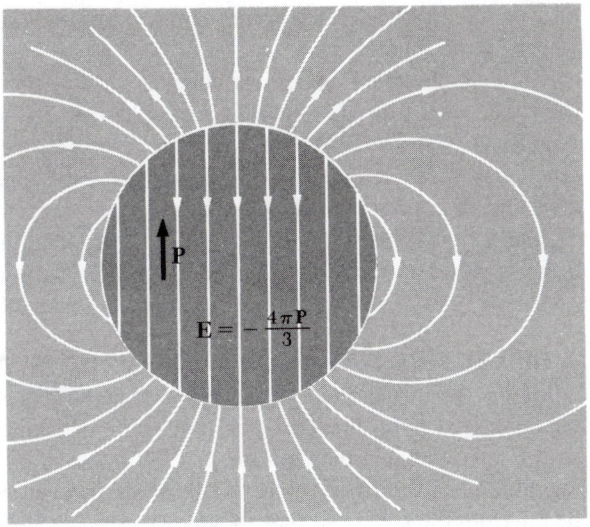

FIGURE 10.24
The field of the uniformly polarized sphere, both inside and outside.

As the direction of **P** was the only thing that distinguished the z axis, we can write our result in more general form:

$$\mathbf{E}_{\text{in}} = -\frac{4\pi\mathbf{P}}{3} \tag{39}$$

This is the macroscopic field **E** in the polarized material.

Figure 10.24 shows both the internal and external field. At the upper pole of the sphere, the strength of the upward-pointing external field is, from Eq. 14 for the field of a dipole,

$$E_z = \frac{2p_0}{r^3} = \frac{2(4\pi r_0^3 P/3)}{r_0^3} = \frac{8\pi P}{3} \qquad \text{(outside)} \tag{40}$$

which is just twice the magnitude of the downward-pointing internal field.

This example illustrates the general rules for the behavior of the field components at the surface of a polarized medium. **E** is discontinuous at the boundary of a polarized medium exactly as it would be at a surface in vacuum which carried a surface charge density $\sigma = P_\perp$. The symbol $P_\perp$ stands for the component of **P** normal to the surface outward. It follows that $E_\perp$, the normal component of **E**, must change abruptly by an amount $4\pi P_\perp$, while $E_\parallel$ the component of **E** parallel to the boundary remains continuous, that is, has the same value on both sides of the boundary (Fig. 10.25). Indeed, at the north pole of our sphere the net change in E_z is $8\pi P/3 - (-4\pi P/3)$ or $4\pi P$. Referring to Eq. 15 for the dipole field, you can check that the component

FIGURE 10.25
The change in E at the boundary of a polarized dielectric. $E_\parallel$ is the same on both sides of the boundary. $E_\perp$ increases by $4\pi P_\perp$ in going from dielectric to vacuum. (E and P are not drawn to the same scale).

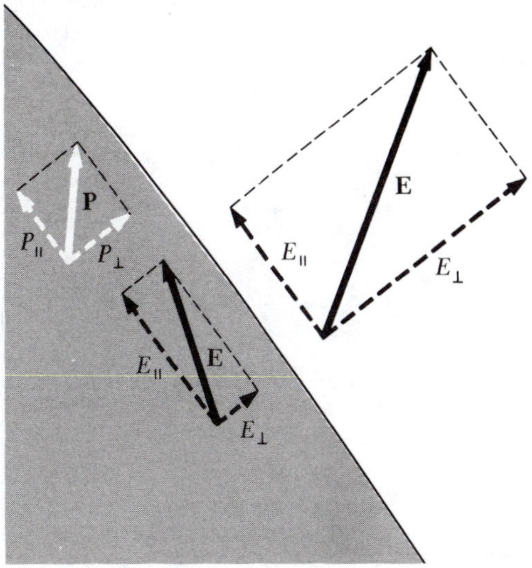

of **E** parallel to the surface is continuous from inside to outside everywhere on the sphere.

None of these conclusions depends on how the polarization of the sphere was caused. Assuming any sphere *is* uniformly polarized, Fig. 10.24 shows *its* field. Onto this can be superposed any field from other sources, thus representing many possible systems. This will not affect the discontinuity in **E** at the boundary of the polarized medium. The rules just stated therefore apply in any system, the discontinuity in **E** being determined solely by the existing polarization.

A DIELECTRIC SPHERE IN A UNIFORM FIELD

10.10 As an example, let us put a sphere of dielectric material characterized by a dielectric constant ϵ into a homogeneous electric field $\mathbf{E}_0$ like the field between the parallel plates of a vacuum capacitor, Fig. 10.26. Let the sources of this field, the charges on the plates, be far from the sphere so that they do not shift as the sphere is introduced. Then whatever the field may be in the vicinity of the sphere, it will remain practically $\mathbf{E}_0$ at a great distance. That is what is meant by putting a sphere into a uniform field. The total field **E** is no longer uniform in the neighborhood of the sphere. It is the *sum* of the uniform field $\mathbf{E}_0$ of the distant sources and a field $\mathbf{E}'$ generated by the polarized matter itself:

$$\mathbf{E} = \mathbf{E}_0 + \mathbf{E}' \tag{41}$$

The field $\mathbf{E}'$ depends on the polarization **P** of the dielectric, which in turn depends on the value of **E** inside the sphere:

$$\mathbf{P} = \chi_e \mathbf{E} = \frac{\epsilon - 1}{4\pi}\mathbf{E} \tag{42}$$

We don't know yet what the total field **E** is; we know only that Eq. 42 has to hold at any point inside the sphere. *If* the sphere becomes uniformly polarized, an assumption that will need to be justified by our results, the relation between the polarization of the sphere and its own field $\mathbf{E}'$, at points inside, was given by Eq. 39. (In Eq. 39 we were using the symbol **E** for this field; in that case it was the only field present.)

$$\mathbf{E}'_{\text{in}} = -\frac{4\pi\mathbf{P}}{3} \tag{43}$$

Now we have enough equations to eliminate **P** and $\mathbf{E}'$, which should give us a relation connecting **E** and $\mathbf{E}_0$. Using Eqs. 41 to 43 we find:

$$\mathbf{E} = \mathbf{E}_0 - \frac{4\pi\mathbf{P}}{3} = \mathbf{E}_0 - \frac{\epsilon - 1}{3}\mathbf{E} \tag{44}$$

FIGURE 10.26
The sources of the field $\mathbf{E}_0$ remain fixed. The dielectric sphere develops some polarization **P.** The total field **E** is the superposition of $\mathbf{E}_0$ and the field of this polarized sphere.

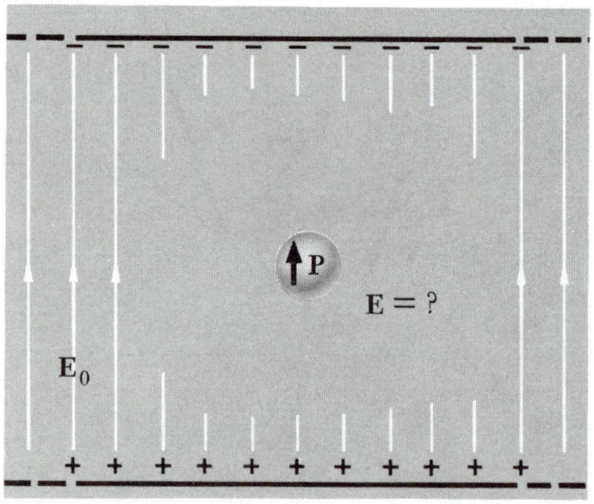

Solving for **E**,

$$\mathbf{E} = \left(\frac{3}{2+\epsilon}\right)\mathbf{E}_0 \tag{45}$$

Because ϵ is greater than one, the factor $3/(2+\epsilon)$ will be less than one; the field inside the dielectric is weaker than $\mathbf{E}_0$. The polarization is

$$\mathbf{P} = \frac{\epsilon - 1}{4\pi}\mathbf{E} = \frac{3}{4\pi}\left(\frac{\epsilon - 1}{\epsilon + 2}\right)\mathbf{E}_0 \tag{46}$$

The assumption of uniform polarization is now seen to be self-consistent.† To compute the total field **E** outside the sphere we must add vectorially to $\mathbf{E}_0$ the field of a central dipole with dipole moment equal to **P** times the volume of the sphere. Some field lines of **E**, both inside and outside the dielectric sphere, are shown in Fig. 10.27.

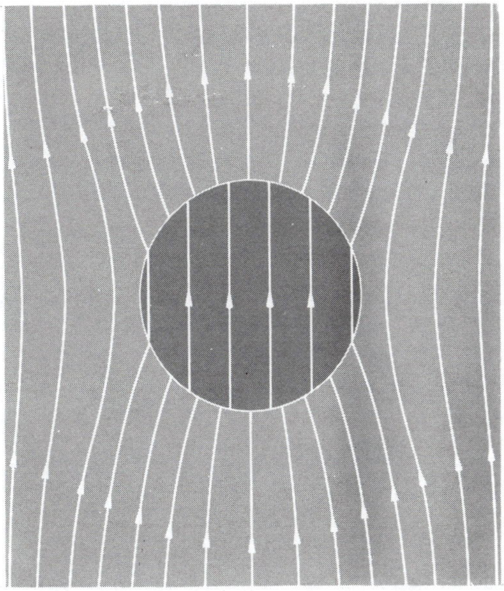

FIGURE 10.27
The total field *E*, both inside and outside the dielectric sphere.

THE FIELD OF A CHARGE IN A DIELECTRIC MEDIUM, AND GAUSS'S LAW

10.11 Suppose that a very large volume of homogeneous dielectric has somewhere within it a concentrated charge Q, not part of the regular molecular structure of the dielectric. Imagine, for instance, that a small metal sphere has been charged and then dropped into a tank of oil. As was stated earlier, the electric field in the oil is simply $1/\epsilon$ times the field that Q would produce in a vacuum.

$$E = \frac{Q}{\epsilon r^2} \tag{47}$$

It is interesting to see how Gauss's law works out. The surface integral of **E** (which is the macroscopic, or space average, field, remember) taken over a sphere surrounding Q, gives $4\pi Q/\epsilon$, if we believe Eq. 47, and *not* $4\pi Q$. Why not? The answer is that Q is not the only charge inside the sphere. There are also all the charges that make up the atoms and molecules of the dielectric. Ordinarily any volume of the oil would be electrically neutral. But now the oil is radially polarized, which means that the charge Q, assuming it is positive, has pulled in toward itself the negative charge in the oil molecules and pushed away

†That is what makes this system easy to deal with. For a dielectric cylinder of finite length in a uniform electric field, the assumption would not work. The field $\mathbf{E}'$ of a uniformly polarized cylinder—for instance one with its length about equal to its diameter—is *not* uniform inside the cylinder. (What must it look like?) Therefore $\mathbf{E} = \mathbf{E}_0 + \mathbf{E}'$ cannot be uniform—but in that case $\mathbf{P} = \chi_e\mathbf{E}$ could not be uniform after all. In fact it is only dielectrics of ellipsoidal shape, of which the sphere is a special case, which acquire uniform polarization in a uniform field.

the positive charges. Although the displacement may be only very slight in each molecule, still on the average any sphere we draw around Q will contain more oil-molecule negative charge than oil-molecule positive charge. Hence the net charge in the sphere, including the "foreign" charge Q at the center, is *less* than Q. In fact it is Q/ϵ.

It is often useful to distinguish between the foreign charge Q and the charges that make up the dielectric itself. Over the former we have some degree of control—charge can be added to or removed from an object, such as the plate of a capacitor. This is often called *free* charge. The other charges, which are integral parts of the atoms or molecules of the dielectric, are usually called *bound* charge. *Structural* charge might be a better name. These charges are not mobile but more or less elastically bound, contributing, by their slight displacement, to the polarization.

One can devise a vector quantity which is related by something like Gauss' law to the free charge only. In the system we have just examined, a point charge Q immersed in a dielectric, the vector $\epsilon\mathbf{E}$ has this property. That is, $\int \epsilon\mathbf{E} \cdot d\mathbf{a}$, taken over some closed surface S, equals $4\pi q$ if S encloses Q, and zero if it does not. By superposition, this must hold for any collection of free charges described by a free-charge density $\rho_{\text{free}}(x, y, z)$ in an infinite homogeneous dielectric medium:

$$\int_S \epsilon\mathbf{E} \cdot d\mathbf{a} = 4\pi \int_V \rho_{\text{free}}\, dv \tag{48}$$

where V is the volume enclosed by the surface S. An integral relation like this implies a "local" relation between the divergence of the vector field $\epsilon\mathbf{E}$ and the free charge density:

$$\text{div}\,(\epsilon\mathbf{E}) = 4\pi\rho_{\text{free}} \tag{49}$$

Since ϵ has been assumed to be constant throughout the medium, Eq. 49 tells us nothing new. However, it can help us to isolate the role of the bound charge. In any system whatever, the fundamental relation between electric field $\mathbf{E}$ and total charge density $\rho_{\text{free}} + \rho_{\text{bound}}$ remains valid:

$$\text{div}\,\mathbf{E} = 4\pi(\rho_{\text{free}} + \rho_{\text{bound}}) \tag{50}$$

From Eqs. 49 and 50 it follows that

$$\text{div}\,(\epsilon - 1)\mathbf{E} = -4\pi\rho_{\text{bound}} \tag{51}$$

According to Eq. 34, $(\epsilon - 1)\mathbf{E} = 4\pi\mathbf{P}$, so Eq. 51 implies that

$$\text{div}\,\mathbf{P} = -\rho_{\text{bound}} \tag{52}$$

Equation 52 states a local relation. It cannot depend on conditions elsewhere in the system, nor on how the particular arrangement of bound charges is maintained. Any arrangement of bound charge which has a certain local excess, per unit volume, of nuclear protons over atomic electrons must represent a polarization with a certain divergence. So Eq. 52 must hold universally, not just in the unbounded dielectric. You can get a feeling for the identity expressed in Eq. 52 by imagining a few polar molecules arranged to give a polarization with a positive divergence (Fig. 10.28). The dipoles point outward, which necessarily leaves a little concentration of negative charge in the middle. Of course, Eq. 52 refers to averages over volume elements so large that **P** and ρ_{bound} can be treated as smoothly varying quantities.

From Eqs. 50 and 52 we get the relation

$$\text{div}\,(\mathbf{E} + 4\pi\mathbf{P}) = 4\pi\rho_{\text{free}} \tag{53}$$

This is quite independent of any relation between **E** and **P**. It is not limited to those materials, which we call dielectrics, in which **P** is proportional to **E**.

It is customary to give the combination $\mathbf{E} + 4\pi\mathbf{P}$ a special name, the *electric displacement* vector, and its own symbol, **D**. That is, we define **D** by

$$\mathbf{D} = \mathbf{E} + 4\pi\mathbf{P} \tag{54}$$

In an isotropic dielectric, **D** is simply $\epsilon\mathbf{E}$, but the relation

$$\text{div}\,\mathbf{D} = 4\pi\rho_{\text{free}} \tag{55}$$

holds in any situation in which the macroscopic quantities **P**, **E**, and ρ can be defined.

The appearance of Eq. 55 may suggest that we should look on **D** as a vector field whose source is the free charge distribution ρ_{free}, in the same sense that the total charge distribution ρ is the source of **E**. That would be wrong. The electrostatic field **E** is uniquely determined—except for the addition of a constant field—by the charge distribution ρ because, supplementing the law $\text{div}\,\mathbf{E} = 4\pi\rho$, there is another universal condition, $\text{curl}\,\mathbf{E} = 0$. It is *not* true, in general, that $\text{curl}\,\mathbf{D} = 0$. Thus the distribution of free charge is not sufficient to determine **D** through Eq. 55. Something else is needed, such as the boundary conditions at various dielectric surfaces. The boundary conditions on **D** are of course merely an alternate way of expressing the boundary conditions involving **E** and **P**, already stated near the end of Section 10.9 and in Fig. 10.25.

In the approach we have taken to electric fields in matter the introduction of **D** is an artifice which is not, on the whole, very helpful. We have mentioned **D** because it is hallowed by tradition, beginning

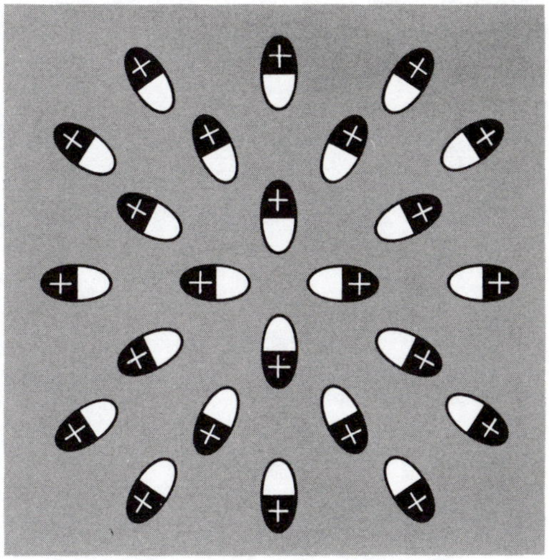

FIGURE 10.28
Molecular dipoles arranged so that div **P** > 0. Note the concentration of negative charge in the middle, consistent with Eq. 52.

with Maxwell,† and the student is sure to encounter it in other books, many of which treat it with more respect than it deserves.

Our essential conclusions about electric fields in matter can be summarized like this:

1. Matter can be polarized, its condition being described completely, so far as the macroscopic field is concerned, by a polarization density **P**, which is the dipole moment per unit volume. The contribution of such matter to the electric field **E** is the same as that of a charge distribution ρ_{bound}, existing in vacuum and having the density $\rho_{\text{bound}} = -\text{div}\,\mathbf{P}$. In particular, at the surface of a polarized substance, where there is a discontinuity in **P**, this reduces to a surface charge of density $\sigma = -P_n$. Add any free charge distribution that may be present, and the electric field is the field that this *total* charge distribution would produce in vacuum. That is the macroscopic field **E** both inside and outside matter, with the understanding that inside matter it is the spatial average of the true microscopic field.

2. If **P** is proportional to **E** in a material, we call the material a dielectric. We define the electric susceptibility χ_e and the dielectric constant ϵ characteristic of that material: $\chi_e = \mathbf{P}/\mathbf{E}$ and $\epsilon = 1 + 4\pi\chi_e$. Free charges immersed in a dielectric give rise to electric fields which are $1/\epsilon$ times as strong as the same charges would produce in vacuum.

A MICROSCOPIC VIEW OF THE DIELECTRIC

10.12 The polarization **P** in the dielectric is simply the large-scale manifestation of the electric dipole moments of the atoms or molecules of which the material is composed. **P** is the mean dipole moment density, the total vector dipole moment per unit volume—averaged, of course, over a region large enough to contain an enormous number of atoms. If there is no electric field to establish a preferred direction, **P** will be zero. That will surely be true for an ordinary liquid or a gas, and for solids too if we ignore the possibility of "frozen-in" polarization mentioned in Section 10.8. In the presence of an electric field in the medium, polarization can arise in two ways. (1) Every atom or molecule will acquire an induced dipole moment proportional to, and in the direction of, the field **E** that acts on that atom or molecule. (2) If molecules with permanent dipole moments are present in the

†The prominence of **D** in Maxwell's formulation of electromagnetic theory, and his choice of the name *displacement* can perhaps be traced to his inclination toward a kind of mechanical model of the "aether." Whittaker has pointed out in his classic, "A History of the Theories of Aether and Electricity," vol. I, Harper, New York, 1960, p. 266, that this inclination may have led Maxwell himself astray at one point in the application of his theory to the problem of reflection of light from a dielectric.

medium, their orientations will no longer be perfectly random; alignment of their dipole moments in the field direction will be favored slightly over alignment in the opposite direction. Both effects (1) and (2) lead to polarization in the direction **E**, that is, to a positive value of **P**/**E**, the electric susceptibility.

Let us consider first the induced atomic moments in a medium in which the atoms or molecules are rather far apart. An example is a gas at atmospheric density, in which there are something like 3×10^{19} molecules per cm^3. We shall assume that the field **E** which acts on an individual molecule is the same as the average, or macroscopic, field **E** in the medium. In making this assumption we are neglecting the field at a molecule which is produced by the induced dipole moment of a nearby molecule. Let α be the polarizability of every molecule and N the mean number of molecules per cubic centimeter. The dipole moment induced in each molecule is $\alpha\mathbf{E}$, and the resulting polarization of the medium, **P**, is simply

$$\mathbf{P} = N\alpha\mathbf{E} \tag{56}$$

This gives us at once the electric susceptibility χ_e:

$$\chi_e = \frac{\mathbf{P}}{\mathbf{E}} = N\alpha \tag{57}$$

and the dielectric constant ϵ:

$$\epsilon = 1 + 4\pi\chi_e = 1 + 4\pi N\alpha \tag{58}$$

The methane molecule in Fig. 10.12 has a polarizability of 2.6×10^{-24} cm^3. At standard conditions of 0°C and atmospheric pressure there are approximately 2.8×10^{19} molecules in 1 cm^3. According to Eq. 58 the dielectric constant of methane at that density ought to have the value

$$\begin{aligned}\epsilon &= 1 + 4\pi N\alpha = 1 + 4\pi \times 2.8 \times 10^{19} \times 2.6 \times 10^{-24} \\ &= 1.00088\end{aligned}$$

This agrees with the value of ϵ listed for methane in Table 10.1. The agreement is hardly surprising, for the value of α given in Fig. 10.12 was probably deduced originally by applying the simple theory we have just developed to an experimentally measured dielectric constant.

We have already noted in Section 10.5 that the atomic polarizability α, which has the dimensions of volume, is in order of magnitude about equal to the volume of an atom. That being so, the product $N\alpha$, which is just χ_e according to Eq. 57, is about equal to the fraction of the volume of the medium which is taken up by atoms. Now the density of a gas under standard conditions is roughly one-thousandth of the density of the same substance condensed to liquid or solid. In the case of methane the ratio is close to 1/1000; in the case of air, 1/700. The gas is about 99.9 percent empty space. In the solid or liquid, on the other hand, the molecules are practically touching one another. The fraction of the volume they occupy is not much less than unity. This tells us that, in condensed matter generally, the induced polarization will result in a susceptibility χ_e of order of magnitude unity. In fact, as our brief list in Table 10.1 suggests and as a more extensive tabulation would confirm, the susceptibility of most nonpolar liquids and solids, that is, the value of $(\epsilon - 1)/4\pi$, ranges from about 0.1 to 1. We can now see why.

We can see, too, why an exact theory of the susceptibility of a solid or liquid is not so easy to develop. When the atoms are crowded together until they almost "touch," the effect of one atom on its neighbors cannot be neglected. The distance b between nearest neighbors is approximately $N^{-1/3}$. Let an electric field E induce a dipole moment $p = E\alpha$ in each atom. This dipole p on one atom will cause a field of strength $E' \approx p/b^3$ at the location of the next atom. But $1/b^3 \approx N$, hence $E' \approx E\alpha N$. As we have just explained, in condensed matter αN is necessarily of order unity. Hence $\mathbf{E}'$ is *not* small, and certainly not negligible, compared with $\mathbf{E}$. Just what the effective field is that polarizes an atom in this situation is a question with no very obvious answer.†

Molecules with permanent electric dipole moments, *polar* molecules, respond to an electric field by trying to line up parallel to it. So long as the dipole moment $\mathbf{p}$ is not pointing in the direction of $\mathbf{E}$, there is a torque $\mathbf{p} \times \mathbf{E}$ tending to turn $\mathbf{p}$ into the direction of $\mathbf{E}$. (Look back at Eq. 18 and Fig. 10.8*b*.) Of course, the torque is zero if $\mathbf{p}$ happens to be pointing exactly opposite to $\mathbf{E}$, but that condition is unstable. Torque on the electric dipole is torque on the molecule itself. A state of lowest energy will have been attained if and when all the polar molecules have rotated to bring their dipole moments into the $\mathbf{E}$ direction. While settling down to that state of perfect alignment they will have given off some energy, through rotational friction, to their surroundings. The resulting polarization would be gigantic. In water there are about 3×10^{22} molecules per cm^3; the dipole moment of each (Fig. 10.14) is 1.84×10^{-18} esu-cm. With complete alignment of the dipoles $\mathbf{P}$ would be 5.6×10^4 esu/cm^2. If Fig. 10.24 were a picture of a water droplet thus polarized, the field strength just outside the drop would exceed 10^5 statvolts/cm!

This does *not* happen. Nothing approaching complete alignment is attained in any reasonable applied field $\mathbf{E}$. Why not? The reason is essentially the same as the reason why the molecules of air in a room are not found all lying on the floor—which is, after all, the arrange-

†An elementary, approximate, treatment of this problem, leading to what is called the Clausius-Mossotti relation, can be found in Section 9.13 in the first edition of this book.

ment of lowest potential energy. We must think about *temperature* and about the energy of thermal agitation which every molecule exhibits at a given absolute temperature T. In magnitude that energy is kT, where k is the universal constant called *Boltzmann's constant*. At room temperature kT amounts to 4×10^{-14} erg. In a system all at temperature T the mean translational energy of a molecule—or for that matter, of any object small or large—is $\frac{3}{2}kT$. More to the point here, the mean rotational energy of a molecule is just kT. Now the air molecules do not all gather near the floor because the change in gravitational potential energy in elevating by a couple of meters a molecule of mass 5×10^{-23} gm is only, as you can readily compute, about 10^{-17} erg, less than $\frac{1}{1000}$ of kT. On the other hand, the air near the floor *is* slightly more dense than the air near the ceiling, even when there is no temperature gradient. That is just the well-known change of barometric pressure with height. Air near the floor is fractionally more dense (when the difference is slight) by just mgh/kT, mgh being the difference in gravitational potential energy between the two levels.

Similarly, in our dielectric we shall find a slight excess of molecular dipoles in the orientation of lower potential energy, that is, pointing in the direction of **E**, or with a component in that direction. The fractional excess in the favored directions will be, in order of magnitude, pE/kT. The numerator represents the difference in potential energy. Actually the work required to turn a dipole from the direction of **E** to the opposite direction is $2pE$ (see Eq. 19) but averaging over angles would bring in other numerical factors that we are leaving out. With N dipoles per unit volume the polarization P, which would be Np if they were totally aligned, will be smaller by something like the factor pE/kT. The polarization to be expected is therefore, in order of magnitude,

$$P \approx Np\left(\frac{pE}{kT}\right) = \frac{Np^2}{kT}E \tag{59}$$

and the susceptibility is

$$\chi_e = \frac{P}{E} \approx \frac{Np^2}{kT} \tag{60}$$

For water at room temperature the quantity on the right in Eq. 60 is 3.0, whereas with $\epsilon = 80$, the actual value of χ_e is 6.3. Evidently a factor of about 2.1 is needed on the right in Eq. 60, in this case, to convert our order-of-magnitude estimate into a correct prediction. Deriving that factor theoretically is quite difficult, for the interactions of neighboring molecules complicate matters even worse than in the case of the nonpolar dielectric.

If you apply an electric field of 1 statvolt/cm to water, the resulting polarization, $P = (80 - 1)/4\pi$ esu/cm^2, is equivalent to the alignment of 3.4×10^{18} H_2O dipoles per cm^3, or about one molecule in 10,000. Even so, this is an order of magnitude greater polarization than the same field would cause in any nonpolar dielectric.

POLARIZATION IN CHANGING FIELDS

10.13 So far we have considered only electrostatic fields in matter. We need to look at the effects of electric fields that are varying in time, like the field in a capacitor used in an alternating-current circuit. The important question is, will the changes in polarization keep up with the changes in the field? Will the ratio of **P** to **E**, at any instant, be the same as in a static electric field? For very slow changes we should expect no difference but, as always, the criterion for slowness depends on the particular physical process. It turns out that induced polarization and the orientation of permanent dipoles are two processes with quite different response times.

The induced polarization of atoms and molecules occurs by the distortion of the electronic structure. Little mass is involved, and the structure is very stiff; its natural frequencies of vibration are extremely high. To put it another way, the motions of the electrons in atoms and molecules are characterized by periods on the order of 10^{-16} second—something like the period of a visible light wave. To an atom, 10^{-14} second is a *long* time. It has no trouble readjusting its electronic structure in a time like that. Because of this, strictly nonpolar substances behave practically the same from direct current (zero frequency) up to frequencies close to those of visible light. The polarization keeps in step with the field, and the susceptibility $\chi_e = P/E$ is independent of frequency.

The orientation of a polar molecule is a process quite different from the mere distortion of the electron cloud. The whole molecular framework has to rotate. On a microscopic scale, it is rather like turning a peanut end for end in a bag of peanuts. The frictional drag tends to make the rotation lag behind the torque and to reduce the amplitude of the resulting polarization. Where on the time scale this effect sets in, varies enormously from one polar substance to another. In water, the "response time" for dipole reorientation is something like 10^{-11} sec. The dielectric constant remains around 80 up to frequencies on the order of 10^{10} Hz. Above 10^{11} Hz ϵ falls to a modest value typical of a nonpolar liquid. The dipoles simply cannot follow so rapid an alternation of the field. In other substances, especially solids, the characteristic time can be much longer. In ice just below the freezing point the response time for electrical polarization is around 10^{-5} sec. Figure 10.29 shows some experimental curves of dielectric constant versus frequency for water and ice.

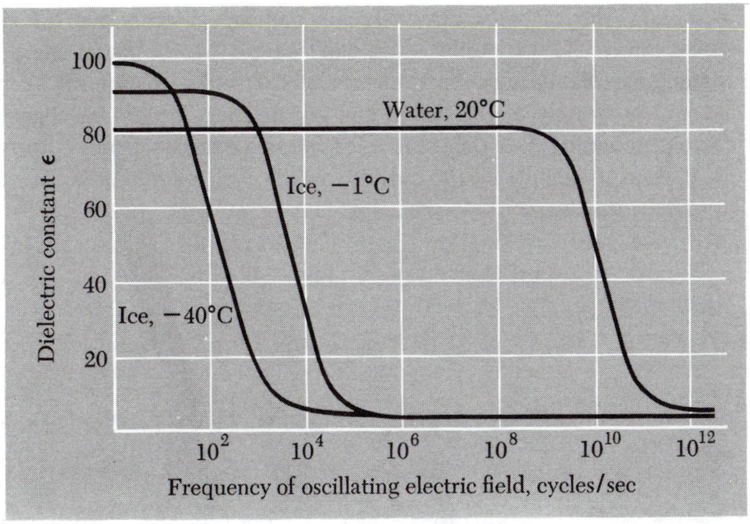

FIGURE 10.29
The variation with frequency of the dielectric constant of water and ice. [Based on information from C. P. Smyth, "Dielectric Behavior and Structure," McGraw-Hill, New York, 1955, for water data; and R. P. Auty and R. H. Cole, *J. Chem. Phys.* **20:** 1309 (1952), for ice data.]

THE BOUND-CHARGE CURRENT

10.14 Wherever the polarization in matter changes with time there is an electric current, a genuine motion of charge. Suppose there are N dipoles in a cubic centimeter of dielectric, and that in the time interval dt each changes from $\mathbf{p}$ to $\mathbf{p} + d\mathbf{p}$. Then the macroscopic polarization density $\mathbf{P}$ changes from $\mathbf{P} = N\mathbf{p}$ to $\mathbf{P} + d\mathbf{P} = N(\mathbf{p} + d\mathbf{p})$. Suppose the change $d\mathbf{p}$ was effected by moving a charge q through a distance $d\mathbf{s}$, in each atom: $q\,d\mathbf{s} = d\mathbf{p}$. Then during the time dt there was actually a charge cloud of density $\mathbf{P} = Nq$, moving with velocity $v = d\mathbf{s}/dt$. That is a conduction current of a certain density $\mathbf{J}$ in esu/sec-cm^2:

$$\mathbf{J} = \rho\mathbf{v} = Nq\frac{d\mathbf{s}}{dt} = N\frac{d\mathbf{p}}{dt} = \frac{d\mathbf{P}}{dt} \tag{61}$$

The connection between rate of change of polarization and current density, $\mathbf{J} = d\mathbf{P}/dt$, is independent of the details of the model. A changing polarization *is* a conduction current, not essentially different from any other.

Naturally, such a current is a source of magnetic field. If there are no other currents around, we should write Maxwell's second equation, curl $\mathbf{B} = (1/c)(\partial\mathbf{E}/\partial t + 4\pi\mathbf{J})$ as

$$\text{curl}\,\mathbf{B} = \frac{1}{c}\left(\frac{\partial\mathbf{E}}{\partial t} + 4\pi\frac{\partial\mathbf{P}}{\partial t}\right) \tag{62}$$

The only difference between an "ordinary" conduction current density and the current density $\partial\mathbf{P}/\partial t$ is that one involves *free* charge in motion, the other *bound* charge in motion. There is one rather obvious practical distinction—you can't have a *steady* bound charge current, one that goes on forever unchanged. Usually we prefer to keep account separately of the bound charge current and the free charge current, retaining **J** as the symbol for the free charge current density only. Then to include all the currents in Maxwell's equation we have to write it this way:

$$\text{curl } \mathbf{B} = \frac{1}{c}\left(\frac{\partial \mathbf{E}}{\partial t} + 4\pi \underbrace{\frac{\partial \mathbf{P}}{\partial t}}_{\substack{\uparrow \\ \text{Bound charge} \\ \text{current density}}} + \underbrace{4\pi\mathbf{J}}_{\substack{\uparrow \\ \text{Free charge} \\ \text{current density}}}\right) \tag{63}$$

In a dielectric medium, $\mathbf{E} + 4\pi\mathbf{P} = \epsilon\mathbf{E}$, allowing a shorter version of Eq. 63.

$$\text{curl } \mathbf{B} = \frac{1}{c}\left(\epsilon \frac{\partial \mathbf{E}}{\partial t} + 4\pi\mathbf{J}\right) \tag{64}$$

More generally, Eq. 63 can also be abbreviated by introducing the vector **D**, previously defined as $\mathbf{E} + 4\pi\mathbf{P}$:

$$\text{curl } \mathbf{B} = \frac{1}{c}\left(\frac{\partial \mathbf{D}}{\partial t} + 4\pi\mathbf{J}\right) \tag{65}$$

The term $\partial\mathbf{D}/\partial t$ is usually referred to as the displacement current. Actually, that part of it which involves $\partial\mathbf{P}/\partial t$ represents, as we have seen, an honest conduction current, real charges in motion. The only part of the total current density that is not simply charge in motion is the $\partial\mathbf{E}/\partial t$ part, the true vacuum displacement current which we discussed in Chapter 9. Incidentally, if we want to express all components of the full current density in units corresponding to those of **J**, we should note that no 4π appears in the first term, and fix that up by writing Eq. 63 as follows:

$$\text{curl } \mathbf{B} = \frac{4\pi}{c}\left(\underbrace{\frac{1}{4\pi}\frac{\partial \mathbf{E}}{\partial t}}_{\substack{\uparrow \\ \text{Vacuum} \\ \text{displacement} \\ \text{current} \\ \text{density}}} + \underbrace{\frac{\partial \mathbf{P}}{\partial t}}_{\substack{\uparrow \\ \text{Bound} \\ \text{charge} \\ \text{current} \\ \text{density}}} + \underbrace{\mathbf{J}}_{\substack{\uparrow \\ \text{Free} \\ \text{charge} \\ \text{current} \\ \text{density}}}\right) \tag{66}$$

Involved in the distinction between bound charge and free charge is a question we haven't squarely faced: Can one always identify unambiguously the "molecular dipole moments" in matter, especially solid matter? The answer is no. Let us take a microscopic view of a thin wafer of sodium chloride crystal. The arrangement of the positive sodium ions and the negative chlorine ions was shown in Fig. 1.7. Figure 10.30 is a cross section through the crystal, which extends on out to the right and the left. If we choose to, we may consider an

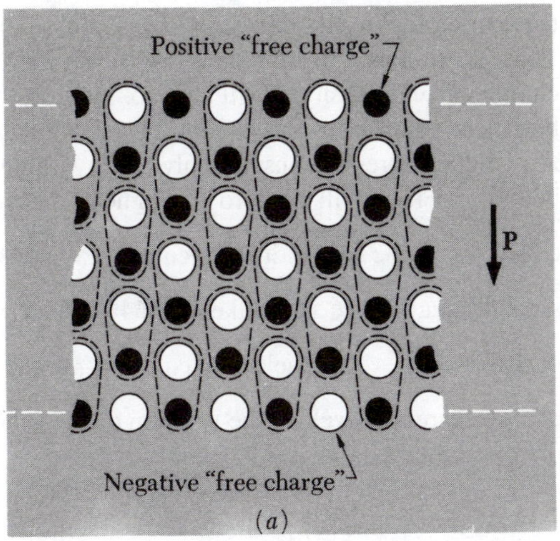

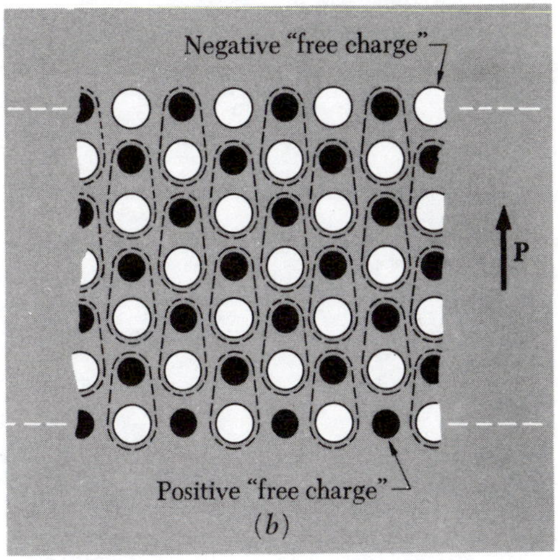

FIGURE 10.30
The same ionic lattice, with charges grouped in pairs as "molecules," in two ways: polarization vector directed downward (*a*), or upward (*b*). The systems are physically identical; the difference is only in the description.

adjacent pair of ions as a neutral molecule with a dipole moment. Grouping them as in Fig. 10.30*a*, we describe the medium as having a uniform macroscopic polarization density **P**, a vector directed downward. At the same time, we observe that there is a layer of positive charge over the top of the crystal, and negative charge over the bottom which, not having been included in our molecules, must be accounted *free charge*.

Now we might just as well have chosen to group the ions as in Fig. 10.30*b*. According to that description, **P** is a vector *upward*, but we have a negative free charge layer on top of the crystal and a positive free charge layer beneath. *Either description is correct*. You will have no trouble finding another one, also correct, in which **P** is zero and there is no free charge. Each description predicts **E** = 0. The macroscopic field **E** is an observable physical quantity. It can depend only on the charge distribution, not on how we choose to *describe* the charge distribution.

This example teaches us that in the real atomic world the distinction between bound charge and free charge is more or less arbitrary, and so, therefore, is the concept of polarization density **P**. The molecular dipole is a well-defined notion only where molecules as such are identifiable—where there is some physical reason for saying, "This atom belongs to this molecule and not to that." In many crystals such an assignment is meaningless. An atom or ion may interact about equally strongly with all its neighbors; one can only speak of the whole crystal as a single molecule.

AN ELECTROMAGNETIC WAVE IN A DIELECTRIC

10.15 In Eqs. 15 of Chapter 9 we wrote out Maxwell's equations for the electric and magnetic fields in vacuum, including source terms, charge density ρ and current density **J**. Now we want to consider an electromagnetic field in an unbounded dielectric medium. The dielectric is a perfect insulator, we shall assume, so there is no free current. That is, the last term on the right in Eqs. 63 through 65, the free charge current density **J**, will be zero. No free charge is present either, but there could be a nonzero density of bound charge if div **E** is not zero. Let us agree to consider only fields with div $\mathbf{E} = 0$. Then ρ, both bound and free, will be zero throughout the medium. No change is called for in the first induction equation, curl $\mathbf{E} = -\frac{1}{c}\frac{\partial \mathbf{B}}{\partial t}$. For the second equation we now take Eq. 64 without the free current term: curl $\mathbf{B} = \frac{\epsilon}{c}\frac{\partial \mathbf{E}}{\partial t}$. The dielectric constant ϵ takes account of the bound charge current as well as the vacuum displacement current. Our complete set of equations has become

$$\begin{aligned} \text{curl } \mathbf{E} &= -\frac{1}{c}\frac{\partial \mathbf{B}}{\partial t} & \text{div } \mathbf{E} &= 0 \\ \text{curl } \mathbf{B} &= \frac{\epsilon}{c}\frac{\partial \mathbf{E}}{\partial t} & \text{div } \mathbf{B} &= 0 \end{aligned} \tag{67}$$

These differ from Eq. 16 of Chapter 9 only in the presence of the constant factor ϵ in the second induction equation.

As we did in Section 9.4, let us construct a wavelike electromagnetic field that can be made to satisfy Maxwell's equations. This time we'll give our trial wave function a slightly more general form:

$$\begin{aligned} \mathbf{E} &= \hat{\mathbf{z}}E_0 \sin(ky - \omega t) \\ \mathbf{B} &= \hat{\mathbf{x}}B_0 \sin(ky - \omega t) \end{aligned} \tag{68}$$

The angle $(ky - \omega t)$ is called the *phase* of the wave. For a point that moves in the positive y direction with speed ω/k, the phase $ky - \omega t$ remains constant. In other words, ω/k is the *phase velocity* of this wave. This term is used when it is necessary to distinguish between two velocities, phase velocity and group velocity. There is no difference in the case we are considering, so we shall call ω/k simply the wave velocity, the same as v in our discussion in Section 9.4. At any fixed location, such as $y = y_0$, the fields oscillate in time with angular frequency ω. At any instant of time, such as $t = t_0$, the phase differs by 2π at planes one wavelength λ apart, where $\lambda = 2\pi/k$.

The space and time derivatives we need are those listed in Eq. 9.19 with small alterations:

$$\begin{aligned} \text{curl } \mathbf{E} &= \hat{\mathbf{x}}E_0 k \cos(ky - \omega t) & \frac{\partial \mathbf{E}}{\partial t} &= -\hat{\mathbf{z}}E_0\omega \cos(ky - \omega t) \\ \text{curl } \mathbf{B} &= -\hat{\mathbf{z}}B_0 k \cos(ky - \omega t) & \frac{\partial \mathbf{B}}{\partial t} &= -\hat{\mathbf{x}}B_0\omega \cos(ky - \omega t) \end{aligned} \tag{69}$$

Substituting these into Eq. 67, we find that the equations are satisfied if

$$\frac{\omega}{k} = \frac{c}{\sqrt{\epsilon}} \qquad \text{and} \qquad B_0 = \sqrt{\epsilon}E_0 \tag{70}$$

The wave velocity ω/k differs from the velocity of light in vacuum by the factor $1/\sqrt{\epsilon}$. The electric and magnetic field amplitudes, E_0 and B_0, which are precisely equal in the wave in vacuum, here differ by the factor $\sqrt{\epsilon}$, the electric amplitude being the smaller. In other respects the wave resembles our plane wave in vacuum: **B** is perpendicular to **E**, and the wave travels in the direction of $\mathbf{E} \times \mathbf{B}$. Of course, if we compare a wave in a dielectric with a wave of the same frequency in vacuum, the wavelength λ in the dielectric will be less than the vacuum wavelength by $1/\sqrt{\epsilon}$, since *frequency* $\times$ *wavelength* = *velocity*.

Light traveling through glass provides an example of the wave just described. In optics it is customary to define *n*, the *index of refraction* of a medium, as the ratio of the speed of light in vacuum to the speed of light in that medium. We have now discovered that n is nothing more than $\sqrt{\epsilon}$. In fact we have now laid most of the foundation for a classical theory of optics.

PROBLEMS

10.1 You have a supply of polyethylene tape, dielectric constant 2.3, 2.25 inches wide, and 0.001 inch thick; also, a supply of aluminum tape 2 inches wide and 0.0005 inch thick. You want to make a capacitor of about 0.05-microfarad capacitance, in the form of a compact cylindrical roll. Describe how you might do this, estimating the amount of tape of each kind that would be needed, and the overall diameter of the finished capacitor.

10.2 In 1746 a Professor Musschenbroek in Leiden charged water in a bottle by touching a wire, projecting from the neck of the bottle, to his electrostatic machine. When his assistant, who was holding the bottle in one hand, tried to remove the wire with the other, he got a violent shock. Thus did the simple capacitor force itself on the attention of electrical scientists. The discovery of the "Leyden jar" revolutionized electrical experimentation. Already in 1747 Benjamin Franklin was writing about his experiments with "Mr. Musschenbroek's wonderful bottle." The jar was really nothing but glass with a conductor on each side of it. To see why it caused such a sensation, estimate the capacitance of a jar made of a 1-liter bottle with walls 2 mm thick, the glass having a dielectric constant 4. What diameter sphere, in air, would have the same capacitance?

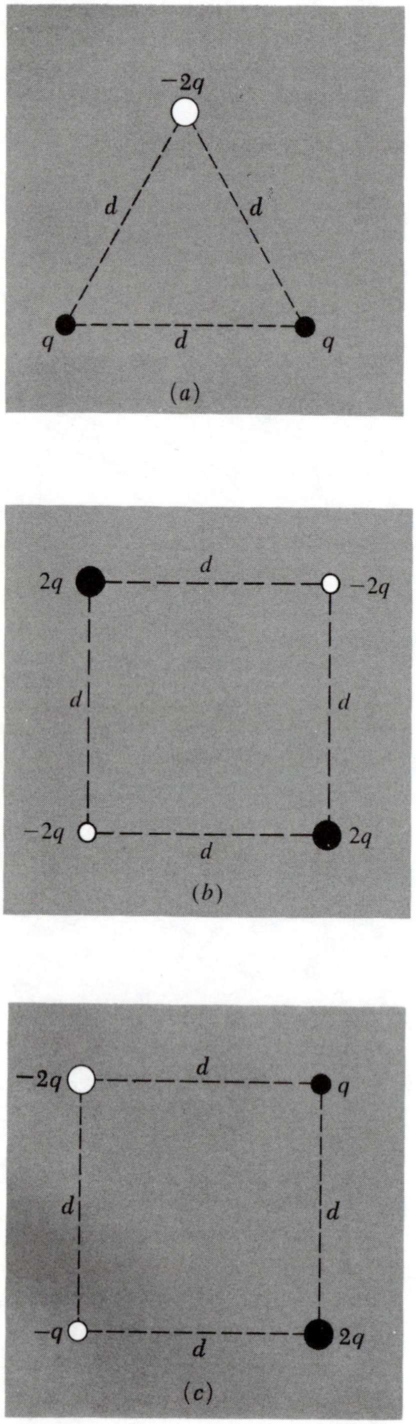

PROBLEM 10-3

10.3 What is the magnitude of the dipole moment of each of the charge distributions in parts (*a*), (*b*), and (*c*) of the figure? What is the direction of the dipole moment vector **p**?

10.4 In the hydrogen chloride molecule the distance between the chlorine nucleus and the proton is 1.28 angstroms. Suppose the electron from the hydrogen atom is transferred entirely to the chlorine atom, joining with the other electrons to form a spherically symmetrical negative charge and centered on the chlorine nucleus. How does the electric dipole moment of this model compare with the actual HCl dipole moment given in Fig. 9.16? Where must the actual center of gravity of the negative charge distribution be located in the real molecule? (The chlorine nucleus has a charge $17e$, the hydrogen nucleus, a charge e.)

10.5 A hydrogen chloride molecule is located at the origin with the H—Cl line along the z axis and Cl uppermost. What is the direction of the electric field, and its strength in statvolts/cm, at a point 10 angstroms up from the origin, on the z axis? At a point 10 angstroms out from the origin, on the y axis?

10.6 A parallel-plate capacitor, with a measured capacitance $C = 250$ cm, is charged to a potential difference of 6 statvolts. The plates are 1.5 cm apart. We are interested in the field outside the capacitor, the "fringing" field which we usually ignore. In particular, we would like to know the field at a distance from the capacitor large compared with the size of the capacitor itself. This can be found by treating the charge distribution on the capacitor as a dipole. Estimate the electric field strength

(*a*) At a point 3 meters from the capacitor in the plane of the plates.

(*b*) At a point the same distance away, in a direction perpendicular to the plates.

10.7 In Section 4.11 we discussed the relaxation time of a capacitor filled with a material having a resistivity ρ. If you will look back at that discussion you will notice that we dodged the question of the dielectric constant of the material. Now you can repair that omission. Introduce ϵ properly into the expression for the time constant. A leaky capacitor important to us all is formed by the wall of a living cell, an insulator (among its many other functions!) that separates two conducting fluids. Its electrical properties are of particular interest in the case of the nerve cell, for the propagation of a nerve impulse is accompanied by rapid changes in the electric potential difference between interior and exterior. The cell membrane typically has a capacitance around 1 microfarad per cm^2 of membrane area. It is believed the membrane consists of material having a dielectric constant about 3. You can now figure out what thickness this implies. Other electrical measurements have indicated that the resistance of 1 cm^2 of cell membrane, measured from the conducting fluid on one side to that on the other, is around 1000 ohms. Show that the time constant of such a leaky capacitor is independent of the area of the capacitor. How large is it in this case? Where would the resistivity ρ of such membrane material fall on the chart of Fig. 4.8?

10.8 How much work is done in moving unit positive charge from A to B in the field of the dipole **p**?

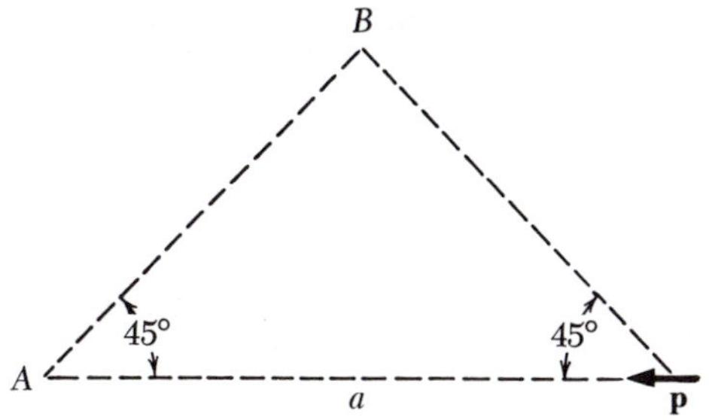

PROBLEM 10-8

10.9 What is the direction of the force on the central dipole caused by the field of the other two dipoles? Calculate the magnitude of the force.

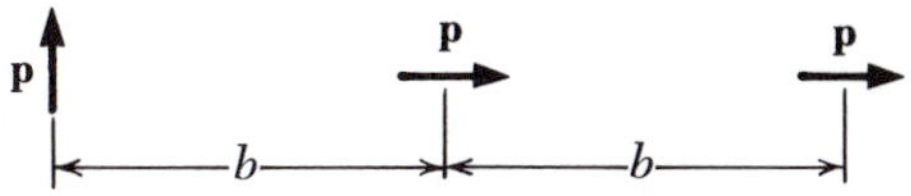

PROBLEM 10-9

10.10 A dipole of strength $p = 200$ esu-cm is located at the origin, pointing in the $\hat{z}$ direction. To its field is added a uniform electric field of strength 5 statvolts/cm in the $\hat{y}$ direction. At how many places, located where, is the total field zero?

Ans. (0, −3.134, +2.216) and (0, +3.134, −2.216).

10.11 A field line in the dipole field is described in polar coordinates by the very simple equation $r = r_0 \sin^2 \theta$, in which r_0 is the radius at which the field line passes through the equatorial plane of the dipole. Show that this is true by demonstrating that at any point on that curve the tangent has the same direction as the dipole field.

10.12 Our formula for the dielectric sphere can actually serve to describe a metal sphere in a uniform field. To demonstrate this, investigate the limiting case, $\epsilon \to \infty$, and show that the external field then takes on a form which satisfies the perfect-conductor boundary conditions. What about the internal field? Make a sketch of some field lines for this limiting case. How large is the dipole moment induced in a conducting sphere of radius a, in a field E_0? What is the radius of a conducting sphere with polarizability equal to that of the hydrogen atom, given in Table 10.2?

10.13 By considering how the introduction of a dielectric changes the energy stored in a capacitor, show that the correct expression for the energy density in a dielectric must be $\epsilon E^2/8\pi$. Then compare the energy stored in the electric field with that stored in the magnetic field in the wave studied in Section 10.15.

10.14 The figure shows three capacitors of the same area and plate separation. Call the capacitance of the vacuum condenser C_0. Each of the others is half-filled with a dielectric, with the same dielectric constant ϵ, but differently disposed, as shown. Find the capacitance of each of these two capacitors. (Neglect edge effects.)

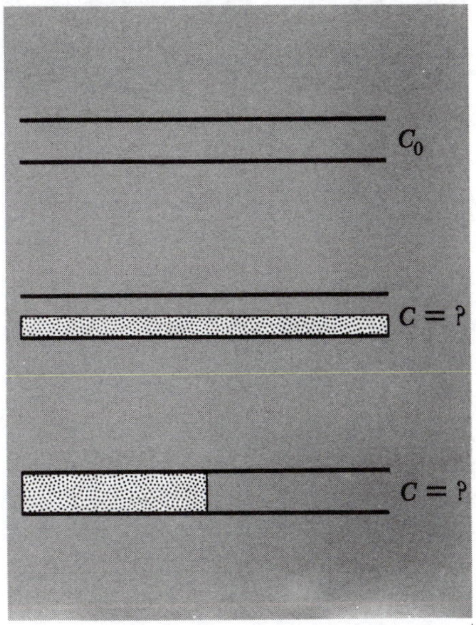

PROBLEM 10-14

10.15 The electric dipole moment of the water molecule is given in Fig. 10.14. Imagine that all the molecular dipoles in a cup of water could be made to point down. Calculate the magnitude of the resulting surface charge density at the upper water surface, and express it in electrons per cm^2.

10.16 In Section 10.10 the fact that the electric field is uniform inside the polarized sphere was deduced from the form of the potential on the boundary. You can also prove it by superposing the internal fields of two balls of charge whose centers are separated.

(*a*) Show that inside a spherical, uniform charge distribution **E** is proportional to **r**.

(*b*) Now take two spherical distributions with density ρ and $-\rho$, centers at C_1 and C_2, and show that the resultant field is constant and parallel to the line from C_1 to C_2.

(*c*) Analyze in the same way the field of a long circularly cylindrical rod which is polarized perpendicular to its axis.

10.17 Shown below are four different arrangements of the electric dipole moments of two neighboring polar molecules. Find the potential energy of each arrangement, the potential energy being defined as the work done in bringing the two molecules together from infinite separation while keeping their moments in the specified orientation. That is not necessarily the easiest way to calculate it. You can always bring them together one way and then rotate them.

PROBLEM 10.17

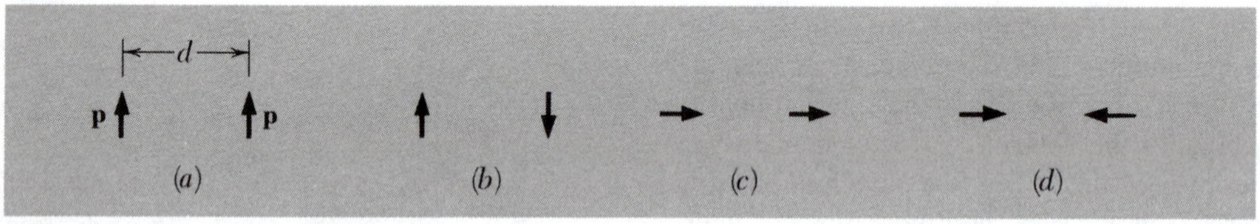

10.18 The phenomenon of *hydration* is important in the chemistry of aqueous solutions. This refers to the fact that an ion in solution gathers around itself a cluster of water molecules, which cling to it rather tightly. The force of attraction between a dipole and a point charge is responsible for this. Estimate the energy required to separate an ion carrying a single charge e from a water molecule, assuming that initially the ion is located 1.5 angstroms from the effective location of the H_2O dipole. (This distance is actually a rather ill-defined quantity, since the water molecule, viewed from close up, is a charge distribution, not an infinitesimal dipole.) Which part of the water molecule will be found nearest to a negative ion?

10.19 Between any ion and any neutral atom there is a force which arises as follows. The electric field of the ion polarizes the atom; the field of that induced dipole reacts on the ion. Show that this force is always attractive, and that it varies with the inverse fifth power of the distance of separation r. Derive an expression for the associated potential energy, with zero energy corresponding to infinite separation. For what distance r is this potential energy of the same magnitude as kT at room temperature, which is 4×10^{-14} erg, if the ion is singly charged and the atom is a sodium atom? (See Table 10.2.)

10.20 Two polarizable atoms A and B are a fixed distance apart. The polarizability of each atom is α. Consider the following intriguing possibility. Atom A is polarized by an electric field, the source of which is the electric dipole moment $\mathbf{p}$ of atom B. This dipole moment is induced in atom B by an electric field, the source of which is the dipole moment of atom A. Can this happen? If so, under what conditions? If not, why not?

10.21 Materials to be used as insulators or dielectrics in capacitors are rated with respect to *dielectric strength,* defined as the maximum internal electric field the material can support without danger of electrical breakdown. It is customary to express the dielectric strength in kilovolts per mil. (One mil is 0.001 inch, or 0.00254 cm.) For example Mylar (a Dupont polyester film) is rated as having a dielectric strength of 14 kilovolts/mil when it is used in a thin sheet—as it would be in a typical capacitor. The dielectric constant ϵ of Mylar is 3.25. Its density is 1.40 gm/cm^3. Calculate the maximum amount of energy that can be stored in a Mylar-filled capacitor, and express it in joules/kg of Mylar. Assuming the electrodes and case account for 25 percent of the capacitor's weight, how high could the capacitor be lifted by the energy stored in it? Compare the capacitor as an energy storage device with the batteries in Problems 4.28 and 4.29.

10.22 From the values of ϵ given for water, ammonia, and methanol in Table 10.1 calculate the electric susceptibility χ_e for each liquid. Our theoretical prediction in Eq. 60 can be written $\chi_e = CNp^2/kT$, with the factor C as yet unknown, but expected to be of order of magnitude unity. The densities of the liquids are 1.00, 0.82, and 1.33 gm/cm^3, respectively; their molecular weights are 18, 17, and 32. Taking the value of the dipole moment from Fig. 10.14, find for each case the value of C required to fit the observed value of χ_e.

10.23 Consider an oscillating electric field, $E_0 \cos \omega t$, inside a dielectric medium that is not a perfect insulator. The medium has dielectric constant ϵ and conductivity σ. This could be the electric field in some leaky capacitor which is part of a resonant circuit, or it could be the electric field at a particular location in an electromagnetic wave. Show that the Q factor, as defined by Eq. 13 of Chapter 8, is

$\epsilon\omega/4\pi\sigma$ for this system, and evaluate it for seawater at a frequency of 1000 MHz. (The conductivity is given in Table 4.1, and the dielectric constant may be assumed to be the same as that of pure water at the same frequency. See Fig. 10.29.) What does your result suggest about the propagation of decimeter waves through seawater?

10.24 A block of glass, refractive index $n = \sqrt{\epsilon}$, fills the space $y > 0$, its surface being the xz plane. A plane wave traveling in the positive y direction through the empty space $y < 0$ is incident upon this surface. The electric field in this wave is $\hat{\mathbf{z}}E_i \sin(ky - \omega t)$. There is a wave inside the glass block, described exactly by Eq. 68. There is also a reflected wave in the space $y < 0$, traveling away from the glass in the negative y direction. Its electric field is $\hat{\mathbf{z}}E_r \sin(ky + \omega t)$. Of course, each wave has its magnetic field, of amplitude, respectively, B_i, B_0, and B_r. The total magnetic field must be continuous at $y = 0$, and the total electric field, being parallel to the surface, must be continuous also. Show that this requirement, and the relation of B_0 to E_0 given in Eq. 70, suffice to determine the ratio of E_r to E_i. When a light wave is incident normally at an air-glass interface, what fraction of the energy is reflected if the index n is 1.6?

11

MAGNETIC FIELDS IN MATTER

HOW VARIOUS SUBSTANCES RESPOND TO A MAGNETIC FIELD

11.1 Imagine doing some experiments with a very intense magnetic field. To be definite, suppose we have built a solenoid of 10-cm inside diameter, 40 cm long, like the one shown in Fig. 11.1. Its outer diameter is 40 cm, most of the space being filled with copper windings. This coil will provide a steady field of 30,000 gauss, or 3.0 teslas, at its center if supplied with 400 kilowatts of electric power—and something like 30 gallons of water per minute, to carry off the heat. We mention these practical details to show that our device, though nothing extraordinary, is a pretty respectable laboratory magnet. The field strength at the center is nearly 10^5 times the earth's field, and probably 5 or 10 times stronger than the field near any iron bar magnet or horseshoe magnet you may have experimented with. The field will be fairly uniform near the center of the solenoid, falling, on the axis at either end, to roughly half its central value. It will be rather less uniform than the field of the solenoid in Fig. 6.18, since our coil is equivalent to a "nested" superposition of solenoids with length-diameter ratio varying from 4:1 to 1:1. In fact, if we analyze our coil in that way and use the formula (Eq. 44 of Chapter 6) which we derived for

FIGURE 11.1
(*a*) A coil designed to produce a strong magnetic field. The water-cooled winding is shown in cross section. (*b*) A graph of the field strength B_z on the axis of the coil.

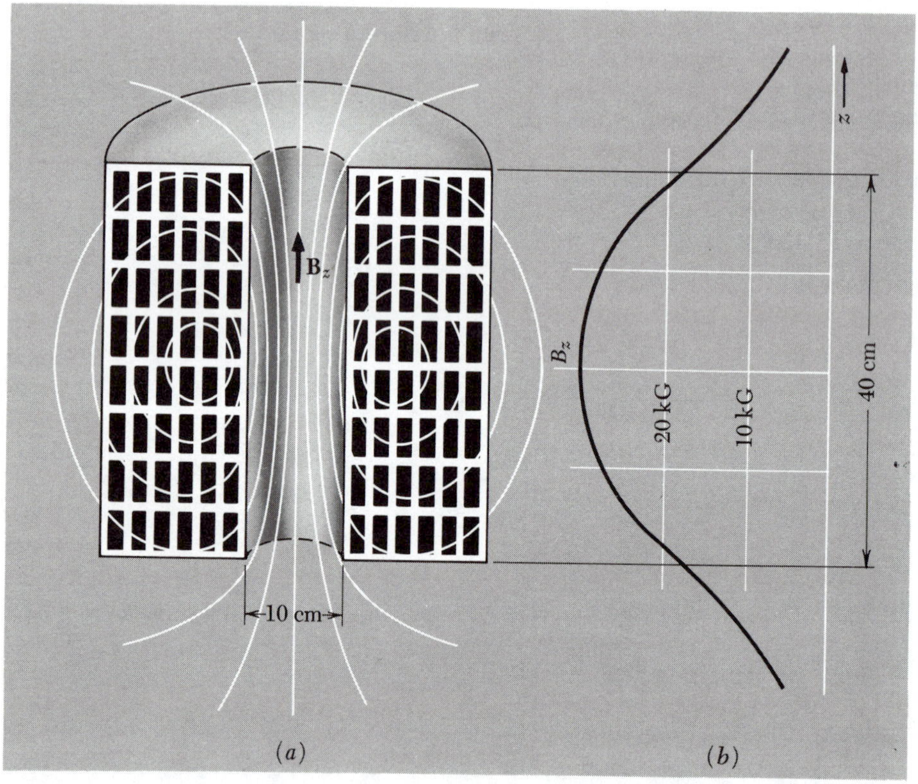

the field on the axis of a solenoid with a single-layer winding, it is not hard to calculate the axial field exactly. A graph of the field strength on the axis, with the central field taken as 30 kilogauss, is included in Fig. 11.1. The intensity just at the end of the coil is 18,000 gauss, and in that neighborhood the field is changing with a gradient of approximately 1700 gauss/cm.

Let's put various substances into this field and see if a force acts on them. Generally, we do detect a force. It vanishes when the current in the coil is switched off. We soon discover that the force is strongest not when our sample of substance is at the center of the coil where the magnetic field B_z is strongest, but when it is located near the end of the coil where the gradient dB_z/dz is large. From now on let us support each sample just inside the upper end of the coil. Figure 11.2 shows one such sample, contained in a test tube suspended by a spring which can be calibrated to indicate the extra force caused by the magnetic field. Naturally we have to do a "blank" experiment with the test tube and suspension alone, to allow for the magnetic force on everything other than the sample.

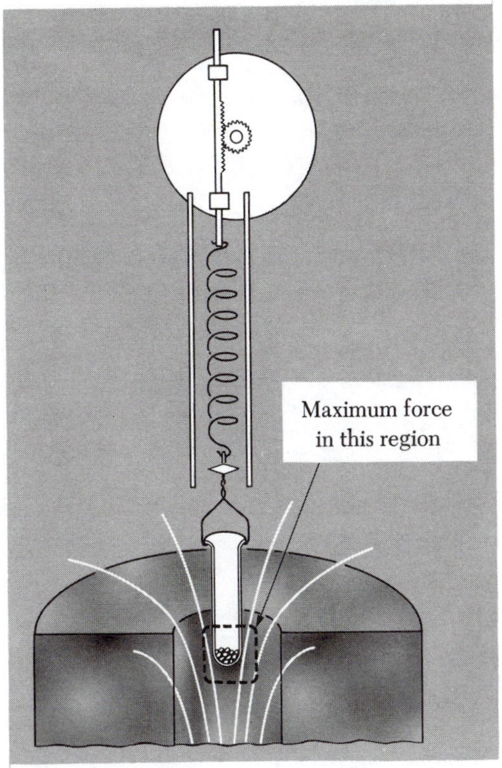
Maximum force
in this region

FIGURE 11.2
An arrangement for measuring the force on a substance in a magnetic field.

We find in such an experiment that the force on a particular substance—metallic aluminum, for instance—is proportional to the mass of the sample and independent of its shape, as long as the sample is not too large. (Experiments with a small sample in this coil show that the force remains practically constant over a region a few centimeters in extent, inside the end of the coil; if we use samples no more than 1 to 2 cm^3 in volume, they can be kept well within this region.) We can express our quantitative results, for a given substance, as so many dynes force per gram of sample, under the conditions B_z = 18,000 gauss, dB_z/dz = 1700 gauss/cm.

But first the qualitative results, which are a bit bewildering: For a large number of quite ordinary pure substances the force observed, although easily measurable, seems after all our effort to provide an intense magnetic field, ridiculously small. It is 10 or 20 dynes/gm, typically, not more than a few percent of the weight of the sample. It is upward for some samples, downward for others. This has nothing to do with the *direction* of the magnetic field, as we can verify by reversing the current in the coil. Instead, it appears that some substances are always pulled in the direction of *increasing* field intensity, others in the direction of *decreasing* field intensity, irrespective of the field direction.

We do find some substances that are attracted to the coil with considerably greater force. For instance, copper chloride crystals are pulled downward with a force of 280 dynes per gram of sample. Liquid oxygen behaves spectacularly in this experiment; it is pulled into the coil with a force nearly 8 times its weight. In fact, if we were to bring an uncovered flask of liquid oxygen up to the bottom end of our coil, the liquid would be lifted right out of the flask. (Where do you think it would end up?) Liquid nitrogen, on the other hand, proves to be quite unexciting; a gram of liquid nitrogen is pushed away from the coil with the feeble force of 10 dynes. In Table 11.1 we have listed some results that one might obtain in such an experiment. The substances, including those already mentioned, have been chosen to suggest, as best one can with a sparse sampling, the wide range of magnetic behavior we find in ordinary materials.

As you know, a few substances, of which the most familiar is metallic iron, seem far more "magnetic" than any others. In Table 11.1 we give the force that would act on a 1-gm piece of iron put in the same position in the field as the other samples. The force is nearly a pound! (We would not have been so naive as to approach our magnet with several grams of iron suspended in a test tube from a delicate spring—a different suspension would have to be used.) Note that there is a factor of more than 10^5 between the force that acts on a gram of iron and the force on a gram of copper, elements not otherwise radically different. Incidentally, this suggests that reliable magnetic measurements on a substance like copper may not be easy. A few parts per million contamination by metallic iron particles would utterly falsify the result.

There is another essential difference between the behavior of the iron and the magnetite and that of the other substances in the table. Suppose we make the obvious test, by varying the field strength of the magnet, to ascertain whether the force on a sample is proportional to the field. For instance, we might reduce the solenoid current by half,

TABLE 11.1

Force on a 1-gm Sample Near the Upper End of the Coil where B_z = 18,000 gauss, dB_z/dz = 1700 gauss/cm

Substance	Formula	Force, dynes†
Diamagnetic		
Water	H_2O	−22
Copper	Cu	−2.6
Sodium chloride	NaCl	−15
Sulfur	S	−16
Diamond	C	−16
Graphite	C	−110
Liquid nitrogen	N_2	−10 (78 K)
Paramagnetic		
Sodium	Na	20
Aluminum	Al	17
Copper chloride	$CuCl_2$	280
Nickel sulfate	$NiSO_4$	830
Liquid oxygen	O_2	7,500 (90 K)
Ferromagnetic		
Iron	Fe	400,000
Magnetite	Fe_3O_4	120,000

†Direction of force: downward (into coil) +; upward −. All measurements made at temperature of 20°C except as noted.

thereby halving both the field intensity B_z and its gradient dB_z/dz. We would find, in the case of every substance above iron in the table, that the force is reduced to *one-fourth* its former value, whereas the force on the iron sample, and that on the magnetite, would be reduced only to one-half or perhaps a bit less. Evidently the force, under these conditions at least, is proportional to the square of the field strength for all the other substances listed, but nearly proportional to the field strength itself for Fe and Fe_3O_4.

It appears that we may be dealing with several different phenomena here, and complicated ones at that. As a small step toward understanding, we can introduce some classification.

First, those substances which are feebly repelled by our magnet, water, sodium chloride, quartz, etc., are called *diamagnetic*. The majority of inorganic compounds and practically all organic compounds are diamagnetic. It turns out, in fact, that diamagnetism is a property of *every* atom and molecule. When the opposite behavior is observed, it is because the diamagnetism is outweighed by a different and stronger effect, one that leads to attraction.

Substances which are attracted toward the region of stronger field are called *paramagnetic*. In some cases, notably metals such as Al, Na, and many others, the paramagnetism is not much stronger than the common diamagnetism. In other materials such as the $NiSO_4$ and the $CuCl_2$ on our list, the paramagnetic effect is much stronger. In these substances also, it *increases* as the temperature is lowered, leading to quite large effects at temperatures near absolute zero. The increase of paramagnetism with lowering temperature is responsible in part for the large force recorded for liquid oxygen. If you think all this is going to be easy to explain, observe that copper is diamagnetic while copper chloride is paramagnetic, but sodium is paramagnetic while sodium chloride is diamagnetic.

Finally, substances that behave like iron and magnetite are called *ferromagnetic*. In addition to the common metals of this class, iron, cobalt, and nickel, quite a number of ferromagnetic alloys and crystalline compounds are known. Indeed current research in ferromagnetism is steadily lengthening the list.

In this chapter we have two tasks. One is to develop a treatment of the large-scale phenomena involving magnetized matter, in which the material itself is characterized by a few parameters and the experimentally determined relations among them. It is like a treatment of dielectrics based on some observed relation between electric field and bulk polarization. We sometimes call such a theory *phenomenological*; it is more of a description than an explanation. Our second task is to try to understand, at least in a general way, the atomic origin of the various magnetic effects. Even more than dielectric phenomena the magnetic effects, once understood, reveal some basic features of atomic structure.

One general fact stands out in the table. Very little energy, on the scale of molecular energies, is involved in diamagnetism and paramagnetism. Take the extreme example of liquid oxygen. To pull 1 gm of liquid oxygen away from our magnet, energy would have to be expended amounting, in ergs, to 7500 dynes times a distance of several centimeters (since the field strength falls off substantially in a few centimeters distance). In order of magnitude, let us say, 10^5 ergs. That is less than 10^{-17} erg per molecule, of which there are 2×10^{22} in 1 gm of the liquid. Just to vaporize 1 gm of liquid oxygen requires 50 calories, or about 10^{-13} erg per molecule. (Most of that energy is used in separating the molecules from one another.) Whatever may be happening in liquid oxygen at the molecular level as a result of the magnetic field, it is apparently a very minor affair in terms of energy.

Even a strong magnetic field has hardly any effect on chemical processes, including biochemical. You could put your hand and forearm into our 30-kilogauss solenoid without experiencing any significant sensation or consequence. It is hard to predict whether your arm would prove to be paramagnetic or diamagnetic, but the force on it would be no more than a fraction of an ounce in any case. Conversely, the presence of someone's hand close to the sample in Fig. 11.2 would perturb the field and change the force on the sample by no more than a few parts in a million. In whole-body imaging with nuclear magnetic resonance, the body is pervaded by magnetic field up to a few kilogauss in strength with no physiological effects whatever. It appears that the only hazard associated with large-scale, strong, steady magnetic fields is the danger that a loose iron object will be snatched by the fringing field and hurled into the magnet.

In its interaction with matter the magnetic field plays a role utterly different from that of the electric field. The reason is simple and fundamental. Atoms and molecules are made of electrically charged particles which move with velocities generally small compared with the speed of light. A magnetic field exerts no force at all on a stationary electric charge; on a moving charged particle the force is proportional to v/c. Electric forces overwhelmingly dominate the atomic scene. As we have remarked before, magnetism appears, in our world at least, to be a relativistic effect. The story would be different if matter were made of magnetically charged particles. We must explain now what *magnetic charge* means and what its apparent absence signifies.

THE ABSENCE OF MAGNETIC "CHARGE"

11.2 The magnetic field outside a magnetized rod such as a compass needle looks very much like the electric field outside an electrically polarized rod, a rod that has an excess of positive charge at one end, negative charge at the other (Fig. 11.3). It is conceivable that the

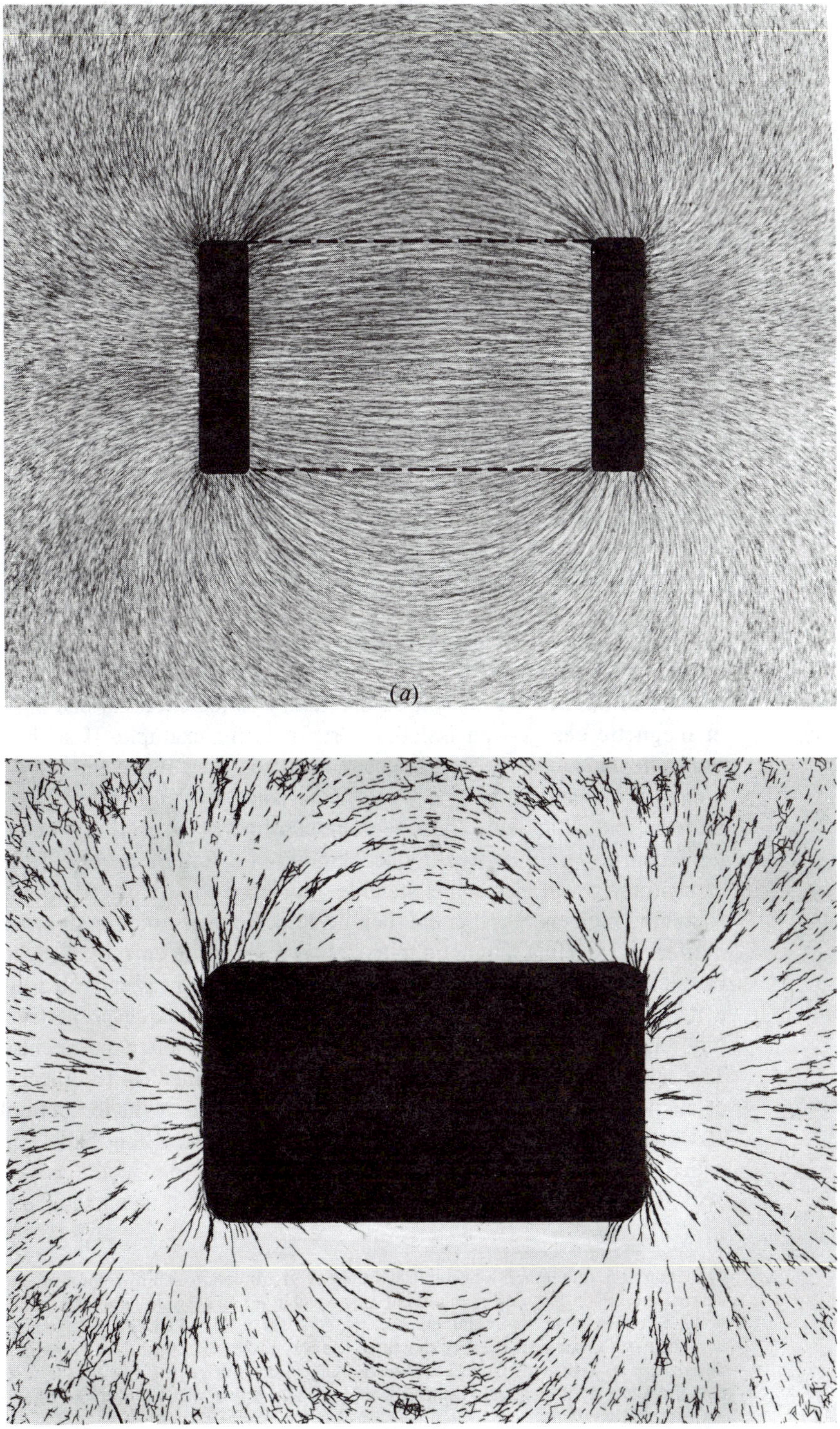

FIGURE 11.3
(*a*) Two oppositely charged disks (the electrodes showing in cross section as solid black bars) have an electric field which is the same as that of a polarized rod. That is, if you imagine such a rod to occupy the region within the dashed boundary, its external field would be like that drawn. The electric field here was made visible by a multitude of tiny black fibers, suspended in oil, which oriented themselves along the field direction. This elegant method of demonstrating electric field configurations is due to Harold M. Waage, Palmer Physical Laboratory, Princeton University, who has kindly prepared the original photograph for this illustration [H. M. Waage, *Am. J. Phys.*, **32:**388 (1964)]. (*b*) The magnetic field around a magnetized cylinder, shown by the orientation of small pieces of nickel wire, immersed in glycerine. (This attempt to improve on the traditional iron filings demonstration by an adaptation of Waage's technique was not very successful—the nickel wires tend to join in long strings which are then pulled in toward the magnet.) Theoretically constructed diagrams of the fields in the two systems are shown later in Fig. 11.21.

magnetic field has sources which are related to it the way electric charge is related to the electric field. Then the north pole of the compass needle would be the location of an excess of one kind of magnetic charge, and the south pole would be the location of an excess of the opposite kind. We might call "north charge" positive and "south charge" negative, with magnetic field directed from positive to negative, a rule like that adopted for electric field and electric charge. Historically, that is how our convention about the positive direction of magnetic field was established.† What we have called *magnetic charge* has usually been called *magnetic pole strength.*

This idea is perfectly sound as far as it goes. It becomes even more plausible when we recall that the fundamental equations of the electromagnetic field are quite symmetrical in **E** and **B**. Why, then, should we not expect to find symmetry in the sources of the field? With magnetic charge as a possible source of the static magnetic field **B**, we would have div $\mathbf{B} = 4\pi\eta$, where η stands for the density of magnetic charge, in complete analogy to the electric charge density ρ. Two positive magnetic charges (or north poles) of unit strength, 1 cm apart, would repel one another with a force of 1 dyne, and so on.

The trouble is, that is not the way things are. Nature for some reason has not made use of this opportunity. The world around us appears totally asymmetrical in the sense that we find *no magnetic charges at all.* No one has yet observed an isolated excess of one kind of magnetic charge—an isolated north pole for example. If such a *magnetic monopole* existed it could be recognized in several ways. Unlike a magnetic dipole, it would experience a force if placed in a uniform magnetic field. Thus an elementary particle carrying a magnetic charge would be steadily accelerated in a static magnetic field, as a proton or an electron is steadily accelerated in an electric field. Reaching high energy, it could then be detected by its interaction with matter. A traveling magnetic monopole is a magnetic current; it must be encircled by an electric field, as an electric current is encircled by a magnetic field. With strategies based on these unique properties, physicists have looked for magnetic monopoles in many experiments. The search was recently renewed when a development in the theory of elementary particles suggested that the universe ought to contain at least a few magnetic monopoles, left over from the "big bang" in which it presumably began. But not one magnetic monopole has yet been detected, and it is now evident that if they exist at all they are

†In Chapter 6, remember, we established the positive direction of **B** by reference to a current direction (direction of motion of positive charge) and a right-hand rule. Now *north pole* means "north-seeking pole" of the compass needle. We know of no reason why the earth's magnetic polarity should be one way rather than the other. Franklin's designation of "positive" electricity had nothing to do with any of this. So the fact that it takes a right-hand rule rather than a left-hand rule to make this all consistent is the purest accident.

exceedingly rare. Of course, the proven existence of even one magnetically charged particle would have profound implications, but it would not alter the fact that in matter as we know, it the only sources of the magnetic field are electric currents. As far as we know,

$$\text{div}\,\mathbf{B} = 0 \qquad (everywhere) \tag{1}$$

This takes us back to the hypothesis of Ampère, his idea that magnetism in matter is to be accounted for by a multitude of tiny rings of electric current distributed through the substance. We'll begin by studying the magnetic field created by a single current loop at points relatively far from the loop.

THE FIELD OF A CURRENT LOOP

11.3 A closed conducting loop, not necessarily circular, lies in the *xy* plane encircling the origin, as in Fig. 11.4*a*. A steady current I, measured in esu/sec, flows around the loop. We are interested in the magnetic field this current creates—not near the loop, but at distant points like P_1 in the figure. We shall assume that r_1, the distance to P_1, is much larger than any dimension of the loop. To simplify the diagram we have located P_1 in the yz plane; it will turn out that this is no restriction. This is a good place to use the vector potential. We shall compute first the vector potential $\mathbf{A}$ at the location P_1, that is, $\mathbf{A}(0, y_1, z_1)$. From this it will be obvious what the vector potential is at any other point (x, y, z) far from the loop. Then by taking the curl of $\mathbf{A}$ we shall get the magnetic field $\mathbf{B}$.

For a current confined to a wire, we had, as Eq. 35 of Chapter 6:

$$\mathbf{A}(0, y_1, z_1) = \frac{I}{c}\int_{\text{loop}} \frac{d\mathbf{l}_2}{r_{12}} \tag{2}$$

At that time we were concerned only with the contribution of a small segment of the circuit; now we have to integrate around the entire loop. Consider the variation in the denominator r_{12} as we go around the loop. If P_1 is far away, the first-order variation in r_{12} depends only on the coordinate y_2 of the segment $d\mathbf{l}_2$, and not on x_2. This should be clear from the side view in Fig. 11.4*b*. Thus, neglecting quantities proportional to $(x_2/r_{12})^2$, we may treat r_{12} and r'_{12}, which lie on top of one another in the side view, as equal. And in general, to first order in the ratio (loop dimension/distance to P_1), we have

$$r_{12} \approx r_1 - y_2 \sin\theta \tag{3}$$

Look now at the two elements of the path $d\mathbf{l}_2$ and $d\mathbf{l}'_2$ shown in

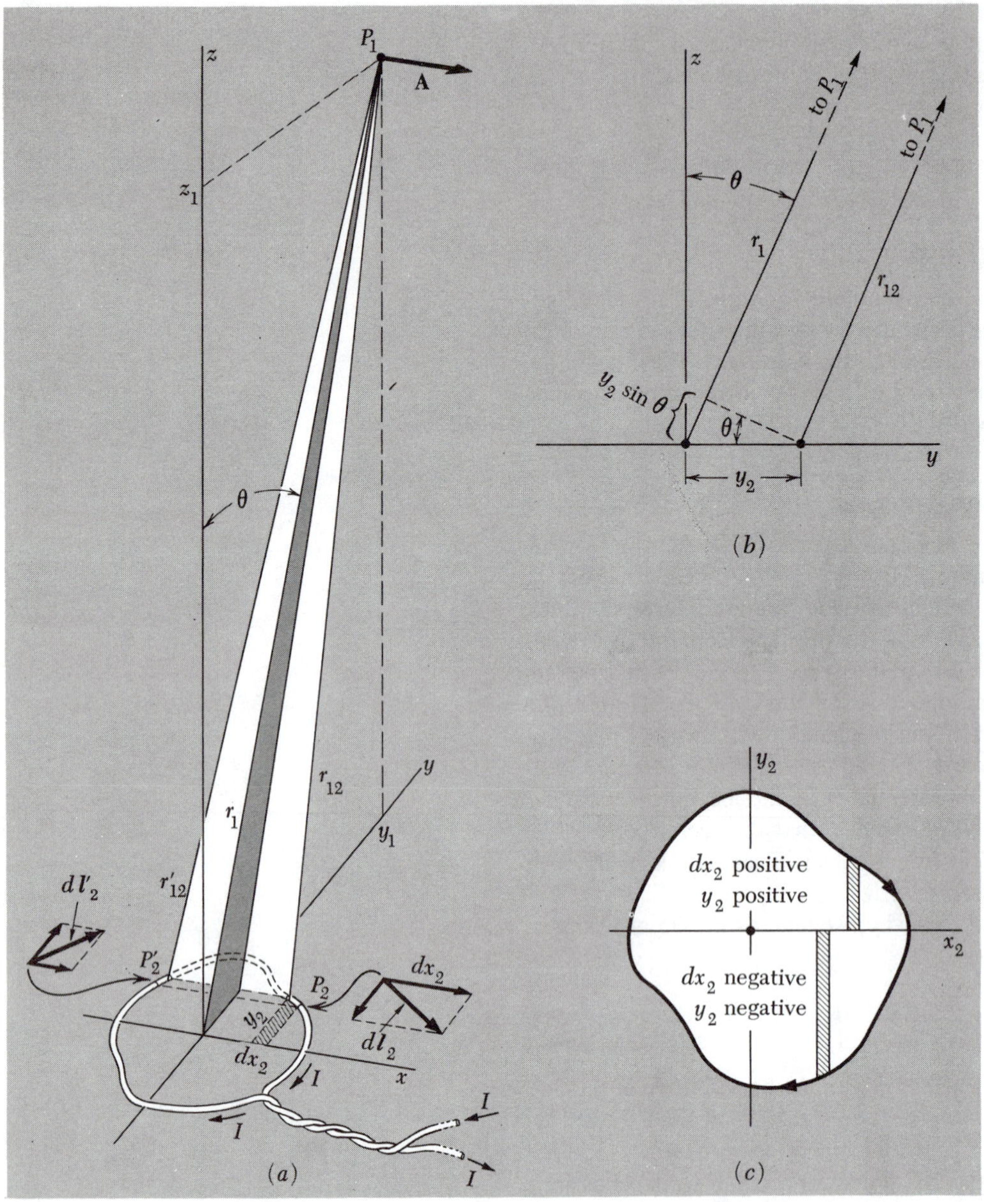

FIGURE 11.4

(a) Calculation of the vector potential **A** at a point far from the current loop. (b) Side view, looking in along the x axis, showing that

$$r_{12} \approx r_1 - y_2 \sin\theta \qquad \text{if } r_1 \gg y_2$$

(c) Top view, to show that $\int_{\text{loop}} y_2\, dx_2$ is the area of the loop.

Fig. 11.4a. For these the dy_2's are equal and opposite, and as we have already pointed out, the r_{12}'s are equal to first order. To this order then, their contributions to the line integral will cancel, and this will be true for the whole loop. Hence **A** at P_1 will not have a y component. Obviously it will not have a z component, for the current path itself has nowhere a z component. The x component of the vector potential comes from the dx part of the path integral. Thus

$$\mathbf{A}(0, y_1, z_1) = \hat{\mathbf{x}} \frac{I}{c} \int \frac{dx_2}{r_{12}} \tag{4}$$

Without spoiling our first-order approximation, we can turn Eq. 3 into

$$\frac{1}{r_{12}} \approx \frac{1}{r_1}\left(1 + \frac{y_2 \sin\theta}{r_1}\right) \tag{5}$$

and using this for the integrand we have

$$\mathbf{A}(0, y_1, z_1) = \hat{\mathbf{x}} \frac{I}{cr_1} \int \left(1 + \frac{y_2 \sin\theta}{r_1}\right) dx_2 \tag{6}$$

In the integration r_1 and θ are constants. Obviously $\int dx_2$ around the loop vanishes. Now $\int y_2\, dx_2$ around the loop is just the area of the loop, regardless of its shape (see Fig. 11.4*c*). So we get finally

$$\mathbf{A}(0, y_1, z_1) = \hat{\mathbf{x}} \frac{I \sin\theta}{cr_1^2} \times (\text{area of loop}) \tag{7}$$

Here is a simple but crucial point: Since the *shape* of the loop hasn't mattered, our restriction on P_1 to the yz plane can't make any essential difference. Therefore we must have in Eq. 7 the general result we seek, if only we *state* it generally: The vector potential of a current loop of any shape, at a distance r from the loop which is much greater than the size of the loop, is a vector perpendicular to the plane containing $\mathbf{r}$ and the normal to the plane of the loop, of magnitude

$$A = \frac{Ia \sin\theta}{cr^2} \tag{8}$$

where a stands for the area of the loop.

This vector potential is symmetrical around the axis of the loop, which implies that the field $\mathbf{B}$ will be symmetrical also. The explanation is that we are considering regions so far from the loop that the details of the shape of the loop have negligible influence. All loops with the same *current* $\times$ *area* product produce the same far field. We call the product Ia/c the *magnetic dipole moment* of the current loop, and denote it by $\mathbf{m}$. The magnetic dipole moment is evidently a vector, its direction being that of the normal to the loop, or that of the vector $\mathbf{a}$, the directed area of the path surrounded by the loop

$$\mathbf{m} = \frac{I}{c}\mathbf{a} \tag{9}$$

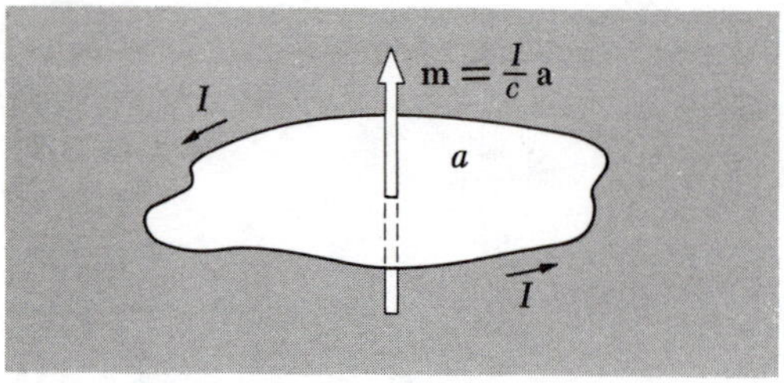

FIGURE 11.5
By definition, the magnetic moment vector is related to the current by a right-hand-screw rule as here shown.

As for sign, let us agree that the direction of **m** and the sense of positive current flow in the loop are to be related by a right-hand-screw rule, illustrated in Fig. 11.5. (The dipole moment of the loop in Fig. 11.4*a* points downward, according to this rule.) The vector potential for the field of a magnetic dipole **m** can now be written neatly with vectors:

$$\mathbf{A} = \frac{\mathbf{m} \times \hat{\mathbf{r}}}{r^2} \tag{10}$$

where $\hat{\mathbf{r}}$ is a unit vector in the direction *from* the loop *to* the point for which **A** is being computed. You can check that this agrees with our convention about sign. Note that the direction of **A** must always be that of the current in the *nearest* part of the loop.

Figure 11.6 shows a magnetic dipole located at the origin, with the dipole moment vector **m** pointed in the positive z direction. To express the vector potential at any point (x, y, z), we observe that $r^2 = x^2 + y^2 + z^2$, and $\sin\theta = \sqrt{x^2 + y^2}/r$. The magnitude A of the vector potential at that point is

$$A = \frac{m \sin\theta}{r^2} = \frac{m\sqrt{x^2 + y^2}}{r^3} \tag{11}$$

Since **A** is tangent to a horizontal circle around the z axis, its components are

$$\begin{aligned} A_x &= A\left(\frac{-y}{\sqrt{x^2 + y^2}}\right) = \frac{-my}{r^3} \\ A_y &= A\left(\frac{x}{\sqrt{x^2 + y^2}}\right) = \frac{mx}{r^3} \\ A_z &= 0 \end{aligned} \tag{12}$$

FIGURE 11.6
A magnetic dipole located at the origin. At every point far from the loop, A is a vector parallel to the xy plane, tangent to a circle around the z axis.

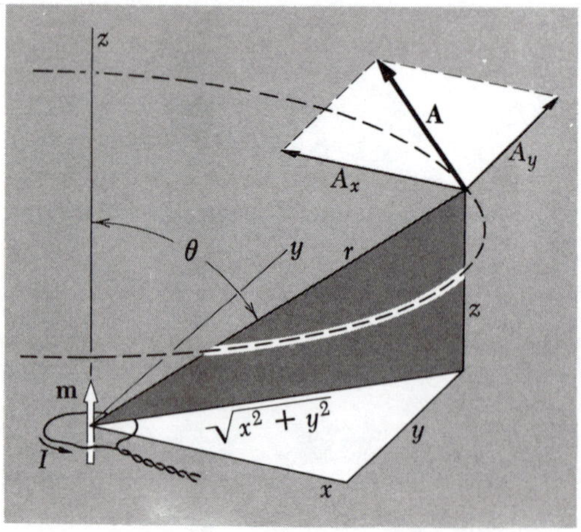

Let's evaluate **B** for a point in the xz plane, by finding the components of curl **A** and then (not before!) setting $y = 0$.

$$B_x = (\nabla \times \mathbf{A})_x = \frac{\partial A_z}{\partial y} - \frac{\partial A_y}{\partial z} = -\frac{\partial}{\partial z}\frac{mx}{(x^2 + y^2 + z^2)^{3/2}} = \frac{3mxz}{r^5}$$

$$B_y = (\nabla \times \mathbf{A})_y = \frac{\partial A_x}{\partial z} - \frac{\partial A_z}{\partial x} = \frac{\partial}{\partial z}\frac{-my}{(x^2 + y^2 + z^2)^{3/2}} = \frac{3myz}{r^5}$$

$$B_z = (\nabla \times \mathbf{A})_z = \frac{\partial A_y}{\partial x} - \frac{\partial A_x}{\partial y}$$

$$= m\left[\frac{-2x^2 + y^2 + z^2}{(x^2 + y^2 + z^2)^{5/2}} + \frac{x^2 - 2y^2 + z^2}{(x^2 + y^2 + z^2)^{5/2}}\right] = \frac{m(3z^2 - r^2)}{r^5} \tag{13}$$

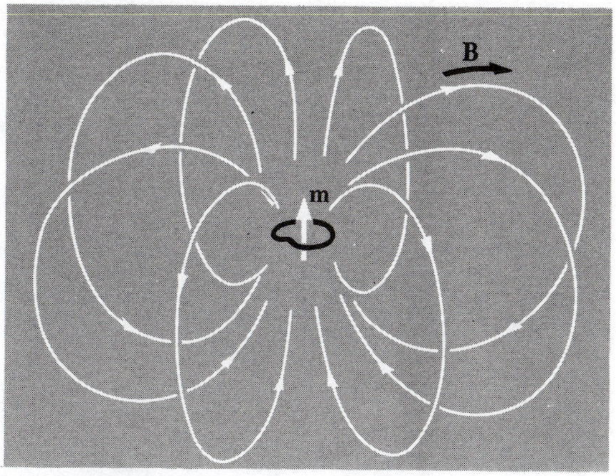

FIGURE 11.7
Some magnetic field lines in the field of a magnetic dipole, that is, a small loop of current.

In the xz plane, $y = 0$, $\sin\theta = x/r$, and $\cos\theta = z/r$. The field components at any point in that plane are thus given by:

$$B_x = \frac{3m\sin\theta\cos\theta}{r^3}$$
$$B_y = 0 \qquad (14)$$
$$B_z = \frac{m(3\cos^2\theta - 1)}{r^3}$$

Now turn back to Section 10.3, where in Eq. 10.14 we expressed the components in the xz plane of the field **E** of an electric dipole **p**, which was situated exactly like our magnetic dipole **m**. The expressions are identical. We have thus found that the magnetic field of a small current loop has at remote points the same form as the electric field of two separated charges. We already know what that field, the electric dipole field, looks like. Figure 11.7 is an attempt to suggest the three-dimensional form of the magnetic field **B** arising from our current loop with dipole moment **m**. As in the case of the electric dipole, the field is described somewhat more simply in spherical polar coordinates:

$$B_r = \frac{2m}{r^3}\cos\theta \qquad B_\theta = \frac{m}{r^3}\sin\theta \qquad B_\varphi = 0 \qquad (15)$$

The magnetic field *close* to a current loop is entirely different from the electric field close to a pair of separated positive and negative charges, as the comparison in Fig. 11.8 shows. Notice that between the charges the electric field points down, while inside the current ring the magnetic field points up, although the far fields are alike. This reflects the fact that our magnetic field satisfies $\nabla \cdot \mathbf{B} = 0$ everywhere, *even inside the source.* The magnetic field lines don't end. By

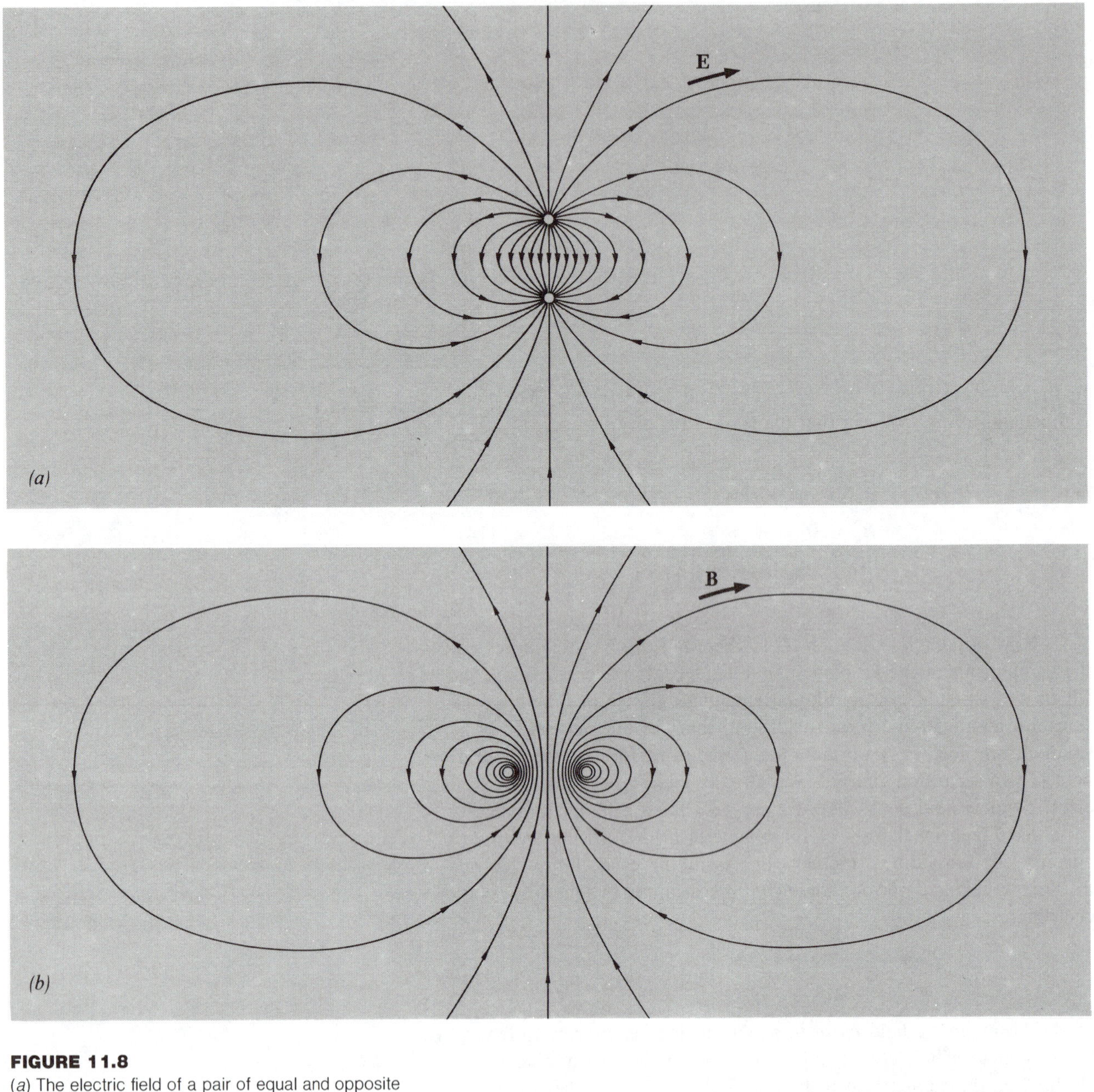

FIGURE 11.8
(*a*) The electric field of a pair of equal and opposite charges. Far away it becomes the field of an electric dipole. (*b*) The magnetic field of a current ring. Far away it becomes the field of a magnetic dipole.

near and *far* we mean, of course, relative to the size of the current loop or the separation of the charges. If we imagine the current ring shrinking in size, the current meanwhile increasing so that the dipole moment $m = Ia/c$ remains constant, we approach the infinitesimal magnetic dipole, the counterpart of the infinitesimal electric dipole described in Chapter 10.

THE FORCE ON A DIPOLE IN AN EXTERNAL FIELD

11.4 Consider a small circular current loop of radius r, placed in the magnetic field of some other current system, such as a solenoid. In Fig. 11.9, a field $\mathbf{B}$ is drawn that is generally in the z direction. It is not a uniform field. Instead, it gets weaker as we proceed in the z direction; that is evident from the fanning out of the field lines. Let us assume, for simplicity, that the field is symmetric about the z axis. Then it resembles the field near the upper end of the solenoid in Fig. 11.1. The field represented in Fig. 11.9 does *not* include the magnetic field of the current ring itself. We want to find the force on the current ring caused by the other field, which we shall call, for want of a better name, the *external field*. The net force on the current ring due to its *own* field is certainly zero, so we are free to ignore its own field in this discussion.

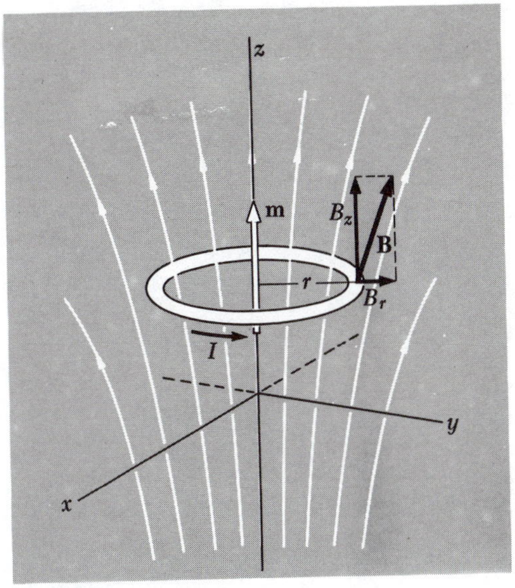

FIGURE 11.9
A current ring in an inhomogeneous magnetic field. (The field of the ring itself is not shown.) Because of the radial component of field B_r, there is a force on the ring as a whole.

If you study the situation in Fig. 11.9, you will soon conclude that there is a net force on the current ring. It arises because the external field $\mathbf{B}$ has an *outward* component B_r everywhere around the ring. Therefore if the current flows in the direction indicated, each element of the loop, dl, must be experiencing a downward force of magnitude $IB_r\,dl/c$. If B_r has the same magnitude at all points on the ring, as it must in the symmetrically spreading field assumed, the total downward force will have the magnitude

$$F = \frac{2\pi r I B_r}{c} \tag{16}$$

Now B_r can be directly related to the gradient of B_z. Since div $\mathbf{B} = 0$ at all points, the net flux of magnetic field out of any volume is zero. Consider the little cylinder of radius r and height Δz (Fig. 11.10). The outward flux from the side is $2\pi r(\Delta z)B_r$ and the net outward flux from the end surfaces is

$$\pi r^2[-B_z(z) + B_z(z + \Delta z)]$$

which to the first order in the small distance Δz is $\pi r^2(\partial B_z/\partial z)\,\Delta z$. Setting the total flux equal to zero: $0 = \pi r^2(\partial B_z/\partial z)\,\Delta z + 2\pi r B_r\,\Delta z$,

$$B_r = -\frac{r}{2}\frac{\partial B_z}{\partial z} \tag{17}$$

FIGURE 11.10
Gauss' theorem can be used to relate B_r and $\partial B_z/\partial z$, leading to Eq. 17.

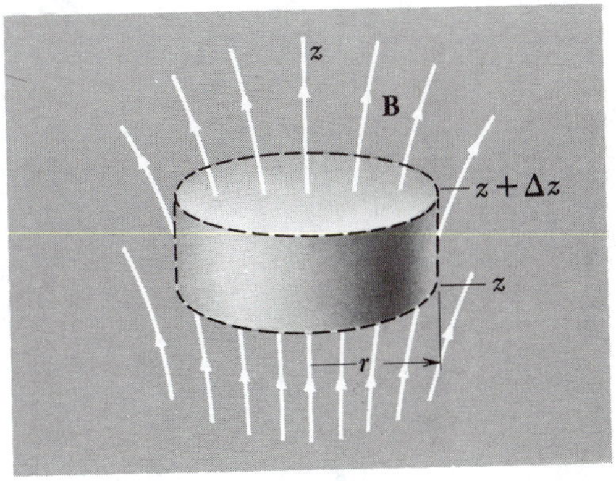

As a check on the sign, notice that, according to Eq. 17, B_r is positive when B_z is decreasing upward; a glance at the figure shows that to be correct.

The force on the dipole can now be expressed in terms of the gradient of the component B_z of the external field:

$$F = \frac{2\pi r I}{c}\frac{r}{2}\frac{\partial B_z}{\partial z} = \frac{\pi r^2 I}{c}\frac{\partial B_z}{\partial z} \tag{18}$$

In the factor $\pi r^2 I/c$ we recognize the magnitude m of the magnetic dipole moment of our current ring. So the force on the ring can be expressed very simply in terms of the dipole moment:

$$F = m \frac{\partial B_z}{\partial z} \tag{19}$$

We haven't proved it, but you will not be surprised to hear that for small loops of any other shape the force depends only on the *current* $\times$ *area* product, that is, on the dipole moment. The shape doesn't matter. Of course, we are discussing only loops small enough so that only the first-order variation of the external field, over the span of the loop, is significant.

Our ring in Fig. 11.9 has a magnetic dipole moment **m** pointing upward, and the force on it is downward. Obviously, if we could reverse the current in the ring, thereby reversing **m,** the force would reverse its direction. The situation can be summarized this way:

Dipole moment *parallel* to external field: Force acts in direction of *increasing* field strength.

Dipole moment *antiparallel* to external field: Force acts in direction of *decreasing* field strength.

Uniform external field: *Zero* force.

Quite obviously, this is not the most general situation. The moment **m** could be pointing at some odd angle with respect to the field **B**, and the different components of **B** could be varying, spatially, in different ways. It is not hard to develop a formula for the force **F** that is experienced in the general case. It would be exactly like the general formula we gave, as Eq. 10.23, for the force on an electric dipole in a nonuniform electric field. That is, the x component of force on any magnetic dipole **m** is given by

$$F_x = \mathbf{m} \cdot \text{grad } B_x \tag{20}$$

with corresponding formulas for F_y and F_z.

In Eqs. 19 and 20 the force is in dynes, with the magnetic field gradient in gauss/cm and the magnetic dipole moment m given by Eq. 9, $m = Ia/c$, where I is in esu/sec, a in cm^2, and c in cm/sec. There are several equivalent ways to express the units of m. We shall adopt ergs/gauss. As you can see from Eq. 19,

$$m = \frac{\text{dynes}}{\text{gauss/cm}} = \frac{\text{dyne-cm}}{\text{gauss}} = \frac{\text{ergs}}{\text{gauss}}$$

Now we can begin to see what must be happening in the experiments described at the beginning of this chapter. A substance located at the position of the sample in Fig. 11.2 would be attracted *into* the solenoid if it contained magnetic dipoles *parallel* to the field **B** of the coil. It would be pushed *out* of the solenoid if it contained dipoles pointing in the opposite direction, *antiparallel* to the field. The force would depend on the gradient of the axial field strength, and would be zero at the midpoint of the solenoid. Also, if the total strength of dipole moments in the sample were proportional to the field strength **B**, then in a given position the force would be proportional to B times $\partial B/\partial z$, and hence to the square of the solenoid current. That is the observed behavior in the case of the diamagnetic and the paramagnetic substances. It looks as if the ferromagnetic samples must have possessed a magnetic moment nearly independent of field strength, but we must set them aside for a special discussion anyway.

How does the application of a magnetic field to a substance evoke in the substance magnetic dipole moments with total strength proportional to the applied field? And why should they be parallel to the field in some substances, and oppositely directed in others? If we can answer these questions, we shall be on the way to understanding the physics of diamagnetism and paramagnetism.

ELECTRIC CURRENTS IN ATOMS

11.5 We know that an atom consists of a positive nucleus surrounded by negative electrons. To describe it fully we would need the concepts of quantum physics. Fortunately, a simple and easily visualized model of an atom can explain diamagnetism very well. It is a planetary model with the electrons in orbits around the nucleus, like the model in Bohr's first quantum theory of the hydrogen atom.

We begin with one electron moving at constant speed on a circular path. Since we are not attempting here to explain atomic structure, we shall not inquire into the reasons why the electron has this particular orbit. We ask only, if it does move in such an orbit, what magnetic effects are to be expected? In Fig. 11.11 we see the electron, visualized as a particle carrying a concentrated electric charge $-e$, moving with speed v on a circular path of radius r. In the middle is a positive nuclear charge, making the system electrically neutral, but the nucleus, because of its relatively great mass, moves so slowly that its magnetic effects can be neglected.

At any instant, the electron and the positive charge would appear as an electric dipole, but on the time average the electric dipole moment is zero, producing no steady electric field at a distance. We discussed this point in Section 10.5. The *magnetic* field of the system, far away, is *not* zero on the time average. Instead, it is just the field of a current ring. For, as concerns the time average, it can't make any difference whether we have all the negative charge gathered into one lump, going around the track, or distributed in bits, as in Fig. 11.11*b*, to make a uniform endless procession. The current is the amount of

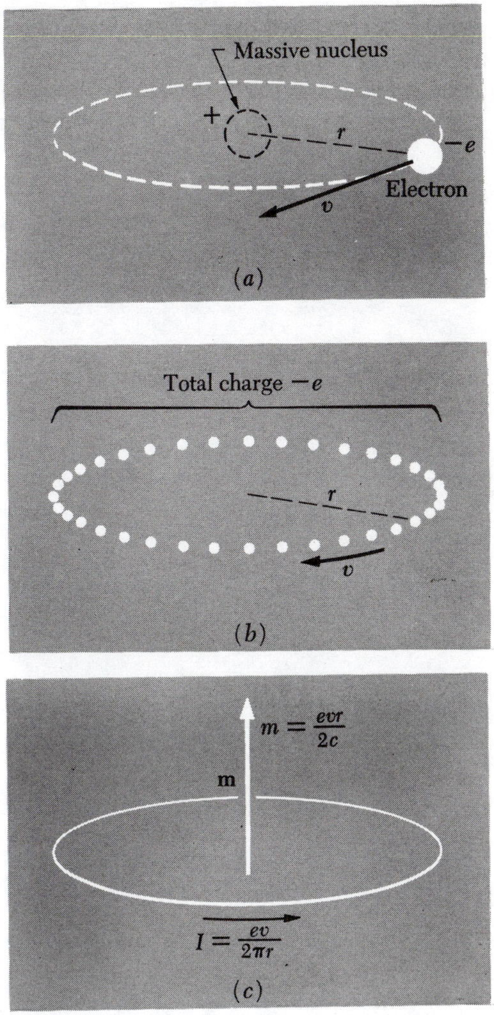

FIGURE 11.11
(*a*) A model of an atom in which one electron moves at speed v on a circular orbit. (*b*) Equivalent procession of charge. The average electric current is the same as if the charge $-e$ were divided into small bits, forming a rotating ring of charge. (*c*) The magnetic moment is 1/c times the product of current and area.

charge that passes a given point on the ring, per second. Since the electron makes $v/2\pi r$ revolutions per sec, the current, in esu/sec if e is in esu, is

$$I = \frac{ev}{2\pi r} \tag{21}$$

The orbiting electron is equivalent to a ring current of this magnitude with the direction of positive flow opposite to v, as shown in Fig. 11.11*c*. Its far field is therefore that of a magnetic dipole, of strength

$$m = \frac{\pi r^2 I}{c} = \frac{evr}{2c} \tag{22}$$

Let us note in passing a simple relation between the magnetic moment **m** associated with the electron orbit, and the orbital angular momentum **L**. The angular momentum is a vector of magnitude $L = m_e vr$, where m_e denotes the mass of the electron,† and it points downward if the electron is revolving in the sense shown in Fig. 11.11*a*. Notice that the product vr occurs in both m and L. With due regard to direction, we can write:

$$\mathbf{m} = \frac{-e}{2m_e c}\mathbf{L} \tag{23}$$

This relation involves nothing but fundamental constants, which should make you suspect that it holds quite generally. Indeed that is the case, although we shall not prove it here. It holds for elliptical orbits, and it holds even for the rosettelike orbits that occur in a central field that is not inverse-square. Remember the important property of any orbit in a central field: Angular momentum is a constant of the motion. It follows then, from the general relation expressed by Eq. 23 (derived by us only for a special case), that wherever angular momentum is conserved, the magnetic moment also remains constant in magnitude and direction. The factor

$$\frac{-e}{2m_e c} \quad \text{or} \quad \frac{\text{magnetic moment}}{\text{angular momentum}}$$

is called the *orbital magnetomechanical ratio* for the electron.‡ The intimate connection between magnetic moment and angular momentum is central to any account of atomic magnetism.

†We shall be dealing with speeds v much less than c, so m_e stands for the rest mass, 9.0×10^{-28} gm. Our choice of the symbol **m** for magnetic moment makes it necessary, in this chapter, to use a different symbol for the electron mass. For angular momentum we choose the symbol **L**, because **L** is traditionally used in atomic physics for orbital angular momentum, which is what we here consider.

‡Many people use the term *gyromagnetic ratio* for this quantity. Some call it the *magnetogyric ratio*. Whatever the name, it is understood that the magnetic moment is the numerator.

Why don't we notice the magnetic fields of all the electrons orbiting in all the atoms of every substance? The answer must be that there is a mutual cancellation. In an ordinary lump of matter there must be as many electrons going one way as the other. This is to be expected, for there is nothing to make one sense of rotation intrinsically easier than another, or otherwise to distinguish any unique axial direction. There would have to be something in the structure of the material to single out not merely an axis, but a *sense of rotation around that axis!*

We may picture a piece of matter, in the absence of any external magnetic field, as containing revolving electrons with their various orbital angular momentum vectors and associated orbital magnetic moments distributed evenly over all directions in space. Consider those orbits which happen to have their planes approximately parallel to the xy plane, of which there will be about equal numbers with **m** up and **m** down. Let's find out what happens to one of these orbits when we switch on an external magnetic field in the z direction.

We'll analyze first an electromechanical system that doesn't look much like an atom. In Fig. 11.12 there is an object of mass M and electric charge q, tethered to a fixed point by a cord of fixed length r. This cord provides the centripetal force that holds the object in its circular orbit. The magnitude of that force F_0 is given, as we know, by

$$F_0 = \frac{Mv_0^2}{r} \tag{24}$$

In the initial state, Fig. 11.12a, there is no external magnetic field. Now, by means of some suitable large solenoid, we begin creating a field **B** in the negative z direction, uniform over the whole region at any given time. While this field is growing at the rate dB/dt, there will be an induced electric field **E** all around the path, as indicated in Fig. 11.12b. To find the magnitude of this field **E** we note that the rate of change of flux through the circular path is

$$\frac{d\Phi}{dt} = \pi r^2 \frac{dB}{dt} \tag{25}$$

This determines the line integral of the electric field, which is really all that matters (we only assume for symmetry and simplicity that it is the same all around the path).

$$\int \mathbf{E} \cdot d\mathbf{l} = \frac{\pi r^2}{c}\frac{dB}{dt} = 2\pi r E \tag{26}$$

Thus we find that

$$E = \frac{r}{2c}\frac{dB}{dt} \tag{27}$$

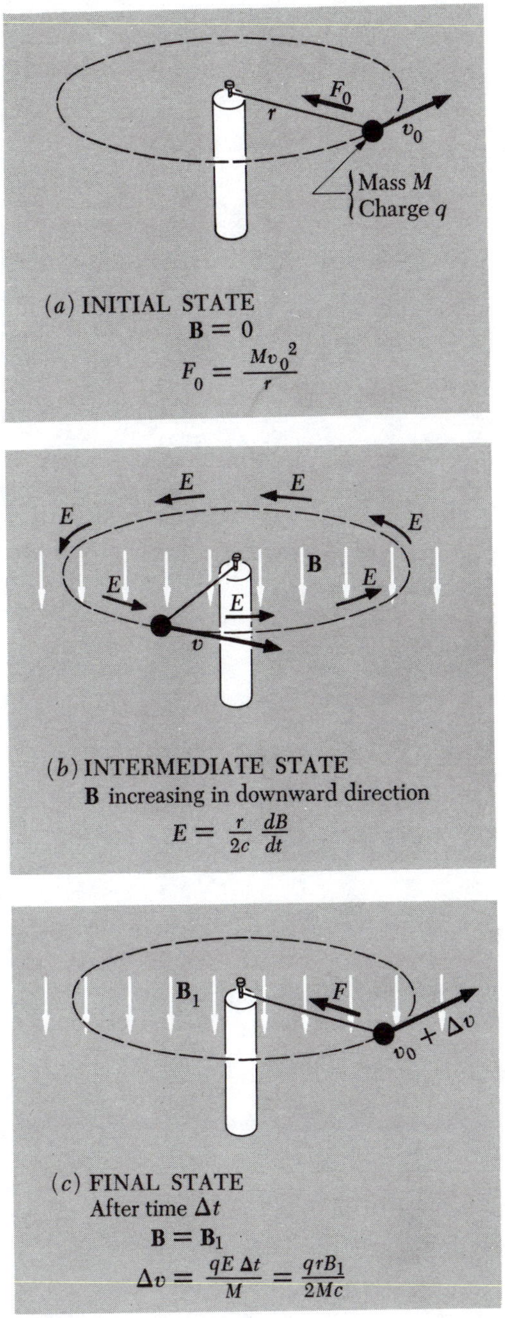

F_0
r
v_0
Mass M
Charge q
(a) INITIAL STATE
$\mathbf{B} = 0$
$F_0 = \frac{Mv_0^2}{r}$
E
E
E
E
E
E
E
B
v
(b) INTERMEDIATE STATE
B increasing in downward direction
$E = \frac{r}{2c}\frac{dB}{dt}$
$\mathbf{B}_1$
F
$v_0 + \Delta v$
(c) FINAL STATE
After time Δt
$\mathbf{B} = \mathbf{B}_1$
$\Delta v = \frac{qE\,\Delta t}{M} = \frac{qrB_1}{2Mc}$

FIGURE 11.12
The growth of the magnetic field **B** induces an electric field **E** that accelerates the revolving charged body.

We have ignored signs so far, but if you apply to Fig. 11.12 your favorite rule for finding the direction of an induced electromotive force, you will see that **E** must be in a direction to accelerate the body, if q is a positive charge. The acceleration along the path, dv/dt, is determined by the force qE:

$$M\frac{dv}{dt} = qE = \frac{qr}{2c}\frac{dB}{dt} \tag{28}$$

so that we have a relation between the change in v and the change in B:

$$dv = \frac{qr}{2Mc}\,dB \tag{29}$$

The radius r being fixed by the length of the cord, the factor $qr/2Mc$ is a constant. Let Δv denote the net change in v in the whole process of bringing the field up to the final value B_1. Then

$$\Delta v = \int_{v_0}^{v_0+\Delta v} dv = \frac{qr}{2Mc}\int_0^{B_1} dB = \frac{qrB_1}{2Mc} \tag{30}$$

Notice that the time has dropped out—the final velocity is the same whether the change is made slowly or quickly.

The increased speed of the charge in the final state means an increase in the upward-directed magnetic moment **m**. A *negatively* charged body would have been *de*celerated under similar circumstances, which would have *decreased* its *downward* moment. In either case, then, the application of the field $\mathbf{B}_1$ has brought about a change in magnetic moment opposite to the field. The magnitude of the change in magnetic moment Δm is

$$\Delta m = \frac{qr}{2c}\Delta v = \frac{q^2r^2}{4Mc^2}B_1 \tag{31}$$

Likewise for charges, either positive or negative, revolving in the other direction, the induced change in magnetic moment is opposite the change in applied magnetic field. Figure 11.13 shows this for a positive charge. It appears that the following relation holds for either sign of charge and either direction of revolution:

$$\Delta\mathbf{m} = -\frac{q^2r^2}{4Mc^2}\mathbf{B}_1 \tag{32}$$

In this example we forced r to be constant by using a cord of fixed length. Let us see how the tension in the cord has changed. We shall assume that B_1 is small enough so that $\Delta v \ll v_0$. In the final state we require a centripetal force of magnitude

$$F_1 = \frac{M(v_0+\Delta v)^2}{r} \approx \frac{Mv_0^2}{r} + \frac{2Mv_0\,\Delta v}{r} \tag{33}$$

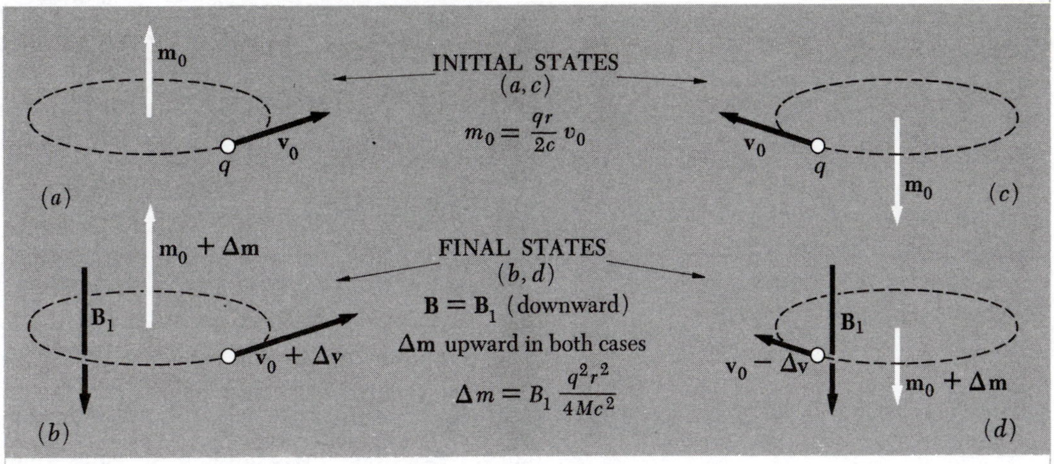

FIGURE 11.13
The change in the magnetic moment vector is opposite to the direction of **B**, for both directions of motion.

neglecting the term proportional to $(\Delta v)^2$. But now the magnetic field itself provides an inward force on the moving charge, given by $q(v_0 + \Delta v)B_1/c$. Using Eq. 30 to express B_1 in terms of Δv, we find that this extra inward force has the magnitude $[(v_0 + \Delta v)/c](2Mc\ \Delta v/r)$ which, to first order in $\Delta v/v_0$, is $2Mv_0\ \Delta v/r$. That is just what is needed, according to Eq. 33, to avoid any extra demand on our cord! Hence the tension in the cord *remains unchanged at the value* F_0.

This points to a surprising conclusion: Our result, Eq. 32, must be valid for *any* kind of tethering force, no matter how it varies with radius. Our cord could be replaced by an elastic spring without affecting the outcome—the radius would still be unchanged in the final state. Or to go at once to a system we are interested in, it could be replaced by the Coulomb attraction of a nucleus for an electron. Or it could be the effective force that acts on one electron in an atom containing many electrons, which has a still different dependence on radius.

Let us apply this to an electron in an atom, substituting the electron mass m_e for M, and e^2 for q^2. Now Δm is the magnetic moment induced by the application of a field B_1 to the atom. In other words $\Delta m/B_1$ is a magnetic polarizability, defined in the same way as the electrical polarizability α we introduced in Section 10.5. Remember that α had the dimensions of volume and turned out to be, in order of magnitude, 10^{-24} cm^3, roughly the volume of an atom. By Eq. 32 the magnetic polarizability due to one electron in an orbit of radius r is

$$\frac{\Delta m}{B_1} = -\frac{e^2 r^2}{4m_e c^2} \tag{34}$$

This too has the dimensions of volume, for $e^2/m_e c^2$ is a length, namely, the *classical electron radius* r_0, a constant with the value 2.8×10^{-13}

cm. For the orbit radius r let us substitute the Bohr radius 0.53×10^{-8} cm. Then Eq. 34 gives $\Delta m/B_1 = 2 \times 10^{-30}$ cm^3. Notice that this is five or six orders of magnitude smaller than typical *electric* polarizabilities, as sampled in Table 10.2. It is smaller by the ratio, roughly, of the classical electron radius r_0 to an atomic radius.

Let us see if this will account for the force on our diamagnetic samples listed in Table 11.1. The total number of electrons is about the same in one gram of almost anything. It is about one electron for every two nucleons, or 3×10^{23} electrons per gm of matter. (Recall that the atomic weight is about twice the atomic number for most of the elements.) Of course, r^2 must now be replaced by a mean square orbit radius $\langle r^2 \rangle$, where the average is taken over all the electrons in the atom, some of which have larger orbits than others. Actually $\langle r^2 \rangle$ varies remarkably little from atom to atom through the whole periodic table, and a_0^2, the square of the Bohr radius which we have just used, remains a surprisingly good estimate. Adopting that, we would predict that a field of 18 kilogauss would induce in 1 gm of substance a magnetic moment of magnitude

$$\begin{aligned} \Delta m &= (3 \times 10^{23})(2 \times 10^{-30})(1.8 \times 10^4) \\ &= 8.4 \times 10^{-3} \text{ cm}^3\text{-gauss} \end{aligned} \tag{35}$$

which in a gradient of 1700 gauss/cm would give rise to a force of magnitude

$$F = \Delta m \frac{\partial B_z}{\partial z} = 8.4 \times 10^{-3} \times 1700 = 14 \text{ dynes} \tag{36}$$

This agrees quite well, indeed, better than we had any right to expect, with the values for the several purely diamagnetic substances listed in Table 11.1.

We can see now why diamagnetism is a universal phenomenon, and a rather inconspicuous one. It is about the same in molecules as in atoms. The fact that a molecule can be a much larger structure than an atom—it may be built of hundreds or thousands of atoms—does not generally increase the effective mean-square orbit radius. The reason is that in a molecule any given electron is pretty well localized on an atom. There are some interesting exceptions and we included one in Table 11.1—graphite. The anomalous diamagnetism of graphite is due to an unusual structure which permits some electrons to circulate rather freely within a planar group of atoms in the crystal lattice. For these electrons $\langle r^2 \rangle$ is extraordinarily large.

ELECTRON SPIN AND MAGNETIC MOMENT

11.6 The electron possesses angular momentum that has nothing to do with its orbital motion. It behaves in many ways as if it were continually rotating around an axis of its own. This property is called *spin.* When the magnitude of the spin angular momentum is measured, the same result is always obtained: $h/4\pi$, where h is Planck's constant, 6.624×10^{-27} gm-cm^2/sec. Electron spin is a quantum phenomenon. Its significance for us now lies in the fact that there is associated with this intrinsic, or "built-in," angular momentum a *magnetic moment,* likewise of invariable magnitude. This magnetic moment points in the direction you would expect if you visualize the electron as a ball of negative charge spinning around its axis. That is, the magnetic moment vector points antiparallel to the spin angular momentum vector, as indicated in Fig. 11.14. The magnetic moment, however, is twice as large, relative to the angular momentum, as is the case in orbital motion.

There is no point in trying to devise a classical model of this object; its properties are essentially quantum mechanical. We need not even go so far as to say it *is* a current loop. What matters is only that it behaves like one in the following respects: (1) it produces a magnetic field which, at a distance, is that of a magnetic dipole; (2) in an external field **B** it experiences a torque equal to that which would act on a current loop of equivalent dipole moment; (3) within the space occupied by the electron, div **B** = 0 everywhere, as in the ordinary sources of magnetic field with which we are already familiar.

Since the magnitude of the spin magnetic moment is always the same, the only thing an external field can influence is its direction. A magnetic dipole in an external field experiences a torque. If you worked through Problem 6.22, you proved that the torque **N** on a current loop of any shape, with dipole moment **m**, in a field **B**, is given by

$$\mathbf{N} = \mathbf{m} \times \mathbf{B} \qquad (37)$$

For those who have not been through that demonstration, let's take time out to calculate the torque in a simple special case. In Fig. 11.15 we see a rectangular loop of wire carrying current I. The loop has a magnetic moment **m**, of magnitude $m = Iab/c$. The torque on the loop arises from the forces $\mathbf{F}_1$ and $\mathbf{F}_2$ that act on the horizontal wires. Each of these forces has the magnitude $F = IbB/c$, and its moment arm is the distance $(a/2)\sin\theta$. We see that the magnitude of the torque on the loop is

$$N = 2\frac{IbB}{c}\frac{a}{2}\sin\theta = \left(\frac{Iab}{c}\right)B\sin\theta = mB\sin\theta \qquad (38)$$

The torque acts in a direction to bring **m** parallel to **B**; it is represented by a vector **N** in the positive x direction, in the situation shown. All this is consistent with the general formula, Eq. 37. Notice that Eq. 37 corresponds exactly to the formula we derived in Chapter 10 for the torque on an electric dipole **p** in an external field **E**, namely, $\mathbf{N} = \mathbf{p} \times \mathbf{E}$. The orientation with **m** in the direction of **B**, like that of the

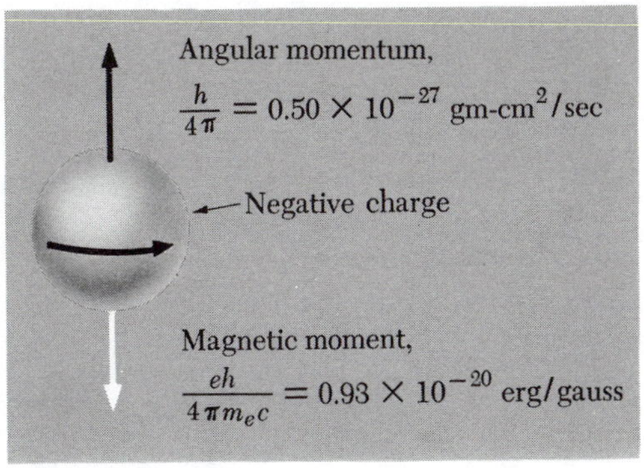
Angular momentum,
$\frac{h}{4\pi} = 0.50 \times 10^{-27}$ gm-cm^2/sec
Negative charge
Magnetic moment,
$\frac{eh}{4\pi m_e c} = 0.93 \times 10^{-20}$ erg/gauss

FIGURE 11.14
The intrinsic angular momentum, or spin, and the associated magnetic moment, of the electron. Note that the ratio of magnetic moment to angular momentum is e/m_ec, not $e/2m_ec$ as it is for orbital motion (Eq. 23). This has no classical explanation.

FIGURE 11.15
Calculation of the torque on a current loop in a magnetic field **B**. The magnetic moment of the current loop is **m**.

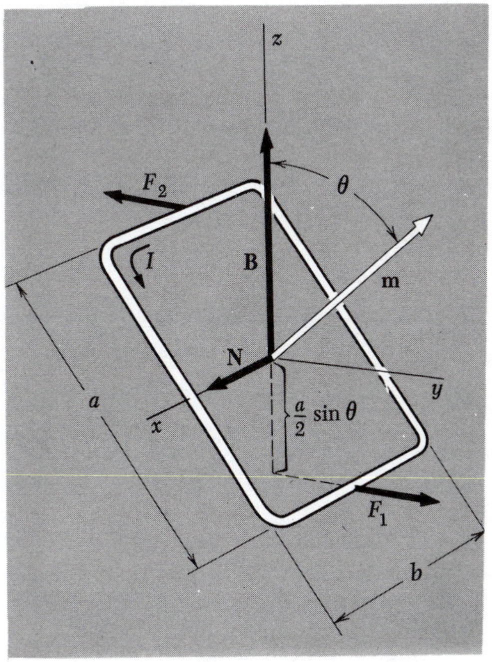
z
F_2
θ
I
B
m
N
a
x
y
$\frac{a}{2}\sin\theta$
F_1
b

electric dipole parallel to **E**, is the position of lowest energy. Similarly, the work required to rotate a dipole **m** from parallel to antiparallel is $2mB$. (See Eq. 19 of Chapter 10; we can simply take over this result for the magnetic case.)

If the electron spin moments in a substance are free to orient themselves, we expect them to prefer the orientation in the direction of any applied field **B**, the orientation of lowest energy. Suppose every electron in a gram of material takes up this orientation. We have already calculated that there are roughly 3×10^{23} electrons in a gram of anything. The spin magnetic moment of an electron, m_s, is given in Fig. 11.14 as 0.93×10^{-20} erg/gauss. The total magnetic moment of our lined-up spins will be $(3 \times 10^{23}) \times (0.9 \times 10^{-20})$ or 2700 ergs/gauss. The force on such a sample, in our coil where the field gradient is 1700 gauss/cm, would be 4.6×10^6 dynes, or a little over 10 pounds!

Obviously this is much greater than the force recorded for any of the paramagnetic samples. Our assumptions were wrong in two ways. First, the electron spin moments are not all free to orient themselves. Second, thermal agitation prevents perfect alignment of any spin moments that are free.

In most atoms and molecules the electrons are associated in pairs, with the spins in each pair constrained to point in opposite directions regardless of the applied magnetic field. As a result, the magnetic moments of such a pair of electrons exactly cancel one another. All that is left is the diamagnetism of the orbital motion which we have already explored. The vast majority of molecules are purely diamagnetic. A few molecules (really *very* few) contain an odd number of electrons. In such a molecule total cancellation of spin moments in pairs is clearly impossible. Nitric oxide, NO, with 15 electrons in the molecule is an example; it is paramagnetic. The oxygen molecule O_2 contains 16 electrons, but its electronic structure happens to favor noncancellation of two of the electron spins. In single atoms the inner electrons are generally paired, and if there is an outer unpaired electron, its spin is often paired off with that of a neighbor when the atom is part of a compound or crystal. Certain atoms, however, do contain unpaired electron spins which remain relatively free to orient in a field even when the atom is packed in with others. Important examples are the elements ranging from chromium to copper in the periodic table, a sequence that includes iron, cobalt, and nickel. Another group of elements with this property is the rare earth sequence around gadolinium. Compounds or alloys of these elements are generally paramagnetic, and in some cases ferromagnetic. The number of free electron spins involved in paramagnetism is typically one or two per atom. We can think of each paramagnetic atom as equipped with one freely swiveling magnetic moment **m**, which in a field **B** would be found pointing, like a tiny compass needle, in the direction of the field—if it were not for thermal disturbances.

Thermal agitation tends always to create a random distribution of spin axis directions. The degree of alignment that eventually prevails represents a compromise between the preference for the direction of lowest energy and the disorienting influence of thermal motion. We have met this problem before. In Section 10.12 we considered the alignment by an electric field **E** of the electric dipole moments of polar molecules. It turned out to depend on the ratio of two energies: pE, the energetic advantage of orientation of a dipole moment **p** parallel to **E** as compared with an average over completely random orientations, and kT, the mean thermal energy associated with any form of molecular motion at absolute temperature T. Only if pE were much larger than kT would nearly complete alignment of the dipole moments be attained. If pE is much smaller than kT, the equilibrium polarization is equivalent to perfect alignment of a small fraction, approximately pE/kT, of the dipoles. We can take this result over directly for paramagnetism. We need only replace pE by mB, the energy involved in the orientation of a magnetic dipole moment **m** in a magnetic field **B**. Providing mB/kT is small, it follows that the total magnetic moment resulting from application of the field **B** to N dipoles will be approximately $(mB/kT)Nm$, or $(Nm^2/kT)B$. The induced moment is proportional to B and inversely proportional to the temperature.

For one electron spin moment in our field of 18 kilogauss, mB is 1.6×10^{-16} erg. For room temperature, kT is 4×10^{-14} erg; in that case mB/kT is indeed small. But if we could lower the temperature to 1 K in the same field, mB/kT would be about unity. With further lowering of the temperature we could expect to approach complete alignment, with total moment approaching Nm. These conditions are quite frequently achieved in low-temperature experiments. Indeed, paramagnetism is both more impressive and more interesting at very low temperatures, in contrast to dielectric polarization. Molecular electric dipoles would be totally frozen in position, incapable of any reorientation. The electron spin moments are still remarkably free.

MAGNETIC SUSCEPTIBILITY

11.7 We have seen that both diamagnetic and paramagnetic substances develop a magnetic moment proportional to the applied field. At least, that is true under most conditions. At very low temperatures, in fairly strong fields, the induced paramagnetic moment can be observed to approach a limiting value as the field strength is increased, as we have noted. Setting this "saturation" effect aside, the relation between moment and applied field is linear, so that we can characterize the magnetic properties of a substance by the ratio of induced moment to applied field. The ratio is called the *magnetic susceptibility*. Depending on whether we choose the moment of 1 gm of material, of 1 cm^3 of material, or of 1 mole, we define the *specific* susceptibility, the *volume* susceptibility, or the *molar* susceptibility. Our discussion in Section 11.5 suggests that for diamagnetic substances the specific susceptibility, based on the induced moment per gram, should be most nearly the same from one substance to another. However, the volume susceptibility, based on the induced magnetic moment per cubic centimeter, is more relevant to our present concerns.

The magnetic moment per unit volume we shall call the *magnetic polarization*, or the *magnetization*, using for it the symbol **M**. Now magnetization **M** and magnetic field **B** have similar dimensions.† To verify that, recall that the field **B** of a magnetic dipole is given by $\frac{\text{magnetic dipole moment}}{(\text{distance})^3}$, while **M**, as we have just defined it, has the dimensions $\frac{\text{magnetic dipole moment}}{\text{volume}}$. If we now define the volume magnetic susceptibility, denoted by χ_m, through the relation

$$\mathbf{M} = \chi_m \mathbf{B} \qquad (\textit{Warning: see remarks below}) \qquad (39)$$

the susceptibility will be a dimensionless number, negative for diamagnetic substances, positive for paramagnetic. This is exactly analogous to the procedure, expressed in Eq. 10.34, by which we defined the electric susceptibility χ_e as the ratio of electric polarization P to electric field E. For the paramagnetic contribution, if any, to the susceptibility, let us denote it χ_{pm}, we shall have a formula analogous to Eq. 60 of the last chapter:

$$\chi_{pm} \approx \frac{Nm^2}{kT} \qquad (40)$$

Of course the full susceptibility χ_m includes the ever-present diamagnetic contribution, which is negative, and derivable from Eq. 34.

Unfortunately, Eq. 39 is *not* the customary definition of volume magnetic susceptibility. In the usual definition another field **H**, which we shall meet in Section 11.10, appears instead of **B**. Although illogical, the definition in terms of **H** has a certain practical justification, and the tradition is so well established that we shall eventually have to bow to it. But in this chapter we want to follow as long as we can a path that naturally and consistently parallels the description of the electric fields in matter. A significant parallel is this: The macroscopic field **B** inside matter will turn out to be the average of the microscopic

†While the dimensions of M and B are the same, it would be confusing to express them in the same *units*, because of a factor 4π that will turn up presently. If and when a name for the units of M is called for, we shall use ergs/gauss-cm^3.

B, just as the macroscopic **E** turned out to be the average of the microscopic **E**.

The difference in definition is of no practical consequence as long as χ_m is a number very small compared with one. The values of χ_m for purely diamagnetic substances, solid or liquid, lie typically between -0.5×10^{-6} and -1.0×10^{-6}. Even for oxygen under the conditions given in Table 11.1, the paramagnetic susceptibility is less than 10^{-3}. This means that the magnetic field caused by the dipole moments in the substance, at least as a large-scale average, is very much weaker than the applied field **B**. That gives us some confidence that in such systems we may assume the field that acts on the atomic dipole to orient them is the same as the field that would exist there in the absence of the sample. However, we shall be interested in other systems in which the field of the magnetic moments is *not* small. Therefore we must study, just as we did in the case of electric polarization, the magnetic fields that magnetized matter itself produces, both inside and outside the material.

THE MAGNETIC FIELD CAUSED BY MAGNETIZED MATTER

11.8 A block of material which contains, evenly distributed through its volume, a large number of atomic magnetic dipoles all pointing in the same direction, is said to be *uniformly magnetized.* The magnetization vector **M** is simply the product of the number of oriented dipoles per unit volume and the magnetic moment **m** of each dipole. We don't care how the alignment of these dipoles is maintained. There may be some field applied from another source, but we are not interested in that. We want to study only the field produced by the dipoles themselves.

Consider first a slab of material of thickness dz, sliced out perpendicular to the direction of magnetization, as shown in Fig. 11.16*a*. The slab can be divided into little tiles. One such tile, which has a top surface of area da, contains a total dipole moment amounting to $M\, da\, dz$, since M is the dipole moment per unit volume (Fig. 11.16*b*). The magnetic field this tile produces at all *distant* points—distant compared to the size of the tile—is just that of any dipole with the same magnetic moment. We could construct a dipole of that strength by bending a conducting ribbon of width dz into the shape of the tile, and sending around this loop a current $I = Mc\, dz$ (Fig. 11.16*c*). That will give the loop a dipole moment:

$$m = \frac{I}{c} \times \text{area} = \frac{Mc\, dz}{c}\, da = M\, da\, dz \qquad (41)$$

which is the same as that of the tile.

Let us substitute such a current loop for every tile in the slab,

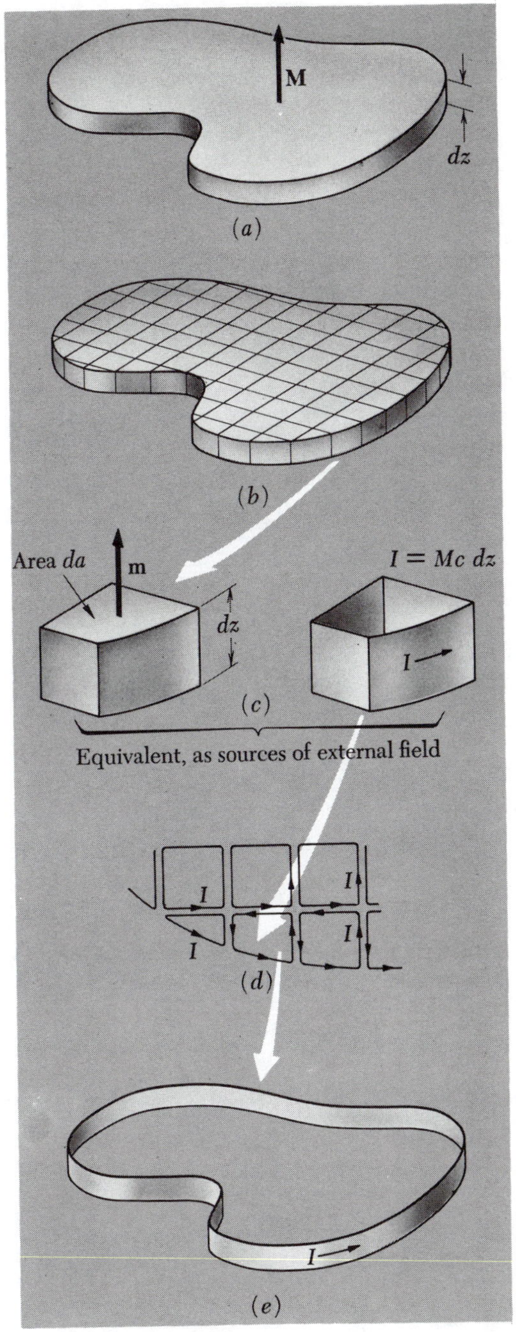

FIGURE 11.16
The thin slab, magnetized perpendicular to its broad surface, is equivalent to a ribbon of current so far as its external field is concerned.

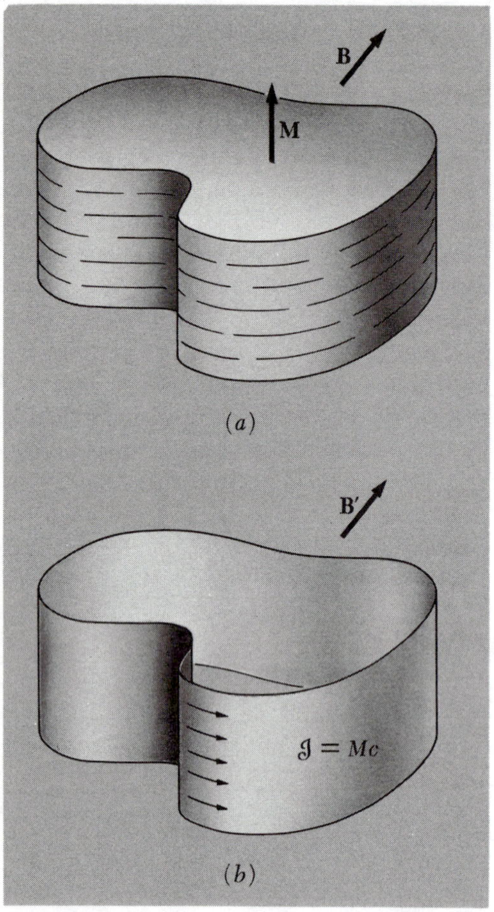

FIGURE 11.17
A uniformly magnetized block is equivalent to a band of surface current.

as indicated in Fig. 11.16*d*. The current is the same in all of these and therefore, at every interior boundary we find equal and opposite currents, equivalent to zero current. Our "egg-crate" of loops is therefore equivalent to a single ribbon running around the outside, carrying the current $Mc\,dz$ (Fig. 11.16*e*). Now these tiles can be made quite small, so long as we don't subdivide clear down to molecular size. They must be large enough so that their magnetization does not vary appreciably from one tile to the next. Within that limitation, we can state that the field at any *external* point, even close to the slab, is the same as that of the current ribbon.

It remains only to reconstruct a whole block from such laminations, or slabs, as in Fig. 11.17*a*. The entire block is then equivalent to the wide ribbon in Fib. 11.17*b* around which flows a current $Mc\,dz$, in esu/sec, in every strip dz, or, stated more simply, a surface current of density $\mathcal{J}$, in esu/sec-cm, given by

$$\mathcal{J} = Mc \tag{42}$$

The magnetic field **B** at any point outside the magnetized block in Fig. 11.17*a,* and even close to the block provided we don't approach within molecular distances, is the same as the field **B′** at the corresponding point in the neighborhood of the wide current ribbon in Fig. 11.17*b*.

But what about the field inside the magnetized block? Here we face a question like the one we met in Chapter 10. Inside matter the magnetic field is not at all uniform if we observe it on the atomic scale which we have been calling microscopic. It varies sharply in both magnitude and direction between points only a few angstroms apart. This *microscopic* field **B** is simply a magnetic field in vacuum, for from the microscopic viewpoint, as we emphasized in Chapter 10, matter is a collection of particles and electric charge in otherwise empty space. The only large-scale field that can be uniquely defined inside matter is the spatial average of the microscopic field.

Because of the absence of effects attributable to magnetic charge, we believe that the microscopic field itself satisfies div **B** = 0. If that is true, it follows quite directly that the spatial average of the internal microscopic field in our block is equal to the field **B′** inside the equivalent hollow cylinder of current.

To demonstrate this, consider the long rod uniformly magnetized parallel to its length, shown in Fig. 11.18*a*. We have just shown that the external field will be the same as that of the long cylinder of current (practically equivalent to a single-layer solenoid) shown in Fig. 11.18*b*. S in Fig. 11.18*a* indicates a closed surface which includes a portion S_1 passing through the interior of the rod. Because div **B** = 0 for the internal microscopic field, as well as for the external field, div **B** is zero throughout the entire volume enclosed by S. It then follows from Gauss's theorem that the surface integral of **B** over S must be zero. The surface integral of **B′** over the closed surface S' is zero

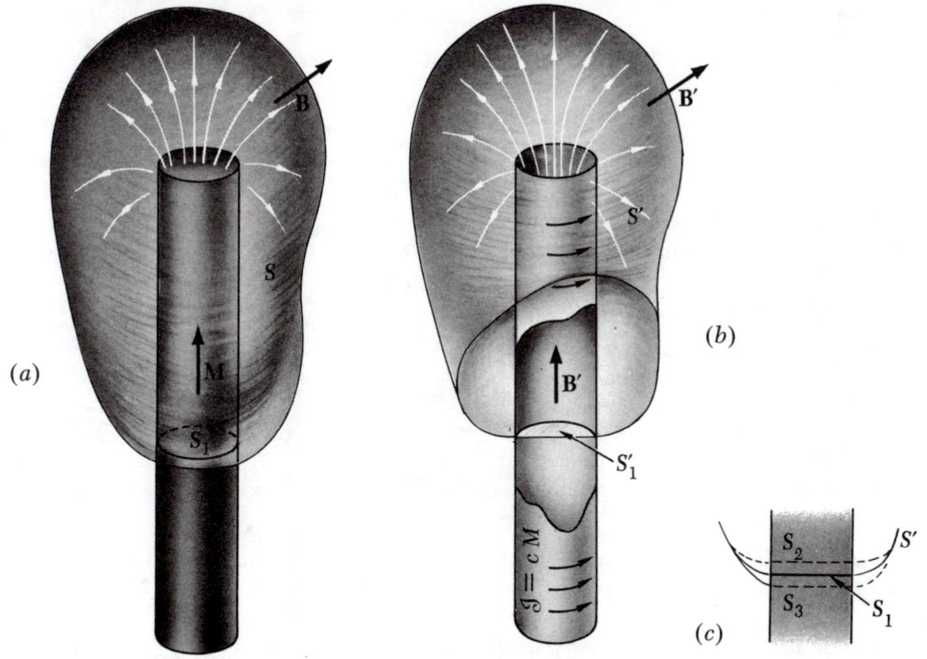

FIGURE 11.18
(*a*) A uniformly magnetized cylindrical rod. (*b*) The equivalent hollow cylinder, or sheath, of current. Its field is $\mathbf{B}'$. (*c*) We can sample the interior of the rod, and thus obtain a spatial average of the microscopic field, by closely spaced parallel surfaces, S_1, S_2, ...

also. Over the portions of S and S' *external* to the cylinders, B and B' are identical. Therefore the surface integral of $\mathbf{B}$ over the internal disk S_1 must be equal to the surface integral of $\mathbf{B}'$ over the internal disk S'_1. This must hold also for any one of a closely spaced set of parallel disks, such as S_2, S_3, etc., indicated in Fig. 11.18*c*, because the field outside the cylinder in this neighborhood is negligibly small, so that the outside parts don't change anything. Now taking the surface integral over a series of equally spaced planes like that is a perfectly good way to compute the volume average of the field $\mathbf{B}$ in that neighborhood, for it samples all volume elements impartially. It follows that the spatial average of the microscopic field $\mathbf{B}$ inside the magnetized rod is equal to the field $\mathbf{B}'$ inside the current sheath of Fig. 11.18*b*.

It is instructive to compare the arguments we have just developed with our analysis of the corresponding questions in Chapter 10. Figure 11.19 displays these developments side by side. You will see that they run logically parallel, but that at each stage there is a difference which reflects the essential asymmetry epitomized in the observation that *electric charges* are the source of *electric fields,* while *moving electric charges* are the source of *magnetic fields.* For example, in the arguments about the average of the microscopic field, the key to the problem in the electric case is the assumption that curl $\mathbf{E} = 0$ for the microscopic electric field. In the magnetic case, the key is the assumption that div $\mathbf{B} = 0$ for the microscopic magnetic field.

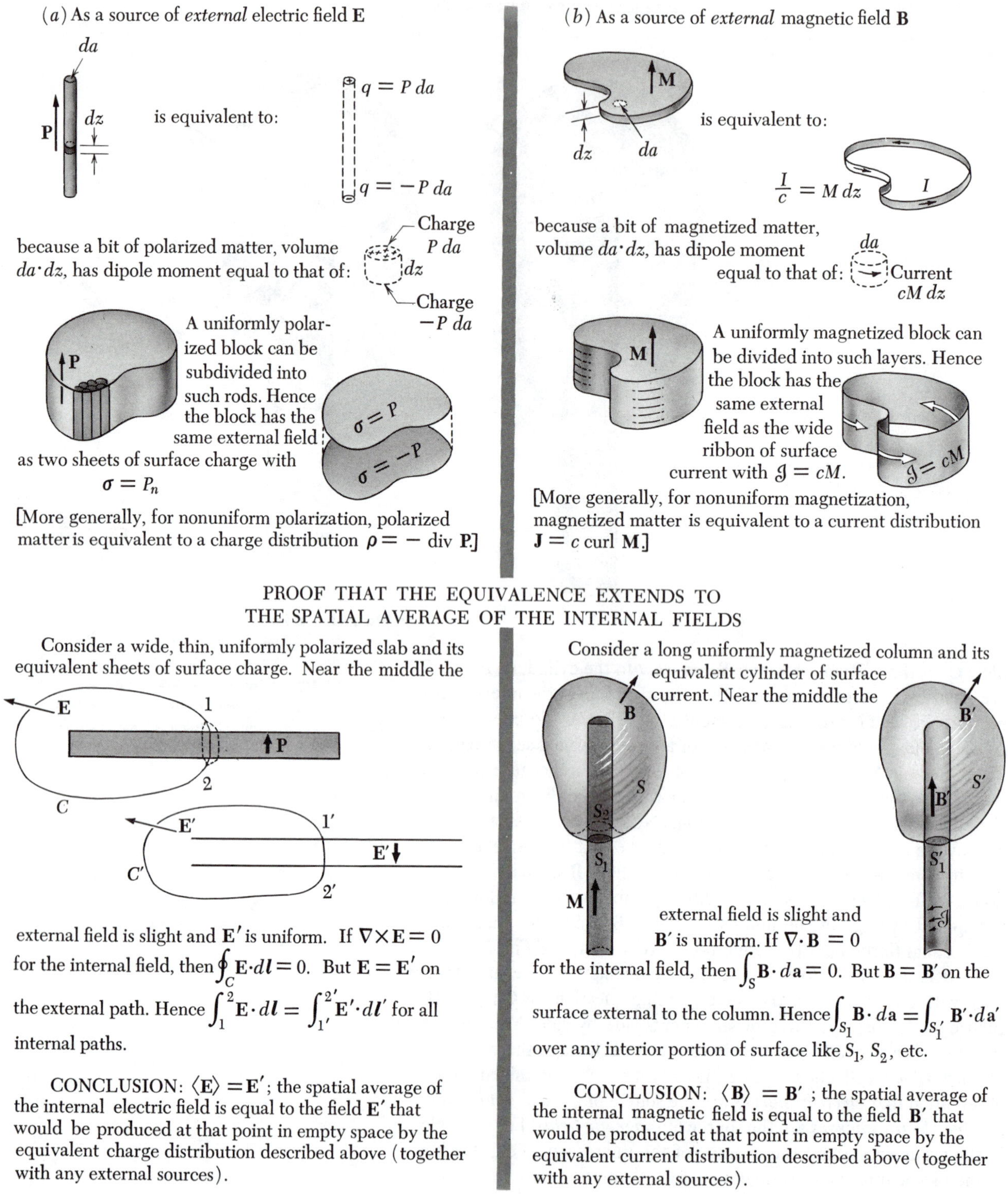
(a) As a source of external electric field E
da
P
dz
is equivalent to:
q = P da
q = −P da
because a bit of polarized matter, volume da·dz, has dipole moment equal to that of:
Charge P da
dz
Charge −P da
A uniformly polarized block can be subdivided into such rods. Hence the block has the same external field as two sheets of surface charge with σ = Pn
σ = P
σ = −P
[More generally, for nonuniform polarization, polarized matter is equivalent to a charge distribution ρ = − div P.]
(b) As a source of external magnetic field B
M
dz
da
is equivalent to:
I/c = M dz
I
because a bit of magnetized matter, volume da·dz, has dipole moment equal to that of:
da
Current cM dz
A uniformly magnetized block can be divided into such layers. Hence the block has the same external field as the wide ribbon of surface current with 𝒥 = cM.
𝒥 = cM
[More generally, for nonuniform magnetization, magnetized matter is equivalent to a current distribution J = c curl M.]
PROOF THAT THE EQUIVALENCE EXTENDS TO THE SPATIAL AVERAGE OF THE INTERNAL FIELDS
Consider a wide, thin, uniformly polarized slab and its equivalent sheets of surface charge. Near the middle the
E
1
P
2
C
E′
1′
E′
C′
2′
external field is slight and E′ is uniform. If ∇×E = 0 for the internal field, then ∮C E·dl = 0. But E = E′ on the external path. Hence ∫₁² E·dl = ∫₁′²′ E′·dl′ for all internal paths.
CONCLUSION: ⟨E⟩ = E′; the spatial average of the internal electric field is equal to the field E′ that would be produced at that point in empty space by the equivalent charge distribution described above (together with any external sources).
Consider a long uniformly magnetized column and its equivalent cylinder of surface current. Near the middle the
B
S
S2
S1
M
B′
S′
B′
S′1
𝒥
external field is slight and B′ is uniform. If ∇·B = 0 for the internal field, then ∫S B·da = 0. But B = B′ on the surface external to the column. Hence ∫S1 B·da = ∫S′1 B′·da′ over any interior portion of surface like S1, S2, etc.
CONCLUSION: ⟨B⟩ = B′; the spatial average of the internal magnetic field is equal to the field B′ that would be produced at that point in empty space by the equivalent current distribution described above (together with any external sources).

FIGURE 11.19

The electric (*a*) and magnetic (*b*) cases compared.

If the magnetization **M** within a volume of material is not uniform but instead varies with position as **M**(x, y, x), the equivalent current distribution is given simply by

$$\mathbf{J} = c \text{ curl } \mathbf{M} \tag{43}$$

Let's see how this comes about in one situation. Suppose there is a magnetization in the z direction, which gets stronger as we proceed in the y direction. This is represented in Fig. 11.20a, which shows a small region in the material subdivided into little blocks. The blocks are supposed to be so small that we may consider the magnetization uniform within a single block. Then we can replace each block by a current

FIGURE 11.20
Nonuniform magnetization is equivalent to a volume current density.

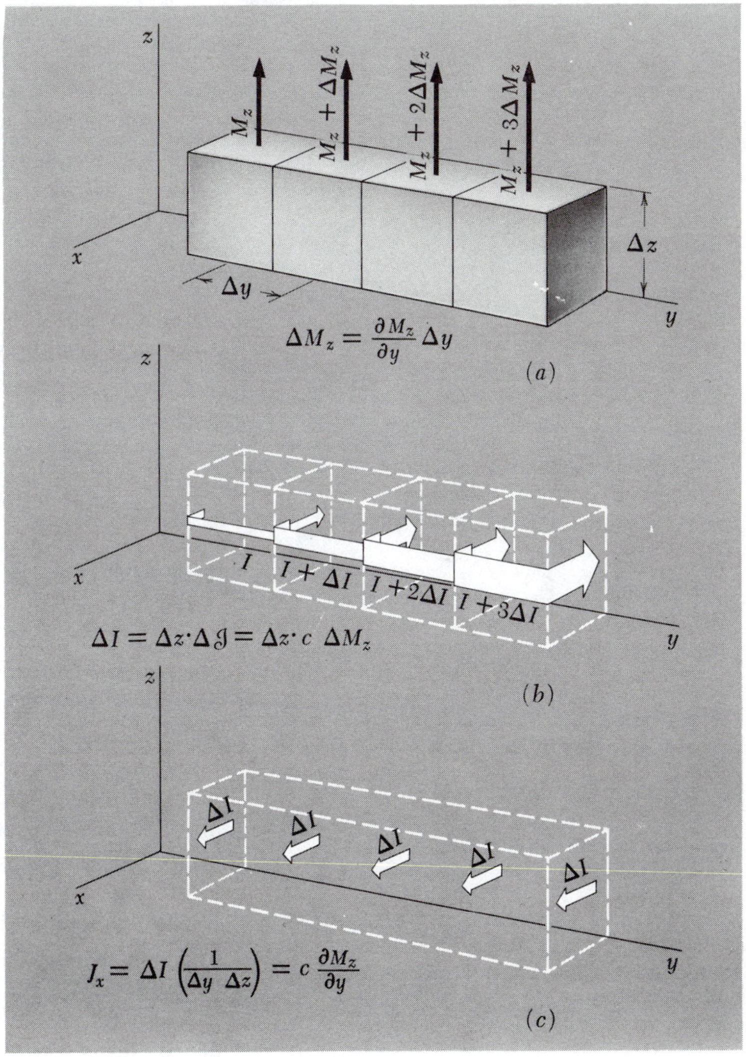

M_z
$M_z + \Delta M_z$
$M_z + 2\Delta M_z$
$M_z + 3\Delta M_z$
Δy
Δz
$\Delta M_z = \frac{\partial M_z}{\partial y} \Delta y$
(a)
I
$I + \Delta I$
$I + 2\Delta I$
$I + 3\Delta I$
$\Delta I = \Delta z \cdot \Delta \mathcal{J} = \Delta z \cdot c\ \Delta M_z$
(b)
ΔI
$J_x = \Delta I \left(\frac{1}{\Delta y\ \Delta z}\right) = c \frac{\partial M_z}{\partial y}$
(c)

ribbon, with surface current density $\mathcal{J} = cM_z$. The current I carried by such a ribbon, if the block is Δz in height, is $\mathcal{J}\,\Delta z$ or $cM_z\,\Delta z$. Now each ribbon has a bit more current density than the one to the left of it. The current in each loop is greater than the current in the loop to the left by

$$\Delta I = c\,\Delta z\,\Delta M_z = c\,\Delta z\,\frac{\partial M_z}{\partial y}\,\Delta y \tag{44}$$

At every interface in this row of blocks there is a net current in the x direction of magnitude ΔI (Fig. 11.20*c*). To get the current per unit area flowing in the x direction we have to multiply by the number of blocks per unit area, which is $1/(\Delta y\,\Delta z)$. Thus

$$J_x = \Delta I\left(\frac{1}{\Delta y\,\Delta z}\right) = c\,\frac{\partial M_z}{\partial y} \tag{45}$$

Another way of getting an x-directed current is to have a y component of magnetization that varies in the z direction. If you trace through that case, using a vertical column of blocks, you will find that the net x-directed current density is given by

$$J_x = -c\,\frac{\partial M_y}{\partial z} \tag{46}$$

In general then, by superposition of these two situations,

$$J_x = c\left(\frac{\partial M_z}{\partial y} - \frac{\partial M_y}{\partial z}\right) = c(\text{curl}\,\mathbf{M})_x \tag{47}$$

which is enough to establish Eq. 43.

THE FIELD OF A PERMANENT MAGNET

11.9 The uniformly polarized spheres and rods we talked about in Chapter 10 are seldom seen, even in the laboratory. Frozen-in electric polarization can occur in some substances, although it is usually disguised by some accumulation of free charge. To make Fig. 11.3*a,* which shows how the field of a polarized rod *would* look, it was necessary to use two charged disks. On the other hand, materials with permanent magnetic polarization, that is, permanent *magnetization,* are familiar and useful. Permanent magnets can be made from many alloys and compounds of ferromagnetic substances. What makes this possible is a question we'll leave for Section 11.11, where we dip briefly into the physics of ferromagnetisim. In this section, taking the existence of permanent magnets for granted, we want to study the magnetic field **B** of a uniformly magnetized cylindrical rod and compare it carefully with the electric field **E** of a uniformly polarized rod of the same shape.

(a)

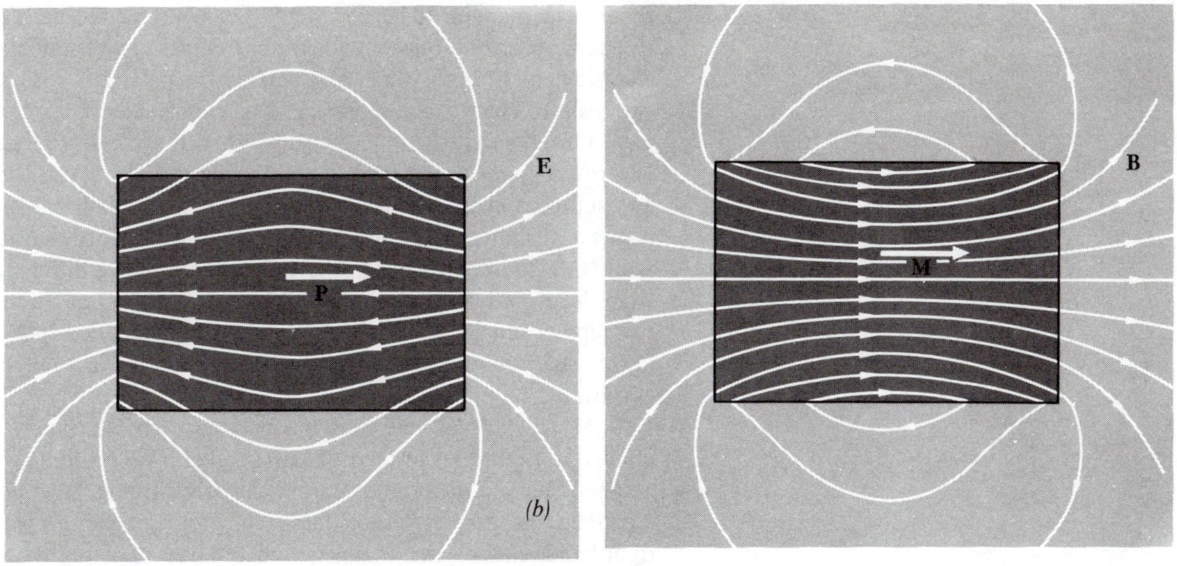

FIGURE 11.21

(*a*) The electric field **E** outside and inside a uniformly polarized cylinder. (*b*) The magnetic field **B** outside and inside a uniformly magnetized cylinder. In each case, the interior field shown is the macroscopic field, that is, the local average of the atomic or microscopic field.

Figure 11.21 shows each of these solid cylinders in cross section. The polarization, in each case, is parallel to the axis, and it is uniform. That is, the polarization **P** and the magnetization **M** have the same magnitude and direction everywhere within their respective cylinders. In the magnetic case this implies that every cubic millimeter of the permanent magnet has the same number of lined-up electron spins, pointing in the same direction. (A very good approximation to this can be achieved with modern permanent magnet materials.)

By the field inside the cylinder we mean, of course, the macroscopic field defined as the space average of the microscopic field. With this understanding, we show in Fig. 11.21 the field lines both inside and outside the rods. By the way, these rods are not supposed to be near one another; we only put the diagrams together for convenient comparison. Each rod is isolated in otherwise field-free space. (Which do you think would more seriously disturb the field of the other, if they *were* close together?)

Outside the rods the fields **E** and **B** *look alike.* In fact the field lines follow precisely the same course. That should not surprise you if you recall that the electric dipole and the magnetic dipole have similar far fields. Each little chunk of the magnet is a magnetic dipole, each little chuck of the polarized rod (sometimes called an *electret*) is an electric dipole, and the field outside is the superposition of all their far fields.

The field **B**, inside and out, is the same as that of a cylindrical sheath of current. In fact if we were to wind very evenly, on a cardboard cylinder, a single-layer solenoid of fine wire, we could hook a battery up to it and duplicate the exterior and interior field **B** of the permanent magnet. (The coil would get hot, and the battery would run down; electron spins provide the current free and frictionless!) The electric field **E**, both inside and outside the polarized rod, is that of two disks of charge, one at each end of the cylinder.

Observe that the *interior* fields **E** and **B** are essentially different in form: **B** points to the right, is continuous at the ends of the cylinder, and suffers a sharp change in direction at the cylindrical surface. **E** points to the left, passes through the cylindrical surface as if it weren't there, but is discontinuous at the end surfaces. These differences arise from the essential difference between the "inside" of the physical electric dipole, and the "inside" of the physical magnetic dipole, seen in Fig. 11.8. By *physical,* we mean the ones Nature has actually provided us with.

If the external field were our only concern, we could use either picture to describe the field of our magnet. We could say that the magnetic field of the permanent magnet arises from a layer of positive magnetic charge—a surface density of north magnetic poles on the right-hand end of the magnet, and a layer of negative magnetic charge, south poles, on the other end. We could adopt a scalar potential function φ_{mag}, such that $\mathbf{B} = -\text{grad}\ \varphi_{\text{mag}}$. The potential function φ_{mag} would be related to the fictitious pole density as the electric potential is related to charge density. The simplicity of the scalar potential compared with the vector potential is rather appealing. Moreover, the magnetic scalar potential can be related in a very neat way to the currents that are the real source of **B**, and thus one can use the scalar potential without any explicit use of the fictitious poles. You may want to use this device if you ever have to design magnets or calculate magnetic fields.

We must abandon the magnetic pole fiction, however, if we want to understand the field inside the magnetic material. That the macroscopic magnetic field inside a permanent magnet is, in a very real sense, like the field in Fig. 11.21*b* rather than the field in Fig. 11.21*a* has been demonstrated experimentally by deflecting energetic charged particles in magnetized iron, as well as by the magnetic effects on slow neutrons, which pass even more easily through the interior of matter.

Figure 11.22*a* shows a small disk-shaped permanent magnet, in which the magnetization is parallel to the axis of symmetry. You are probably more familiar with permanent magnets in the shape of long bars. However, flat disk magnets of considerable strength can be made with certain new materials. The magnetization M is given as 150 in CGS units. The magnetic moment of the electron is 0.93×10^{-20} erg/gauss, so this value of M corresponds to 1.6×10^{22} lined-up electron spins per cm^3. The disk is equivalent to a band of current around its rim, of surface density $\mathcal{J} = cM$. The rim being 0.3 cm wide, the current I amounts to

$$0.3cM = (0.3)(3 \times 10^{10})(150)$$

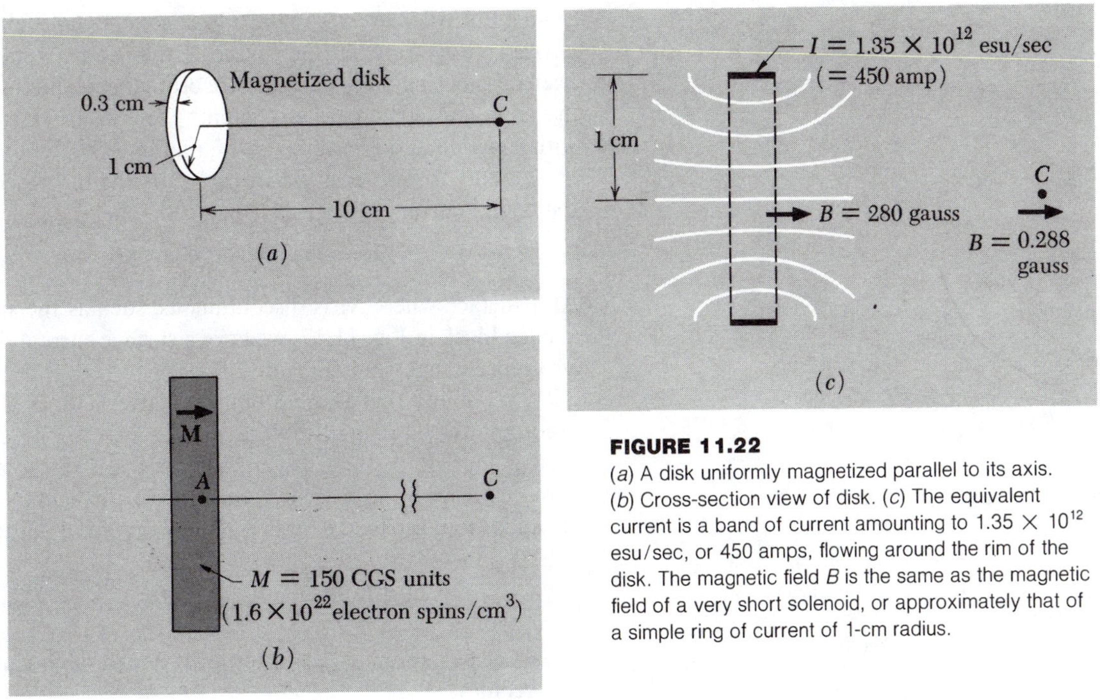

FIGURE 11.22
(*a*) A disk uniformly magnetized parallel to its axis. (*b*) Cross-section view of disk. (*c*) The equivalent current is a band of current amounting to 1.35×10^{12} esu/sec, or 450 amps, flowing around the rim of the disk. The magnetic field B is the same as the magnetic field of a very short solenoid, or approximately that of a simple ring of current of 1-cm radius.

or 1.35×10^{12} esu/sec. This is 450 amps—rather more current than you draw by short-circuiting an automobile battery! The field **B** at any point in space, including points inside the disk, is simply the field of this band of current. For instance, near the center of the disk, **B** is approximately

$$B = \frac{2\pi I}{rc} = \frac{2\pi(0.3cM)}{rc} = \frac{2\pi(0.3)(150)}{(1.0)} = 280 \text{ gauss} \qquad (48)$$

The approximation consists in treating the 0.3-cm-wide band of current as if it were concentrated in a single thin ring. (In the corresponding approximation in the electrical example we treated the equivalent charge sheets as large compared with their separation.) As for the field at a distant point, it would be easy to compute it for the ring current, but we could also, for an approximate calculation, proceed as we did in the electrical example. That is, we could find the total magnetic moment of the object, and find the distant field of a single dipole of that strength.

FREE CURRENTS, AND THE FIELD H

11.10 It is often useful to distinguish between bound currents and free currents. *Bound* currents are currents associated with molecular or atomic magnetic moments, including the intrinsic magnetic moment of particles with spin. These are the molecular current loops envisioned by Ampère, the source of the magnetization we have just been considering. *Free* currents are ordinary conduction currents flowing on macroscopic paths—currents that can be started and stopped with a switch and measured with an ammeter.

The current density $\mathbf{J}$ in Eq. 43 is the macroscopic average of the bound currents, so let us henceforth label it $\mathbf{J}_{\text{bound}}$:

$$\mathbf{J}_{\text{bound}} = c \text{ curl } \mathbf{M} \tag{49}$$

At a surface where $\mathbf{M}$ is discontinuous, such as the side of the magnetized block in Fig. 11.17, we have a surface current density $\mathcal{J}$ which also represents bound current.

We found that $\mathbf{B}$, both outside matter and, as a space average, inside matter, is related to $\mathbf{J}_{\text{bound}}$ just as it is to any current density. That is, curl $\mathbf{B} = (4\pi/c)\mathbf{J}_{\text{bound}}$. But that was in the absence of free currents. If we bring these into the picture, the field they produce simply adds on to the field caused by the magnetized matter and we have

$$\text{curl } \mathbf{B} = \frac{4\pi}{c}(\mathbf{J}_{\text{bound}} + \mathbf{J}_{\text{free}}) = \frac{4\pi}{c}\mathbf{J}_{\text{total}} \tag{50}$$

Let us express $\mathbf{J}_{\text{bound}}$ in terms of $\mathbf{M}$, through Eq. 49. Then Eq. 50 becomes

$$\text{curl } \mathbf{B} = \frac{4\pi}{c}(c \text{ curl } \mathbf{M}) + \frac{4\pi}{c}\mathbf{J}_{\text{free}}$$

which can be rearranged as

$$\text{curl}(\mathbf{B} - 4\pi\mathbf{M}) = \frac{4\pi}{c}\mathbf{J}_{\text{free}} \tag{51}$$

If we now *define* a vector function $\mathbf{H}(x, y, z)$ at every point in space by the relation

$$\mathbf{H} = \mathbf{B} - 4\pi\mathbf{M} \tag{52}$$

Eq. 51 can be written

$$\text{curl } \mathbf{H} = \frac{4\pi}{c}\mathbf{J}_{\text{free}} \tag{53}$$

In other words, the vector $\mathbf{H}$, defined by Eq. 52, is related to the *free* current in the way $\mathbf{B}$ is related to the total current, *bound* plus *free*. The parallel is not complete, however, for we always have div $\mathbf{B} = 0$, whereas our vector function $\mathbf{H}$ does not necessarily have zero divergence.

This surely has reminded you of the vector $\mathbf{D}$ which we introduced, a bit grudgingly, in the last chapter. $\mathbf{D}$, remember, was related to the free charge as $\mathbf{E}$ is related to the total charge. Although we rather disparaged **D**, the vector **H** is really useful, for a practical reason that is worth understanding. In electrical systems, what we can easily control and measure are the potential differences of bodies, and not the amounts of free charge on them. Thus we control the electric field **E** directly. **D** is out of our direct control, and since it is not a fundamental quantity in any sense, what happens to it is not of much concern. In magnetic systems, however, it is precisely the free currents that we can most readily control. We lead them through wires, measure them with ammeters, channel them in well-defined paths with insulation, and so on. We have much less direct control, as a rule, over magnetization, and hence over **B**. So the auxiliary vector **H** *is* useful, even if **D** is not.

The integral relation equivalent to Eq. 53 is

$$\int_C \mathbf{H} \cdot d\mathbf{l} = \frac{4\pi}{c} \int_S \mathbf{J}_{\text{free}} \cdot d\mathbf{a} = \frac{4\pi}{c} I_{\text{free}} \tag{54}$$

where I_{free} is the total free current enclosed by the path C. Suppose we wind a coil around a piece of iron and send through this coil a certain current I which we can measure by connecting an ammeter in series with the coil. This is the free current, and it is the only free current in the system. Therefore one thing we know for sure is the line integral of **H** around any closed path, whether that path goes through the iron or not. The integral depends only on the number of turns of our coil that are linked by the path, and not on the magnetization in the iron. The determination of **M** and **B** in this system may be rather complicated. It helps to have singled out one quantity that we can determine quite directly.

Figure 11.23 illustrates this property of **H** by an example, and is a reminder of the units we may use in a practical case. H has the same dimensions as B; in the Gaussian CGS system they are related in exactly the same way to current in esu/sec. As you know, the unit of magnetic field strength B in this system is named the gauss. There was no compelling need for a different name for the unit of H. Nevertheless, people who like to name things have given the unit of H a name all its own, the *oersted*. Because you will find this name used elsewhere, we have introduced it in Fig. 11.23.

We consider **B** the fundamental magnetic field vector because the absence of magnetic charge, which we discussed in Section 10.2, implies div **B** = 0 everywhere, even inside atoms and molecules. From div **B** = 0 it follows, as we showed in Section 11.8, that the average macroscopic field inside matter is **B**, not **H**. The implications of this have not always been understood or heeded in the past. However, **H** has the practical advantage we have already explained. In some older books you will find **H** introduced as the primary magnetic field. **B** is then defined as **H** + 4π**M**, and given the name *magnetic induction.* Even some modern writers who treat **B** as the primary field feel

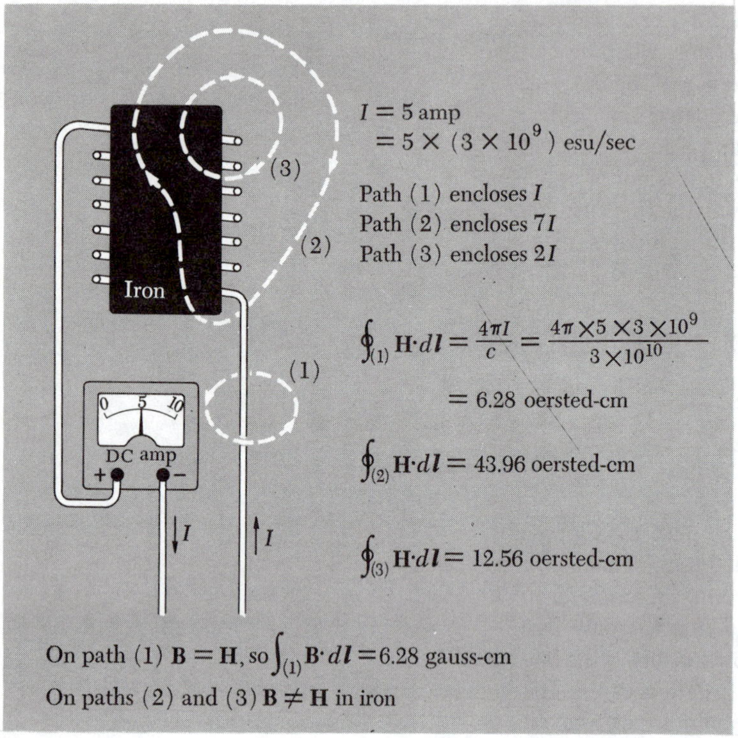

Iron
(3)
(2)
(1)
DC amp
0 5 10
+ −
I
I
$I = 5$ amp
$= 5 \times (3 \times 10^9)$ esu/sec
Path (1) encloses I
Path (2) encloses $7I$
Path (3) encloses $2I$
$\oint_{(1)} \mathbf{H} \cdot d\boldsymbol{l} = \frac{4\pi I}{c} = \frac{4\pi \times 5 \times 3 \times 10^9}{3 \times 10^{10}}$
$= 6.28$ oersted-cm
$\oint_{(2)} \mathbf{H} \cdot d\boldsymbol{l} = 43.96$ oersted-cm
$\oint_{(3)} \mathbf{H} \cdot d\boldsymbol{l} = 12.56$ oersted-cm
On path (1) $\mathbf{B} = \mathbf{H}$, so $\int_{(1)} \mathbf{B} \cdot d\boldsymbol{l} = 6.28$ gauss-cm
On paths (2) and (3) $\mathbf{B} \neq \mathbf{H}$ in iron

FIGURE 11.23
Illustrating the relation between free current and the line integral of **H**.

obliged to call it the magnetic induction because the name *magnetic field* was historically preempted by **H**. This seems clumsy and pedantic. If you go into the laboratory and ask a physicist what causes the pion trajectories in his bubble chamber to curve, he'll probably answer "magnetic field," not "magnetic induction." You will seldom hear a geophysicist refer to the earth's magnetic induction, or an astrophysicist talk about the magnetic induction in the galaxy. We propose to keep on calling **B** the magnetic field. As for **H**, although other names have been invented for it, we shall call it *the field **H***, or even, *the magnetic field **H***.

It is only the names that give trouble, not the symbols. Everyone agrees that in the Gaussian CGS system the relation connecting **B**, **M**, and **H** is that stated in Eq. 52. In vacuum there is no essential distinction between **B** and **H**, for **M** must be zero where there is no matter. You will often see Maxwell's equations for the vacuum fields written with **E** and **H**, rather than **E** and **B**.

In SI units the relation of **H** to the free current is written

$$\text{curl}\, \mathbf{H} = \mathbf{J}_{\text{free}} \tag{53'}$$

with its integral equivalent:

$$\int_C \mathbf{H} \cdot d\mathbf{l} = \int_S \mathbf{J}_{\text{free}} \cdot d\mathbf{a} = I_{\text{free}} \tag{54'}$$

Since $\mathbf{J}_{\text{free}}$ is expressed in amps/m^2, the unit of **H** is 1 amp/meter. In empty space $\mathbf{H} = \mathbf{B}/\mu_0$. When SI units are used in the description of an electromagnetic wave, it is customary to use **H** and **E**, rather than **B** and **E**, for the magnetic and electric fields. For the plane wave in free space that we studied in Section 9.4 the relation between the magnetic amplitude H_0 in amps/meter and the electric amplitude E_0 in volts/meter involves the constant $\sqrt{\mu_0/e_0}$ which has the dimensions of resistance and the approximate value 377 ohms. For its exact value, see Appendix E. We met this constant before in Section 9.6, where it appeared in the expression for the power density in the plane wave, Eq. 28. The condition that corresponds to E_0 and B_0, as stated for CGS units by Eq. 20 in Section 9.4, becomes in SI units

$$E_0(\text{volt/meter}) = H_0(\text{amps/meter}) \times 377 \text{ ohms} \tag{55}$$

This makes a convenient system of units for dealing with electromagnetic fields in vacuum whose sources are macroscopic alternating currents and voltages. But remember that the basic magnetic field *inside* matter is **B** not **H** as we found in the last section. That is not a matter of mere definition, but a consequence of the absence of magnetic charge.

The way in which **H** is related to **B** and **M** is reviewed in Fig. 11.24, for both systems of units. These relations hold whether **M** is proportional to **B** or not. However, if **M** is proportional to **B**, then it will also be proportional to **H**. In fact, the traditional definition of the volume magnetic susceptibility χ_m is not the logically preferable one given in Eq. 39, but rather:

$$\mathbf{M} = \chi_m \mathbf{H} \tag{56}$$

which we shall reluctantly adopt from here on.

The permanent magnet in Fig. 11.21*b* is an instructive example of the relation of **H** to **B** and **M**. To obtain **H** at some point inside the magnetized material, we have to add vectorially to the magnetic field **B** at that point the vector $-4\pi\mathbf{M}$. Figure 11.25 depicts this for a particular point *P*. It turns out that the lines of **H** inside the magnet look just like the lines of **E** inside the polarized cylinder of Fig. 11.21*a*. That is as it should be, for if magnetic poles really were the source of the magnetization, rather than electric currents, the macroscopic magnetic field inside the material would be **H**, not **B**, and the similarity of magnetic polarization and electric polarization would be complete.

In the permanent magnet there are no free currents at all. Consequently, the line integral of **H**, according to Eq. 54, must be zero around any closed path. You can see that it will be if the **H** lines really look like the **E** lines in Fig. 11.21*a*, for we know the line integral of that electrostatic field is zero around any closed path. In this example of the permanent magnet, Eq. 56 does not apply. The magnetization vector **M** is not proportional to **H** but is determined, instead, by the

esu-sec $^{-1}$
cm ^{2}

$\mathbf{m} = \dfrac{I\mathbf{A}}{c}$

erg-gauss $^{-1}$

$B = \dfrac{2m}{r^3}$

cm^3

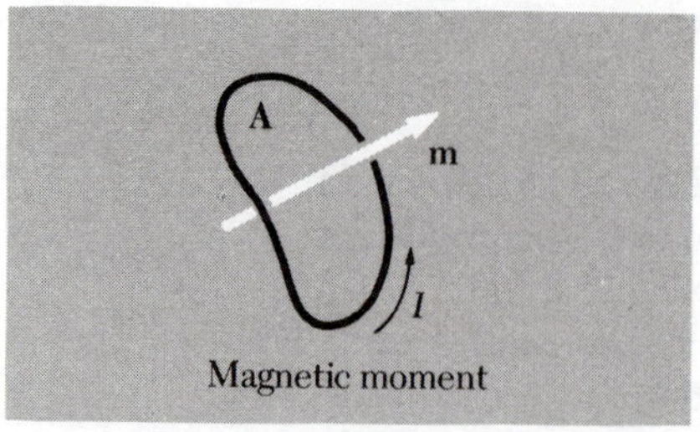

A
m
I
Magnetic moment

ampere

$\mathbf{m} = I\mathbf{A}$

meter ^{2}

joule-tesla $^{-1}$

$B = \left(\dfrac{\mu_0}{4\pi}\right)\dfrac{2m}{r^3}$

meter^3

$\mathbf{H} = \mathbf{B}$

oersted

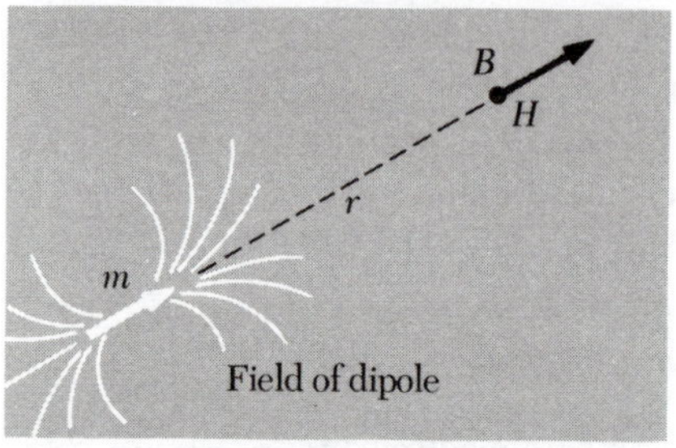

B
H
r
m
Field of dipole

$\mathbf{H} = \dfrac{\mathbf{B}}{\mu_0}$

amp-meter^{-1}

Torque $= \mathbf{m} \times \mathbf{B}_0$

dyne-cm

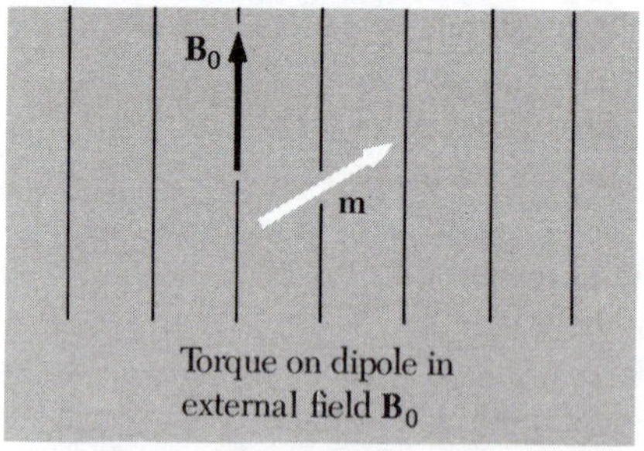

B$_0$
m
Torque on dipole in
external field B$_0$

Torque $= \mathbf{m} \times \mathbf{B}_0$

newton-meter

$\mathbf{M}$ = moment in 1 cm^3

erg-gauss^{-1} cm^{-3}

esu-sec^{-1}cm^{-2}

curl $\mathbf{M} = \frac{1}{c}\mathbf{J}_{\text{bound}}$

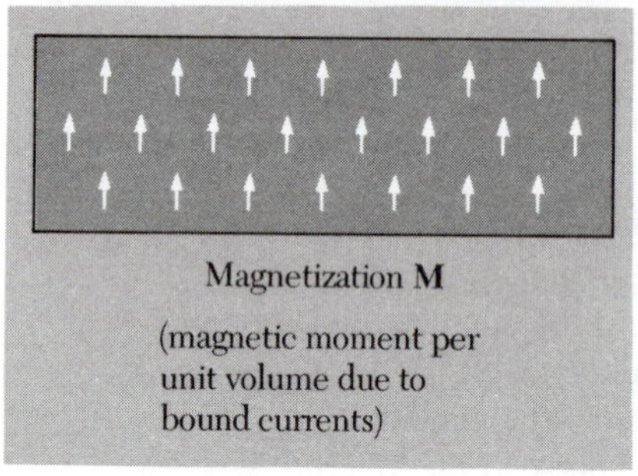

Magnetization M
(magnetic moment per
unit volume due to
bound currents)

$\mathbf{M}$ = moment in 1 meter^3

joule-tesla^{-1} meter^{-3}

amp-meter^{-2}

curl $\mathbf{M} = \mathbf{J}_{\text{bound}}$

curl $\mathbf{B} = \frac{4\pi}{c}(\mathbf{J}_{\text{free}} + \mathbf{J}_{\text{bound}})$

Define: $\mathbf{H} = \mathbf{B} - 4\pi\mathbf{M}$

Then: curl $\mathbf{H} = \frac{4\pi}{c}\mathbf{J}_{\text{free}}$

esu-sec^{-1}

or $\oint \mathbf{H} \cdot d\mathbf{l} = \frac{4\pi}{c} I_{\text{free}}$

curl $\mathbf{B} = \mu_o\,(\mathbf{J}_{\text{free}} + \mathbf{J}_{\text{bound}})$

Define: $\mathbf{H} = \mathbf{B}/\mu_o - \mathbf{M}$

Then: curl $\mathbf{H} = \mathbf{J}_{\text{free}}$

amperes

or $\oint \mathbf{H} \cdot d\mathbf{l} = I_{\text{free}}$

FIGURE 11.24
Summary of relations involving **B**, **H**, **M**, **m**, $\mathbf{J}_{\text{free}}$, and $\mathbf{J}_{\text{bound}}$.

previous treatment of the material. How this can come about will be explained in the next section.

For any material in which **M** is proportional to **H**, so that Eq. 56 applies as well as the basic relation, Eq. 52, we have:

$$\mathbf{B} = \mathbf{H} + 4\pi\mathbf{M} = (1 + 4\pi\chi_m)\mathbf{H} \tag{57}$$

B is then proportional to **H**. The factor of proportionality, $(1 + 4\pi\chi_m)$, is called the *magnetic permeability* and denoted usually by μ:

$$\mathbf{B} = \mu\mathbf{H} \tag{58}$$

The permeability μ, rather than the susceptibility χ, is customarily used in describing ferromagnetism.

FERROMAGNETISM

11.11 Ferromagnetism has served and puzzled man for a long time. The *lodestone* (magnetite) was known in antiquity, and the influence on history of iron in the shape of compass needles was perhaps second only to that of iron in the shape of swords. For nearly a century our electrical technology has depended heavily on the circumstance that one abundant metal happens to possess this peculiar property. Nevertheless, it is only in recent years that anything like a fundamental understanding of ferromagnetism has been achieved.

We have already described some properties of ferromagnets. In a very strong magnetic field the force on a ferromagnetic substance is in such a direction as to pull it into a stronger field, as for paramagnetic materials, but instead of being proportional to the product of the field **B** and its gradient, the force is proportional to the gradient itself. As we remarked at the end of Section 11.4, this suggests that, if the field is strong enough, the magnetic moment acquired by the ferromagnet reaches some limiting magnitude. The direction of the magnetic moment vector must still be controlled by the field, for otherwise the force would not always act in the direction of increasing field intensity.

In permanent magnets we observe a magnetic moment even in the absence of any externally applied field, and it maintains its magnitude and direction even when external fields are applied, if they are not too strong. The field of the permanent magnet itself is always present of course, and you may wonder whether it could not keep its own sources lined up. However, if you look again at Fig. 11.21*b* and Fig. 11.25, you will notice that **M** is generally not parallel to either **B** or **H**. This suggests that the magnetic dipoles must be clamped in direction by something other than purely magnetic forces.

The magnetization observed in ferromagnetic materials is much larger than we are used to in paramagnetic substances. Permanent

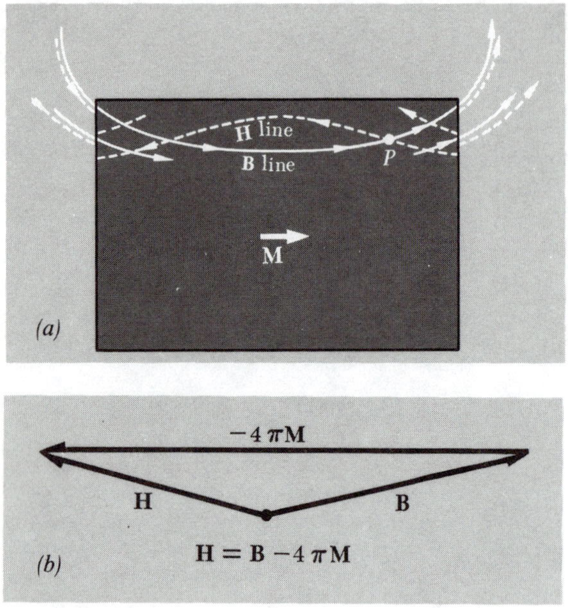

FIGURE 11.25
(*a*) The relation of **B**, **H**, and **M** at a point inside the magnetized cylinder of Fig. 11.21*b*. (*b*) Relation of vectors at point *P*.

magnets quite commonly have fields in the range of a few thousand gauss. A more characteristic quantity is the limiting value of the magnetization, the magnetic moment per unit volume, which the material acquires in a very strong field. This is called the *saturation* magnetization. We can deduce the saturation magnetization of iron from the data in Table 11.1. In a field with a gradient of 1700 gauss/cm, the force on 1 gm of iron was 4×10^5 dynes. From Eq. 19, which relates the force on a dipole to the field gradient, we find

$$m = \frac{F}{dB/dz} = \frac{4 \times 10^5 \text{ dynes}}{1700 \text{ gauss/cm}} = 235 \text{ ergs/gauss} \qquad \text{(for 1 gm)} \tag{59}$$

To get the moment per cubic centimeter we multiply m by the density of iron, 7.8 gm/cm^3. The magnetization M is thus

$$M = 235 \times 7.8 = 1830 \text{ ergs/gauss-cm}^3 \tag{60}$$

It is $4\pi M$, not M, that we should compare with field strengths in gauss.

It is more interesting to see how many electron spin moments this magnetization corresponds to. Dividing M by the electron moment given in Fig. 11.14, 0.93×10^{-20} erg/gauss, we get about 2×10^{23} spin moments per cm^3. Now 1 cm^3 of iron contains about 10^{23} atoms. The limiting magnetization seems to correspond to about two lined-up spins per atom. As most of the electrons in the atom are paired off and have no magnetic effect at all, this indicates that we are dealing with substantially complete alignment of those few electron spins in the atom's structure that are at liberty to point in the same direction.

A very suggestive fact about ferromagnets is this: A given ferromagnetic substance, pure iron for example, loses its ferromagnetic properties quite abruptly if heated to a certain temperature. Above 770°C, pure iron acts like a paramagnetic substance. Cooled below 770°C, it immediately recovers its ferromagnetic properties. This transition temperature, called the *Curie point* after Pierre Curie who was one of its early investigators, is different for different substances. For pure nickel it is 358°C.

What is this ferromagnetic behavior which so sharply distinguishes iron below 770°C from iron above 770°C, and from copper at any temperature? It is the *spontaneous* lining up in one direction of the atomic magnetic moments, which implies alignment of the spin axes of certain electrons in each iron atom. By *spontaneous,* we mean that no external magnetic field need be involved. Over a region in the iron large enough to contain millions of atoms, the spins and magnetic moments of nearly all the atoms are pointing in the same direction. Well below the Curie point—at room temperature, for instance, in the case of iron—the alignment is nearly perfect. If you could magically look into the interior of a crystal of metallic iron and see the elementary magnetic moments as vectors with arrowheads on them, you might see something like Fig. 11.26.

It is hardly surprising that a high temperature should destroy this neat arrangement. Thermal energy is the enemy of order, so to speak. A crystal, an orderly arrangement of atoms, changes to a liquid, a much less orderly arrangement, at a sharply defined temperature, the melting point. The melting point, like the Curie point, is different for different substances. Let us concentrate here on the ordered state itself. Two or three questions are obvious:

FIGURE 11.26
The orderliness of the spin directions in a small region in a crystal of iron. Each arrow represents the magnetic moment of one iron atom.

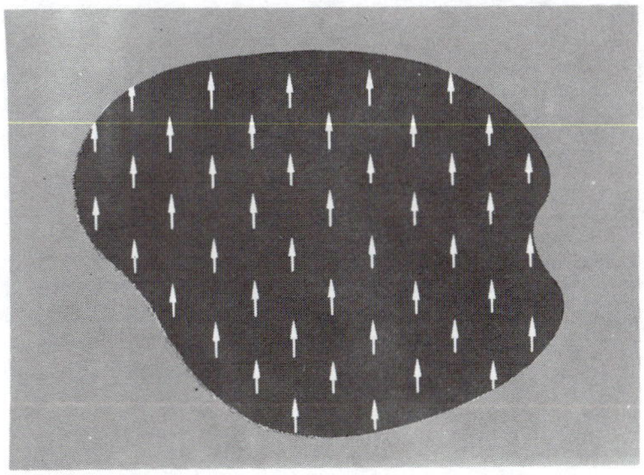

Question 1 What makes the spins line up and keeps them lined up?

Question 2 How, if there is no external field present, can the spins choose one direction rather than another? Why didn't all the moments in Fig. 11.26 point down, or to the right, or to the left?

Question 3 If the atomic moments *are* all lined up, why isn't every piece of iron at room temperature a strong magnet?

The answers to these three questions will help us to understand, in a general way at least, the behavior of ferromagnetic materials when an external field, neither very strong nor very weak, is applied. That includes a very rich variety of phenomena which we haven't even described yet.

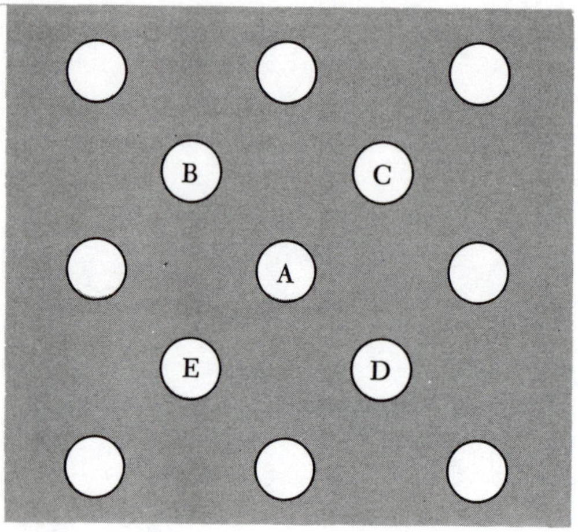

FIGURE 11.27
An atom *A* and its nearest neighbors in the crystal lattice. (Of course, the lattice is really three-dimensional.)

FIGURE 11.28
In iron the energetically preferred direction of magnetization is along a cubic axis of the crystal.

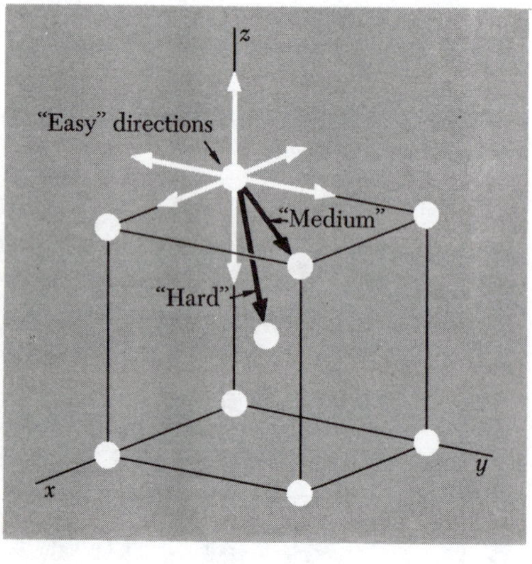

Answer 1 For some reason connected with the quantum mechanics of the structure of the iron atom, it is energetically favorable for the spins of adjacent iron atoms to be parallel. This is *not* due to their magnetic interaction. It is a stronger effect than that, and moreover, it favors parallel spins whether like this ↑↑ or like this →→ (dipole interactions don't work that way—see Problem 10.17). Now if atom A (Fig. 11.27) wants to have its spin in the same direction as that of its neighbors, atoms B, C, D, and E, and each of *them* prefers to have its spin in the same direction as the spin of *its* neighbors, including atom A, you can readily imagine that if a local majority ever develops there will be a strong tendency to "make it unanimous," and then the fad will spread.

Answer 2 Accident somehow determines which of the various equivalent directions in the crystal is chosen, if we commence from a disordered state—as, for example, if the iron is cooled through its Curie point without any external field applied. Pure iron consists of body-centered cubic crystals. Each atom has eight nearest neighbors. The symmetry of the environment imposes itself on every physical aspect of the atom, including the coupling between spins. In iron the cubic axes happen to be the axes of easiest magnetization. That is, the spins like to point in the same direction, but they like it even better if that direction is one of the six directions $\pm\hat{\mathbf{x}}$, $\pm\hat{\mathbf{y}}$, $\pm\hat{\mathbf{z}}$ (Fig. 11.28). This is important because it means that the spins cannot easily swivel around *en masse* from one of the easy directions to an equivalent one at right angles. To do so, they would have to swing through *less* favorable orientations on the way. It is just this hindrance that makes permanent magnets possible.

Answer 3 An apparently unmagnetized piece of iron is actually composed of many *domains,* in each of which the spins are all lined up one way, but in a direction different from that of the spins in neighboring domains. On the average over the whole piece of "unmagnetized" iron, all directions are equally represented, so no large-scale magnetic field results. Even in a single crystal the magnetic domains establish themselves. The domains are usually microscopic in the everyday sense of the word. In fact they can be made visible under a low-power microscope. That is still enormous, of course, on an atomic scale, so a magnetic domain typically includes billions of elementary magnetic moments. Figure 11.29 depicts a division into domains. The division comes about because it is cheaper in energy than an arrangement with all the spins pointing in one direction. The latter arrangement would be a permanent magnet with a strong field extending out into the space around it. The energy stored in this exterior field is larger than the energy needed to turn some small fraction of the spins in the crystal, namely, those at a domain boundary, out of line with their immediate neighbors. The domain structure is thus the outcome of an energy-minimization contest.

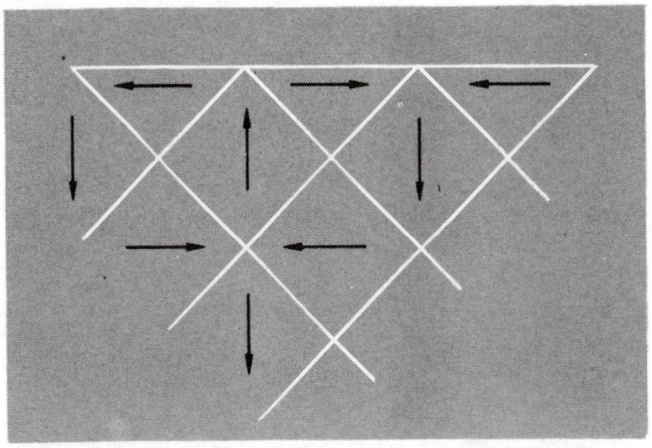

FIGURE 11.29
Possible arrangement of magnetic domains in a single uniform crystal of iron.

If we wind a coil of wire around an iron rod, we can apply a magnetic field to the material by passing a current through the wire. In this field, moments pointing parallel to the field will have a lower energy than those pointing antiparallel, or in some other direction. This favors some domains over others; those that happen to have a favorably oriented moment direction† will tend to grow at the expense of the others, if that is possible. A domain grows like a club, that is, by expanding its membership. This happens at the boundaries. Spins belonging to an unfavored domain but located next to the boundary with a favored domain, simply switch allegiance by adopting the favored direction. That merely shifts the domain boundary, which is nothing more than the dividing surface between the two classes of spins. This happens rather easily in single crystals. That is, a very weak applied field can bring about, through boundary movement, a very large domain growth, and hence a large overall change in magnetization. Depending on the grain structure of the material, however, the movement of domain boundaries can be difficult.

If the applied field does not happen to lie along one of the "easy" directions (in the case of a cubic crystal, for example), the exhaustion of the unfavored domains still leaves the moments not pointing exactly parallel to the field. It may now take a considerably stronger field to pull them into line with the field direction so as to create, finally, the maximum magnetization possible.

Let us look at the large-scale consequences of this, as they appear in the magnetic behavior of a piece of iron under various applied fields. A convenient experimental arrangement is an iron torus, around which we wound two coils (Fig. 11.30). This affords a practically uniform field within the iron, with no end effects to complicate matters. By measuring the voltage induced in one of the coils we can determine changes in flux Φ, and hence in $\mathbf{B}$ inside the iron. If we keep track of the *changes* in $\mathbf{B}$, starting from $B = 0$, we always know what $\mathbf{B}$ is. A current through the other coil establishes $\mathbf{H}$, which we take as the independent variable. If we know $\mathbf{B}$ and $\mathbf{H}$, we can always compute $\mathbf{M}$. It is more usual to plot $\mathbf{B}$ rather than $\mathbf{M}$, as a function of $\mathbf{H}$. A typical B-H curve for iron is shown in Fig. 11.31. Notice that the scales on abscissa and ordinate are vastly different. If there were no iron in the coil, 1 oersted would be worth exactly 1

FIGURE 11.30
Arrangement for investigating the relation between **B** and **M**, or **B** and **H**, in a ferromagnetic material.

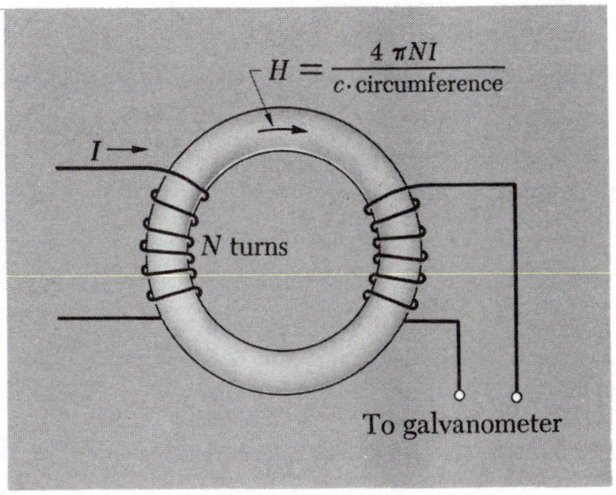
$H = \frac{4\pi NI}{c \cdot \text{circumference}}$
I
N turns
To galvanometer

†We tend to use *spins* and *moments* almost interchangeably in this discussion. The moment is an intrinsic aspect of the spin, and if one is lined up so is the other. To be meticulous, we should remind the reader that in the case of the electron the magnetic moment and angular momentum vectors point in opposite directions (Fig. 11.14).

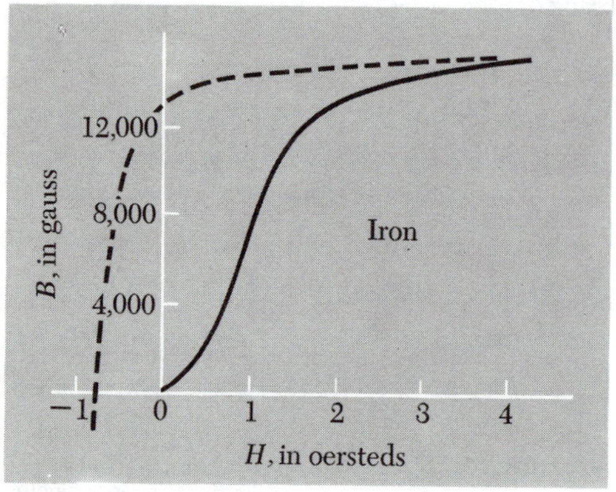

FIGURE 11.31
Magnetization curve for fairly pure iron. The dashed curve is obtained as H is reduced from a high positive value.

gauss. Instead, when the field H is only a few oersteds, B has risen to thousands of gauss. Of course B and H here refer to an average throughout the whole iron ring; the fine domain structure as such never exhibits itself.

Starting with unmagnetized iron, $B = 0$ and $H = 0$, increasing H causes B to rise in a conspicuously nonlinear way, slowly at first, then more rapidly, then very slowly, finally flattening off. What actually becomes constant in the limit is not B but M. In this graph however, since $M = (B - H)/4\pi$, and $H \ll B$, the difference between B and $4\pi M$ is not appreciable.

The lower part of the B-H curve is governed by the motion of domain boundaries, that is, by the growth of "right-pointing" domains at the expense of "wrong-pointing" domains. In the flattening part of the curve, the atomic moments are being pulled by "brute force" into line with the field. The iron here is an ordinary polycrystalline metal, so only a small fraction of the microcrystals will be fortunate enough to have an easy direction lined up with the field direction.

If we now slowly decrease the current in the coil, thus lowering H, the curve *does not retrace itself.* Instead, we find the behavior given by the dashed curve in Fig. 11.31. This irreversibility is called *hysteresis.* It is largely due to the domain boundary movements being partially irreversible. The reasons are not obvious from anything we have said, but are well understood by physicists who work on ferromagnetism. The irreversibility is a nuisance, and a cause of energy loss in many technical applications of ferromagnetic materials—for instance, in alternating-current transformers. But it is indispensable for permanent magnetization, and for such applications, one wants to enhance the irreversibility. Figure 11.32 shows the corresponding portion of the B-H curve for a good permanent magnet alloy. Notice that H has to become 600 oersteds in the *reverse* direction before B is reduced to zero. If the coil is simply switched off and removed, we are left with B at 13,000 gauss, called the *remanence.* Since H is zero, this is essentially the same as the magnetization M, except for the factor 4π. The alloy has acquired a permanent magnetization, that is, one that will persist indefinitely if it is exposed only to weak magnetic fields. All the information that is stored on magnetic tapes and disks owes its permanence to this physical phenomenon.

FIGURE 11.32
Alnico V is an alloy of aluminum, nickel, and cobalt, which is used for permanent magnets. Compare this portion of its magnetization curve with the corresponding portion of the characteristic for a "soft" magnetic material, shown in Fig. 11.31.

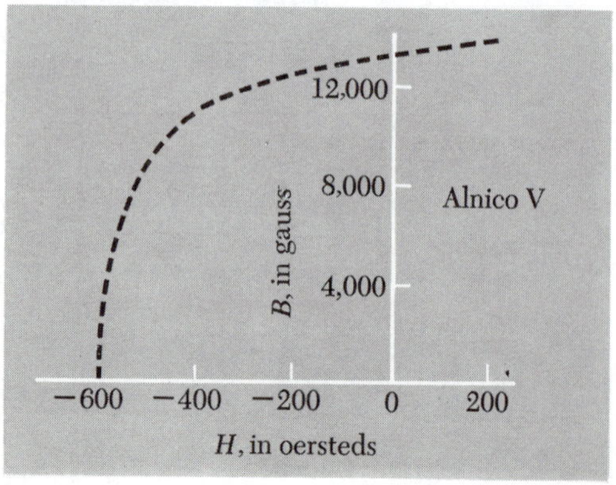

PROBLEMS

11.1 From the data in Table 11.1 determine the diamagnetic susceptibility of water.

11.2 In Chapter 6 we calculated the field at a point on the axis of a current ring of radius b. (See Eq. 41 of Chapter 6.) Show that for z

$\gg b$ this approaches the field of a magnetic dipole, and find how far out on the axis one has to go before the field has come within 1 percent of the field that an infinitesimal dipole of the same dipole moment would produce at that point.

11.3 How large is the magnetic moment of 1 gm of liquid oxygen in a field of 18 kilogauss, according to the data in Table 11.1? Given that the density of liquid oxygen is 0.85 gm/cm^3 at 90 K, what is its magnetic susceptibility χ_m?

11.4 At the north magnetic pole the earth's magnetic field is vertical and has a strength of 0.62 gauss. The earth's field at the surface and further out is approximately that of a central dipole.

(*a*) What is the magnitude of the dipole moment in ergs/gauss?

(*b*) In joules/tesla?

(*c*) Imagine that the source of the field is a current ring on the "equator" of the earth's metallic core, which has a radius of 3000 km, about half the earth's radius. How large would the current have to be?

11.5 A solenoid like the one described in Section 11.1 is located in the basement of a physics laboratory. A physicist on the top floor of the building, 60 feet higher and displaced horizontally 80 feet, complains that its field is disturbing his measurements. Assuming that the solenoid is operating under the conditions described, and treating it as a simple magnetic dipole, compute the field strength at the location of the complaining physicist. Comment, if you see any grounds for doing so, on the merit of his complaint.

11.6 A cube of magnetite 5 cm on an edge is magnetized to saturation in a direction perpendicular to two of its faces. Find the magnitude in amperes of the ribbon of bound-charge current that flows around the circuit consisting of the other four faces of the cube. The saturation magnetization in magnetite is 4.8×10^5 joules/tesla-m^3. Would the field of this cubical magnet seriously disturb a compass 2 meters away?

11.7 A sphere of radius R carries the charge Q which is distributed uniformly over the surface of the sphere with the density $\sigma = Q/4\pi R^2$. This shell of charge is rotating about an axis of the sphere with the angular velocity ω, in radians/sec. Find its magnetic moment. (Divide the sphere into narrow bands of rotating charge; find the current to which each band is equivalent, and its dipole moment, and integrate over all bands.)

Ans. $QR^2\omega/3c$.

11.8 Show that the work done in pulling 1 gm of paramagnetic material from a region where the magnetic field strength is B to a region where the field strength is negligibly small is $\frac{1}{2}\chi B^2$, χ being the specific susceptibility. Then calculate exactly how much work, per gram, would be required to remove the liquid oxygen from the position referred to in Sec. 11.1. (Of course, this only applies if χ is a constant over the range of field strengths involved.)

11.9 A cylindrical solenoid has a single-layer winding of radius r_0. It is so long that near one end the field may be taken to be that of a semi-infinite solenoid. Show that the point on the axis of the solenoid where a small paramagnetic sample will experience the greatest force is located a distance $r_0/\sqrt{15}$ in from the end.

11.10 In the case of an electric dipole made of two charges Q and $-Q$ separated by a distance s, the volume of the near region, where the field is essentially different from the ideal dipole field, is proportional to s^3. The field strength in this region is proportional to Q/s^2, at similar points as s is varied. The dipole moment $p = Qs$, so that if we shrink s while holding p constant, the product of volume and field strength does what? Carry through the corresponding argument for the magnetic field of a current loop. The moral is: If we are concerned with the space average field in any volume containing dipoles, the essential difference between the insides of electric and magnetic dipoles *cannot* be ignored, even when we are treating the dipoles otherwise as infinitesimal.

11.11 Write out Maxwell's equations as they would appear if we had magnetic charge and magnetic charge currents as well as electric charge and electric currents. Invent any new symbols you need and define carefully what they stand for. Be particularly careful about + and − signs.

11.12 We want to find the energy required to bring two dipoles from infinite separation into the configuration shown in (*a*) below, defined by the distance apart r and the angles θ_1 and θ_2. Both dipoles lie in the plane of the paper. Perhaps the simplest way to compute the energy is this: Bring the dipoles in from infinity while keeping them in the orientation shown in (*b*). This takes no work, for the force on each dipole is zero. Now calculate the work done in rotating $\mathbf{m}_1$ into

PROBLEM 11.12

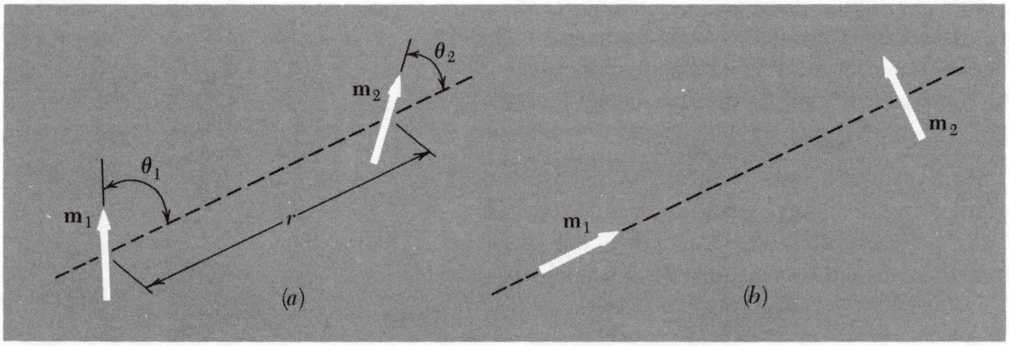

its final orientation while holding $\mathbf{m}_2$ fixed. Then calculate the work required to rotate $\mathbf{m}_2$ into its final orientation. Thus show that the total work done, which we may call the potential energy of the system, is $(\sin\theta_1 \sin\theta_2 - 2\cos\theta_1 \cos\theta_2)m_1 m_2/r^3$.

11.13 Two opposite vertices of a regular octahedron of edge length b are located on the z axis. At each of these vertices, and also at each of the other four vertices, is a dipole of strength m pointing in the $\hat{\mathbf{z}}$ direction. Using the result for Problem 11.12, calculate the potential energy of this system.

11.14 Let us denote by χ_m' the magnetic susceptibility defined by Eq. 39, to distinguish it from the susceptibility χ_m of the conventional definition, Eq. 56. Show that

$$\chi_m = \chi_m'/(1 - 4\pi\chi_m')$$

11.15 In magnetite, Fe_3O_4, the saturation magnetization M_0, in CGS units, is 480 erg/gauss-cm^3. The magnetic bacteria discovered in 1975 by R. P. Blakemore contain crystals of magnetite, approximately cubical, of dimension 5×10^{-6} cm. A bacterium, itself about 10^{-4} cm in size, may contain from 10 to 20 such crystals strung out as a chain. This magnet keeps the whole cell aligned with the earth's magnetic field, and thus controls the direction in which the bacterium swims. See "Magnetic Navigation in Bacteria" by R. P. Blakemore and R. B. Frankel, *Scientific American*, December 1981. Calculate the energy involved in rotating a cell containing such a magnet through 90° in the earth's field, and compare it with the energy of thermal agitation, kT.

11.16 A remarkable new permanent magnet alloy of samarium and cobalt has a saturation magnetization of 750 erg/gauss-cm^3, which it retains undiminished in external fields up to 15 kilogauss. It is the nearest thing yet to rigidly frozen magnetization. Consider a sphere of uniformly magnetized samarium-cobalt 1 cm in radius. (*a*) What is the strength of its magnetic field **B** just outside the sphere at one of its poles? (*b*) At its magnetic equator? (*c*) Imagine two such spheres magnetically stuck together with unlike poles touching. How much force must be applied to separate them?

11.17 An iron plate 20 cm thick is magnetized to saturation in a direction parallel to the surface of the plate. A 10-Gev muon moving perpendicular to that surface enters the plate and passes through it with relatively little loss of energy. Calculate approximately the angular deflection of the muon's trajectory, given that the rest mass of the muon is 200 Mev and that the saturation magnetization in iron is equivalent to 1.5×10^{23} electron moments per cm^3.

11.18 Three magnetic compasses are placed at the corners of a horizontal equilateral triangle. As in any ordinary compass, each compass needle is a magnetic dipole constrained to rotate in a horizontal plane. In this case the earth's magnetic field has been precisely annulled. The only field that acts on each dipole is that of the other two dipoles. What orientation will they eventually assume? (Use symmetry arguments!) Can your answer be generalized for N compasses at the vertices of an N-gon?

11.19 The electric dipole moment of a polar molecule is typically 10^{-18} esu-cm in order of magnitude (Fig. 10.14). The magnetic moment of an atom or molecule with an unpaired electron spin is 10^{-20} erg/gauss. Although we are using different names for the units, the dimensions of electric dipole moment and magnetic dipole moment are actually the same. The numerical comparison is therefore significant (as it would not be in SI units) and provides another reminder that on the atomic scale magnetism is a relatively feeble effect. But consider large-scale polarization P and magnetization M. What limits the practically attainable ratio of P in a dielectric to M in a ferromagnetic material, and how large do you think it could be?

11.20 Imagine that a magnetic dipole of strength m is located at the center of every square on a chessboard, with dipoles on white squares pointing up, dipoles on black squares pointing down. The side of a square is s.

(*a*) Compute the work required to remove any particular one of the dipoles to infinity, leaving the other 63 fixed in position and orientation. Thus determine which of the dipoles are in this respect most tightly bound.

(*b*) How much work must be done to disperse all 64 dipoles to infinite separation from one another? To answer these questions you will have to write and run a little program.

11.21 The magnetic dipole $\mathbf{m}$ in the diagram oscillates at frequency ω. Some of its flux links the nearby circuit C_1, inducing in C_1 an electromotive force, $\mathscr{E}_1 \sin \omega t$. It would be easy to compute $\mathscr{E}_1$ if we knew how much flux from the dipole C_1 encloses, but that might be hard to calculate. Suppose that all we know about C_1 is this: If a current I_1 were flowing in C_1, it would produce a magnetic field $\mathbf{B}_1$ at the location of $\mathbf{m}$. We are told the value of $\mathbf{B}_1/I_1$, but nothing more about C_1, not even its shape or location. Show that this information suffices to relate $\mathscr{E}_1$ to $\mathbf{m}_0$ by the simple formula $\mathscr{E} = (\omega/I_1)\mathbf{B}_1 \cdot \mathbf{m}_0$. *Hint:* Represent $\mathbf{m}$ as a small loop of area A carrying current I_2. Call this circuit C_2. Consider the voltage induced in C_2 by a varying current in C_1; then invoke the reciprocity of mutual inductance which we proved in Section 7.7. (Note that this formula works in either CGS or SI units. In the former $\mathscr{E}I$ is in ergs/sec and m_0 is in ergs/gauss; in the latter $\mathscr{E}I$ is in watts and m_0 is in joules/tesla.)

PROBLEM 11.21

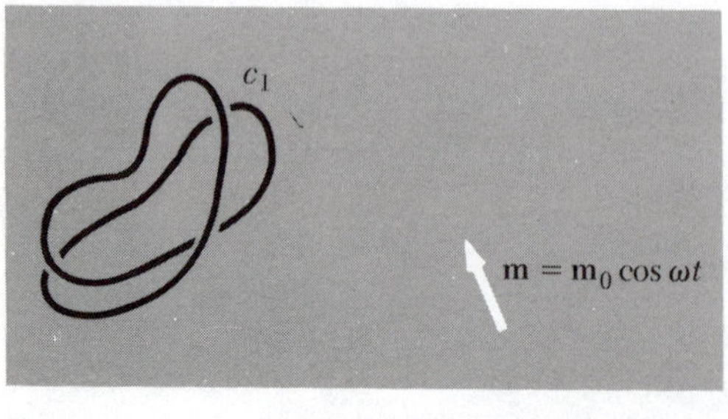

11.22 The direction of the earth's magnetic field in geological ages past can be deduced by studying the remanent magnetization in rocks. The magnetic moment of a rock specimen can be determined by rotating it inside a coil and measuring the alternating voltage thereby induced. The two coils in the diagram are connected in series. Each has 1500 turns and a mean radius of 6 cm. The rock is rotated at 1740 revolutions per minute by a shaft perpendicular to the plane of the diagram.

(*a*) How large is the magnetic moment of the rock if the amplitude of the induced electromotive force is 1 millivolt? The formula derived in Problem 11.21 is useful here.

(*b*) In order of magnitude, what is the minimum amount of ferromagnetic material required to produce an effect that large?

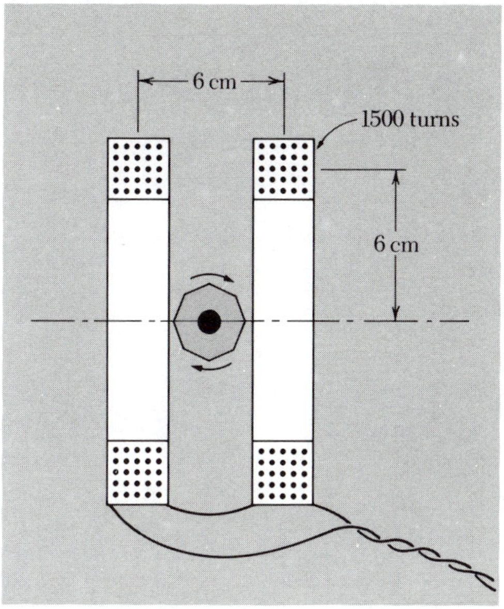

PROBLEM 11.22

11.23 A magnetic dipole of strength m is placed in a homogeneous magnetic field of strength B_0, with the dipole moment directed opposite to the field. Show that, in the combined field, there is a certain spherical surface, centered on the dipole, through which no field lines pass. The external field, one may say, has been "pushed out" of this sphere. The field lines outside the sphere have been plotted in the figure. What do the field lines inside the sphere look like? What is the strength of the field immediately outside the sphere, at the equator?

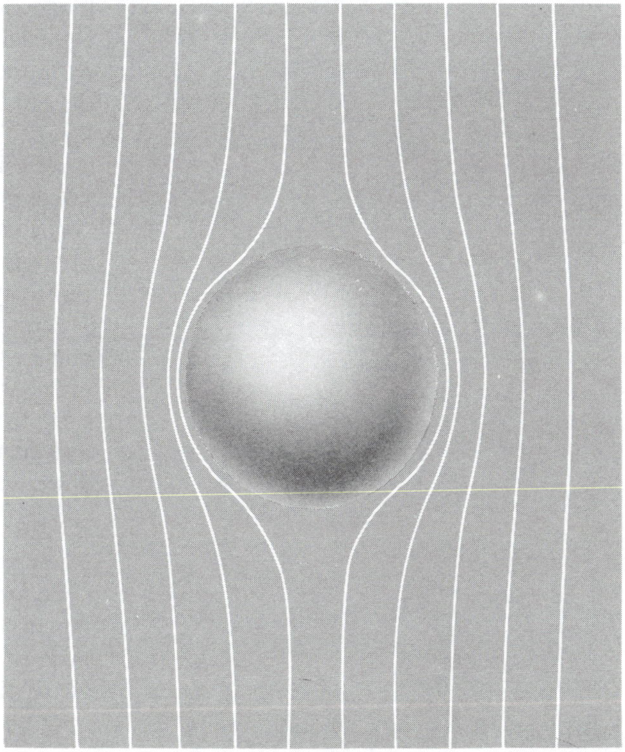

PROBLEM 11.23

So far as its effect on the external field is concerned, the dipole could be replaced by currents flowing in the spherical surface, if we could provide just the right current distribution. What is the field inside the sphere in this case? Why can you be sure? (This is an important configuration in the study of superconductivity. A superconducting sphere, in fact, does push out all field from its interior.)

11.24 An iron torus of inner diameter 10 cm, outer diameter 12 cm, has 20 turns of wire wound on it. Use the *B-H* curve in Fig. 11.31 to estimate the current required to produce a field in the iron of 12,000 gauss.

11.25 For deflecting a beam of high-energy particles in a certain experiment one requires a magnetic field of 16,000-gauss intensity, maintained over a rectangular region 3 meters long in the beam direction, 60 cm wide, and 20 cm high. A suitable magnet might be designed along the lines indicated in parts (*a*) and (*b*) of the figure. Taking the dimensions as given, determine (1) the total amount of ampere-turns required in the two coils to produce a 16-kilogauss field in the gap; (2) the power in kilowatts that must be supplied; (3) the number of turns that each coil should contain, and the corresponding cross-sectional area of the wire, so that the desired field will be attained when the coils are connected in series to a 400-volt dc power supply. For use in (1), a portion of the *B-H* curve for Armco magnet iron is shown in part (*d*) of the figure. All that you need to determine is the line integral of H around a path like *abcdea*. In the gap, $H = B$. In the iron, you may assume that B has the same intensity as in the gap. The field lines will look something like those in part (*c*) of the figure. You can estimate roughly the length of path in the iron. This is not very critical, for you will find that the long path *bcdea* contributes a relatively small amount to the line integral, compared with the contribution of the air path *ab*. (In fact, it is not a bad approximation, at lower field strengths, to neglect H in the iron.) For (2) assume the resistivity of copper $\rho = 2.0 \times 10^{-6}$ ohm-cm, and let each coil contain N turns. You will find that the power required for a given number of ampere turns is independent of N; that is, it is the same for many turns of fine wire or a few turns of thick wire, providing the total cross section of copper is fixed as specified. The designer therefore selects N and conductor cross section to match the magnet to the intended power source.

11.26 The water molecule H_2O contains 10 electrons with spins paired off and, consequently, zero magnetic moment. Its electronic structure is purely diamagnetic. However, the hydrogen nucleus, the proton, is a particle with intrinsic spin and magnetic moment. The magnetic moment of the proton is about 700 times smaller than that of the electron. In water the two proton spins in a molecule are not

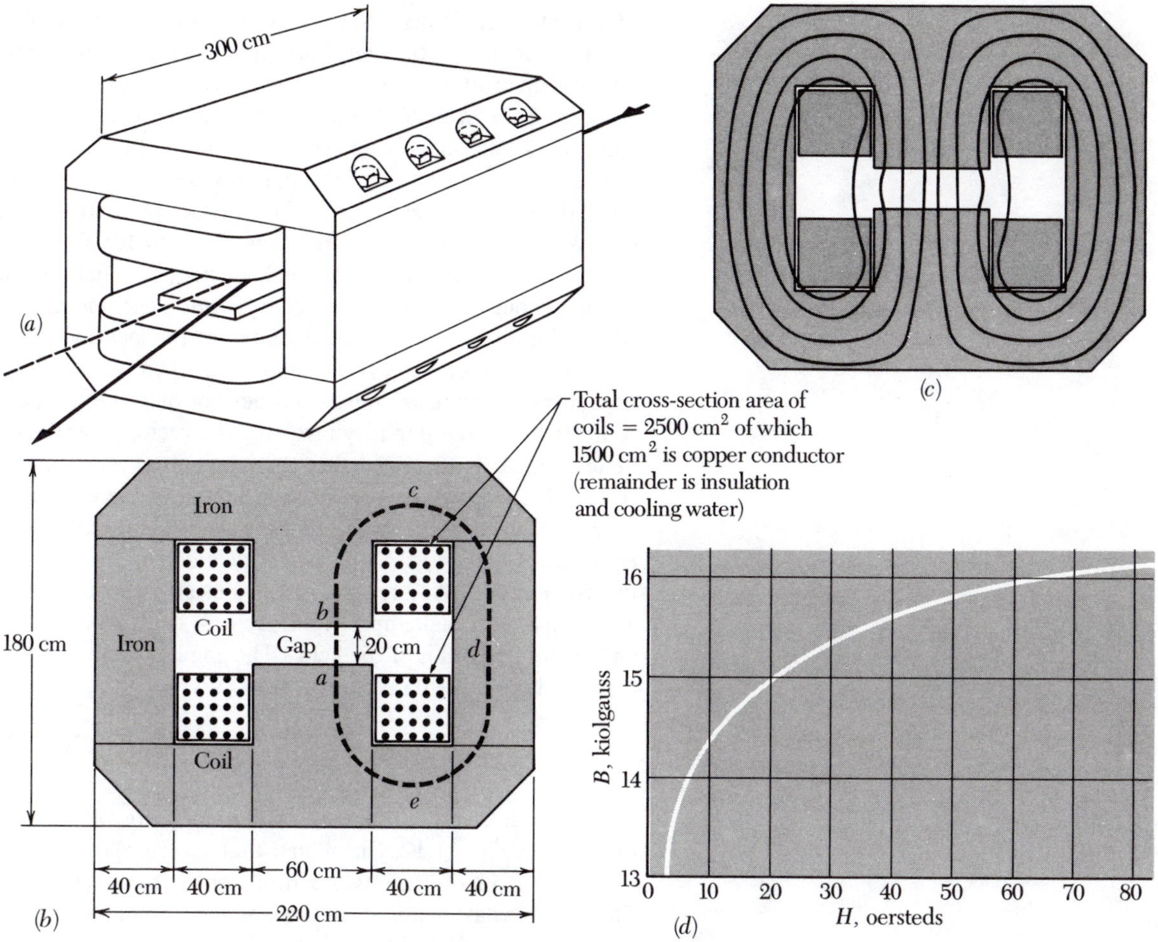

PROBLEM 11.25

locked antiparallel but are practically free to orient individually, subject only to thermal agitation.

(*a*) Using Eq. 40, calculate the resulting paramagnetic susceptibility of water at 20°C.

(*b*) How large is the magnetic moment induced in 1 liter of water in a field of 15 kilogauss?

(*c*) If you wrapped a single turn of wire around a 1-liter flask, about how large a current, in microamps, would produce an equivalent magnetic moment?

Ans. (*c*). 800 microamp.

11.27 Someone who knows a little about the quantum theory of the atom might be troubled by one point in our analysis in Section 11.5 of the effect of a magnetic field on the orbital velocity of an atomic electron. When the velocity changes, while r remains constant, the angular momentum mvr changes. But the angular momentum of an electron orbit is supposed to be precisely an integral multiple of the constant $h/2\pi$, h being the universal quantum constant, Planck's constant. How can mvr change without violating this fundamental quantum law? The resolution of this paradox is important for the quantum mechanics of charged particles, but it is not peculiar to quantum theory. When we consider conservation of energy for a particle carrying charge q, moving in an external electrostatic field $\mathbf{E}$, we always include, along with the kinetic energy $\frac{1}{2}mv^2$, the potential energy $q\varphi$, where φ is the scalar electric potential at the location of the particle. We should not be surprised to find that, when we consider conservation of momentum, we must consider not only the ordinary momentum $M\mathbf{v}$ but also a quantity involving the vector potential of the magnetic field, $\mathbf{A}$. It turns out that the momentum must be taken as $M\mathbf{v} + (q/c)\mathbf{A}$, where $\mathbf{A}$ is the vector potential of the external field evaluated at the location of the particle. We might call $M\mathbf{v}$ the *kinetic momentum* and $(q/c)\mathbf{A}$ the *potential momentum.* (In relativity the inclusion of the $q\mathbf{A}/c$ term is an obvious step because, just as energy and momentum make up a "four-vector," so do φ and $\mathbf{A}/c$, the scalar and vector potentials of the field.) The angular momentum which concerns us here must then be, not just

$$\mathbf{r} \times (M\mathbf{v}) \qquad \text{but} \qquad \mathbf{r} \times \left(M\mathbf{v} + \frac{q}{c}\mathbf{A} \right)$$

Go back now to the case of the charge revolving at the end of the cord, in Fig. 11.12. Check first that a vector potential appropriate to a field $\mathbf{B}$ in the negative z direction is $\mathbf{A} = (B/2)\,(\hat{\mathbf{x}}y - \hat{\mathbf{y}}x)$. Then find what happens to the angular momentum $\mathbf{r} \times [M\mathbf{v} + (q/c)\mathbf{A}]$ as the field is turned on.

A

A SHORT REVIEW OF SPECIAL RELATIVITY

We assume the reader has already been introduced to special relativity. Here we shall review the principal ideas and the formulas that are used in the text beginning in Chapter 5. Most essential is the concept of an inertial frame of reference for space-time events and the transformation of the coordinates of an event from one inertial frame to another.

A frame of reference is a coordinate system laid out with measuring rods and provided with clocks. Clocks are everywhere. When something happens at a certain place, the time of its occurrence is read from a clock that was at, and stays at, that place. That is, time is measured by a *local* clock that is *stationary* in the frame. The clocks belonging to the frame are all *synchronized.* One way to accomplish this (not the only way) was described by Einstein in his great paper of 1905. Light signals are used. From a point A, at time t_A, a short pulse of light is sent out toward a remote point B. It arrives at B at the time t_B, as read on a clock at B, and is immediately reflected back toward A, where it arrives at t'_A. If $t_B = (t_A + t'_A)/2$, the clocks at A and B are synchronized. If not, one of them requires adjustment. In this way, all clocks in the frame can be synchronized. Note that the job of observers in this procedure is merely to record local clock readings for subsequent comparison.

An *event* is located in space and time by its coordinates x, y, z, t in some chosen reference frame. The event might be the passage of a particle at time t_1, through the space point (x_1, y_1, z_1). The history of the particle's motion is a sequence of such events. Suppose the sequence has the special property that $x = v_x t$, $y = v_y t$, $z = v_z t$, at every time t, with v_x, v_y, and v_z constant. That describes motion in a straight line at constant speed with respect to this frame. An *inertial frame of reference* is a frame in which an isolated body, free from external influences, moves in this way. An inertial frame, in other words, is one in which Newton's first law is obeyed. Behind all of this, including the synchronization of clocks, are two assumptions about empty space: It is *homogeneous;* that is, all locations in space are equivalent. It is *isotropic;* that is, all directions in space are equivalent.

Two frames, let us call them F and F', can differ in several ways. One can simply be displaced with respect to the other, the origin of coordinates in F' being fixed at a point in F which is not at the F coordinate origin. Or the axes in F' might not be parallel to the axes in F. As for the timing of events, if F and F' are not moving with respect to one another, a clock stationary in F is stationary also in F'. In that case we can set all F' clocks to agree with the F clocks and then ignore the distinction. Differences in frame location and frame orientation only have no interesting consequences if space is homogeneous and isotropic. Suppose now that the origin of frame F' is *moving* relative to the origin of frame F. The description of a sequence of events by coordinate values and clock times in F can differ from the description of the same events by space coordinate values in F' and times measured by clocks in F'. How must the two descriptions be related? In answering that we shall be concerned only with the case in which F is an inertial frame and F' is a frame which is moving relative to F at constant velocity and without rotating. In that case F' is also an inertial frame.

Special relativity is based on the postulate that physical phenomena observed in different inertial frames of reference appear to obey exactly the same laws. In that respect one frame is as good as another; no frame is unique. If true, this relativity postulate is enough to determine the way a description of events in one frame is related to the description in a different frame of the same events. In that relation there appears a universal speed, the same in all frames, whose value must be found by experiment. Sometimes added as a second postulate is the statement that a measurement of the velocity of light in any frame of reference gives the same result whether the light's source is stationary in that frame or not. One may regard this as a statement about the nature of light rather than an independent postulate. It asserts that electromagnetic waves in fact travel with the limiting speed implied by the relativity postulate. Other things travel at that speed, too—neutrinos, for example—but we are accustomed to call the limiting speed the "speed of light." The deductions from the relativity postulate, expressed in the formulas of special relativity, have been precisely verified by countless experiments. Nothing in physics rests on a firmer foundation.

Consider two events, A and B, observed in an inertial frame F. *Observed,* in this usage, is short for "whose space-time coordinates are determined with the measuring rods and clocks of frame F." (Remember that our observers are equipped merely with pencil and paper, and we must post an observer at the location of every event!) The displacement of one event from the other is given by the four numbers

$$x_B - x_A, \qquad y_B - y_A, \qquad z_B - z_A, \qquad t_B - t_A \tag{1}$$

The *same two events* could have been located by giving their coordinates in some other frame F'. Suppose F' is moving with respect to F in the manner indicated in Fig. A.1. The spatial axes of F' remain parallel to those in F, while, as seen from F, the frame F' moves with

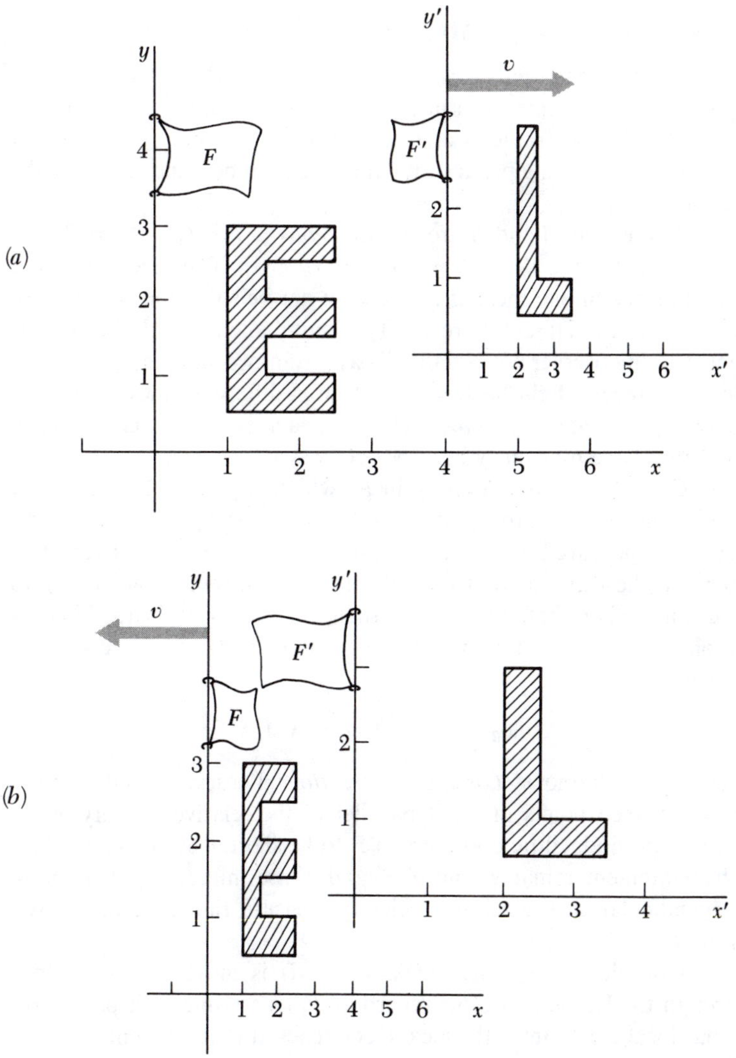

FIGURE A.1
Two frames moving with relative speed v. The "E" is stationary in frame F. The "L" is stationary in frame F'. In this example $\beta = v/c = 0.866$; $\gamma = 2$. (*a*) Where everything was, as determined by observers in F at a particular instant of time t according to clocks in F. (*b*) Where everything was, as determined by observers in F' at a particular instant of time t' according to clocks in F'.

Question: Suppose the clocks in the two frames happened to be set so that the left edge of the E touched the left edge of the L at $t = 0$ according to a local clock in F and at $t' = 0$ according to a local clock in F'. Let the distances be in feet and take c as 1 foot/nanosecond. What is the reading t of all the F clocks in (*a*)? What is the reading t' of all the F' clocks in (*b*)?

Answer: $t = 4.62$ nanoseconds; $t' = 4.04$ nanoseconds. If you don't agree, study the example again.

speed v in the positive x direction. This is a special case, obviously, but it contains most of the interesting physics.

Event A, as observed in F', occurred at x'_A, y'_A, z'_A, t'_A, the last of these numbers being the reading of a clock belonging to (that is, *stationary in*) F'. The space-time displacement, or *interval* between events A and B in F' is not the same as in F. Its components are related to those in F by the *Lorentz transformation:*

$$\begin{aligned} x'_B - x'_A &= \gamma(x_B - x_A) - \beta\gamma c(t_B - t_A) \\ y'_B - y'_A &= y_B - y_A \\ z'_B - z'_A &= z_B - z_A \\ t'_B - t'_A &= \gamma(t_B - t_A) - \beta\gamma(x_B - x_A)/c \end{aligned} \tag{2}$$

In these equations c is the speed of light, $\beta = v/c$, and $\gamma = 1/\sqrt{1 - \beta^2}$. The inverse transformation has a similar appearance—as it should if no frame is unique. It can be obtained from Eqs. (2) simply by exchanging primed and unprimed symbols and reversing the sign of β.

Two events A and B are *simultaneous* in F if $t_B - t_A = 0$. But that does not make $t'_B - t'_A = 0$ unless $x_B = x_A$. Thus events that are simultaneous in one inertial frame may not be so in another. Do not confuse this fundamental "relativity of simultaneity" with the obvious fact that an observer not equally distant from two simultaneous explosions will receive light flashes from them at different times. The times t'_A and t'_B are recorded by *local* clocks at each event, clocks stationary in F' that have previously been perfectly synchronized.

Consider a rod stationary in F', which is parallel to the x' axis and extends from x'_A to x'_B. Its length in F' is just $x'_B - x'_A$. The rod's length as measured in frame F is the distance $x_B - x_A$ between two points in the frame F that its ends pass *simultaneously* according to clocks in F. For these two events, then, $t_B - t_A = 0$. With this condition the first of the Lorentz transformation equations above gives us at once

$$x_B - x_A = (x'_B - x'_A)/\gamma \tag{3}$$

This is the famous *Lorentz contraction.* Loosely stated, lengths between fixed points in F', if parallel to the relative velocity of the frames, are judged by observers in F to be shorter by the factor $1/\gamma$. This statement remains true if F' and F are interchanged. Lengths perpendicular to the relative velocity measure the same in the two frames.

Consider one of the clocks in F'. It is moving with speed v through the frame F. Let us record as t'_A its reading as it passes one of our local clocks in F; the local clock reads at that moment t_A. Later this moving clock passes another F clock. At that event the local F clock reads t_B, and the reading of the moving clock is recorded as t'_B.

The two events are separated in the F frame by a distance $x_B - x_A = v(t_B - t_A)$. Substituting this into the fourth equation of the Lorentz transformation, Eq. 2, we obtain at once

$$t'_B - t'_A = \gamma(t_B - t_A)(1 - \beta^2) = (t_B - t_A)/\gamma \tag{4}$$

According to the moving clock, less time has elapsed between the two events than is indicated by the stationary clocks in F. This is the *time dilation* that figures in the "twin paradox." It has been verified in many experiments, including one in which an atomic clock was flown around the world.

Remembering that "moving clocks run slow, by the factor $1/\gamma$," and that "moving graph paper is shortened parallel to its motion by the factor $1/\gamma$," you can often figure out the consequences of a Lorentz transformation without writing out the equations. This behavior, it must be emphasized, is not a peculiar physical property of our clocks and paper, but is intrinsic in space and time measurement under the relativity postulate.

The formula for the addition of velocities, which we use in Chapter 5, is easily derived from the Lorentz transformation equations. Suppose an object is moving in the positive x direction in frame F with velocity u_x. What is its velocity in the frame F'? To simplify matters let the moving object pass the origin at $t = 0$. Then its position in F at any time t is simply $x = u_x t$. To simplify further, let the space and time origins of F and F' coincide. Then the first and last of the Lorentz transformation equations become

$$x' = \gamma x - \beta\gamma ct \qquad \text{and} \qquad t' = \gamma t - \beta\gamma x/c$$

By substituting $u_x t$ for x on the right side of each and dividing the first by the second, we get

$$\frac{x'}{t'} = \frac{u_x - \beta c}{1 - \beta u_x/c} \tag{5}$$

On the left we have the velocity of the object in the F' frame, u'_x. The formula is usually written with v instead of βc.

$$u'_x = \frac{u_x - v}{1 - u_x v/c^2} \tag{6}$$

By solving Eq. 6 for u'_x you can verify that the inverse is

$$u_x = \frac{u'_x + v}{1 + u'_x v/c^2} \tag{7}$$

and that in no case will these relations lead to a velocity, either u_x or u'_x, larger than c.

A velocity component perpendicular to v, the relative velocity of the frames, transforms differently, of course. Remembering that $y' = y$, but moving clocks are slow, we must get $u'_y = u_y/\gamma$. The inverse transformation is in this case $u_y = u'_y/\gamma$, and *not* $u_y = \gamma u'_y$. Whether our frame is F or F', it is *always the other fellow's clocks* that are slow compared with clocks at rest in our frame.

A dynamical consequence of special relativity can be stated as follows. Consider a particle moving with velocity $\mathbf{u}$ in an inertial frame F. We find that energy and momentum are conserved in the interactions of this particle with others if we attribute to the particle momentum $\mathbf{p} = \gamma m_0 \mathbf{u}$ and energy $\gamma m_0 c^2$, where m_0 is a constant characteristic of that particle. We call m_0 the *rest mass* of the particle. It could have been determined in a frame in which the particle is moving so slowly that newtonian mechanics applies—for instance, by bouncing the particle against some standard mass. The factor γ multiplying m_0 is $(1 - u^2/c^2)^{-1/2}$, where u is the speed of the particle as observed in our frame F.

Given $\mathbf{p}$ and E, the momentum and energy of a particle as observed in F, what is the momentum of that particle, and its energy, as observed in another frame F'? As before, we'll assume F' is moving in the positive x direction, with speed v, as seen from F. The transformation turns out to be this:

$$
\begin{aligned}
p'_x &= \gamma p_x - \beta\gamma E/c \\
p'_y &= p_y \\
p'_z &= p_z \\
E' &= \gamma E - \beta\gamma c p_x
\end{aligned}
\tag{8}
$$

Note that βc is here the relative velocity of the two frames, as it was in Eqs. 2, not the particle velocity.

Compare this transformation with Eqs. 2. The resemblance would be perfect if we considered cp instead of p in Eqs. (8), and ct rather than t in Eqs. 2. A set of four quantities that transform in this way is called a *four-vector*.

The meaning of *force* is rate of change of momentum. The force acting on an object is simply $d\mathbf{p}/dt$, where $\mathbf{p}$ is the object's momentum in the chosen frame of reference, and t is measured by clocks in that frame. To find how forces transform, consider a particle of mass m_0 initially at rest at the origin in frame F upon which a force f acts for a short time Δt. We want to find the rate of change of momentum dp'/dt', observed in a frame F'. As before, we shall let F' move in the x direction as seen from F. Consider first the effect of the force component f_x. In time Δt, p_x will increase from zero to $f_x\,\Delta t$, while x increases by

$$\Delta x = \frac{1}{2}\left(\frac{f_x}{m_0}\right)(\Delta t)^2 \tag{9}$$

and the particle's energy increases by $\Delta E = (f_x\,\Delta t)^2/2m_0$, the kinetic energy it acquires, as observed in F. (The particle's speed in F is still so slight that newtonian mechanics applies there.) Using the first of Eqs. (8) we find the change in p'_x:

$$\Delta p'_x = \gamma\,\Delta p_x - \beta\gamma\,\Delta E/c \tag{10}$$

and using the fourth of Eqs. (2) gives

$$\Delta t' = \gamma\Delta t - \beta\gamma\,\Delta x/c \tag{11}$$

Now both ΔE and Δx are proportional to $(\Delta t)^2$, so when we take the limit $\Delta t \to 0$, the last term in each of these equations will drop out, giving

$$\frac{dp'_x}{dt'} = \lim_{\Delta t' \to 0} \frac{\Delta p'_x}{\Delta t'} = \frac{\gamma f_x}{\gamma} = f_x \tag{12}$$

Conclusion: the force component *parallel* to the relative frame motion has the same value in the moving frame as in the rest frame of the particle.

A transverse force component behaves differently. In frame F, $\Delta p_y = f_y\,\Delta t$. But now $\Delta p'_y = \Delta p_y$, and $\Delta t' = \gamma\,\Delta t$, so we get

$$\frac{dp'_y}{dt'} = \frac{f_y\,\Delta t}{\gamma\,\Delta t} = \frac{f_y}{\gamma} \tag{13}$$

A force component perpendicular to the relative frame motion, observed in F', is *smaller* by the factor $1/\gamma$ than the value determined by observers in the rest frame of the particle.

The transformation of a force from F' to some other moving frame F'' would be a little more complicated. We can always work it out, if we have to, by transforming to the rest frame of the particle and then back to the other moving frame.

We'll conclude our review with a remark about Lorentz invariance. If you square both sides of Eq. 8 and remember that $\gamma^2 - \beta^2\gamma^2 = 1$, you can easily show that

$$c^2(p'^2_x + p'^2_y + p'^2_z) - E'^2 = c^2(p^2_x + p^2_y + p^2_z) - E^2 \tag{14}$$

Evidently this quantity $c^2p^2 - E^2$ is *not changed* by a Lorentz transformation. It is often called the *invariant four-momentum* (even though it has dimensions of energy squared). It has the same value in every frame of reference, including the particle's rest frame. In the rest frame the particle's momentum is zero and its energy E is just m_0c^2. The invariant four-momentum is therefore $-m_0^2c^4$. It follows that in any other frame

$$E^2 = c^2p^2 + m_0^2c^4 \tag{15}$$

The invariant constructed in the same way with Eqs. 2 is $(x_B - x_A)^2 + (y_B - y_A)^2 + (z_B - z_A)^2 - c^2(t_B - t_A)^2$. Two events, A and

B, for which this quantity is positive are said to have a *spacelike* separation. It is always possible to find a frame in which they are simultaneous. If the invariant is negative, the events have a *timelike* separation. In that case a frame exists in which they occur at different times, but at the same place. If this "invariant interval" is zero, the two events can be connected by a flash of light.

B

RADIATION BY AN ACCELERATED CHARGE

A particle with charge q has been moving in a straight line at constant speed v_0 for a long time. It runs into something, let us imagine, and in a short period of constant deceleration, of duration τ, the particle is brought to rest. The graph of velocity versus time in Fig. B.1 describes its motion. What must the electric field of this particle look like after that? Figure B.2 shows how to derive it.

We shall assume that v_0 is small compared with c. Let $t = 0$ be the instant the deceleration began, and let $x = 0$ be the position of the particle at that instant. By the time the particle has completely stopped it will have moved a little farther on, to $x = \frac{1}{2}v_0\tau$. That distance, although we tried to indicate it on our diagram, is small compared with the other distances that will be involved.

We now examine the electric field at a time $t = T \gg \tau$. Observers farther away from the origin than $R = cT$ cannot have learned that the particle was decelerated. Throughout that region, region I in Fig. B.2, the field must be that of a charge which has been moving *and is still moving* at the constant speed v_0. That field, as we discovered in Section 5.7, appears to emanate from the present position of the charge, which for an observer anywhere in region I is the point $x = v_0T$ on the x axis. That is where the particle would be now if it hadn't been decelerated. On the other hand, for any observer whose distance from the origin is less than $c(T - \tau)$, that is to say, for any observer in region II, the field is that of a charge at rest close to the origin (actually at $x = \frac{1}{2}v_0\tau$).

What must the field be like in the transition region, the spherical shell of thickness $c\tau$? Gauss's law provides the key. A field line such as AB lies on a cone around the x axis which includes a certain

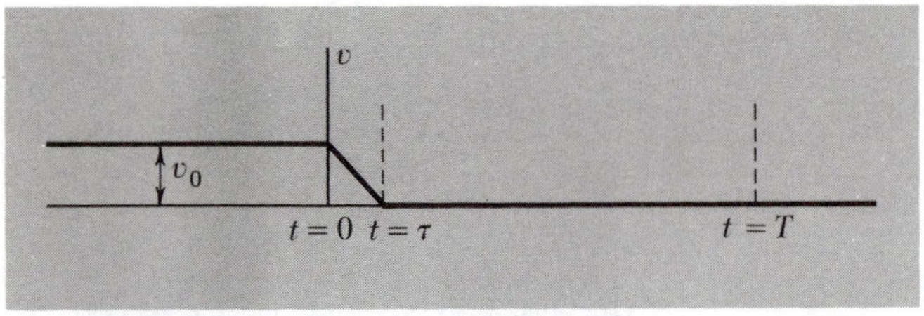

FIGURE B.1
Velocity-time diagram for a particle which traveled at constant speed v_0 until $t = 0$. It then experienced a constant negative acceleration of magnitude $a = v_0/\tau$, which brought it to rest at time $t = \tau$, We assume v_0 is small compared to c.

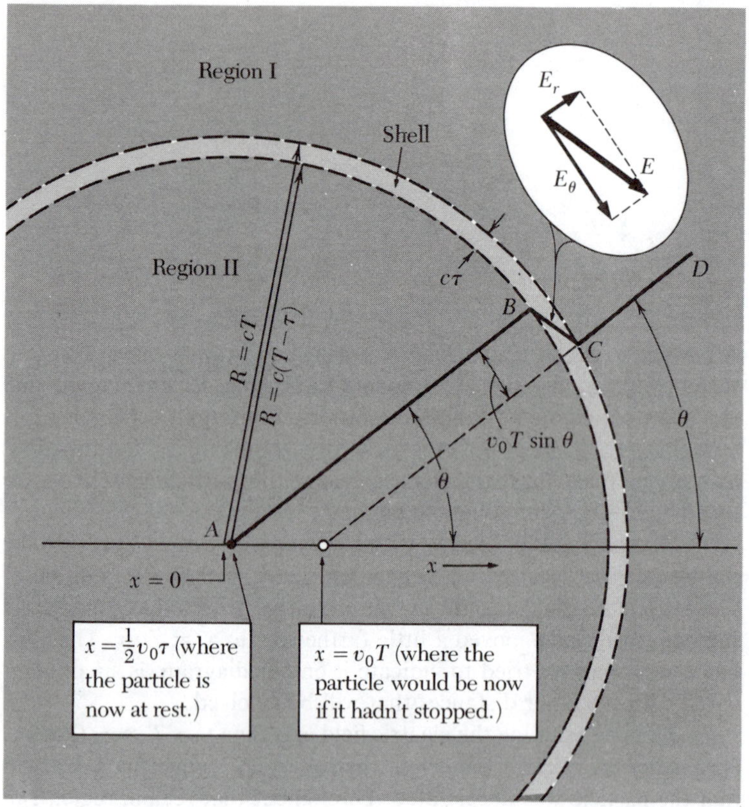

FIGURE B.2
Space diagram for the instant $t = T \gg \tau$, a long time after the particle has stopped. For observers in region *I*, the field must be that of a charge located at the position $x = v_0T$; for observers in region II, it is that of a particle at rest close to the origin. The transition region is a shell of thickness $c\tau$.

amount of flux from the charge q. If CD makes the same angle θ with the axis, the cone on which it lies includes that same amount of flux. (Because v_0 is small, the relativistic compression of field lines visible in Fig. 5.13 and 5.17 is here negligible.) Hence AB and CD must be parts of the same field line, connected by a segment BC. This tells us the *direction* of the field **E** within the shell; it is the direction of the line segment BC. This field **E** within the shell has both a radial component E_r and a transverse component E_θ. From the geometry of the figure their ratio is easily found.

$$\frac{E_\theta}{E_r} = \frac{v_0 T \sin\theta}{c\tau} \tag{1}$$

Now E_r must have the same value within the shell thickness that it does in region II near B. (Gauss's law again!) Therefore $E_r = q/R^2 = q/c^2T^2$, and substituting this in Eq. 1 we obtain

$$E_\theta = \frac{v_0 T \sin\theta}{c\tau} E_r = \frac{q v_0 \sin\theta}{c^3 T \tau} \tag{2}$$

But $v_0/\tau = a$, the magnitude of the (negative) acceleration, and $cT = R$, so our result can be written

$$E_\theta = \frac{qa \sin\theta}{c^2 R} \tag{3}$$

A remarkable fact is here revealed: E_θ is proportional to $1/R$, *not* to $1/R^2$! As time goes on and R increases, the transverse field E_θ will eventually become very much stronger than E_r. Accompanying this transverse (that is, perpendicular to **R**) electric field will be a magnetic field of equal strength perpendicular to both **R** and **E**. This is a general property of an electromagnetic wave, explained in Chapter 9.

Let us calculate the energy stored in the transverse electric field above, in the whole spherical shell. The energy density is

$$\frac{E_\theta^2}{8\pi} = \frac{q^2 a^2 \sin^2\theta}{8\pi R^2 c^4} \tag{4}$$

The volume of the shell is $4\pi R^2 c\tau$, and the average value of $\sin^2\theta$ over a sphere† is ⅔. The total energy of the transverse electric field is therefore

$$\frac{2}{3} 4\pi R^2 c\tau \frac{q^2 a^2}{8\pi R^2 c^4} = \frac{1}{3}\frac{q^2 a^2 \tau}{c^3}$$

To this we must add an equal amount for the energy stored in the transverse magnetic field:

$$\text{Total energy in transverse electromagnetic field} = \frac{2}{3}\frac{q^2 a^2 \tau}{c^3} \tag{5}$$

The radius R has canceled out. This amount of energy simply travels outward, undiminished, with speed c from the site of the deceleration. Since τ is the duration of the deceleration, and is also the duration of the electromagnetic pulse a distant observer measures, we can say that the *power* radiated during the acceleration process was

$$P_{\text{rad}} = \frac{2}{3}\frac{q^2 a^2}{c^3} \tag{6}$$

---

†Our polar axis in this figure is the x axis: $\cos^2\theta = x^2/R^2$. With a bar denoting an average over the sphere, $\overline{x^2} = \overline{y^2} = \overline{z^2} = \frac{1}{3}R^2$. Hence $\overline{\cos^2\theta} = \frac{1}{3}$, and $\overline{\sin^2\theta} = 1 - \overline{\cos^2\theta} = \frac{2}{3}$.

As it is the square of the instantaneous acceleration that appears in Eq. 6, it doesn't matter whether a is positive or negative. Of course it ought not to, for stopping in one inertial frame could be starting in another. Speaking of different frames, P_{rad} itself turns out to be Lorentz-invariant, which is sometimes very handy. That is because P_{rad} is *energy/time,* and energy transforms like time, each being the fourth component of a four-vector, as noted in Appendix A.

We have here a more general result than we might have expected. Equation 6 correctly gives the instantaneous rate of radiation of energy by a charged particle moving with variable acceleration—for instance, a particle vibrating in simple harmonic motion. It applies to a wide variety of radiating systems from radio antennas to atoms and nuclei.

PROBLEMS

B.1 An electron moving initially at constant speed v is brought to rest with uniform deceleration a lasting for a time $t = v/a$. Compare the electromagnetic energy radiated during the deceleration with the electron's initial kinetic energy. Express the ratio in terms of two lengths, the distance light travels in time t and the classical electron radius r_0, defined as e^2/mc^2.

B.2 An elastically bound electron vibrates in simple harmonic motion at frequency ω with amplitude A.

(a) Find the average rate of loss of energy by radiation.

(b) If no energy is supplied to make up the loss, how long will it take for the oscillator's energy to fall to $1/e$ of its initial value?

Ans. (b): $3mc^3/2e^2\omega^2$.

B.3 A plane electromagnetic wave with frequency ω and electric field amplitude E_0 is incident on an isolated electron. In the resulting sinusoidal oscillation of the electron the maximum acceleration is E_0e/m. How much power is radiated by this oscillating charge, averaged over many cycles? (Note that it is independent of the frequency ω.) Divide this average radiated power by $E_0^2c/8\pi$, the average power density (power per unit area of wavefront) in the incident wave. This gives a constant σ with the dimensions of area, called a *scattering cross section.* The energy radiated, or scattered, by the electron, and thus lost from the plane wave, is equivalent to that falling on an area σ. (The case here considered, involving a free electron moving nonrelativistically is often called *Thomson scattering* after J. J. Thomson, the discoverer of the electron, who first calculated it.)

B.4 Our master formula, Eq. 6, is useful for relativistically moving particles, even though we assumed $v_0 \ll c$ in the derivation. All we have to do is transform to an inertial frame F' in which the particle in question is, at least temporarily, moving slowly, apply Eq. 6 in that frame, then transform back to any frame we choose. Consider a highly relativistic electron ($\gamma \gg 1$) moving perpendicular to a magnetic field **B.** It is continually accelerated perpendicular to the field, and must radiate. At what rate does it lose energy? To answer this, transform to a frame F' moving momentarily along with the electron, find E' in that frame, and P'_{rad}. Now show that, because power is *energy/time*, $P_{rad} = P'_{rad}$. This radiation is generally called *synchrotron radiation.*

Ans. $P_{rad} = \frac{2}{3}\gamma^2 B^2 e^4 / m^2 c^3$.

C

SUPER-CONDUCTIVITY

The metal lead is a moderately good conductor at room temperature. Its resistivity, like that of other pure metals, varies approximately in proportion to the absolute temperature. As a lead wire is cooled to 15 K its resistance falls to about $\frac{1}{20}$ of its value at room temperature, and the resistance continues to decrease as the temperature is lowered further. But as the temperature 7.22 K is passed, there occurs without forewarning a startling change: the electrical resistance of the lead wire vanishes! So small does it become that a current flowing in a closed ring of lead wire colder than 7.22 K—a current which would ordinarily die out in much less than a microsecond—will flow for *years* without measurably decreasing. That has been directly demonstrated. Other experiments indicate that such a current could persist for billions of years. One can hardly quibble with the flat statement that the resistivity is zero. Evidently something quite different from ordinary electrical conduction occurs in lead below 7.22 K. We call it *superconductivity*.

Superconductivity was discovered in 1911 by the great Dutch low-temperature experimenter Kamerlingh Onnes. He observed it first in mercury, for which the critical temperature is 4.16 K. Since then dozens of pure metals and alloys have been found to become superconductors. Their individual critical temperatures range from a few hundredths of a degree up to the highest yet discovered, 23.2 K for a certain compound of niobium and germanium. Curiously, among the elements which do *not* become superconducting are some of the best normal conductors such as silver, copper, and the alkali metals.

Only recently has superconductivity been satisfactorily explained. It is essentially a quantum-mechanical phenomenon, and a rather subtle one at that. The freely flowing electric current consists of electrons in perfectly orderly motion. Like the motion of an electron in an atom, this electron flow is immune to small disturbances—and for a similar reason: A finite amount of energy would be required to make any change in the state of motion. It is something like the situation in an insulator in which all the levels in the valence band are occupied and separated by an energy gap from the higher energy levels in the conduction band. But unlike electrons filling the valence band, which must in total give exactly zero net flow, the lowest energy state of the superconducting electrons can have a net electron velocity, hence current flow, in some direction. Why should such a strange state become possible below a certain critical temperature? We can't explain that here. †It involves the interaction of the conduction electrons not only with each other but with the whole lattice of positive ions through which they are moving. That is why different substances can have different critical temperatures, and why some substances are expected to remain normal conductors right down to absolute zero.

In the physics of superconductivity magnetic fields are even more important than you might expect. We must state at once that the phenomena of superconductivity *in no way* violate Maxwell's equations. Thus the persistent current that can flow in a ring of superconducting wire is a direct consequence of Faraday's law of induction, given that the resistance of the ring is really zero. For if we start with a certain amount of flux Φ_0 threading the ring, then because $\int \mathbf{E} \cdot d\mathbf{s}$ around the ring remains always zero, $d\Phi/dt$ must be zero. The flux cannot change; the current I in the ring will automatically assume whatever value is necessary to maintain the flux at Φ_0. Figure C.1 outlines a simple demonstration of this, and shows how a persistent current can be established in an isolated superconducting circuit.

The magnetic field *inside* superconducting material itself (except very near the surface) is always zero. That is *not* a consequence of Maxwell's equations but a property of the superconducting state, as fundamental, and once as baffling, a puzzle as the absence of resistance. The condition $\mathbf{B} = 0$ inside the bulk of the superconductor is automatically maintained by currents flowing in a thin surface layer.

A strong magnetic field destroys superconductivity. None of the superconductors known before 1957 could stand more than a few hundred gauss. That discouraged practical applications of zero-resis-

†The abrupt emergence of a state of order at a certain critical temperature reminds us of the spontaneous alignment of electron spins which occurs in iron below its Curie temperature (mentioned in Section 11.11). Such *cooperative* phenomena always involve a large number of mutually interacting particles. A more familiar cooperative phenomenon is the freezing of water, also characterized by a well-defined critical temperature.

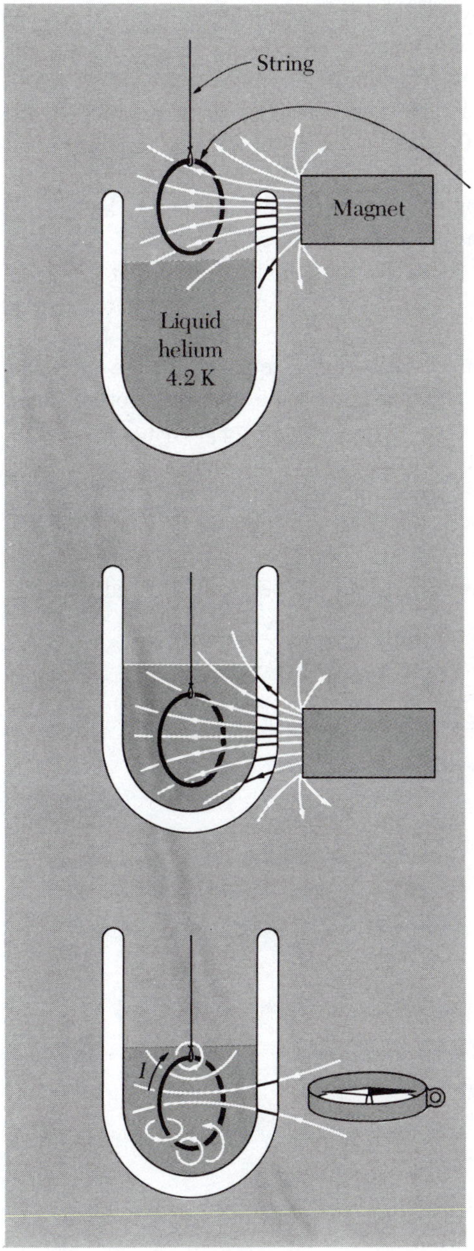
String
Magnet
Liquid
helium
4.2 K
I

Ring of solder (lead-tin alloy); normal conductor; current zero; permanent magnet causes flux Φ_0 through ring.

Ring cooled below its critical temperature. (Some helium has boiled away.) Flux through ring unchanged. Ring is now a superconductor.

Magnet removed. Persistent current I now flows in ring to maintain flux at value Φ_0. Compass needle responds to field of persistent current.

FIGURE C.1
Establishing a persistent current in a superconducting ring. The ring is made of ordinary solder, a lead-tin alloy. (*a*) The ring, not yet cooled, is a normal conductor with ohmic resistance. Bringing up the permanent magnet will induce a current in the ring, which will quickly die out, leaving the magnetic flux from the magnet, in amount Φ, passing through the ring. (*b*) The helium bath is raised without altering the relative position of the ring and the permanent magnet. The ring, now cooled below its critical temperature, is a superconductor with resistance zero. (*c*) The magnet is removed. The flux through the zero-resistance ring cannot change. It is maintained at the value Φ by a current in the ring which will flow as long as the ring remains below the critical temperature. The magnetic field of the persistent current can be demonstrated with the compass.

tance conductors. One could not pass a large current through a superconducting wire because the magnetic field of the current itself would destroy the superconducting state. But then another type of superconductor was discovered which can preserve zero resistance in fields up to 10^5 gauss or more. A widely used superconductor of this type is an alloy of niobium and tin, which has a critical temperature of 18 K and if cooled to 4 K remains superconducting in fields up to 200 kilogauss. Superconducting solenoids are now common which produce steady magnetic fields of 50 to 100 kilogauss without any cost in power other than that incident to their refrigeration. There are good prospects for the use of superconductors in large electrical machinery and in the long-distance transmission of electrical energy.

At the other end of the scale, the quantum physics of superconductivity makes possible electrical measurements of unprecedented sensitivity and accuracy—including the standardization of the volt in terms of an easily measured oscillation frequency. To the physicist, superconductivity is a fascinating large-scale manifestation of quantum mechanics. We can trace the permanent magnetism of the magnet in Fig. C.1 (down to the intrinsic magnetic moment of a spinning electron—a kind of supercurrent in a circuit less than 10^{-8} in size. The ring of solder wire with the persistent current flowing in it is in some sense like a gigantic atom, the motion of its associated electrons, numerous as they are, marshaled into the perfectly ordered behavior of a single quantum state.

D

MAGNETIC RESONANCE

The electron has angular momentum of spin, **J.** Its magnitude is always the same, $h/4\pi$, or 5.271×10^{-28} gm-cm^2/sec. Associated with the axis of spin is a magnetic dipole moment μ of magnitude 0.9273×10^{-20} erg/gauss (Section 11.6). An electron in a magnetic field experiences a torque tending to align the magnetic dipole in the field direction. It responds like any rapidly spinning gyroscope: Instead of lining up with the field, the spin axis *precesses* around the field direction. Let us see why any spinning magnet does this. In Fig. D.1 the magnetic moment $\boldsymbol{\mu}$ is shown pointing opposite to the angular momentum **J**, as it would for a negatively charged body like an electron. The magnetic field **B** (the field of some solenoid or magnet not shown) causes a torque tending to rotate $\boldsymbol{\mu}$ into the direction of **B**. This torque is a vector in the negative $\hat{\mathbf{x}}$ direction at the time of our picture. Its magnitude is given by Eq. 38 in Chapter 11; it is $\mu B \sin\theta$. In a short time Δt the torque adds to the angular momentum of our top a vector increment $\Delta\mathbf{J}$ in the direction of the torque vector and of magnitude $\mu B \sin\theta\, \Delta t$. The horizontal component of **J**, in magnitude J $\sin\theta$, is thereby rotated through a small angle $\Delta\psi$ given by

$$\Delta\psi = \frac{\Delta J}{J \sin\theta} = \frac{\mu \mathrm{B}\, \Delta t}{J} \qquad (1)$$

As this continues the upper end of the vector **J** will simply move around the circle with constant angular velocity ω_p:

$$\omega_p = \frac{\Delta\psi}{\Delta t} = \frac{\mu B}{J} \qquad (2)$$

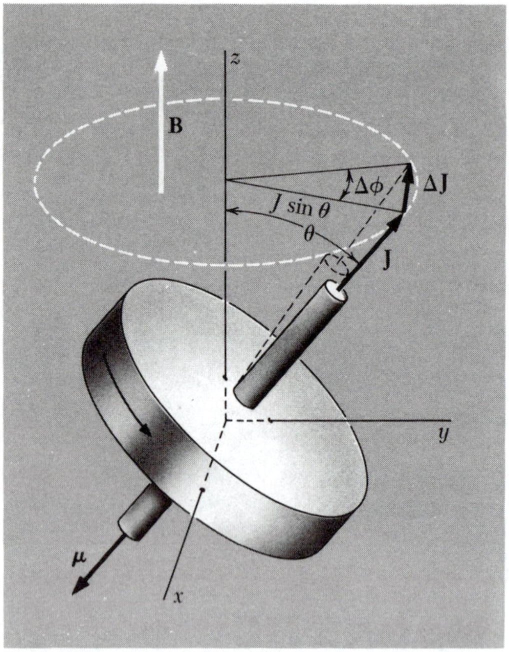

FIGURE D.1
The precession of a magnetic top in an external field. The angular momentum of spin **J** and the magnetic dipole moment μ are oppositely directed, as they would be for a negatively charged rotator.

This is the rate of precession of the axis of spin. Notice that it is the same for any angle of tip; sin θ has cancelled out.

For the electron μ/J has the value $1.759 \times 10 \text{ sec}^{-1} \text{ gauss}^{-1}$. In a field of 1 gauss the spin vector precesses at 1.759×10^7 radians/sec, or 2.800×10^6 revolutions per sec. The proton has exactly the same intrinsic spin angular momentum as the electron, $h/4\pi$, but the associated magnetic moment is smaller. That was to be expected since the mass of the proton is 1836 times the mass of the electron. As in the case of orbital angular momentum (Eq. 23 in Chapter 11) the magnetic moment of an elementary particle with spin ought to be inversely proportional to its mass, other things being equal. Actually the proton's magnetic moment is 1.411×10^{-23} erg/gauss, only about 660 times smaller than the electron moment, which shows that the proton is in some way a composite particle. In a field of 1 gauss the proton spin precesses at 4258 revolutions per sec. About 40 percent of the stable atomic nuclei have intrinsic angular momentum and associated magnetic dipole moments.

We can detect the precession of magnetic dipole moments through their influence on an electric circuit. Imagine a proton in a magnetic field B, with its spin axis perpendicular to the field, and surrounded by a small coil of wire, as in Fig. D.2. The precession of the proton causes some alternating flux through the coil, as would the end-over-end rotation of a little bar magnet. A voltage alternating at the precession frequency will be induced in the coil. As you might expect, the voltage thus induced by a single proton would be much too feeble to detect. But it is easy to provide more protons—1 cm^3 of water contains about 7×10^{22} protons, and all of them will precess at the same frequency. Unfortunately they will not all be pointing in the same direction at the same instant. In fact, their spin axes and magnetic moments will be distributed so uniformly over all possible directions that their fields will very nearly cancel one another. But not quite, if we introduce another step. If we apply a strong magnetic field B to water, for several seconds there will develop a slight excess of proton moments pointing in the direction of **B**, the direction they energetically favor. The fractional excess will be $\mu B/kT$ in order of magnitude, as in ordinary paramagnetism. It may be no more than one in a million, but these uncanceled moments, if they are now caused to precess in our coil, will induce an observable signal. A simple method for observing nuclear spin precession in weak fields such as the earth's field, is described in Fig. D.3. Many other schemes are used to observe the spin precession of electrons and of nuclei. They generally involve a combination of a steady magnetic field and oscillating magnetic fields with frequency in the neighborhood of ω_p. For electron spins (*electron paramagnetic resonance,* or EPR) the frequencies are typically several thousand megahertz, while for nuclear spins (*nuclear magnetic resonance,* or NMR) they are several tens of megahertz. The

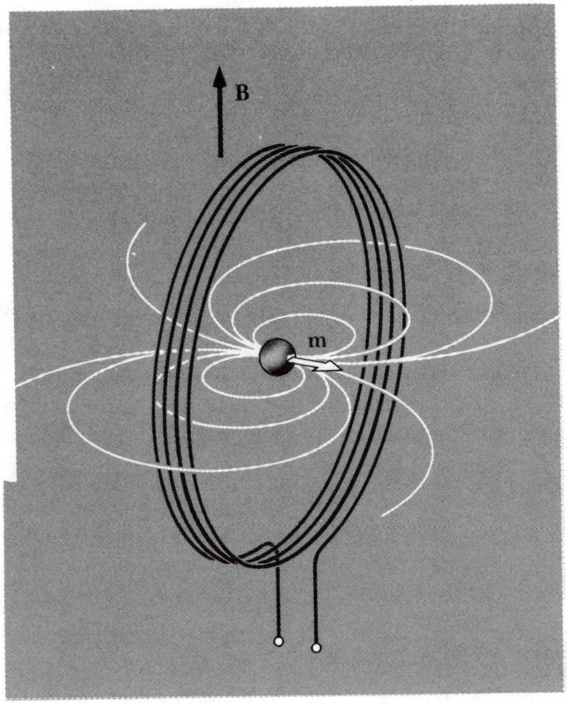

FIGURE D.2
A precessing magnetic dipole moment at the center of a coil causes a periodic change in the flux through the coil, inducing an alternating electromotive force in the coil. Notice that the flux from the dipole **m** which links the coil is that which loops around outside it. See Problem D.1.

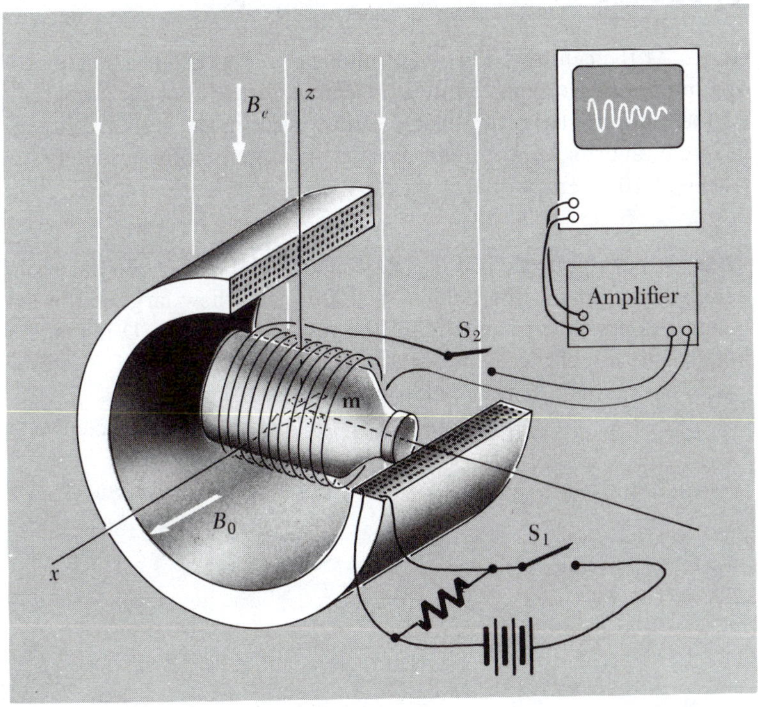

FIGURE D.3
Apparatus for observing proton spin precession in the earth's field B_e. A bottle of water is surrounded by two orthogonal coils. With switch S_2 open and switch S_1 closed, the large solenoid creates a strong magnetic field B_0. As in ordinary paramagnetism (Section 11.6) the energy is lowered if the dipoles point in the direction of the field but thermal agitation causes disorder. Our dipoles here are the protons (hydrogen nuclei) in the molecules of water. When thermal equilibrium has been attained, which in this case takes several seconds, the magnetization is what you would get by lining up with the magnetic field the small fraction $\mu B_0/kT$ of all the proton moments. We now switch off the strong field B_0 and close switch S_2 to connect the coil around the bottle to the amplifier. The magnetic moment **m** now precesses in the x-y plane around the remaining, relatively weak, magnetic field B_e, with precession frequency given by Eq. 2. The alternating y component of the rotating vector **m** induces an alternating voltage in the coil which can be amplified and observed. From its frequency B_e can be very precisely determined. This signal itself will die away in a few seconds as thermal agitation destroys the magnetization the strong field B_0 had brought about. Magnetic resonance magnetometers of this and other types are used by geophysicists to explore the earth's field, and even by archaeologists to locate buried artifacts.

exact frequency of precession, or resonance, in a given applied field can be slightly shifted by magnetic interactions within a molecule. This has made NMR, in particular, useful in chemistry. The position of a proton in a complex molecule can often be deduced from the small shift in its precession frequency.

Magnetic fields easily penetrate ordinary nonmagnetic materials, and that includes alternating magnetic fields if their frequency or the electric conductivity of the material is not too great. A steady field of 2000 gauss applied to the bottle of water in our example would cause any proton polarization to precess at a frequency of 8.516×10^6 revolutions per sec. The field of the precessing moments would induce a signal of 8.516 MHz frequency in the coil outside the bottle. This applies as well to the human body, which, viewed as a dielectric, is simply an assembly of more or less watery obejcts. In *NMR imaging* the interior of the body is mapped by means of nuclear magnetic resonance. The concentration of hydrogen atoms at a particular location is revealed by the radiofrequency signal induced in an external coil by the precessing protons. The location of the source within the body can be inferred from the precise frequency of the signal if the steady field B, which determines the frequency according to Eq. 2, varies spatially with a known gradient.

PROBLEMS

D.1 At the center of the coil of radius a in Fig. D.2 is a single proton, precessing at angular rate ω_p. Derive a formula for the amplitude of the induced alternating electromotive force in the coil in volts, for a in cm, and for ω_p in radians/sec, given that the proton moment is 1.411×10^{-23} erg/gauss.

Ans. $3.55 \times 10^{-30}\ \omega/a$ volts.

D.2 (*a*) If the bottle in Fig. D.2 contains 200 cm^3 of H_2O at room temperature, and if the field B_0 is 1000 gauss, how large is the net magnetic moment **m**? (*b*) Using the result of Problem D.1, make a rough estimate of the signal voltage available from a coil of 500 turns and 4-cm radius when the field strength B_e is 0.4 gauss.

Ans. (*a*) 2.5×10^{-4} erg/gauss; (*b*) 10 microvolts.

E

EXACT RELATIONS AMONG SI AND CGS UNITS

In 1983 the General Conference on Weights and Measures officially redefined the meter as the distance that light travels in vacuum in 1/299,792,458 of a second. The second is defined in terms of a certain atomic frequency in a way that does not concern us here. The nine-digit integer was chosen to make the assigned value of c agree with the most accurate measured value to well within the uncertainty in the latter. Henceforth the velocity of light is, by *definition,* 299,792,458 meters/sec. An experiment in which the passage of a light pulse from point A to point B is timed is to be regarded as a measurement of the distance from A to B, not a measurement of the speed of light.

While this step has no immediate practical consequences, it does bring a welcome simplification of the exact relations connecting various electromagnetic units. As we learn in Chapter 9, Maxwell's equations for the vacuum fields, formulated in SI units, have a solution in the form of a traveling wave with velocity $c = (\mu_0\epsilon_0)^{-1/2}$. The SI constant μ_0 has always been defined exactly as $4\pi \times 10^{-7}$, whereas the value of ϵ_0 has depended on the experimentally determined value of the speed of light, any refinement of which called for adjustment of the value of ϵ_0. But now ϵ_0 acquires a permanent and perfectly precise value of its own, through the requirement that

$$(\mu_0\epsilon_0)^{-1/2} = 299{,}792{,}458 \text{ meters/sec} \tag{1}$$

In our CGS system no such question arises. Wherever c is involved, it appears in plain view, and all other quantities are defined exactly, beginning with the electrostatic unit of charge, the esu, whose definition by Coulomb's law involves no arbitrary factor.

With the adoption of Eq. 1 in consequence of the redefinition of the meter, the relations among the units in the systems we have been using can be stated with unlimited precision. These relations are given in Table E.1 for the principal quantities we deal with. In the table the symbol "3" stands for the precise decimal 2.99792458; the symbol "9" stands for the 17-digit square of that number, 8.9875517873681764.

The exact numbers are uninteresting and for our work quite unnecessary. That "3" happens to be so close to 3 is sheer luck, an accidental consequence of the length of the meter and the second. When 0.1 percent accuracy is good enough we need only remember that "300 volts is a statvolt" and "3×10^9 esu is a coulomb." Much less precisely, but still within 12 percent, a capacitance of 1 cm is equivalent to 1 picofarad.

An important SI constant is $(\mu_0/\epsilon_0)^{1/2}$, which is a resistance in ohms. Its precise value is stated below the table. One tends to remember it, and even refer to it, as "377 ohms." It is the ratio of the electric field strength E, in volts/meter, in a plane wave in vacuum, to the strength in amperes/meter of the accompanying magnetic field H. For this reason the constant $(\mu_0/\epsilon_0)^{1/2}$ is sometimes denoted by Z_0 and called, rather cryptically, the *impedance of the vacuum.* In a plane wave in vacuum in which E_{rms} is the rms electric field in volts/meter, the mean density of power transmitted, in watts/m^2, is E^2_{rms}/Z_0.

The logical relation of the SI electrical units to one another takes on now a slightly different aspect. Before the redefinition of the meter it was customary to designate one of the electrical units as *primary,* in this sense: Its precise value could, at least in principle, be established by a procedure involving the SI mechanical and metrical units only. Thus the ampere, to which this role has usually been assigned, was defined in terms of the force in newtons between parallel currents, using the relation Eq. 7′ of Chapter 6. This was possible because the constant μ_0 in that relation has the precise value $4\pi \times 10^{-7}$. Then with the ampere as the primary electrical unit, the coulomb was defined precisely as 1 ampere-second. The coulomb itself, owing to the presence of ϵ_0 in Coulomb's law, was not eligible to serve as the primary unit. Now with ϵ_0 as well as μ_0 assigned an exact numerical value, the system can be built up with any unit as the starting point. All quantities are in this sense on an equal footing, and the choice of a primary unit loses its significance. Never a very interesting question anyway, it can now be relegated to history.

TABLE E.1

	In SI units		In CGS units
Energy	1 joule	=	10^7 erg
Force	1 newton	=	10^5 dyne
Electric charge	1 coulomb	=	"3" $\times$ 10^9 esu
Electric current	1 ampere	=	"3" $\times$ 10^9 esu/sec
Electric potential	"3" $\times$ 10^2 volts	=	1 statvolt (1 erg/esu)
Electric field E	"3" $\times$ 10^4 volts/m	=	1 statvolt/cm (1 dyne/esu)
Magnetic field B	1 tesla	=	10^4 gauss (10^4 dynes/esu)
Magnetic field H	1 ampere/m	=	$4\pi \times 10^{-3}$ oersted
Capacitance	1 farad	=	"9" $\times$ 10^{11} cm
Inductance	1 henry	=	("9" $\times$ $10^{11})^{-1}$ sec^2/cm
Resistance	1 ohm	=	("9" $\times$ $10^{11})^{-1}$ sec/cm

$\mu_0 = 4\pi \times 10^{-7}$ ohm-sec/m $\qquad$ $\epsilon_0 = (4\pi \times$ "9" $\times 10^9)^{-1}$ sec/ohm-m

$(\mu_0/\epsilon_0)^{1/2} = 40\pi \times$ "3" ohms $= 376.73 \cdots$ ohms

"3" = 2.9979245800000 $\cdots$ $\qquad$ "9" = "3" $\times$ "3"

INDEX

INDEX

		CGS unit	SI unit
distance	s	centimeter	meter
mass	m	gram	kilogram
time	t	second	second
velocity	v	cm sec^{-1}	m sec^{-1}
momentum	p	gm cm sec^{-1}	kg m sec^{-1}
force	F	dyne (gm cm sec^{-2})	newton (kg m sec^{-2})
energy, work	W	erg (dyne cm)	joule (newton m)
power	P	erg sec^{-1}	watt (joule sec^{-1})
electric charge	q	esu	coulomb (amp sec)
charge density	ρ	esu cm^{-3}	coulomb m^{-3}
electric potential	φ	statvolt (erg esu^{-1})	volt (joule $coulomb^{-1}$)
electric field	E	statvolt cm^{-1} (dyne esu^{-1})	volt m^{-1} (newton $coulomb^{-1}$)
electric current	I	esu sec^{-1}	ampere (coulomb sec^{-1})
current density	J	esu sec^{-1} cm^{-2}	amp m^{-2}
resistance	R	cm^{-1} sec	ohm (volt amp^{-1})
resistivity	ρ	sec	ohm m
magnetic field	B	gauss (dyne esu^{-1})	tesla (newton m^{-1} amp^{-1})
magnetic field	H	oersted (dyne esu^{-1})	amp m^{-1}
magnetic flux	Φ	gauss cm^2	weber (tesla m^2)
capacitance	C	cm	farad (ohm^{-1} sec)
inductance	L, M	cm^{-1} sec^2	henry (ohm sec)

1 coulomb $= 3 \times 10^9$ esu

300 volts = 1 statvolt

1 tesla $= 10^4$ gauss

1 farad $= 9 \times 10^{11}$ cm

1 ampere $= 3 \times 10^9$ esu sec^{-1}

3×10^4 volt m^{-1} = 1 statvolt cm^{-1}

1 ohm $= 1.113 \times 10^{-12}$ cm^{-1} sec

1 henry $= 1.113 \times 10^{-12}$ cm^{-1} sec^2

Introduction to Philosophical Thinking

Readings and Commentary